STP 1367

# *Fretting Fatigue: Current Technology and Practices*

*David W. Hoeppner, V. Chandrasekaran,*
*and Charles B. Elliott III, editors*

ASTM Stock Numer: STP 1367

ASTM
100 Barr Harbor Drive
West Conshohocken, PA 19428-2959

Printed in the U.S.A.

## Library of Congress Cataloging-in-Publication Data

Fretting fatigue: current technology and practices/David W. Hoeppner, V.
Chandrasekaran, and Charles B. Elliott III, editors.
    p. cm. -- (STP; 1367)
    ASTM Stock Number: STP1367.
    Includes bibliographical references and index.
    ISBN 0-8031-2851-7
    1. Metals--Fatigue. 2. Fretting corrosion. 3. Contact mechanics. I. Hoeppner, David W.
II. Chandrasekaran, V., 1964- III. Elliott, Charles B., 1941- IV. International Symposium
on Fretting Fatigue (2nd: 1998: University of Utah) V. ASTM special technical publication; 1367.

TA460 .F699 2000
620.1'66--dc21                                                                99-059181

## Photocopy Rights

## Peer Review Policy

    Each paper published in this volume was evaluated by two peer reviewers and at least one
editor. The authors addressed all of the reviewers' comments to the satisfaction of both the technical
editor(s) and the ASTM Committee on Publications.
    To make technical information available as quickly as possible, the peer-reviewed papers in this
publication were prepared "camera-ready" as submitted by the authors.
    The quality of the papers in this publication reflects not only the obvious efforts of the authors
and the technical editor(s), but also the work of the peer reviewers. In keeping with long standing
publication practices, ASTM maintains the anonymity of the peer reviewers. The ASTM Committee on
Publications acknowledges with appreciation their dedication and contribution of time and effort on
behalf of ASTM.

Printed in Baltimore, MD
January 2000

# Foreword

This publication, Fretting Fatigue: Current Technology and Practices, contains papers presented at the symposium held at the University of Utah, Salt Lake City, Utah on Aug. 31, 1998. The symposium was sponsored by University of Utah, United Technologies Research Center, MTS Systems Corporation, FASIDE International, INC. and co-sponsored by Committee E8 on Fatigue and Fracture. The symposium was chaired by David W. Heoppner, V. Chandrasekaran, and Charles B. Elliott III served as co-chairmen. They all served as STP editors of this publication.

# Contents

# Overview

The Second International Symposium on Fretting Fatigue was held at the University of Utah August 31–September 2, 1998. This symposium was held to continue the exchange of information on the subject of fretting fatigue that was accelerated within the ASTM Symposium on *Standardization of Fretting Fatigue Methods and Equipment* held in San Antonio, TX on November 12–13, 1990 (see *ASTM STP1159* edited by Attia and Waterhouse, ASTM, 1992) and the International Symposium on Fretting Fatigue held at the University of Sheffield in April, 1993 (see *Fretting Fatigue*, ESIS Publication 18, edited by Waterhouse and Lindley, 1994). The contribution of fretting to nucleating fatigue failures, often well before they were expected to occur is well known now even though the phenomenon had not been formally identified until the 20th century. A great deal of progress dedicated to understanding the phenomenon of fretting fatigue has occurred within the past century. Thus, this symposium was organized to focus on the progress and to continue the extensive interchange of ideas that has occurred—particularly within the past 50 years.

Fifty-six delegates from ten countries attended the symposium to present papers and participate in lively discussions on the subject of fretting fatigue. The attendees included Dr. Waterhouse and Dr. Hirakawa who did pioneering research and development from the 1960's to the present. Technical leaders in the area of fretting fatigue were in attendance from most of the leading countries that are currently involved in fretting fatigue research, development, and engineering design related matters as well as failure analysis and maintenance engineering issues. ASTM Committee E08 provided the ASTM organizational support for the symposium. The collection of papers contained in this volume will serve as an update to a great deal of information on fretting fatigue. It contains additional contributions that may prove useful in life estimation. More applications of these methods are required. The damage mapping approach presented in some of the papers should assist the community in developing more understanding of fretting fatigue and also provide significant guidance to developing fretting fatigue design methods, and prevention and alleviation schemes. This volume thus serves engineers that have need to develop an understanding of fretting fatigue and also serves the fretting fatigue community including both newcomers and those that have been involved for some time.

The Symposium was sponsored by the following organizations: 1) The Quality and Integrity Engineering Design Center at the Department of Mechanical Engineering at the University of Utah—Dr. David Hoeppner—contact. 2) MTS Systems Corporation—Mr. Arthur Braun—contact. 3) United Technologies Research Center (UTRC)—Dr. Donald Anton—contact and 4) FASIDE International Inc.—Dr. David Hoeppner—contact.

All of the above organizations provided valuable technical assistance as well as financial support. The Symposium was held at the University Park Hotel adjacent to the University of Utah campus. Many of the delegates took part in pre- and post-symposium tours of area National Parks and other sites. Sally Elliott of Utah Escapades, Part City, UT, coordinated the activities and program.

The organizing committee was formed at the conclusion of the International Symposium of Fretting Fatigue held at the University of Sheffield in Sheffield, England April 19–22, 1993. The committee members were: Dr. David Hoeppner, P.E., Chair (USA), Dr. Leo Vincent (France), Dr. Toshio Hattori (Japan), Dr. Trevor Lindley (England), and Dr. Helmi Attia (Canada). Forty papers were presented and this volume contains 36 of those papers.

At the conclusion of the symposium the planning committee for the next two symposia was formed. Dr. Mutoh of Japan will coordinate and chair the next meeting with support from the fretting fatigue community of Japan. Another symposium will be held a few years after the Japan symposium in France with Dr. Vincent as coordinator and chair.

Editing and review coordination of the symposium was done with the outstanding coordination of Ms. Annette Adams of ASTM. The editors are very grateful to her for her extensive effort in assisting in concluding the paper reviews and issuing this volume in a timely manner.

The symposium opened with remarks by the symposium chair. Subsequently, Dr. Robert Waterhouse gave the Distinguished Keynote Lecture. A session of six keynote papers followed the paper of Dr. Waterhouse and is included as the Background Section in this volume.

The papers enclosed in this volume cover the following topics: Fretting fatigue parameter effects, environmental effects, fretting fatigue crack nucleation, material and microstructural effects, fretting damage analysis, fracture mechanics applied to fretting fatigue, life prediction, experimental studies, surface treatments, and applications.

The symposium involved the presentation of methods for studying the phenomenon and for analyzing the damage that fretting produces. It is now very clear that fretting is a process that may occur conjointly with fatigue and the fretting damage acts to nucleate cracks prematurely. More evidence of this is presented in the papers presented in this volume. Although a few laboratories are expending significant efforts on the utilization of fracture mechanics to estimate both the occurrence of fretting fatigue and its progression, there was lively discussion of when cracks are actually nucleated during the fretting fatigue process. As with many of the symposia held on topics related to fatigue over the past 40 years, part of the problem stems from the use of the conceptual view on "initiation of cracks" rather than on the processes by which cracks nucleate (e.g., fretting), and grow in their "short or small" stage and in their long stages where LEFM, EPFM, or FPFM are directly applicable. Even though ASTM committee E 8 has attempted to have the community use the term crack formation or nucleation rather than initiation, this symposium had several papers that persist in this conceptual framework and thus a great deal of discussion centered on this issue. As well, some investigators simply substitute the word nucleation or formation for "initiation." This also resulted in lively discussion at the Symposium, and readers of this volume will find this aspect most interesting. The papers will, when taken as a whole, assist the community in expanding our understanding of fretting fatigue a great deal. This will undoubtedly assist engineers in both the prevention and control of fretting fatigue and in formulating standards to deal with experimentation related to it in the future.

Extensive progress has been made in understanding the phenomenon of fretting fatigue. Even though analytical techniques have emerged to assist in life estimation for fretting fatigue and the analytical techniques also provide guidance for alleviation of fretting fatigue, it is still necessary to conduct experiments to attempt to simulate the fretting fatigue behavior of joints. New experimental techniques have emerged that allow characterization of fretting fatigue in much greater detail than at any time previous to this and new testing techniques are emerging. A standard to assist in development of fretting fatigue data still has not emerged, but one of the participating countries has made an effort to attempt to develop a standard. As well, a manual of standard terminology for fretting fatigue still has not emerged. ASTM E 8 was asked by the planning committee to ask their fretting fatigue subcommittee to undertake to develop the list of terms and phrases and come up with a manual of these within the next two years—hopefully, before the next symposium in Japan.

Several papers dealt with the application of fracture mechanics to fretting fatigue. This is not new but some newer computational models are discussed, and these applications provide a means by which to manage the occurrence of fretting fatigue induced cracks in practice. Thus, the crack propagation portion of cracks induced by fretting is manageable as was shown in works as early as 1975. Some papers herein provide additional insight into the application of fracture mechanics to fretting fatigue. One of the areas that has not received as much interest and study as it should is the area of surface treatments (coatings, self-stresses, diffusion layers, and implanted layers, etc.). This is regrettable since one of the most important ways to prevent fretting degradation is to provide a change in the surface behavior. Hopefully, more effort will be expended on this aspect, and more results will be presented at the next symposium. It is suspected that the scientific community of the USA, for example, does not view this as a new science area to be studied. If this is true and extends to other countries, this would slow the development of fretting fatigue prevention schemes. Another area that has not received anywhere near the attention needed, even though Waterhouse and Hoeppner both have emphasized the need for additional effort and study to adequately understand the phenomenon, is the area of environmental effects on fretting fatigue. The review of this subject by D. Taylor in the 1993 discussed this issue in depth but little progress seems to have occurred in this area. This is regrettable since it is very likely that the environmental (both chemical and thermal) contribution to fretting fatigue is substantial. Thus, more effort needs to be directed at this area in the future.

Work in France, Japan, and two US laboratories (UTRC and the University of Utah) is progressing on a more holistic, systems oriented approach to fretting fatigue. This includes damage characterization during the process, the development of fretting maps and/or damage maps, attempting to characterize the physics of the crack nucleation and propagation processes as well formulate mechanics based formulations of life estimation. These papers are reflected in this volume. It is clear that additional progress will be made in the next several years to assist the engineering and science community in understanding and dealing with fretting fatigue. The papers contained herein will assist in this endeavor.

*David W. Hoeppner, P.E., Ph.D.*
*V. Chandrasekaran, Ph.D.*
*Charles Elliott III, P.E. Ph.D.*

University of Utah
Symposium Chairman, Co-chairmen, and STP Editors

**Background and Critical Issues
Related to Fretting Fatigue**

R B Waterhouse[1]

## Plastic Deformation in Fretting Processes – a Review

**REFERENCE:** Waterhouse, R. B., **"Plastic Deformation in Fretting Processes—a Review,"** *Fretting Fatigue: Current Technology and Practices, ASTM STP 1367,* D. W. Hoeppner, V. Chandrasekaran, and C. B. Elliott, Eds., American Society of Testing and Materials, West Conshohocken, PA, 2000.

**ABSTRACT:** In recent years, analytical treatments of contacting surfaces and resultant fretting, the initiation and early propagation of fatigue cracks, have been the subject of elastic stress analysis. However, direct observations of fretting damage in optical and scanning electron microscopes indicates that plastic deformation of the contacting surfaces is usually an important feature. In this respect it has some similarity with other surface deformation processes, such as shot-peening and surface rolling, in that residual stresses are developed or existing stresses are modified. Surface films which are there as a result of oxidation or applied as an anti-fretting palliative can be seriously disrupted by plastic deformations of the substrate, resulting in a "tribologically transformed layer" or third-body intervention. Consideration of these factors can play a role in the development of methods to counteract the effect of fretting, and is the basis of this review.

**KEYWORDS:** plastic deformation, adhesion, work hardening, residual stress, fretting debris, surface films

### Introduction

Recent developments in the study of fretting, and particularly fretting fatigue, have concentrated on finite element analysis of the contact, locating the site of initiation of a crack. If the configuration is that often used in experimental studies i.e. a square-edged fretting pad applied to a flat fatigue specimen, the crack initiates at a singularity created by the sharp edge of the pad. Even with the less severe contact of a cylindrical pad these analyses usually locate the crack at the edge of the contact region [1]. Further analysis allows the oblique course of the crack to be predicted and its velocity calculated under mixed-mode stress conditions. Once the crack has migrated out of the region affected by the contact stresses, the crack propagates in a plane perpendicular to the alternating fatigue stress.

Direct observation of fretting fatigue failures often indicates that the initial crack is generated in the boundary between the slip region and the non-slip region in the partial slip regime. As the slip region usually migrates further into the contact, by the time the eventual failure occurs the crack appears to be well within the slip region. An empirical analysis developed by Ruiz [2] predicts that this will be the case. It depends on the

[1]Professor, Department of Materials Engineering and Materials Design, University of Nottingham, University Park, Nottingham, NG7 2RD, United Kingdom.

product of the amplitude of slip δ, the shear stress τ and the tensile stress in the surface σ reaching a critical value.

Close examinations of the fretted surface reveals that surface films, usually oxide, are disrupted, intimate metal-to-metal contact occurs and local welds are formed which result in material being plucked above the original surface [3] and even the formation of macroscopic welds [4]. These events can be followed by the measurement of contact resistance [5] between the surfaces and finally by topographical analysis [6] and SEM examination [7]. These observations confirm that the disruption of the surface material is rather more violent than appears in the FE analysis.

**Direct Observation**

The development of the scanning electron microscope (SEM) has proved invaluable in examining surfaces and surface damage. Figure 1 shows damage in the early stages of fretting fatigue in a 0.2C steel. Material has been pulled up above the original surface and has resulted in the creation of a very visible crack. However, the optical microscope also is still a very useful means of investigating such damage. By protecting the surface by nickel or other plating and then sectioning, the superficial and subsurface changes can be readily seen. Figure 2 shows evidence of plastic deformation and the initiation of a crack in the same steel. Figure 3 is a different form of damage where shallow cracks have joined and where a loose particle could result. Intimate intermetallic contact in the early stage is indicated by very low electrical contact resistance and the formation of local welds. Figure 4 is a section through the specimen and bridge foot showing that the deformation occurs in both surfaces. Eventually a wedge of material can develop, Figure 5. The welds can be quite strong – in some cases a normal tensile force of 4kg has had to be applied to remove the bridge. If the weld is 0.2 mm in diameter which is apparent from such photographs and there four such welds, two on each bridge foot [8], the strength of the welds is about 330MPa. If it is likely that one or two welds break first and the remainder are prized off then the individual welds could be stronger and of the same order as the strength of the steel.

Sections which have been polished but not electroplated can be examined in the SEM to show both the surface and the underlying material, Figure 6, which shows subsurface cavities.

Occasionally somewhat bizarre features are seen in the SEM. Figure 7 shows the result of fretting on a pure copper specimen. A thin surface layer has broken open to reveal a series of parallel tubular holes just below the surface and apparently parallel to the surface. This has been termed the "zip fastener effect" and is related to the pile-up of dislocations in the surface region [9].

**Profilometer Observations**

In the early days a single traverse by the profilometer would confirm the features already seen in the SEM e.g. the undamaged plateau in the centre of the fretting scar in the partial slip regime, Figure 8. The course of the damage as fretting proceeded could also be assessed as in Figure 9. Nowadays, with computers, orthogonal projections of the scar can be produced and the volume of material both raised above and missing below

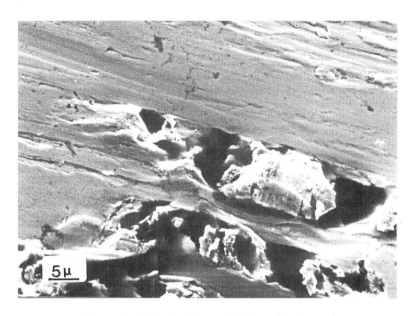

Figure 1 - *Adhesive damage in the early stages of
fretting fatigue on a 0.2C steel – slip amplitude 12 µm*

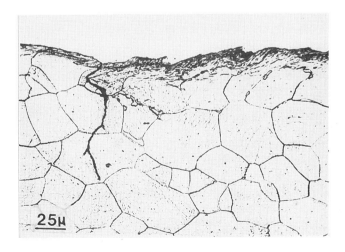

Figure 2 - *Fretting fatigue damage on mild steel 10000 cycles x 400*

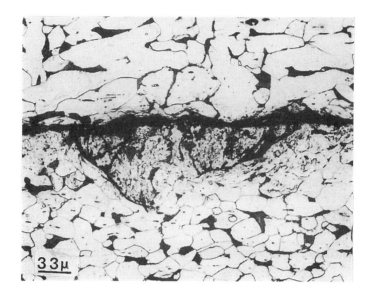

Figure 3 - *Confluent fatigue cracks on mild steel x 200*

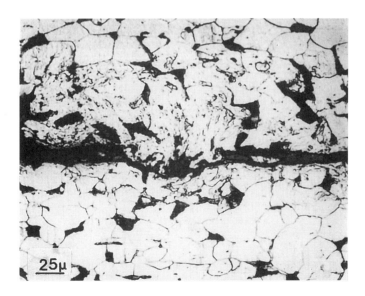

Figure 4 - *Local weld between mild steel surfaces 25000 cycles x 400*

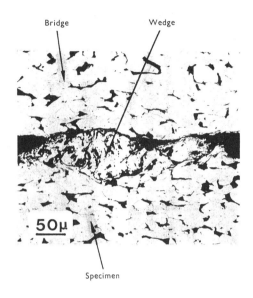

Figure 5 - *Development of a wedge of material x 150*

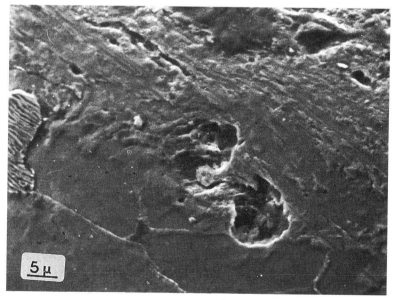

Figure 6 - *Subsurface cavities developed after $10^5$ cycles of fretting fatigue –0.2C steel*

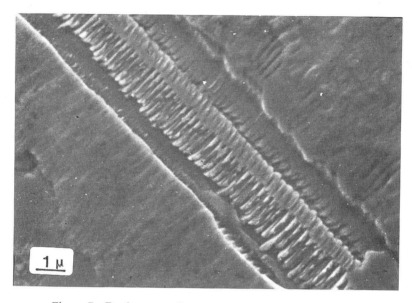

Figure 7 - *Zip fastener effect in fretting fatigue of pure copper*

Figure 8 - *Profilometer trace across damage caused by bridge foot showing non-slip region. Vertical magnification/horizontal x 50*

the original surface can immediately be calculated, a valuable contribution in studying fretting wear. Figure 10 shows such an example from the fretting fatigue of steel wires.

## Metallurgical Evidence

The structural changes in the fretted area can be investigated by x-ray diffraction and microhardness measurements. Back reflection patterns from the fretting scar on a bright drawn 0.2 carbon steel showed that the (310) reflection with Co Ka radiation from the ferrite was clearly resolved whereas the doublet from outside the fretting scar was not resolved, the lines being broadened by the cold-worked structure. This was attributed to a possible rise in temperature which has since been observed by other investigators [10, 11]. In a few cases, formation of the so-called white layer has been seen [12, 13].

   Micro hardness traverses on sections through the fretting scar have shown evidence of work hardening on steels and work softening on age-hardened aluminium alloys [14], all the result of plastic deformation. The extensive work in France has promoted the term "tribologically transformed layer" to represent the region affected by fretting action [15].

## Residual Stresses

One of the principal effects of surface working is the development of residual stresses, usually compressive in such processes as shot-peening and surface rolling, but can also be tensile as in wire drawing. Similarly fretting has been shown to develop residual compressive stresses in the surface of a 0.5C steel which are a maximum in a direction perpendicular to the fretting direction [16]. Conversely fretting can cause the fading of existing residual stresses e.g. induced by shot-peening, if they are sufficiently high. These stresses fade under the influence of normal fatigue, but the fading is much more rapid under fretting conditions, Figure 11 [17]. It was found in this investigation that if the Al-Cu-Mg alloy was shot-peened in the solution heat treated condition and then aged, a surface structure was produced which was completely resistant to fretting fatigue damage, Figure 12.

## Fretting Debris

Although the debris has been removed from the surface, it still has an influence on the process. In many cases it has been observed that the debris itself shows evidence of plastic deformation. On shot-peened surfaces of an aluminium alloy spherical debris has been found, Figure 13. The debris on this alloy was found to contain 22% metallic material surrounded by a coating of amorphous $Al_2O_3$, and black in colour. Incidentally this debris was shown to be pyrophoric [18]. The metallic particles originally removed from the surfaces, were rolled up by the oscillatory movement.

## Surface Films

If surface films are not capable of being plastically deformed, which is the case with most oxide films, then deformation of the substrate leads to fracture and breakdown of the film. Figure 14 shows the multiple cracking of such an oxide film. Where films and coatings are applied to prevent fretting damage, it is important that the substrate is

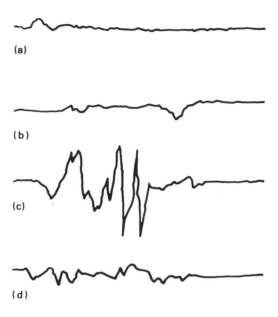

Figure 9 - *Changes in surface roughness as a function of fretting cycles –0.2C steel (a) as received (b) $10^3$ cycles (c) 5 x $10^4$ cycles (d) 3 x $10^5$ cycles. Vertical magnification/horizontal x 100*

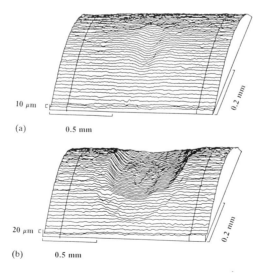

Figure 10 - *Profilometer survey of scar in cross-cylinder fretting*

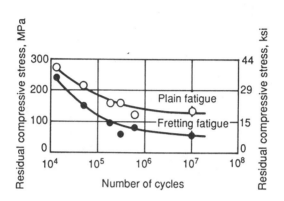

Figure 11 - *Fading of surface residual stress in fatigue and fretting fatigue with number of cycles*

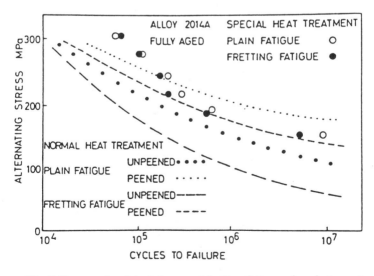

Figure 12 - *S-N curves in plain fatigue and fretting fatigue after shot-peening in SHT condition and re-ageing*

Figure 13 - *Spherical debris due to fretting of shot-peened Al-4Cu-1Mg alloy*

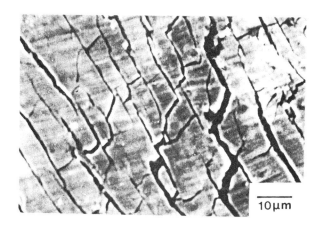

Figure 14 - *Cracking of brittle oxide film due to creep of substrate Inconel 718 at 600 °C*

sufficiently strong to withstand plastic deformation. This is particularly relevant in the application of TiN. Breakdown of this film on an insufficiently strong substrate is particularly dangerous since the material has a hardness in the region of 3000 VHN and therefore introduces a highly abrasive material between the surfaces resulting in enhanced wear [19].

In recent work on the behaviour of TiN under fretting the failure of the coating is related to the mechanical properties of a number of steels, an aluminium alloy and a titanium alloy [20].

**Discussion**

Despite the considerable evidence from experimental observations that plastic deformation is nearly always detected in the early stage of fretting fatigue and fretting wear, it has received little attention from those investigators who have concentrated on the analytical approach to the problem. When Johnson [21] used fretting damage to verify the Mindlin analysis of partial slip between a sphere and flat he was aware that the energy consumption was rather greater than predicted by the theory and suggested that this was due to the plastic deformation of asperities, although the overall behaviour of the contact was dictated by the elastic stress field in the surrounding bulk material. This was taken up by Odvalk and Vingsbo [22] who showed that the apparent amplitude of incipient gross slip was considerably higher than predicted and therefore developed an elastic-plastic model for fretting. The question has been raised recently by Giannakopoulos and Suresh [23]. Their three-dimensional analysis of the fretting of a sphere on flat accords well with previously published experimental observations. However, they have extended it to consider the possibility of plastic yielding in the contact in the fretting fatigue of such a contact. They concluded that "for typical displacements commonly encountered in a Ti alloy, a high coefficient of friction between the fretted surfaces may be sufficient to cause plastic yielding at the surface so that crack initiation can be facilitated by a ductile fracture process".

An early paper by Collins and Marco [24] has proved to provide explanations for recent observations. Their original work showed that if fretting damage was caused on a specimen under static compressive stress, when that stress was removed the fretted region exhibited a local tensile residual stress which resulted in a much reduced fatigue strength in subsequent fatigue testing. In fretting fatigue testing with a square-ended bridge on a flat on the specimen in both push-pull and rotating bending, lifting away at the outer edge of the bridge foot occurs during the tensile part of the cycle and digging in the compressive so that the fretting damage is produced when the surface is in compression [25]. The reverse situation can possibly occur at the inner edge of the foot. It is in these regions that the fatigue crack is generated. In tests where the normal stress on the fretting bridge was cycled in phase with the alternating stress in the specimen (Figure 15) in the case where the normal stress was only applied in the tensile part of the fatigue cycle the fatigue life was longer than when the normal stress was constant (Figure 16) [26].

It is, of course, expected that soft materials will suffer considerable plastic deformation when involved in a fretting contact. This is usually encountered where a soft metal such as tin or lead is applied as a coating on steel to prevent damage. Extrusion of the coating from the contact is a possibility, Figure 17. This is particularly

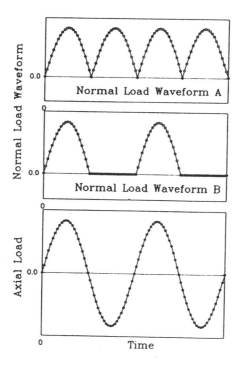

Figure 15 - *Normal load wave forms*

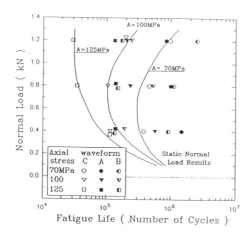

Figure 16 - *The influence of normal load on fretting fatigue life*
*A and B variable load*
*C constant load*

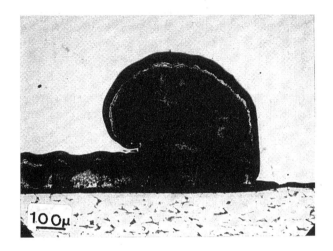

Figure 17 - *Extrusion of lead coating (75 μm thick) from fretting contact x 100*

important in silver plated copper electrical contacts where massive plastic deformation is necessary to achieve their function. This situation has been designated as "gross plastic fretting" since the fretting mechanism is so very different from the more usual fretting which is largely determined by elastic behaviour. Satisfactory contact resistance is only achieved with low slip amplitude and high normal forces where contact area growth of the order of 1000% occurs [27].

With the possible exception of electrical contacts, plastic deformation of the surfaces in a fretting contact is to be avoided if possible. This suggests that hardening of the surface is a possible solution. Hardening by diffusion processes such as carburising nitriding and carbo-nitriding has been a well established process. Laser beam quenching of the surface is a more recent technique [28] as is the use of a laser to apply hard material such as stellite [29]. High hardness has also been achieved by nitrogen treatment of titanium alloys [30]. Shot-peening also produces hardening depending on the work hardenability the material which is a function of the stacking fault energy. These and other treatments are dealt with in a recent review [31].

## Conclusions

Plastic deformation in fretting contacts is a frequent occurrence in the early stages of fretting fatigue. It leads to roughening of the surfaces, adhesion and the early initiation of fatigue cracks. It also results in modification of existing residual stresses arising from mechanical working processes and shot-peening. Deformation of the substrate can lead to the early breakdown of coatings such as TiN. Recent analytical studies have taken into account the probability of plastic deformation. Methods of hardening the surface are recommended for overcoming it.

## References

[1]    Hills, D.A., and Nowell, D., *"Mechanics of Fretting"* Kluwer (1994).

[2]    Ruiz, C., Boddington, P.H.B., and Chen, K.C., "An investigation of fatigue and fretting in a dovetail joint" *Experimental Mechanics* 24(3) 1984 208-217.

[3]    Waterhouse, R.B., *"Fretting Wear"* ASM Handbook Vol 18 *"Friction, Lubrication and Wear Technology"* ASM International, New York, 1992 242-256.

[4]    Waterhouse, R.B., "The role of adhesion and delamination in the fretting wear of metallic materials" *Wear* 45 1977 355-364.

[5]    Pendlebury, R.E., "Unlubricated fretting of mild steel in air at room temperature, Pt II Electrical contact resistance measurements and the effect of wear on intermittent loading" *Wear* 118 1987 341-364.

[6]    Marui, E., Endo, H., Hasegawa, N., and Mizuno, H., "Prototype fretting-wear testing machine and some experimental results" *Wear* 214 1998 221-230.

[7]    Waterhouse, R.B., "The effect of environment in wear processes and the mechanisms of fretting wear" *Fundamentals of Tribology* ed Suh, N.P., and Saka, N., The MIT Press, Cambridge, Mass 1980 567-584.

[8]    Waterhouse, R.B., *"Fretting Corrosion"* Pergamon 1972 p179.

[9]    Doeser, B.A., and Waterhouse, R.B., "Metallographic features of fretting corrosion damage" *Microstructural Science* ed I le May, I., Elsevier, New York 7 1979 205-215.

[10]   Vodopivec, F., Vizintin, J., and Sustarsic, B., "Effect of fretting amplitude on microstructure of 1C-1.5Cr steel" *Material Science and Technology* 12 1996 355-360.

[11]   Attia, M.H., "Friction-induced thermo-elastic effects in the contact zone due to fretting action" *"Fretting Fatigue"* ed Waterhouse, R.B., and Lindley, T.C., MEP Ltd London 1994 307-319.

[12]   Beard, J., and Thomason, P.F., "The mechanics of fatigue crack nucleation in a low alloy steel under fretting conditions" *Conference Proceedings*, Fatigue 87, 28 June-3 July Charlottesville VA Vol 1 1987 53-62.

[13]   Dobromirksi, J., and Smith, I.O., "Metallographic aspects of surface damage, surface temperature and crack initiation in fretting fatigue" *Wear* 117 1987 347-357.

[14]   Waterhouse, R.B., *"Fretting Corrosion"* Pergamon 1972 p104.

[15]   Zhou, Z.R., Sauger, E., and Vincent, L., "Nucleation and early growth of tribologically transformed structure (TTS) induced by fretting" *Wear* 212(4) 1997 50-58.

[16]   Ferrahi, G.H., and Maeder, G., "An experimental study of fretting by means of X-ray diffraction" *Fatigue, Fracture of Engineering Materials and Structures* 15(1) 1992, 91-102.

[17]   Fair, G., Noble, B., and Waterhouse, R.B., "The stability of compressive stresses induced by shot peening under conditions of fatigue and fretting fatigue" *Advances in Surface Treatment*, ed, Niku-Lari, A., Pergamon 1984 1 3-8.

[18]   Waterhouse, R.B., "Fretting corrosion and fretting fatigue in the transport industry" *Proceedings 6th European Congress on Metallic Corrosion*, London 19-23 Sept 1977.

[19]   Cobb, R.C., and Waterhouse, R.B., "The fretting wear of certain hard coatings including TiN applied to a 0.4C steel" *Proceedings International Conference Tribology, Friction, Lubrication and Wear 50 years on*, 1-3 July 1987 London

303-310.

[20]    Shima, M., Okado, J., McColl, I.R., Waterhouse, R.B., Hasegawa, T., and
        Kasaya, M., "The influence of substrate material and hardness on the fretting
        behaviour of TiN", *Wear* 225-229 Pt1 (1999) 38-45.

[21]    Johnson, K.L., "Surface interactions between elastically loaded bodies under
        tangential forces" *Proceedings of the Royal Society* (London) A230 1955 531.

[22]    Ödfalk, M., and Vingsbo, O., "An elastic-plastic model for fretting contact" *Wear
        of Materials 1991*, ASME, New York (1991) 41-47.

[23]    Giannakopoulos, A.E., and Suresh, S., "A three-dimensional analysis of fretting
        fatigue" *Acta Mater* 46(1) 1998 177-192.

[24]    Collins, J.A., and Marco, S.M., "The effect of stress direction during fretting on
        subsequent fatigue life" Proceedings ASTM 64 1964 547-560.

[25]    Bramhall, R., "Studies in Fretting Fatigue" PhD Thesis Oxford University 1973.

[26]    U.S. Fernando, Brown, M.W., Miller, K.J., Cook, R., and Rayaprolu, D.,
        "Fretting fatigue behaviour of BS L65 4 percent copper aluminium alloy
        under variable normal load" *Fretting Fatigue*, ed, Waterhouse, R.B., and
        Lindley, T.C., MEP Ltd London 1994 197-209.

[27]    Rudolphi, A.K., and Jacobson, S., "Gross plastic fretting – examination of the
        gross weld regime" *Wear* 201(1996) 255-264.

[28]    Dai, Z.D., Pan, S.C., Wang, M., Yang, S.R., Zhang, X.S., and Xue, Q.J.,
        "Improving the fretting wear resistance of titanium alloy by laser beam
        quenching" *Wear* 213(1-2) 1997 135-139.

[29]    Le Hosson, J.Th.M.D., van Otterloo, L.De.Mol., Boerstal, B.M., and Huis int
        Veld, A.J., "Surface engineering with lasers: an application to Co-based
        materials", *Euromat 97: 5th European Conference on Advanced Materials*,
        Maastricht, Netherlands, 21-23 April 1997, Netherlands Soc for Material
        Science Vol 3, 85-90.

[30]    Lebrun, J.P., Corre, Y., and Douet, M., "The influence of nitrogen on the
        tribological behaviour of titanium alloys" *Bull Cerole Etudes Metaux*, 16(10)
        1995 28.1-28.11.

[31]    Waterhouse, R.B., and Lindley, T.C., "Prevention of fatigue by surface
        engineering" in the press.

Toshio Hattori,[1] Masayuki Nakamura,[2] and Takashi Watanabe[3]

## A New Approach to the Prediction of the Fretting Fatigue Life that Considers the Shifting of the Contact Edge by Wear

**REFERENCE:** Hattori, T., Nakamura, M., and Watanabe, T., **"A New Approach to the Prediction of the Fretting Fatigue Life that Considers the Shifting of the Contact Edge by Wear,"** *Fretting Fatigue: Current Technology and Practices, ASTM STP 1367,* D. W. Hoeppner, V. Chandrasekaran, and C. B. Elliott, Eds., American Society for Testing and Materials, West Conshohocken, PA, 2000.

**ABSTRACT:** In fretting fatigue the contact edge will shift inward due to wear. The effect of this shifting on crack propagation behavior is considered here for accurate prediction of fretting fatigue life. The stress intensity factor($K_1$) for cracking due to fretting fatigue was calculated by using contact pressure and frictional stress distributions, which were analyzed by the finite element method. The S-N curves for fretting fatigue were predicted by using the relationship between the calculated stress intensity factor range ($\triangle K_1$) and the crack propagation rate (da/dN) obtained using CT specimens. Fretting fatigue tests were performed on Ni-Mo-V steel specimens. The S-N curves of experimental results were in good agreement with the analytical results obtained by considering the shifting of the contact edge.

**KEYWORDS:** crack initiation, wear extension, crack propagation, stress intensity factor, stress singularity parameters

### Introduction

Fretting can occur when a pair of structural elements are in contact under a normal load while cyclic stress and relative displacement are forced along the contact surface. This condition can be seen in bolted or riveted joints [1,2], in shrink-fitted shafts [3,4], in the BL

---

[1]Chief Researcher, Mechanical Engineering Research Laboratory, Hitachi Ltd., Tsuchiura, Ibaraki, Japan.

[2]Researcher, Mechanical Engineering Research Laboratory, Hitachi Ltd., Tsuchiura, Ibaraki, Japan.

[3]Chief Engineer, Hitachi Works, Hitachi Ltd., Hitachi, Ibaraki, Japan.

19

blade dovetail region of turbo machinery [5,6], etc. During fretting the fatigue strength decreases to less than one-third of that without fretting [7,8]. The strength is reduced because of concentrations of contact stresses such as contact pressure and tangential stress at the contact edge, where fretting fatigue cracks initiate and propagate.

This concentration of stress can be calculated using the finite element method [9] or boundary element method. Methods for estimating the strength of fretting fatigue have been developed that use values of this stress concentration on a contact surface [3,5]. However, the stress fields near the contact edges show singularity behavior, where the stress at contact edges are infinite. Thus, maximum stresses cannot be used to evaluate fretting fatigue strength. Accordingly, a recent method for estimating the initiation and propagation of fretting fatigue cracks used the stress singularity parameter at the contact edge [10] and the stress intensity factor range of the crack [11-13]. Here we improve these estimation methods by considering the shifting of the contact edge by wear.

**Fretting Fatigue Process**

Cracking due to fretting fatigue starts very early in fretting fatigue life. During this early period, fretting fatigue cracks tend to close and  propagate very little, due to the high contact pressure acting near this contact edge. But wear on the contact surface reduces the contact pressure near the contact edge, and cracks gradually start to propagate. Hence, fretting fatigue life is dominated by the propagation of small cracks initiated at the contact edge. Here, we form estimates using various fretting fatigue procedures, as illustrated   in Fig. 1. First, we used stress singularity parameters at the contact edge to estimate the initiation of cracks, and then fretting fatigue crack propagation behavior is estimated using stress intensity factor range considering the shifting   of the contact edge from wear.

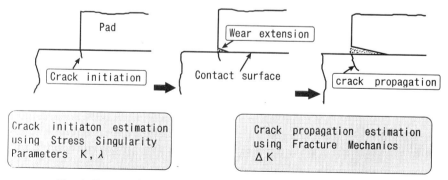

*Fig. 1. Fretting fatigue mechanisms in various processes.*

**Stress Singularity Parameters Approach for Evaluating the Initiation of Fretting Fatigue Cracks**

The stress fields near the contact edges show singularity behavior (Fig. 2). These stress distributions near the contact edges can be expressed using the two stress singularity parameters, H and $\lambda$, as follows,

$$\sigma(r) = H / r^{\lambda} \qquad (1)$$

where $\sigma$ is the stress (MPa), r is the distance from the singularity point (mm), H is the intensity of stress singularity, and $\lambda$ is order of stress singularity. Parameters $\lambda$ and H are calculated as follows. The order of stress singularity ($\lambda$) for the contact edge is calculated analytically [14] using wedge angles $\theta_1$ and $\theta_2$, Young's moduli $E_1$ and $E_2$, Poisson's ratio $\nu_1$ and $\nu_2$, and the frictional coefficient $\mu$ (Fig. 3). However, in this paper $\lambda$ is calculated by finding the best fit of Eq. (1) to the stress distributions near the contact edge. Stress distribution is calculated by numerical stress analysis, such as by the finite element method or the boundary element method. The intensity of the stress singularity (H) is also calculated by best fit of Eq. (1) to the numerically analyzed stress distribution. By comparing these calculated parameters (H and $\lambda$) with the crack

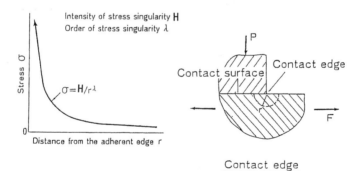

Fig. 2. Stress singularity behavior at contact edge.

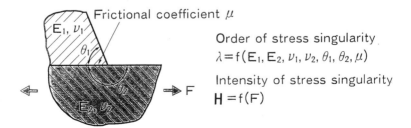

Fig. 3. Geometry of contact edge.

initiation criterion Hc, fretting crack initiation conditions can be estimated for each wedge angle and pad stiffness.   This criterion Hc is derived using stress distributions on both critical conditions, the plain fatigue limit ($\sigma_{wo}$) and the threshold stress intensity factor range ($\Delta K_{th}$) of the rotor materials. On both critical conditions the stress distributions cross at point 0.118mm depth from singular point as shown in Fig. 4. And we expect that on whole critical conditions with each order of stress singularity the stress distributions cross at this point. From these assumptions we can estimate the critical intensity value of stress singularity Hc for each order of stress singularity $\lambda$ as follows(see Fig. 5).

$$Hc(\lambda) = 360 \times (0.118)^{\lambda}$$

$$(2)$$

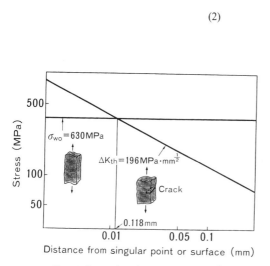

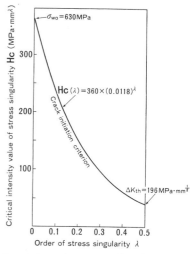

Fig. 4. Stress distributions near singular point on critical conditions.

Fig. 5. Critical intensity value of stress singularity.

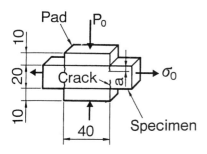

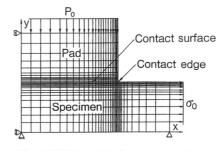

Fig. 6. Model of contact specimen.

Fig. 7. FEM model of contact specimen.

## Fracture Mechanics Approach

*Fretting Fatigue Threshold*

The fretting fatigue threshold can be defined as the point at which the small crack that started at the contact edge ceases to propagate[7]. Hence we analyzed the fretting fatigue threshold by comparing the stress intensity factor range of the small fretting cracks that started at the contact edge and the range of the material's threshold stress intensity factor range.

The stress intensity factor of such cracks can be calculated using the distributions of tangential stress and contact pressure on the contact surface near the contact edge. Additionally, the threshold stress intensity range $\Delta K_{th}$ is influenced by the stress intensity factor ratio R, where

$$R = Kmin/Kmax \tag{3}$$

and by crack length a, especially in the short crack region. In estimating the fretting fatigue limit using fracture mechanics, the precise determination of $\Delta K_{th}$ is essential. Hence, we consider the influence of the stress intensity factor ratio R and crack length a, on $\Delta K_{th}$. Using various experimental results, we determined the relation between $\Delta K_{th}$ and the stress intensity factor ratio R as follows:

$$\Delta K_{th(R)} = \Delta K_{th(R=0)}(1-R)^{0.7} \quad (R<0) \tag{4}$$

$$\Delta K_{th(R)} = \Delta K_{th(R=0)}(1-R)^{\{0.5+\log(1-R)/10\}} \quad (R>0) \tag{4'}$$

To predict $\Delta K_{th}$ in the short crack region, Ohuchida [14], Tanaka [15] and El Haddad [16] derived the following equation:

$$\Delta K_{th(a)} = \Delta K_{th(a=\infty)} \sqrt{\frac{a}{a+a_0}} \tag{5}$$

In this equation, the critical crack length $a_0$ is determined by using the fatigue limit of a plain specimen $\sigma_{w0}$ and the threshold stress intensity factor range for a long crack $\Delta K_{th}(a = \infty)$ as follows:

$$a_0 = \frac{1}{\pi} \left( \frac{\Delta K_{th(a=\infty)}}{\Delta \sigma_{wo}} \right)^2 \tag{6}$$

Using Eqs. (4) and (5), we derive the threshold stress intensity factor range $\Delta K_{th}$, considering both R and a, as follows:

$$\Delta K_{th(R,a)} = \Delta K_{th(R=0,a=\infty)}(1-R)^{0.7} \sqrt{\frac{a}{a+a_0}} \cdot (R<0) \tag{7}$$

$$\Delta K_{th(R,a)} = \Delta K_{th(R=0,a=\infty)}(1-R)^{\{0.5+\log(1-R)/10\}} \cdot \sqrt{\frac{a}{a+a_0}} \cdot (R>0) \tag{7'}$$

Next we will show a sample calculation for the range of the stress intensity factor of fretting fatigue cracks. Using the finite element models under fretting conditions(see Figs. 6 and 7), we calculate the distributions of the tangential stress and contact pressure on the contact surface. Also using these stress distributions, the stress intensity factor of the fretting crack can be obtained as shown in Fig.8, and can be expressed approximately as:

$$K = A + (\sigma_0 - B)D \qquad (8)$$

In this equation A and B are constants for a given contact pressure $P_0$. When $P_0$ = 196 MPa, A is -4 MPa$\sqrt{}$ m and B is 43 MPa, as shown in Fig. 8. The relation between D and crack length a is shown in Fig. 9. Using this equation, the stress intensity factor range $\triangle$ K and stress intensity factor ratio can be calculated as follows.

$$\triangle K = \triangle \sigma_0 D \qquad (9)$$

$$R = \frac{A + (\sigma_{min} - B)D}{A + (\sigma_{max} - B)D} \qquad (10)$$

The threshold stress intensity factor range $\triangle K_{th}$ for the fretting condition can be calculated by substituting Eq. (10) into Eq. (7) or Eq. (7').

The analytical model for the stress intensity factor range when the contact edge shifts inward because of wear is shown in Fig. 10. In this case the small crack started at the initial contact edge and then the contact edge shifted to point e due to wear. As wear region b increased in size the crack closure effect arising from the contact pressure decreased. This effect can be confirmed by the calculations shown in Fig. 10.

*Fretting Fatigue Crack Propagation Behavior*

The results of experiments on crack propagation behavior under fretting and non fretting conditions are shown in Fig. 11. The non fretting condition test was performed using flat specimens with shoulder fillets with a corner radius of 0.5 mm and a stress concentration factor of 3.87. In both cases the stress amplitude was 147 MPa. These results show that almost all the fatigue life under fretting conditions of just the crack propagation process, while under non fretting conditions the fatigue life is dominated by the crack initiation process. Consequently for a precise estimate of the fretting fatigue life, we must analyze the crack propagation behavior. The crack propagation behavior of Ni-Mo-V steel is shown in Fig. 12. This is just an experimental result under pulsating loading conditions (R = 0), but under fretting conditions the stress ratio R becomes negative. So we introduce equivalent stress intensity factor range $\triangle K_e$, which considers stress ratio R and crack length a, just as explained in Eqs. (7) and (7') for the threshold stress intensity factor range $\triangle K_{th}$, as follows:

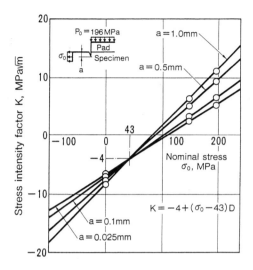

*Fig. 8. Stress intensity factor vs. nominal stress.*

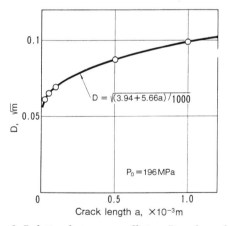

*Fig. 9. Relation between coefficient D and crack length a.*

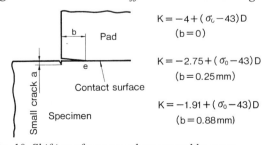

*Fig. 10. Shifting of contact edges caused by wear.*

$$\Delta K_e = \frac{\Delta K_{th(R=0,\ a=\infty)}}{\Delta K_{th(R,\ a)}} \cdot \Delta K_{(R,\ a)} \tag{11}$$

Using this equivalent stress intensity factor range $\Delta K_e$ and the crack propagation behavior of the material as shown in Fig. 12, we can analyze the crack propagation behavior under fretting conditions. Moreover, fretting fatigue life can be estimated by using this crack propagation behavior, ignoring the fretting fatigue initiation life, and assuming initial crack length $a_i$ to be 10 $\mu$ m. Our estimation of the fretting fatigue life, without consideration of wear (b = 0 mm), is shown in Fig. 13. From this we can estimate the fretting fatigue threshold to be 141 MPa and the non propagation crack length to be 80 $\mu$ m. The resulting estimates of fretting fatigue life considering wear (b = 0.88 mm) are shown in Fig. 14. From this we can estimate the fretting fatigue threshold to be 114 MPa and the non propagating crack length to be 90 $\mu$ m. Comparison of these estimated results show that the fretting fatigue strength decreases due to the wear near the contact edge, and converges to a certain level in the b > 0.88 mm region. Hence, to estimate the fretting fatigue strength precisely we must now consider the change in contact conditions due to wear. The crack propagation rate estimated for each crack length is shown in Fig. 15. From this we can see that the crack propagation rate is at its lowest level when the crack length is just about 0.15 mm. Qualitatively these estimated results coincide well with the experimental results obtained by K. Sato [17] using an aluminum alloy (A2024-T3) specimen, as shown in Fig. 16. Although the crack propagation behavior differs quantitatively, due to the difference in material and contact pressure, the validity of our method for estimating fretting fatigue crack propagation behavior using fracture mechanics can be confirmed.

**Fretting Fatigue Strength: Estimated vs. Experimental Results**

The estimated results for the fretting fatigue strength are shown in Fig. 16. In this figure the chained line shows the estimations that do not consider wear (b = 0 mm), and the dotted line and solid line show those that consider wear of b = 0.25 mm and b = 0.88 mm, respectively. The experimental results are shown in this figure by an open circle. Fretting fatigue specimens are shown in Fig. 6 and contact pressure $P_0$ is 196MPa and surface roughness of specimens and pads are both about Ra=0.7 $\mu$ m. The experimental results agree well with the estimated results considering the wear (b = 0.88 mm) especially in the fretting threshold region. This wear region b = 0.88 mm was estimated by observed results as shown in Fig. 18.

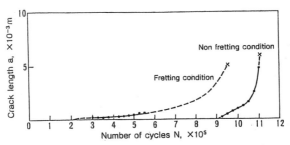

*Fig. 11. Crack propagation behavior under fretting and non fretting conditions.*

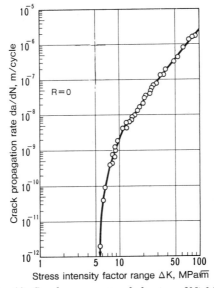

*Fig. 12. Crack propagation behavior of Ni-Mo-V steel.*

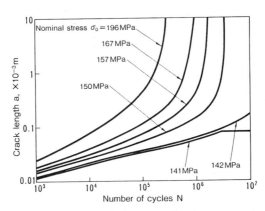

*Fig. 13. Crack propagation behavior under fretting condition (b=0).*

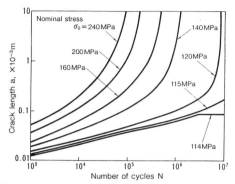

*Fig. 14. Crack propagation behavior under fretting condition (b=0.88mm).*

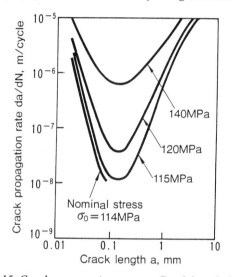

*Fig. 15. Crack propagation rate vs. Crack length (b=0.88mm).*

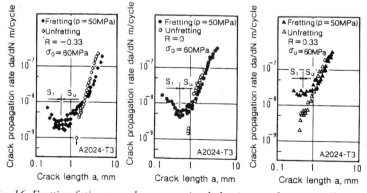

*Fig. 16. Fretting fatigue crack propagation behavior on aluminum alloy.*

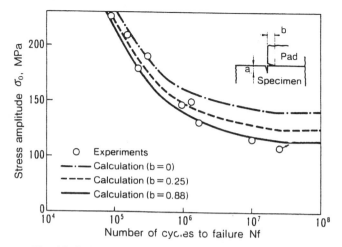

Fig. 17. S-N curves under fretting conditions (R=-1).

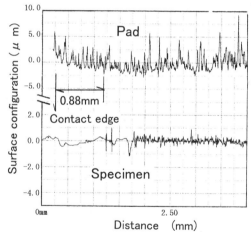

Fig. 18. Fretting damaged region observed on the contact surface
($\sigma_a$=110MPa,N=2×10$^7$).

## Conclusions

1. We simulated the entire fretting fatigue process, including crack initiation, wear, and crack propagation, using stress singularity parameters at the contact edge and fracture mechanics that considered the change in contact conditions due to wear.

2. The fretting fatigue strength or life decreased as the wear near the contact edge increased, and converged at a certain level in the b >0.88 mm region. This estimated convergence coincides well with the experimental results.

3. The estimated fretting fatigue crack propagation rate (da/dN) showed its minimum at a crack length of about a = 0.15 mm, which coincides well with the experimental results.

**References**

1. Gassner, E.,   The value of surface-protective media against fretting corrosion on the basis of fatigue strength tests, *Laboratorium fur Betriebsfestigkeit TM19/67*, 1967.
2. Buch, A., Fatigue and fretting of pin-lug joints with and without interference fit, *Wear*, 1977, 43, p. 9.
3. Hattori, T., Kawai, S., Okamoto, N. and Sonobe, T.,   Torsional fatigue strength of a shrink- fitted shaft, *Bulletin of the JSME*, 1981, 24, 197, p. 1893.
4. Cornelius, E. A. and Contag, D., Die Festigkeits-minderung von Wellen unter dem Einflu $\beta$   von Wellen-Naben- Verbindungen durch Lotung, Nut und Pa $\beta$ feder, Kerbverzahnungen und Keilprofile bei wechselnder Drehung, *Konstruktion,* 1962, 14, 9, p. 337.
5. Hattori, T., Sakata, S. and Ohnishi, H., Slipping behavior and fretting fatigue in the disk/blade dovetail region, *Proceedings, 1983 Tokyo Int.   Gas Turbine Cong.,* 1984, p. 945.
6. Johnson, R. L. and Bill, R. C., Fretting in aircraft turbine engines, *NASA TM X-71606,* 1974.
7. Hattori, T., Nakamura, M. and Watanabe, T., Fretting fatigue analysis by using fracture mechanics, *ASME Paper No.84-WA/DE-10,* 1984.
8. King, R. N.and Lindley, T. C., Fretting fatigue in a 3 $^1/_2$ Ni-Cr-Mo-V rotor steel, *Proc. ICF5,* 1980, p. 631.
9. Okamoto N. and Nakazawa, M., Finite element incremental contact analysis with various frictional conditions, *Int. J. Numer. Methods Eng,* 1979 14, p. 377.
10. Hattori, T., Sakata, H. and Watanabe, T., A stress singularity parameter approach for evaluating adhesive and fretting strength, *ASME Book No. G00485, MD-vol.6,* 1988, p. 43
11. Hattori, T. et al., Fretting fatigue analysis using fracture mechanics, *JSME Int. J, Ser. I,* 1988, 31, p. 100.
12. Hattori, T. et al., Fretting fatigue analysis of strength improvement models with grooving or knurling on a contact surface, *ASTM STP 1159,* ASTM, Philadelphia, 1992, p. 101.
13. Hattori, T., Fretting fatigue problems in structural design, *Fretting Fatigue,* ESIS Publication 18, 1994, p. 437.
14. Ohuchida, H., Usami, S. and Nishioka, A., Initiation and propagation behavior of fatigue crack, *Trans. Japan Soc. Mech. Eng.*, 1975, p.703.
15. Tanaka, T., Nakai, Y. and Yamashita, M., Fatigue crack growth threshold of small cracks, *Int. J. Fracture*, 1981, p.519.
16. El Haddad, M. H., Smith, K. N. and Topper, T. H., Fatigue crack propagation of short cracks, *Trans. ASME. J. Eng. Mater. Technol.* ,1979, p.42.
17. Sato, K., Fujii, H. and Kodama, S., *Fatigue Prevention and Design,* 1986, p.161.

M. Helmi Attia[1]

## On the Standardization of Fretting Fatigue Test Method—Modeling Issues Related to the Thermal Constriction Phenomenon and Prediction of Contact Temperature

**REFERENCE:** Attia, M. H., "**On the Standardization of Fretting Fatigue Test Method-Modeling Issues Related to the Thermal Constriction Phenomenon and Prediction of Contact Temperature,**" *Fretting Fatigue: Current Technology and Practices, ASTM STP 1367,* D. W. Hoeppner, V. Chandrasekaran, and C. B. Elliott, Eds., American Society for Testing and Materials, West Conshohocken, PA, 2000.

**ABSTRACT:** The temperature field in the contact zone has a significant effect on the material microstructure, its properties, the oxidation process, and the thermal contact stresses. To standardize fretting fatigue tests, one has to be able to predict and control the contact temperature. Since direct temperature measurement is practically impossible, analytical models are required to estimate the friction-induced temperatures rise under fretting conditions. The main objective of the present work is to model the thermal constriction phenomenon in fretting fatigue and wear processes, considering the roughness and waviness of contacting surfaces. These asperity-scale models can be combined with large scale analyses, e.g. finite element method, to account for the thermal characteristics of the whole tribo-system, its boundary conditions, as well as the spatial variation in the slip amplitude and coefficient of friction over the interface. The debatable question on whether the contact temperature in fretting fatigue is significant is addressed, considering a wide range of materials and applied loads. The analysis showed that the randomness of the contact size may substantially increase the micro-constriction impedance of the fretting interface. The paper is concluded with recommendations for future work to experimentally validate these models, and to examine the effect of the spatial maldistribution of the micro-contacts, and the effect of surface oxide on the contact temperature prediction.

**KEYWORDS:** fretting fatigue, fretting wear, modeling, thermal contact resistance, thermal constriction, contact temperature, surface topography

### Introduction
It is well established that the temperature field at the contact interface has a decisive effect on the mechanical and chemical aspects of the fretting process [1-5]. To standardize and improve the repeatability of the fretting fatigue test results, it is imperative to ensure that the contact temperatures of the original component and the fretting fatigue and specimen are kept

---

[1] Adjunct Professor, Mechanical Engineering Dept., McMaster University, 1280 Main Street West, Hamilton, Ontario, Canada L8S 4L7.

31

as nearly identical as possible *[6,7]*. Temperature measurement at asperity contacts is practically impossible due to the nature of the temperature field (both in the space and time domains), and the limitations of measurement techniques. Computer simulation results reported by Attia et al. *[8]* showed that the temperature gradient in the contact plane is very steep over a small area of the order 5-15 μm around the center of the asperity contact. The temperature rise is also shown to be contained within a very shallow subsurface layer of approximately 50-100 μm. In addition, this analysis indicated that the temperature field changes rapidly within one-quarter of the oscillation cycle, which is of the order of 2-20 ms. This demonstrates that with the physical dimensions and the response time of conventional temperature sensing elements, the desired spatial and temporal resolutions cannot be achieved. While dynamic thermocouples are severely limited by the presence of wear debris and the material work hardening, radiation and thermographic techniques are also not suitable for measuring contact temperatures of opaque solids. Theoretical and numerical models emerge, therefore, as the most viable and practical approach for predicting the contact temperatures and the related thermoelastic effects.

Debris analysis and metallurgical examination of the microstructure of the fretting zone suggest that the contact temperature may exceed 700 K *[1]*. While similar results were obtained by direct measurement of the temperature gradient in the subsurface layer *[9]*, contradictory results were also reported by others *[10,11]*, who concluded that the contact temperature rise is only in the order of 20 K.

The main objective of this work is three-fold. First, to present an analytical description of the thermal constriction phenomenon during the fretting action, and to quantitatively relate the friction-induced contact temperature rise to the variables of the fretting fatigue process. These asperity-scale models recognizes the stochastic nature of the micro- and macro-irregularities of contacting surfaces and their response to external loads. Second, to propose integrating these models with large scale analysis, e.g. finite element method, to predict the thermal response behaviour of real structures, e.g. the fretting fatigue testing set-up. This approach will allow us to consider realistic boundary conditions, the nonuniformity of the contact stresses, and the variation in the slip amplitude over the fretting interface. The third objective is to revisit the debatable question on whether the contact temperature in fretting fatigue is significant. The paper is concluded with recommendations for future work to experimentally validate these constriction models and to provide estimates for the uncertainties associated with their predictions.

## Micro- and Macroscopic Features of the Contact Interface

Complete modelling of the thermal constriction phenomenon requires a suitable and adequate specification of the surface texture, a knowledge of the mode of deformation under applied load, and the ability to estimate the mechanical and thermal response of the interface within a known confidence limits. Due to the roughness and waviness of real surfaces, point to point contact is observed under mechanical and thermal loading. A number M of these micro-contact areas $A_{mic}$ are usually clustered within a smaller number of bounded zones known as the contour areas $A_{cn}$ or the macro-contacts. The size and spacing of the contour areas depend on the amplitude and wavelength of the surface waviness. The ratio $\epsilon^2$ between the real and apparent contact areas $A_r$ and $A_a$, respectively, is usually < 1-5%, depending on the material properties, surface topography, and applied load. The contact configuration is defined by the following parameters:

a- the average radius of the micro-contact area, $a_{mic}$.
b- the constriction ratio, $\epsilon^2$.
c- the flow pressure over the micro-contact area, $p_m$, which is a function of the coefficient of friction, the slope of the surface asperities [12] and strain hardening. As a first approximation, it is usually taken as the material hardness H at the surface.
d- the density of the micro-contact areas per unit area, $\gamma = M/A_a$.

Mathematically, the contact between two rough surfaces can be reduced to that of an equivalent rough surface in contact with a perfectly smooth semi-infinite body [13,14]. The standard deviation of the asperity heights $\bar{\sigma}$ and the mean slope of the asperities $\bar{m}$ of the equivalent rough surface are related to that of the original surfaces $S_1(\sigma_1,m_1)$ and $S_2(\sigma_2,m_2)$ by the following relations:

$$\bar{\sigma} = \sqrt{\sigma_1^2 + \sigma_2^2}, \quad \text{and} \quad \bar{m} = \sqrt{m_1^2 + m_2^2} \tag{1}$$

For normally distributed surface asperities, the parameters defining the micro-contact configuration are determined by the following relations [13]:

$$\bar{a}_{mic} = \frac{\bar{\sigma}}{\bar{u}.\bar{m}} \tag{2}$$

$$\gamma = \frac{M}{A_a} = \left(\frac{\bar{m}}{\bar{\sigma}}\right)^2 . \frac{\bar{u}\ \phi(\bar{u})}{2\pi\ \Phi(-\varsigma)} \tag{3}$$

where, $\bar{a}_{mic}$ is the average radius of the asperity contacts, $\bar{u}$ is the ratio between the separation 'u' between the median plane of the equivalent rough surface and the smooth semi-infinite body and its standard deviation, $\bar{u} = u/\bar{\sigma}$, and

$$\phi(\bar{u}) = \frac{1}{\sqrt{2\pi}}\ e^{-\bar{u}^2/2}, \quad \text{and} \quad \Phi(-\varsigma) = \int_{-\varsigma}^{\varsigma} \phi(\bar{u})\ d\bar{u} \tag{4}$$

The effect of the applied pressure $p_a$ on $\bar{u}$ is defined by the following relation [12]:

$$\frac{\phi(\bar{u})}{\Phi(-\varsigma)} = \frac{p_a}{p_m} \tag{5}$$

The effect of the dimensionless applied load $p_a/p_m$ on the parameters $\bar{u}$, $\epsilon$, $\bar{a}_{mic}$, and $\gamma$ is shown in Figure 1. The figure shows that while an order of magnitude increase in the applied load results in a small increase in $\bar{a}_{mic}$ (by only a factor of 1.3-1.8), it causes a significant increase in $\gamma$ (by a factor of 4-5). This suggests that the increase in the real contact area with load is mainly due to the increase in the number of micro-contacts rather than the increase in its average radius. It is important, however, to realize that although $\bar{a}_{mic}$ is a weak function of the applied normal load, $a_{mic}$ exhibits a strong random variation around its mean value. The probability density function $\Psi\{\bar{a}\}$ of the dimensionless micro-contact radius, $\bar{a} = a_{mic} (\bar{m}/\bar{\sigma})$, can be described by the following equation [13]:

$$\Psi(\bar{a}) = \frac{[(\bar{u}+\bar{a})^2 - 1]\ \Psi(\bar{u}+\bar{a})}{\bar{u}\ \Psi(\bar{u})} \tag{6}$$

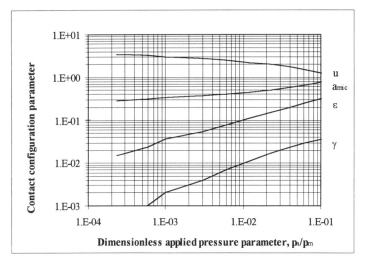

Figure 1  *Effect of the applied contact pressure $p_a$ on the parameters of the micro-contact configuration: $\bar{u}$, $\bar{a}_{mic}$, $\epsilon$, and $\gamma$.*

A three-dimensional presentation of the dependance of the distribution of asperity contact radius $\Psi\{\bar{a}\}$ on the applied load (though its effect on the surface separation $\bar{u}$) and the surface roughness parameters $\bar{\sigma}$ and $\bar{m}$ (though the dimensionless contact radius $\bar{a} = (a_{mic}\,\bar{m})/\bar{\sigma}$) is shown in Figure 2. It will be shown later that the variability in $a_{mic}$ has a significant effect on the uncertainty in modelling the thermal constriction phenomenon.

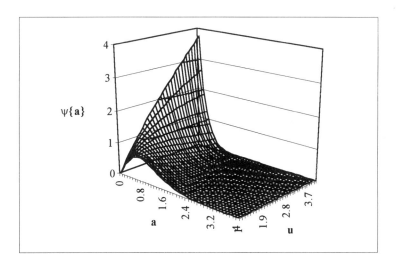

Figure 2  *The probability density function of the dimensionless micro-contact radius as a function of the separation between contacting surfaces*

**Thermal Constriction Impedance**

In tribo-systems, as in fretting fatigue and fretting wear processes, the micro-contact areas can be looked at as micro-frictional heat sources. As the frictional heat flow lines dissipate from the asperity micro-contact area into the solid material, they spread out rather than taking the least resistance, straight path. This gives rise to the so-called "thermal constriction (or spreading) resistance". To overcome this resistance, a steep temperature gradient has to be established in a subsurface layer, known as the thermal disturbance zone. The volume which encompasses a single micro-contact and extends some distance into either solid is defined as the elemental heat flow channel HFC [8,19].

Under static contact and time-independent heat flow conditions, Yovanovich et al. demonstrated [15] that the temperature drop due to the micro-constriction resistance is the difference between the average temperature rise at the contact area $\overline{T}_c$ and the mean temperature $\overline{T}_m$ of the **HFC** cross section at the contact plane. If $\theta$ is defined as the difference between the temperature T at any point and the average bulk interface temperature $\overline{T}_b$, due to external heat sources, then:

$$R_{mic} = \frac{\overline{\theta}_c - \overline{\theta}_m}{Q_{hfc}} \qquad (7)$$

where, $Q_{hfc}$ is the rate of heat flow into a single micro-contact, which varies sinusoidally with time. Non-dimensionalization of the constriction resistance $R_{mic}$ can be made by using the dimensionless constriction parameter $\psi_{mic}$ [16]:

$$\psi_{mic} = 2k\sqrt{A_{mic}}\ R_{mic} \qquad (8)$$

where, k is the material thermal conductivity, and $A_{mic}$ is the micro-contact area. Yovanovich et al. [17] concluded that the $\psi_{mic}$ is insensitive to the shape of the contact area, when $\sqrt{A_{mic}}$ is used as the characteristic length of the constriction resistance. It is clear from Eq. 7 that the definition of the microscopic constriction resistance excludes the thermal resistance of the heat flow channel material, which should be overcome by the temperature drop $\overline{\theta}_m$.

When the contact is static and the heat flow rate is constant, the constriction resistance is only a function of the material thermal conductivity, the size of the contact area in relation to that of the HFC cross section, and the thermal boundary condition at the contact area. However, when the heat flow varies periodically with time, the thermal constriction phenomenon is additionally dependent on the frequency f of the cyclic variation of the heat input, and the thermal diffusivity $\alpha$ of the material. Therefore, one should approach and think of the thermal constriction phenomenon under fretting conditions in terms of constriction impedance **Z** rather than resistance **R**. The constriction impedance will then be the vector sum of both the resistive **R** and capacitive $X_c$ reactance of the lumped thermal disturbance layer, allowing us to use the generalized Ohm's law for a series R-C circuit, which is excited by an alternating heat source:

$$Z = \frac{|\theta|}{|Q|} = \sqrt{R^2 + X_c^2} = \sqrt{R^2 + \left(\frac{1}{\omega C}\right)^2} \qquad (9)$$

where, $\omega$ is circular frequency, $\omega = 2\pi f$, and $|Q|$ and $|\theta|$ are the amplitudes of the sinusoidal excitation (heat flow rate) and the response (temperature rise) of the lumped R-C system, respectively. Equation 9 suggests that the increase in the frequency will lead to a reduction in the impedance of the thermal disturbed zone in the contact region.

## Modelling of the Thermal Constriction Phenomenon and Friction-induced Temperature Rise at the Fretting Interface

To predict the response behaviour of a real fretting tribo-system, with complicated geometry and non-uniform thermal boundary conditions, the thermal characteristic of the whole system should be considered. This can only be achieved by solving the problem on two stages. First, using numerical methods, e.g. finite element or finite difference, one can determine the distribution of the average bulk temperature, micro-slip, and contact stresses over the contact interface. Second, with these values as input, along with the localized coefficient of friction $\mu$, separate models that are developed to deal with the microscopic features of the surface topography should be used to estimate: (a) the division of friction heat between contacting solids, (b) the maximum temperature rise over the micro-contact area, (c) the temperature gradient in the subsurface layer and (d) the associated thermoelastic effects. The mechanical aspect of the second part of the solution is addressed in [18], where a three-dimensional interface element was developed to model the normal and shear compliance of the interface and to deal with the material and geometric nonlinearities. The thermal aspect of this problem is also nonlinear, since the final solution depends on the interface conditions that are unknown in advance. Therefore, the solution has to proceed in iterative fashion. In the first iteration, the micro-slippage at the interface is ignored. Therefore, the conventional thermal contact resistance correlations for static contact can be used for all interfacial nodal points. Once the interface bulk temperatures are determined for various nodal positions, the conditions at the interface are updated and frictional heat sources are attached to those nodes that undergo micro-slip. The proper thermal constriction correlations, to be discussed in subsequent sections, should then be applied to these nodes. Through iterations, the bulk interface temperatures and the coefficients of frictional heat partitioning are tracked and updated until a pre-determined convergence criterion is met. At this point, the solution representing the state of equilibrium will be reached.

Figure 3a shows the finite element mesh idealization of a fretting fatigue test specimen and the wear pad. Because of symmetry, only one-quarter of the model is shown in the figure. A close-up of the fretting interface, modelled with interface elements is shown in Figure 3b. The interface area surrounding a single nodal point, represents a single or multiple contour area(s) $A_{cn}$, depending on the waviness of contacting surfaces. Each of these areas contains M heat flow channels HFC that are connected in parallel (Figure 3c). The idealization of a single contour area, in which the asperity micro-contacts are uniformly distributed in a square mech is shown in Figure 3d. The equivalent thermal network of the fretting interface, which is based on voltage-temperature analogy, is shown in Figure 4. It is evident that the analysis of the heat transfer process at a single micro-contact spot constitutes the basic cell for predicting the overall thermal resistance of the interface.

The computation time of the method presented by Attia et al. [8] to predict the temperature field at and around the micro-contact area is very intensive, since very small time steps are required to ensure adequate convergence to the correct solution. This is particularly

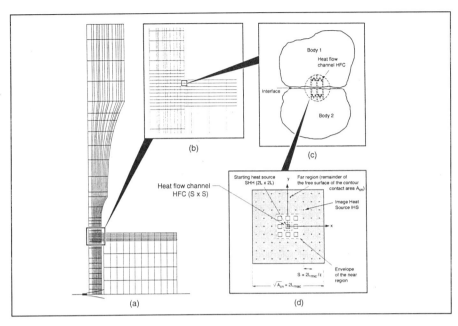

Figure 3  *Modelling of the thermal problem and constriction phenomenon in fretting fatigue testing: (a) finite element idealization of the test set up, (b) close-up of the fretting interface, (c) the equivalent heat flow channels, and (d) the idealized contact configuration of the frictional heat sources.*

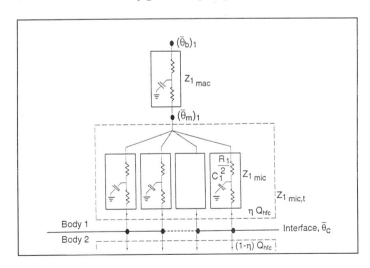

Figure 4  *Thermal network representing the micro- and macroscopic components of the interface thermal constriction impedance.*

true when the size of the contour contact area $A_{cn}$ is large. As $A_{cn}$ approaches infinity, the steady state solution does not exist. This problem can be overcome by tracking the time variation of the temperature differences, instead of absolute temperatures, which converge very rapidly. Therefore, by adopting the concept of thermal impedance (or resistance), which is proportional to the temperature difference, this numerical difficulty is eliminated. In addition, expressing the thermal constriction impedance in Ohm's law form (Eq. 9) allows us to use the electrical network theory to obtain the total micro-constriction impedance $Z_{mic,t}$ of the individual HFC's (which are connected in parallel), and to combine it with the macroscopic constriction impedance $Z_{mac}$ (which is connected in series with $R_{mic,t}$). The overall constriction resistance of the interface $Z_{interface}$ can thus be calculated and used to partition the total frictional heat $Q_f$ generated over the contour area in inverse proportion to the relative impedance of contacting solids.

*Micro-constriction Impedance in Fretting*

As mentioned earlier, the analysis of the thermal constriction phenomenon can be made amenable by replacing the contact between two rough surfaces by a perfectly smooth, rigid and stationary surface in contact with an equivalent rough, oscillating semi-infinite body. The analysis can further be simplified by assuming that the contact asperities have constant square cross section ($2L_{mic} \times 2L_{mic}$) and are arranged in a square pattern with a spacing distance $S$ over the free surface of the half-space (Figure 3d). The following assumptions are also imposed:

a- The relative displacement between the two semi-infinite bodies is described by a simple harmonic motion; $x = \delta \sin(\omega t)$, where $\delta$ and $\omega$ are the amplitude and the circular frequency of oscillation, respectively.

b- Except for the micro-contacts, the free surface of the semi-infinite body is adiabatic.

c- The thermo-mechanical properties are temperature-independent.

d- The materials in contact are homogeneous and isotropic, and their properties are temperature- independent.

The approach followed by Attia et al. *[19]* to estimate average temperatures $\bar{\theta}_c$ and $\bar{\theta}_m$ (Eq. 7) is based on the method of infinite images (Figure 3d), in which the average temperature rise over the micro-contact area is the superposition of the contribution of the frictional heat generated at the micro-contact attached to this HFC (denoted as the starting heat source 'shs') $\theta_{shs}$, and due to the contributions from all other neighbouring image heat sources $\theta_{ihs}$. The latter can be approximately evaluated by grouping the neighbouring image sources into two regions; namely, the near- and far- regions. In the near region, which surrounds the starting heat source and covers a small portion of the contact interface, only a finite number of frictional heat sources $[N_{nr}-1]$ is contained. The contribution of the far region to the average temperature rise over the area $A_{mic}$ is the difference between the average temperature rise $\theta_{c,\infty}$ produced by heating the free surface of the semi-infinite body by the effective heat flux $q_e$ and the average temperature drop $\theta_{c,nr}$ due to a negative heat flux ($-q_e$) over the square area of the *near region* bounded by $x,y = \pm [ (\sqrt{N_{nr}}) / 2S]$. Therefore,

$$\bar{\theta}_c = \left[ \bar{\theta}_{c,shs} \right] + \left[ \sum_{ihs=1}^{N_{nr}-1} \bar{\theta}_{c,ihs} \right] + \left[ \bar{\theta}_{c,\infty} - \bar{\theta}_{c,nr} \right] \qquad (10)$$

The effective heat flux $q_e$ is defined by the following relation:

$$q_e = q_f \epsilon^2 , \quad q_f = \mu \, p_m \, |v(t)| = \mu \, p_m \, \delta \, \omega \, | \cos(\omega t) | \qquad (11)$$

where $q_f$ is the heat flux over the micro-contact area $A_{mic}$, $\mu$ is the coefficient of friction, and $\delta$ is the slip amplitude. To achieve the desired degree of accuracy, only nine image heat sources in the near region are required [8]. Due to the cyclic nature of the thermal load, the instantaneous thermal constriction impedance should be determined:

$$Z_{mic, inst} = \frac{(\bar{\theta}_c - \bar{\theta}_m)}{q_f \cdot (4L_{mic}^2)} \qquad (12)$$

where $q_f$ is defined by Eq. 11. Substitution of the impedance $Z_{mic,ins}$ for $\mathbf{R}$ into Eqs. 7 and 8, the time variation of the micro-constriction impedance parameter $\psi$ is obtained:

$$\psi_{mic, inst} = 4\,k\,L_{mic}\,Z_{mic, inst} \qquad (13)$$

At the ends of the oscillation stroke, $\bar{t} = 0.25$ and $0.75$, the parameter $\psi_{mic,inst}$ approaches infinity, since the relative velocity of between the contacting solids and consequently the frictional heat generation reach zero. Fortunately, the analysis presented in [19] showed that for the practical range of $\underline{Fo} > 10^3$, the variation in the $\psi$ with time is insignificant. Apart from the singularity points at $\bar{t} = (0.25 \text{ and } 0.75) \pm 0.05$, the constriction parameter varies only within $\pm 4.5\%$ of its average value. For all practical purposes, the constriction parameter is assumed to be time-independent, and only its average value $\bar{\psi}_{mic}$ is to be used.

Analysis of the results presented in [19] showed that the increase in the constriction ratio $\epsilon^2$ results in a nearly linear reduction in the constriction resistance. As expected, as $\epsilon$ approaches unity, the constriction (or spreading-out) phenomenon will no longer exist, and the constriction resistance approaches zero. The increase in Fourier modulus, $Fo = \alpha/(fL_{mic}^2)$, leads to an increase in the average constriction parameter $\bar{\psi}_{mic}$, particularly at lower levels of Fourier modulus, due to the increase in the thermal capacitive reactance $\mathbf{X}_c$ of the interface layer (Eq. 9). The results also indicated that the mode of motion (moving versus stationary heat sources) and the amplitude ratio $\mathbf{A}$, which represents the ratio between the slip amplitude $\delta$ and the characteristic length of the micro-contact area $L_{mic}$, have no significant effect on the micro-constriction parameter for $0.5 \leq \mathbf{A} \leq 10$. Therefore, to establish a correlation between $\bar{\psi}_{mic}$ and the process variables, only the effects of Fourier modulus Fo and the constriction ratio $\epsilon^2$ need to be considered. A non-linear least square iterative procedure was used to curve fit the surface $\bar{\psi}\{\epsilon, Fo\}$ shown in Figure 5 [19]:

$$\bar{\psi}_{mic} = A + B\sqrt{Fo} + \frac{C}{\sqrt{Fo}} + D\,\epsilon \qquad (14)$$

where, $A = 0.953$, $B = 0.00074$, $C = 0.0955$, and $D = -1.256$, for $\epsilon^2 \leq 0.0625$, $250 \leq Fo \leq 10^5$, $0.5 \leq \mathbf{A} \leq 10$, which cover the typical fretting conditions encountered in practice.

*Macro-constriction Impedance in Fretting*

At this point, equations 12 to 14 allow us to estimate quantity $(\bar{\theta}_c - \bar{\theta}_m)$. To determine the average temperature rise at the micro-contact area $\theta_c$, one has calculate the average

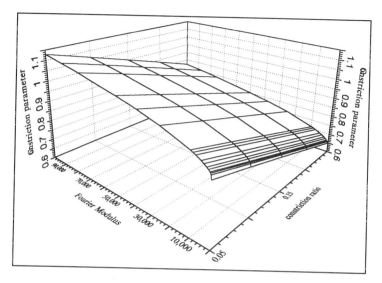

Figure 5    *Effect of Fourier modulus Fo and the constriction ratio $\epsilon$ on the average micro-constriction parameter $\bar{\psi}_{mic}$.*

temperature over the HFC cross section in the contact plane $\bar{\theta}_m$, taking into consideration the extent of the contour area $A_{cn}$. Defining the macro-constriction impedance $\mathbf{Z}_{mac}$ as:

$$\mathbf{Z}_{mac} = \bar{\theta}_m / (q_f \cdot 4L_{mic}^2) = \bar{\theta}_m / Q_{hfc} \tag{15}$$

the total interface constriction impedance $\mathbf{Z}_{interface}$ is thus:

$$\mathbf{Z}_{interface} = \mathbf{Z}_{mic} + \mathbf{Z}_{mac} = \bar{\theta}_c / Q_{hfc} \tag{16}$$

where, $Q_{hfc}$ is the portion of the total frictional heat $Q_f$, which passes through each of the M micro-contact areas; $Q_{hfc} = Q_f/M$. Strictly speaking, the addition of the micro- and macro-impedances (Eq. 16) should be treated as a vectorial sum since the temperatures $\bar{\theta}_c$ and $\bar{\theta}_m$ are not in phase with the heat input $Q_{hfc}$. This point will be addressed in the next section. When the contour area $A_{cn}$ has a finite size, the contribution of the far region heat source $\bar{\theta}_{c,\infty}$ (in Eq. 10) should be replaced by $\bar{\theta}_{c,cn}$. Equation 10 can then be rewritten as:

$$\bar{\theta}_c = Q_{hfc} \cdot \mathbf{Z}_{interface} = \left[ \bar{\theta}_{c,shs} + \sum_{ihs=1}^{N_{nr}-1} \bar{\theta}_{c,ihs} + \left( \bar{\theta}_{c,cn} - \bar{\theta}_{c,nr} \right) \right] \pm \left( \bar{\theta}_{m,\infty} \right) \pm \left( \bar{\theta}_{c,\infty} \right) \tag{17}$$

The average temperature $\bar{\theta}_m$ is estimated using a relationship similar to Eq. 10. Upon substituting $\bar{\theta}_m$ and Eq. 12 into Eq. 17, the following relation is obtained:

$$\bar{\theta}_c = Q_{hfc} \cdot \left[ \mathbf{Z}_{mic} + \frac{\left( \sum_{i=1}^{N_{nr}} \bar{\theta}_{m,ihs} - \bar{\theta}_{m,nr} \right) + \left( \bar{\theta}_{m,\infty} - \bar{\theta}_{c,\infty} \right) + \bar{\theta}_{c,cn}}{Q_{hfc}} \right] \tag{18}$$

Examination of Eq. 18, using the formulation developed in [19], indicates that the difference between the average temperature of the HFC contact plane due to the N discrete starting and image heat sources $\sum \bar{\theta}_{m, ihs}$, $1 < ihs < N_{nr}$, and that due to the negative heat flux $q_e$ over the near region square $\bar{\theta}_{m,nr}$ is two orders of magnitude smaller than $\bar{\theta}_{c, cn}$ produced by the far region heat source. For an infinite far region heat source, the temperature distribution over the mean contact area is uniform and the term $(\bar{\theta}_{m, \infty} - \bar{\theta}_{c, \infty})$ becomes zero. Therefore, Eq. 18 can further be simplified to the following expression:

$$\bar{\theta}_c = Q_{hfc} \cdot \left[ Z_{mic} + \frac{\bar{\theta}_{c, cn}}{Q_{hfc}} \right] \tag{19}$$

With reference to Eq. 16, equation 19 states that the macroscopic constriction impedance $Z_{mac}$ can simply be obtained by calculating the average temperature over the micro-contact area due to the average effective heat flux $q_e$ covering the whole contour area:

$$Z_{mac} = \frac{\bar{\theta}_{c, cn}}{Q_{hfc}} = \frac{\bar{\theta}_{c, cn}}{(4 \mu p_m) (\delta \omega) (L_{mic}^2) \cdot |\cos(\omega t)|} \tag{20}$$

Assuming that the 'M' micro-contacts are identical and uniformly distributed over the entire contour area, then the total micro-constriction impedance $Z_{mic,t}$ is simply: $Z_{mic,t} = Z_{mic}/M$. The ratio $A$ between the slip amplitude $\delta$ and the characteristic length of the contour area $L_{mac} = (\sqrt{A_{cn}})/2$, is typically of the order of 0.01-0.1. Examination of the effect of the ratio $A$ on the average contact temperature $\bar{\theta}_{c, cn}$, indicated that $\bar{\theta}_{c, cn}$ is independent of $A$, for $A < 0.1$, with variation of less than 1%. Therefore, $Z_{mac}$ can accurately be estimated by considering the periodic variation of the frictional heat flux in the time domain only, assuming that the tribo-system is stationary; $A = 0$. Even with this simplification, an analytical closed-form solution for $Z_{mac}$ is difficult to obtain since the time variation of the frictional heat generation is described by the absolute value of cos ($\omega t$). To circumvent this difficulty, a simple harmonic analysis is used to express $|\cos(\omega t)|$ as a Fourier series of cosine terms [20]:

$$|\cos(\omega t)| = \frac{2}{\pi} - \frac{4}{\pi} \left[ \sum_{n=1}^{\infty} \frac{\cos\left(2n\left(\omega t + \frac{\pi}{2}\right)\right)}{n^2 - 1} \right] \tag{21}$$

Equation 21 indicates that the average micro-contact temperature $\bar{\theta}_{c, cn}$, and consequently the macro-constriction impedance $Z_{mac}$ can be assembled from the solutions of two fundamental problems in which the heat flux over a square area ($A_{cn} = 2L_{mac} \times 2L_{mac}$) is (a) constant (first term in Eq. 21), and (b) a simple harmonic (the second summation term).

Using the models developed in [8,19], the quasi-steady state temperature field over the asperity contact, due to the far region heat source was examined for a wide range of Fourier modulus. Analysis of the results showed that the difference between the centroidal temperature $\theta_{max,cn}$ and the average temperature $\bar{\theta}_{c, cn}$ is < 1%. Therefore, to simplify the analysis, without loss of accuracy, the average temperature $\bar{\theta}_{c, cn}$ can be substituted by the temperature rise $\theta_{max,cn}$ at the center of a circular heat source with sinusoidally varying strength. The only condition to be fulfilled is ensure that the area of the circular heat source is equal to that of the contour area $A_{cn}$, in order to satisfy Eq. 8.

The solution for the temperature rise at any point 'o' in a semi-infinite body due to a uniform simple harmonic heat flux $q_{sin}$ spread over an area 'dA' on its surface [21] was used to find the sought-for centroidal temperature $\theta_{o,sin}$:

$$\theta_{o,sin} = \left[ \frac{q_o \ L_{mac}}{\sqrt{\pi} \ k \ \eta} \right] \left\{ \left[ \cos(\omega t)(1 - e^{-\eta}(\cos\eta - \sin\eta)) \right] - \left[ \sin(\omega t)(-1 + e^{-\eta}(\cos\eta - \sin\eta)) \right] \right\} \quad (22)$$

where, $q_o$ is the amplitude of the sinusoidal heat source, and $\eta = 2L_{mac} \cdot \sqrt{f/\alpha}$. The solution for the temperature rise $\theta_{o,const}$ due to a circular constant heat flux ($q_{const}=2/\pi$ in this case, as indicated by Eq. 21) over a semi-infinite body is readily available [22]:

$$\overline{\theta}_{const} = \frac{2 \ q_{const} \ L_{mac}}{\sqrt{\pi} \ k} \quad (23)$$

From Eqs. 22 and 23, the centroidal temperature $\theta_{max,cn}$, which is equivalent to the average temperature $\theta_{c,cn}$, can now be estimated and used to obtain the instantaneous macroscopic constriction impedance:

$$Z_{mac,inst} = \frac{\overline{\theta}_{c,cn}}{Q_{hfc}} \approx \frac{\theta_{max,cn}}{Q_{hfc}} \approx \frac{1}{Q_{hfc}} \cdot \left[ \overline{\theta}_{const} + \sum_{i=1}^{n} \overline{\theta}_{sin} \right]_{cn} \quad (24)$$

Again, the singularity points near the ends of the oscillation stroke, $\overline{t} = 0.25$ and $0.75 \pm 0.015$, were removed and the average constriction parameter $\overline{\psi}_{mac}$ was estimated for various Fo modulus, based on the characteristic length $L_{mac}$ of the contour area $A_{cn}$:

$$\overline{\psi}_{mac} = 2 k \sqrt{A_{cn}} \ Z_{mac} = 4 k L_{mac} \ Z_{mac} \quad (25)$$

Analysis of the results obtained in this investigation indicated that for Fo <100, the phase angle between the heat flow and the temperature rise at various locations in the contact plane varies between 0 and $\pi/6$. However, in calculating the average temperature rise $\theta_{c,cn}$ and the constriction impedance $Z_{mac}$, the assumption that $\phi=0$ results in a relative error of <2% due to the small changes in the amplitude of the instantaneous temperature level during this period of the cycle. In relation to the micro-constriction impedance $Z_{mic}$, the results presented in [19] showed that the phase difference between the quantity $(\theta_c - \theta_m)$ and the sinusoidal friction heat generation is ≤ 3°. Therefore, one can simply add $Z_{mic}$ and $Z_{mac}$ algebraically, instead of treating them vectorially, to obtain the overall impedance $Z_{interface}$:

$$Z_{interface} = Z_{mic} + Z_{mac} \quad (26)$$

Considering the thermal network shown in Figure 4, the following expression can be obtained for the constriction impedance of the interface $Z_{interface}$:

$$Z_{interface} = \frac{1}{k \ L_{mic} \ M} \cdot \left[ \psi_{mic} + \epsilon \ \psi_{mac} \cdot \sqrt{M} \right] \quad (27)$$

The average temperature over the micro-contact area can now be estimated:

$$\bar{\Theta} = \frac{\pi}{2} \zeta \left[ \psi_{mic} + \epsilon \ \psi_{mac} \sqrt{M} \right] \tag{28}$$

where, the dimensionless temperature rise $\Theta$ is related to $\theta$ by the following relation:

$$\bar{\Theta} = \theta \cdot \frac{k}{\left(4L_{mic}^2 \cdot \bar{q}_f\right).L_{mic}} = \theta \cdot \frac{k}{\left(4 \ \mu \ p_m \ \delta \ f\right).L_{mic}} \tag{29}$$

$\bar{q}_f$ is the average heat flux generated over the micro-contact area during the fretting cycle, and $\zeta$ is heat partition coefficient, which is defined in terms of the fraction $\zeta Q_f$ of the total frictional heat $Q_f$ that flows into body 1, which oscillates with respect to the heat source. To determine the coefficient $\zeta$, two coupling conditions should be satisfied at the interface of bodies 1 and 2: (a) the continuity of the average temperature on the micro-contact areas: $\theta_{c,1} = \theta_{c,2} = \theta_c$, and (b) the conservation of energy: $q_f = k_1 \ |\partial\theta_{c,1}/\partial z| + |\partial\theta_{c,2}/\partial z|$.

Applying Eq. 28 to each of the contacting bodies, the constriction impedance of the whole system $Z_{system}$ is obtained:

$$Z_{system} = \frac{[Z_{interface}]_1 \ [Z_{interface}]_2}{[Z_{interface}]_1 + [Z_{interface}]_2} \tag{30}$$

Therefore, the partition coefficient is readily determined from the following relation:

$$\zeta = \frac{[Z_{interface}]_2}{[Z_{interface}]_1 + [Z_{interface}]_2} \tag{31}$$

While the average temperature is needed to estimate the friction heat partitioning between contacting solids, the maximum temperature rise $\theta_{c,max}$ over the contact area is required to predict the tribological reactions of the system under consideration. Investigation of the relationship between the maximum and average temperature rise due to micro-constriction impedance $Z_{mic}$ was carried out using the formulation presented in [8,19]. The results indicated that the ratio $\Re_{\theta,mic} = \theta_{c,max}/\theta_c$ is $1.189 \pm 0.25\%$, for $250 < Fo < 10^5$. As stated earlier, the ratio between the maximum and average temperature rise due to macro-constriction impedance $Z_{mac}$ was shown to be nearly unity, for $Fo < 200$.

## Significance of the Temperature Rise in Fretting Fatigue

*Effect of Process Parameters on Contact Temperature*

To examine the significance of the friction-induced temperature rise in fretting fatigue, a number of materials is considered (see Table 1), for applied pressure-to-hardness ratio $p_a/p_m <$ 0.16. The following fixed conditions were assumed: Slip amplitude $\delta = 10 \ \mu m$, frequency $f = $ 100 Hz, equivalent rough surface, $\bar{\sigma} = 1 \ \mu m$, and $\bar{m} = 0.12$, apparent contact area $A_a = 225$ mm², heat partition coefficient $\zeta = 0.5$, and coefficient of friction $\mu = 1.0$.

Figure 6 shows the effect of applied load on the contact temperature rise. It can be seen that steels, for example, can experience temperature rise of nearly 100-140 °C, with contact pressure of 230-350 MPa, which correspond to $\epsilon^2 = 0.04$-$0.06$. Obviously, the change in surface conditions, particularly the waviness which affects the extent of the contour area will

Table 1- *Thermal and Mechanical Properties of the Selected Materials*

| Case No. | Material | Thermal conductivity, k, W/m-K | Thermal diffusivity α, m²/s | Hardness, $p_m$ N/m² |
|---|---|---|---|---|
| 1 | Aluminum alloy 6061-T6 | 179.9 | $6.89 \times 10^{-5}$ | $9.32 \times 10^{8}$ |
| 2 | Steel, AISI-1095 | 43.24 | $1.26 \times 10^{-5}$ | $5.88 \times 10^{9}$ |
| 3 | Stainless steel, SS-410 | 24.91 | $6.98 \times 10^{-6}$ | $3.82 \times 10^{9}$ |
| 4 | Ti-5Al-2.5Sn | 7.78 | $3.32 \times 10^{-6}$ | $2.94 \times 10^{9}$ |
| 5 | Ceramic- Silicon Nitride | 32 | $7.42 \times 10^{-4}$ | $1.96 \times 10^{10}$ |
| 6 | Polyimide | 0.36 | $2.30 \times 10^{-7}$ | $7.85 \times 10^{7}$ |
| 7 | Fluorocarbon | 10 | $3.16 \times 10^{-4}$ | $3.30 \times 10^{9}$ |

result in significant variation in the contact temperature. It is worth noting that although the contact pressure in fretting wear situations is usually much smaller than that in fretting fatigue, one still expects high contact temperatures, due to the high relative slip amplitude.

In comparison with steels, titanium alloys are characterized by their low conductivity k and thermal diffusivity α ($k_{Ti}/k_{st} \sim 0.18$, and $\alpha_{Ti}/\alpha_{st} \sim 0.26$) as well as low hardness ($H_{Ti}/H_{st} \sim 0.5$). Although this combination results in nearly the same constriction parameters $\psi_{mic}$, and $\psi_{mac}$, the total impedance $Z_{interface}$ is much higher in the case of titanium alloys due to the low thermal conductivity (Eq. 28). Considering the fact that Ti-alloys experience a transition temperature in the region of 400-500 °C *[2]*, where the fretting mechanism and wear rate change considerably, it becomes imperative to consider the effect of the friction-induced temperature rise in fretting fatigue and wear testing.

Fretting fatigue of stainless steels, which have intermediate thermal conductivity, produces contact temperature that fall between steels and Ti-alloys, as shown in Figure 6. Aluminum and its alloys present the other extreme, where the thermal conductivity and diffusivity are quite high ($k_{Al}/k_{st} \sim 4$, and $\alpha_{Al}/\alpha_{st} \sim 5.5$). Due to this thermal characteristics, the contact temperature rise is reduced by more than two orders of magnitude. It is interesting to compare Silicon nitride with steel, since the thermal conductivity is almost the same but the flow pressure (or hardness) is dramatically increased ($H_{SiN}/H_{st} \sim 3.3$). The resultant effect is much higher contact temperature rise (Figure 6). It should, however, be noted that due to sharp increase in the material hardness, applied pressure of 118 and 392 MPa, is required for steel and Silicon nitride, respectively, to obtain the same constriction effect, $\epsilon^2 = 0.02$.

*Uncertainty Analysis of Contact Temperature Prediction*

The models presented above assume that the asperity contacts are of the same size. To estimate the effect of randomness of the contact size on the micro-constriction impedance, the square pattern shown in Figure 3d is maintained with a fixed spacing of $S = 2L_{mic}/\epsilon$.

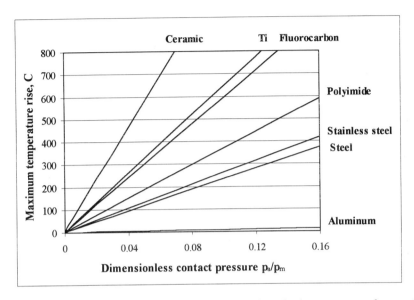

Figure 6    *Effect of the material properties and appl.—d pressure on the maximum temperature Rise during fretting fatigue, for the given operating conditions.*

The characteristic length $L_{mic}$ of each asperity contact varies randomly according to the density distribution function expressed by Eq. 6 (Figure 2). This function was found to be accurately described by the following polynomial function:

$$\Psi\{L_{mic}, \overline{u}\} = \sum_{j=1}^{j=5} c_j(\overline{u}) L_{mic}^j \qquad (32)$$

The coefficients $c_1$ to $c_6$ were estimated for various values of the dimensionless separation parameter $\overline{u}$. Using Eq. 32, one can establish a simple mathematical expression for the relation between the dimensionless separation parameter $\overline{u}$ and the standard deviation and mean value of the probability density function $\Psi\{\overline{a}\}$, $\sigma_a$ and $\overline{a}_{mic} = L_{mic}$, respectively (plotted in Figure 7). Using Eqs. 13, 14 and 32, the expected value of the micro-scopic constriction impedance $E[(Z_{mic})]$ is estimated by the following relation:

$$E[(Z)_{mic}] = \frac{1}{4k} \cdot \int_{L_1}^{L_2} \frac{1}{L_{mic}} \left( A + B\sqrt{Fo} + \frac{C}{\sqrt{Fo}} + D\epsilon \right) \cdot \sum_{j=1}^{j=6} c_j L_{mic}^j \, dL_{mic} \qquad (33)$$

where the practical values of the integration limits fall within the range $0 < \overline{L}_{mic} < 4$. Carrying out the integration after substituting for Fo in terms of $L_{mic}$, and dividing by $Z_{mic}$, one can estimate the ratio $E[(Z)_{mic}]/Z_{mic}$.

For stainless steel with $\sigma_e = 0.6$ μm, $m_e = 0.14$, $f = 20$ Hz, $Fo = 10^4$, and $p_a/p_m = \epsilon^2 = 0.006$, the separation parameter $\overline{u} = 2.5$, and the ratio $E[(Z)_{mic}]/Z_{mic}$ is estimated to be 2. With a rougher surface, $\sigma_e = 0.6$ μm, the ratio becomes 2.4. This demonstrates that the random variation in the micro-contact radius may substantially increase the thermal constriction impedance and the contact temperatures higher than the predictions shown in Figure 6.

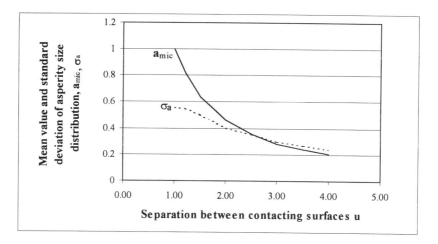

Figure 7    *Variation of the mean value $\bar{a}_{mic}$ and the standard deviation $\sigma_a$ of the probability density function $\psi\{\bar{a}\}$ with surface separation $\bar{u}$.*

It is recommended to investigate this problem further to establish the uncertainty limits on the predictions of the constriction models. The effect of the spatial maldistribution of the micro-contacts needs also to be examined using Monte Carlo simulation technique. Other issues that should be investigated include the effect of surface oxide on the constriction impedance and the contact temperature rise. This problem is quite relevant to fretting fatigue testing, since oxide formation and destruction processes are self-induced tribological reactions that cannot be separated from the fretting action. It is also recommended to carry out controlled experiments to validate the models presented in this investigation.

**Conclusions**

The micro-and macro-scopic components of the constriction impedance at the fretting interface are modelled, to estimate the maximum and average contact temperature rise. It is shown that these asperity-scale models can be integrated with large scale analyses, e.g. finite element method, to accurately model the thermal characteristics of the whole tribo-system. Analysis of the results indicated that the contact temperature rise can be quite significant when the contact pressure-to-hardness ratio is high. This effect is magnified when the material thermal conductivity is relatively low and /or its flow stress is high. Therefore, it is imperative to consider the effect of the friction-induced temperature rise in fretting fatigue and wear testing of such materials as titanium alloys and ceramics. Analysis of the results showed that the random variation in the micro-contact radius may lead to a significant increase in the thermal contact impedance of the fretting interface. Some recommendations are made for future investigation of the effect of the spatial maldistribution of the micro-contacts, as well as the effect of the surface films and oxides on the constriction impedance and the contact temperature predictions.

*Acknowledgment*

The work presented in this paper was conducted under the support of the Natural Sciences and Engineering Research Council of Canada, which the author greatly appreciates.

**References**

1.  Waterhouse, R.B., *Fretting Corrosion*, Pergamon Press, New York, 1st edition, 1972.

2.  Bill, R.C., "The Role of Oxidation in the Fretting Wear Process", *Proceedings*, International Conference on Wear of Materials, San Francisco, Ca., American Society of Mechanical Engineers, N.Y., 1981, pp. 238-250.

3.  Hurricks, P.L. and Ashford, K.S., "The Effect of Temperature on the Fretting Wear of Mild Steel", *Journal of the Institution of Mechanical Engineers*, London, 184(3L), 1969-1970.

4.  Kayaba, T., and Iwabuchi, A., "The Fretting wear of 0.45% Carbon steel and Austenitic Stainless Steel from 20°C up to 650°C in Air", *Proceedings*, International Conference on Wear of Materials, San Francisco, Ca., American Society of Mechanical Engineers, N.Y., 1981, pp. 229-237.

5.  Attia, M.H., "Friction-Induced Thermoelastic Effects due to Fretting Action", *Proceedings*, International Conference on Fretting Fatigue, R. Waterhouse and T.C. Lindley, Eds., sponsored by ESIS, Sheffield, U.K., April 1993, pp.307-319.

6.  Attia, M.H.,"Fretting Fatigue Testing- Current Practice and Future Prospects for Standardization", American Society for Testing and Materials, *ASTM STP-1159*, M.H. Attia, and R.B. Waterhouse, Eds., 1992, pp. 263-275.

7.  Attia, M.H., " A Thermally Controlled Fretting Wear Tribometer- A Step Towards Standardization of Test Equipment and Methods", *Wear*, Vol. 136, 1990, pp.423-440.

8.  Attia, M.H., and Camacho, F., "Temperature Field in the Vicinity of a Contact Asperity during Fretting", *Proceedings*, ASME Symposium on Contact Problems and Surface Interactions in Manufacturing and Tribological Systems, M.H. Attia and R. Komanduri, Eds., ASME Winter Annual Meeting, New Orleans, Louisiana, 1993, pp. 51-61.

9.  Alyabev, M.Y., Kazimirichnik, Y.A., and Onoprienko, V.P., "Determination of Temperature in the Zone of Fretting Corrosion", Fiz. Khim. Mekh. Mater., Vol. 6, No. 3, 1970, pp. 12-15.

10. Wright, K.H.R., "An Investigation of Fretting Corrosion", *Proceedings*,, Institution of Mechanical Engineers, London B, 1 , 1952-1953, pp. 556-563.

11. Sproles Jr., E.S., and Duquette, D.J., "Interface Temperature Measurement in the Fretting of a Medium Carbon Steel", *Wear*, Vol. 47, 1978, pp. 387-396.

12. Hisakado, T., "On the Mechanism of Contact Between Solid Surfaces- 2nd Report: The Real Area of Contact, the Separation and the Penetration Depth", *Bulletin of the Japanese Society of Mechanical Engineering*, Vol. 12, No. 54, 1969, pp. 1528-1536.

13. Hisakado, T., "On the Mechanism of Contact Between Solid Surfaces- 3rd Report: The Number and Distribution of Radii of Contact Points",*Bull. Japanese Society of Mechanical Engineering*, Vol. 12, No. 54, 1969, pp. 1537-1545.

14. Tsukada, T. and Anno, Y., "An Analysis of the Deformation of Contacting Rough Surfaces- 3rd Report: Introduction of a New Contact Theory of Rough Surfaces", *Bull. Japanese Society of Mechanical Engineering*, Vol. 15, No. 86, 1972, pp. 996-1003.

15. Yovanovich, M.M., "General Expression for Circular Constriction Resistance for Arbitrary Flux Distribution", AIAA 13th Aerospace Science Meeting, Pasadena, CA, Jan. 20-22, 1975, Paper No. 75-188.

16. Yovanovich, M.M., Brude, S.S., and Thompson, J.C., "Thermal Constriction Resistance of Arbitrary Planar Contact With Constant Flux", AIAA 11th Thermodynamics Conference, San Diego, Ca, July 14-16, 1976, paper No. 76-440.

17. Negus, K.J., Yovanovich, M.M., and Beck, J.V., "On the Nondimensionalization of Constriction Resistance for Semi-Infinite Heat Flux Tubes", *Transactions of the ASME, Journal of Heat Transfer*, Vol. 111, 1998, pp. 804-807.

18. Attia, M.H., and Fraser, S., "Prediction and Control of the State of Stresses and the Extent of the Partial-Slip Zone During Fretting", *Proceedings*, International Symposium on Fretting, Chengdu, China, Nov., 1997.

19. Attia, M.H., and Yovanovich, M.M., "A Model for Predicting the Thermal Constriction Resistance in Fretting", *Proceedings*, ASME Symposium on Contact Problems and Surface Interactions in Manufacturing and Tribological Systems, PED-Vol.67/TRIB-Vol.4, M.H. Attia and R. Komanduri, Eds., Nov. 1993, pp. 63-74.

20. Hanna, J.R., and Rowland, J.H., *"Fourier Series, Transforms, and Boundary Value Problems"*, John Wiley & Sons, Inc., N.Y., 2nd edition, 1990.

21. Podolsky, B., "A Problem in Heat Transfer", *Journal of Applied Physics*, Vol. 22, No. 5, 1951, pp.581-585.

22. Blok, H., "Theoretical Study of Temperature Rise at Surfaces of Actual Contact Under Oiliness Lubricating Conditions", *Proceedings*, General Discussion on Lubrication and Lubricants, London, Institution. of Mechanical Engineers, Vol. 2, 1937, pp. 222-235.

Siegfried Fouvry,[1] Philippe Kapsa,[1] and Léo Vincent [2]

**Fretting-Wear and Fretting-Fatigue: Relation Through a Mapping Concept**

**REFERENCE:** Fouvry, S., Kapsa, P., and Vincent, L., **"Fretting-Wear and Fretting-Fatigue: Relation Through a Mapping Concept,"** *Fretting Fatigue: Current Technology and Practices, ASTM STP 1367,* D. W. Hoeppner, V. Chandrasekaran, and C. B. Elliott, Eds., American Society for Testing and Materials, West Conshohocken, PA, 2000.

**ABSTRACT:** Developments in wear mapping recently described frontiers between partial and gross slip and between the main damages i.e. cracking and particle detachment both from experiences and from theoretical analysis. These maps, known as running condition fretting maps (RCFM) and material response fretting maps (MRFM), are powerful tools to analyze the contact and to predict failures. Especially MRFM is a major contribution for designers and engineers. Mapping concept enables a new approach of fretting-fatigue as it considers crack nucleation for ranges of contact conditions. Effects of displacement and normal load on cracking are now well admitted under fretting wear conditions. The question is still related to the effect of the cyclic external loading that it is superimposed to the contact loading. Theoretical and experimental results related to homogeneous metallic contacts are discussed to attempt an objective analysis on the interest of fretting fatigue tests.

**KEYWORDS:** fretting wear, fretting wear on pre-stressed specimen, fretting fatigue, crack nucleation, high-cycle fatigue, size effect, fretting map

**Introduction**

Fretting damage is often the origin of catastrophic failures or loss of functionality in many industrial applications. Considered to be a plague for modern industry, fretting is encountered in all quasi-static loadings submitted to vibration and thus concerns many industrial branches [1, 2]. Specifically, fretting-fatigue damage was reported by Hoeppner [3] to occur in parts found in helicopters, fixed wing aircraft, trains, ships, automotive, trucks and buses, farm machinery, engines, construction equipment, orthopedic implants, artificial hearts, rocket motor cases, wire ropes, etc. Two different damage states are usually observed, cracking and wear associated with debris formation.

---

[1] UMR CNRS 5513, Ecole Centrale de Lyon, BP 163, 69131, Ecully, France.
[2] UMR CNRS 5621, Ecole Centrale de Lyon, BP 163, 69131, Ecully, France.

Since the contact analysis of industrial situations is relatively complex, scientists have primarily analyzed simple geometries such as cylinder/plane and sphere/plane contacts. They are well described by the Hertzian pressure and stress formulations whereas the tangential stress field imposed during the alternating tangential loading is expressed through the Cattaneo & Mindlin's formalism [4, 5]. To analyze the degradation two different tests have been developed:

- the fretting wear test where a contact load is generated by a relative motion between two surfaces. This corresponds to very small displacement amplitudes of a classical tribological wear test called reciprocating testing. This can explain why such a configuration is usually employed to analyze wear [6],

- the fretting fatigue test corresponds to the adaptation of a fatigue test where two pads are pressed on the fatigue specimen. Submitted to external bulk loading the strain generated through the fatigue specimen induced a relative displacement between the two contacted surfaces and consequently a cyclic tangential contact loading. This configuration is used to analyze cracking because after crack nucleation the external bulk loading favors the propagation and then the breaking of the specimen [7].

Numerous studies have nevertheless demonstrated that a direct association between fretting fatigue test and cracking or fretting wear test and wear is not systematic. Wear can be observed under fretting fatigue configurations whereas severe examples of cracking are encountered under fretting wear situations (Fig. 1) [1, 8, 9].

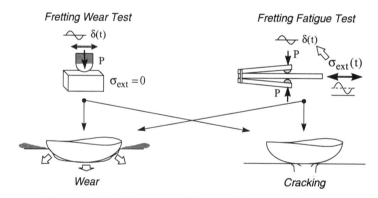

Figure 1 – *Fretting test conditions and damage evolutions.*

In fact, damage evolution is directly related to the contact sliding condition (Fig. 2). For very small amplitudes the tangential force presents a quasi proportional evolution versus displacement which leads to a closed partial slip hysteresis loop characterizing the partial slip condition. Sphere/plane or cylinder/plane contacts are then composed of a central stick domain surrounded by a sliding domain [4,5,10]. Such a condition favors the appearance of cracks [1,8]. For larger displacement amplitudes, the tangential force reaches a constant value leading to a full sliding over the whole contacted area. The Q-$\delta$ fretting hysteresis loop displays a quadratic shape defined as the gross slip condition. Such a sliding condition favors the wear induced by the debris formation [1, 8].

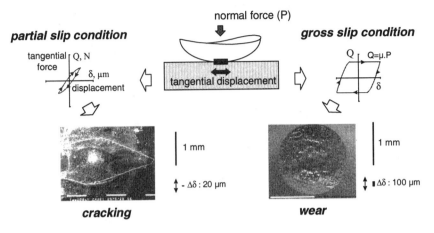

Figure 2 - *Damage evolution as a function of the fretting sliding condition.*

The sliding condition can nevertheless change during the test. Therefore, taking into account the loading history in the contact, recent developments have introduced the fretting regime concept [8,11]. The sliding condition defined by the fretting loop is then represented through a 3D representation integrating the time evolution along a logarithmic scale. Three fretting regimes have been identified:
- the Partial Slip Regime (PSR), observed for the smallest amplitudes is defined by a constant partial slip condition,
- the Gross Slip Regime (GSR), observed for the largest amplitudes is defined by a constant gross slip condition,
- the Mixed Fretting Regime (MFR), observed for intermediate amplitudes, is characterized by an evolution from one sliding condition to another (Fig. 3).

Generally, the gross slip condition is observed at the beginning of the test followed by a partial slip.

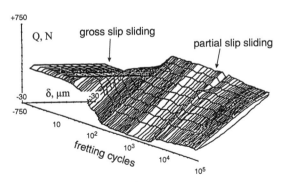

Figure 3 - *Mixed Fretting Regime (MFR) observed for an aluminum contact [15] (7075 vs. 7351 (R=300 mm); δ\*: ± 30 μm; f: 1 Hz; P:500 N).*

To identify the fretting response of the system, different fretting maps have been introduced. Usually the sliding conditions and the damage evolutions are represented in a two-dimensional graph with in abscise the displacement amplitude and in ordinate the normal force (Fig. 4). The sliding domains were mapped by Vingsbo et al. [12] and the fretting regime was successively introduced by Vincent et al [8] (RCFM : Running Condition Fretting Map).

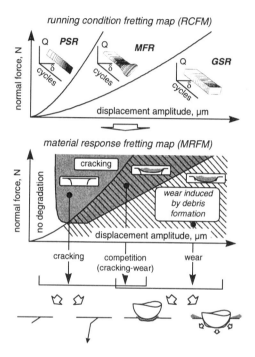

Figure 4 - *Representation of the fretting chart which combines the fretting regime analysis (RCFM) with the material response (MRFM).*

The experimental mapping of the material response (MRFM : Material response fretting mapping) was originally introduced by P. Blanchard et al. [13]. Due to this mapping analysis, a better overview of interactions between the sliding regimes and damage evolution has been identified. It has been shown for instance that the cracking phenomenon is more detrimental under the mixed slip regime. Based on this mapping representation, recent developments have introduced various types of modeling to explain the fretting contact behavior (Fig. 5) [14]. The sliding conditions and more particularly the transition between partial and gross slip have been quantified through the introduction of sliding criteria. Crack nucleation under partial slip was quantified through the introduction of multiaxial fatigue criteria [15, 16], whereas the wear induced by debris formation was quantified by introducing an energy approach [14]. Even if such developments give us the opportunity to model both the RCFM and MRFM, numerous questions still arise.

Mainly developed for the fretting wear configuration, such a mapping approach has not yet been transposed to other loading situations. This paper will demonstrate that for any fretting test situations, crack nucleation can be experimentally analyzed in terms of fretting maps and quantified through an appropriate multiaxial analysis.

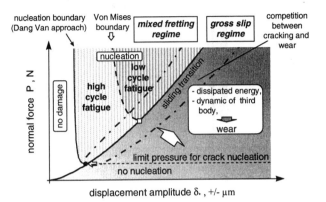

Figure 5 – *Schematic view of Mapping Concept involved to quantify fretting damages.*

A recent fretting wear test crack nucleation analysis of a well defined tempered aeronautical steel (30NiCrMo) has shown that crack nucleation can be predicted from classical fatigue variables, if the size effect is considered [17]. Previous works performed on similar tempered aeronautical steels (32Cr-1MoV [15], 35NiCrMo [16]) tested respectively under fretting wear on pre-stressed specimens and fretting fatigue configurations, will be associated in light of this advancement. Including all the fretting crack nucleation analysis, a normalized crack nucleation fretting map is expressed as a function of the shear stress amplitude and the tensile state.

## Methodology of Crack Nucleation Modeling

This fretting crack nucleation analysis is conducted for high cycle fatigue conditions with the number of cycles greater than $10^6$. Conducted under partial slip or mixed slip regimes, the surface damage remains limited which facilitates the stress field description. The loading associated with this crack nucleation analysis is significantly below the plastic value. This permits the application of Mindlin and Hamilton contact stress descriptions [18]. The methodology used here, first consists of identifying experimentally the lower load just inducing the crack nucleation. The damage mechanism was determined via optical and SEM observations of the contact area and metallographic cross section examinations. The resolution of microcrack observation is around 10-20 µm. The analysis under such high cycle conditions concludes that the first crack nucleation is always observed next to the contact border along the sliding axis. Considering the aspect of various test configurations, the different stress paths imposed below the surface are determined for each experimental condition. Based on the calculated loading path, a post treatment permits the determination of the cracking risk, whereas the critical point where the first crack is predicted to appear is determined. The applied multiaxial fatigue approach is the Dang Van crack nucleation criterion. It was shown to work well for the steel alloys studied where it predicted the location of the first crack nucleation through contact. In the Dang Van multiaxial fatigue criterion, detailed in [19], the loading path is described by the history of two parameters: $\hat{\tau}$, local microscopic shear stress amplitude (see Refs. [19,17] for more precise definition) and $\hat{p}$, local hydrostatic pressure describing a mean tensile state.

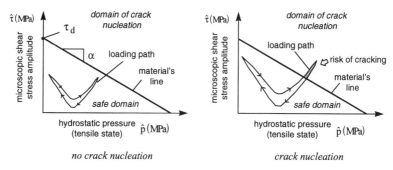

Figure 6 – *Example of Dang Van diagram.*

If the loading path remains beneath the cracking limit defined by the classical bending and shear alternated fatigue variables, fatigue failure will not occur. Otherwise, rupture is predicted (Fig. 6). To quantify the cracking risk, a scalar variable "d" is introduced

$$d = \max_{\bar{n},t}\left\{\frac{\|\hat{\tau}(\bar{n},t)\|}{\tau_d - \alpha \cdot \hat{p}(t)}\right\} \quad \text{with} \quad \alpha = \frac{\tau_d - \sigma_d/2}{\sigma_d/3} \tag{1}$$

Where :

$\hat{\tau}(\bar{n},t)$ : Time evolution of the local shear stress on the plane defined by the normal $\bar{n}$.

$\hat{p}(t)$ : Time evolution of the local hydrostatic pressure.

> *If "d" is greater than or equal to 1 there is a cracking risk,*
> *If "d" remains smaller than 1 there is no cracking risk.*

## Fretting Wear Test

The sphere/plane cracking analysis is fully described in Ref. [17]. Only the elementary information required by the analysis is presented.

### Test Conditions

Fretting tests were conducted using a tension/compression servo hydraulic machine already described elsewhere [8]. During the test, the normal force (P), the tangential force (Q) and the displacement (δ) are recorded. The magnitude of the tangential force versus the displacement is recorded for each cycle. The partial slip condition is maintained during all the test duration (i.e. Partial Slip Regime). A 50 mm radius chromium 52100 steel ball (Table 1) with a very low surface roughness (Ra<0.02 μm) is applied on an aeronautical steel plane which after a mirror polishing also displays a very low surface roughness (Ra<0.01). The crack analysis was done after $10^6$ cycles at a frequency of 10 Hz.

*Material*

The 30NiCrMo steel, presents the following weight composition 0.31% C, 4.08% Ni, 1.39% Cr, 0.25% Si, 0.49% Mn, 0.5% Mo. It was heat treated at 850°C for 1 hour, oil quenched and tempered at 585°C for 1 hour to obtain the mechanical properties given in Table 1 [20].

Table 1 - *Mechanical properties of 30NiCrMo [20] and 52100 [15] steels*

| Materials | Hardness | E, GPa | v | $Re_{0.2}$, MPa | Rm, MPa | $\sigma_d$, MPa | $\tau_d$, MPa |
|---|---|---|---|---|---|---|---|
| 30NiCrMo | 400 Hv | 200 | 0.3 | 1080 | 1200 | 690 | 428 |
| 52100 | 62 HRc | 210 | 0.3 | 1700 | - | - | - |

$\sigma_d$ (MPa)  : Alternated bending fatigue limit ($10^6$ cycles).
$\tau_d$ (MPa)  : Alternated shear fatigue limit ($10^6$ cycles).
$Re_{0.2}$ (MPa)  : Yield stress for 0.2% residual plastic strain.
Rm (MPa)  : Maximum stress.

*Fretting Analysis: Identification of the Crack Nucleation Boundary*

The objective of this study is to determine, as accurately as possible, the limit tangential force amplitude required for a given normal force which induces crack nucleation after $10^6$ cycles. For each normal force, several fretting tests have been performed maintaining a constant tangential force ($Q_*$). The critical value ($Q_{c*}$) is progressively extrapolated by increasing or decreasing the applied tangential value, if no crack or a crack is observed respectively (Fig. 7a, Table 2). At least more than 10 fretting tests are required to bracket the critical tangential force ($Q_{c*}$) with a precision inferior to 5 Newton.

*Crack Nucleation Prediction*

Fatigue analysis has shown that it is essential to take into account the size effect [7, 17]. Good correlation is achieved with the Dang Van model if, instead of considering the point stress description, the loading path is averaged through a critical quantity of material [17] (Fig. 7b). The stress averaging considers a cubic volume defined by the length side l. The stress components are calculated on the 27 points defining the cubic volume (8 at the cubic apexes, 12 at the center of the cubic edges, 6 at the center of the cubic faces and 1 at the center of the cube) and averaged to define the mean loading state. Comparison between experiments and theoretical predictions concludes that for the studied steel alloy, the critical length parameter corresponds to a 5$\mu$m edge cubic volume (l=5$\mu$m) [17]. It appears related to the tempered microstructure and independent of the contact dimension [17]. Therefore, if the stress analysis considers such a size effect, stress condition leading to the crack nucleation can be determined from the experimental parameters (P, $Q_{c*}$ and c). It can be observed that independently of the normal force variation, the critical stresses ($\hat{\tau}$; $\hat{p}$) required for the crack nucleation are restricted to a very narrow domain (Fig. 7a, Table 2).

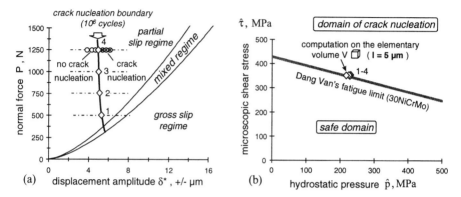

Figure 7 – *Fretting analysis of the 30NiCrMo steel : (a) Identification of the crack nucleation boundary through the MRFM (1 : limit condition); (b) Dang Van diagram of crack nucleation taking into account the size effect.*

Table 2 - *Fretting crack nucleation conditions of the 30NiCrMo steel : The loading path required for the fatigue analysis is averaged through a 5 μm cubic edge volume*

| Conditions | (1) | (2) | (3) | (4) |
|---|---|---|---|---|
| P, N | 500 | 750 | 1000 | 1250 |
| $p_0$, MPa | 788 | 902 | 993 | 1070 |
| a, μm | 550 | 630 | 696 | 747 |
| $Q_{c*}$, N | 260±5 | 325±5 | 385±5 | 435±5 |
| k=c/a | 0.6 | 0.65 | 0.68 | 0.7 |
| $\hat{\tau}$, MPa | 344 | 343 | 343 | 342 |
| $\hat{p}$, MPa | 248 | 245 | 243 | 242 |
| D | 1.018 | 1.009 | 1.007 | 1.004 |

$p_0$ : maximum Hertzian pressure; a : contact radius; c: measured stick radius.

To extend this fatigue analysis, different loading conditions have to be considered. For instance, a higher level of hydrostatic pressure induced by an external static loading, or at higher shear stress amplitude generated through the fretting fatigue test can be studied.

## Fretting Wear on Pre Stressed Specimen (FSS)

### Test Condition

In a previous study, the crack nucleation and propagation in a similar 35NiCrMo aeronautical steel was performed through a specific configuration, where in addition to the conventional fretting wear loading an external static stress is superimposed (FSS) (Fig. 8) [16]. Through this configuration, the influence of a homogeneous static stress on the crack nucleation in the contact can be measured. This is a useful way of analyzing the residual stress influence on fretting cracking.

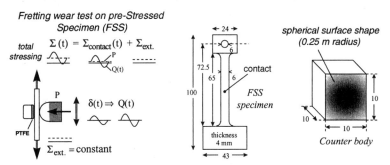

Figure 8 – *Fretting Wear Test on Pre Stressed Specimen (FSS) : loading conditions and design of specimens.*

As for the fretting wear test, both contact displacement and tangential force amplitudes can be monitored. The external stress and the normal force are imposed before starting the test. A homogeneous contact (35NiCrMo/35NiCrMo) with surface roughness, Ra less than 0.1 $\mu$m was studied. The counter body was fixed with one face machined to a 0.25 m radius spherical surface.

*Material*

The 35NiCrMo steel, presents the following weight composition 0.35% C, 3.6% Ni, 1.6% Cr, 0.3% Si, 0.5% Mn, 0.4% Mo. The material was heat treated at 850°C for 1 hour, oil quenched and tempered at 610°C for 1 hour to obtain the mechanical properties given in Table 3.

Table 3 - *Mechanical properties of 35NiCrMo steel*

| Materials | Hardness | E, GPa | $\nu$ | $Re_{0.2}$, MPa | Rm, MPa | $\sigma_d$, MPa* | $\tau_d$, MPa* |
|---|---|---|---|---|---|---|---|
| 35NiCrMo | 39HRc | 210 | 0.3 | 1020 | 1150 | 585 | 375 |

\* Extrapolated values from the measured value of Rm

Metallographic observations have concluded that the microstructure is very close to the previous 30NiCrMo alloy. It displays the same degenerated martensitic structure with a fine dispersion of ferrite and cementite domains.

*Fretting Analysis: Influence of a Static Tensile Stress on Crack Nucleation*

The normal force was maintained constant at 1500 Newtons with a 396 MPa Hertzian pressure and a 1.35 mm contact radius. Three different external tensile stresses, $\sigma_{ext}$, were applied (100, 250 and 350 MPa). For each value the contact was studied for three different sliding amplitudes. Two are under mixed fretting regime stabilized in partial slip condition, the last, next to gross slip regime boundary, is stabilized in gross slip condition after a few thousand cycles (Fig. 9a). Even if $10^6$ fretting cycles are imposed, only the first 50000 cycles are here considered to calculate the representative tangential force amplitude according that next to the sliding transition the friction coefficient tends to decrease with the test duration. The friction coefficient value which is measured at the sliding transition is around 0.85. Because the stick radius was not measured, this value will be considered to determine the loading path and computerize the cracking risk.

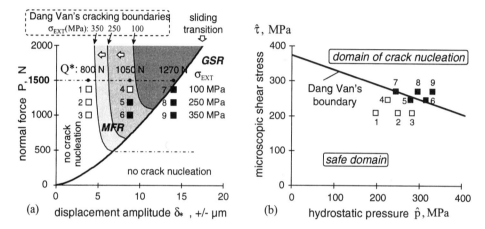

Figure 9 – FSS analysis of the 35NiCrMo steel ($10^6$ cycles; ■ : crack nucleation; □ : no crack nucleation ): (a) Identification of the damage evolution through the MRFM (1 : studied condition); (b) Dang Van diagram of crack nucleation taking into account the size effect (l=5μm).

Table 4 – Damage evolution of 35NiCrMo steel under FSS loadings : The loading path required in the fatigue analysis is averaged through a 5 μm cubic edge volume

| case | Damage | Q*, N | k=c/a | $\sigma_{ext}$, MPa | $\hat{p}$, MPa | $\hat{\tau}$, MPa | d |
|---|---|---|---|---|---|---|---|
| 1 | no nucleation | 800 ±20 | 0.72 | 100 ±5 | 196 | 209 | 0.72 |
| 2 | no nucleation | 800 ±20 | 0.72 | 250 ±5 | 248 | 209 | 0.78 |
| 3 | no nucleation | 800 ±20 | 0.72 | 350 ±5 | 282 | 209 | 0.82 |
| 4 | no nucleation | 1050 ±20 | 0.55 | 100 ±5 | 224 | 247 | 0.88 |
| 5 | Nucleation | 1050 ±20 | 0.55 | 250 ±5 | 278 | 247 | 0.96 |
| 6 | Nucleation | 1050 ±20 | 0.55 | 350 ±5 | 314 | 247 | 1.02 |
| 7 | Nucleation | 1270 ±20 | 0.0 | 100 ±5 | 243 | 272 | 1.00 |
| 8 | Nucleation | 1270 ±20 | 0.0 | 250 ±5 | 296 | 272 | 1.09 |
| 9 | Nucleation | 1270 ±20 | 0.0 | 350 ±5 | 330 | 272 | 1.15 |

The stress analysis of the ball-on-plane configuration is obtained by combining the Cattaneo-Mindlin contact description with the Hamilton formulation. The steel displays a similar microstructure to the 30NiCrMo steel, the size effect was also averaged through a $5\mu m$ edge cubic volume. The comparison of Table 4 indicates that the multiaxial fatigue approach permits a rather good estimation of the cracking risks. All the nucleations events have been predicted except the limit condition ($Q^*=1050$; $\sigma_{ext} =250$ N). The difference with the thresold value ($d=1$) is nevertheless very small (less than 10%) which confirms the stability of the fatigue approach. This also demonstrates the influence of a static stress level on the crack nucleation phenomenon, crack nucleation is extended with an increase of the static loading.

**Fretting Fatigue Test**

This analysis considers previous results published by Petiot who analyzed the crack nucleation of a similar aeronautical steel under fretting fatigue loading [15].

*Test Condition*

A specific set-up was developed to simulate fretting-fatigue. Two cylindrical fretting pads (diameter, 10 mm) are clamped against both surfaces of flat uniaxial fatigue specimen tested under constant loading ($\sigma_{min}/\sigma_{max} = 0.1$) (Fig. 10). The geometry is a cylinder on plane with a 2 mm length contact. The pads are made of 52100 bearing steel. Specimen and pads are machined with a surface finish with an Ra of 0.4 and 0.1 $\mu m$ respectively. The imposed oscillatory motions between the pads and the specimen are linked to the imposed oscillatory fatigue stress $\sigma$ in the specimen. For a maximum stress $\sigma_{max} = 500$MPa the total displacement of the half part of the fatigue specimen is 55 $\mu m$. Considering the elastic response of the system, a compliance constant (K) can be extrapolated which relates the maximum bulk stress such that :

$$\delta_*(\pm\mu m) = K.\sigma_{max} \text{ with } K = \frac{0.9}{2} \cdot \frac{55}{500} = 4.95 \, 10^{-2} \, \mu m.MPa^{-1}$$

The fretting-fatigue loops defined by the operating parameters($Q(t)$-$\sigma(t)$) are plotted to identify the different sliding conditions and fretting regimes by varying the normal force for each $\sigma_{max}$ investigated (Fig. 10).

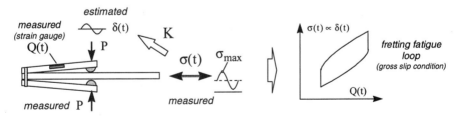

Figure 10 - *Fretting Fatigue Test.*

*Material*

The alloy studied was 32Cr-1MoV steel, which is composed of the following weight composition 0.33% C, 2.93% Cr, 0.83% Mo, 0.55% Mn, 0.3% V, 0.05% Ni, 0.3% Si. It was heat treated at 950°C, oil quenched and tempered at 635°C for 1 hour to obtain the mechanical properties given in Table 6.

Table 6 - *Mechanical properties of 32Cr-1MoV steel [15]*

| Materials | Hardness Hv 50g | E, GPa | $\nu$ | $Re_{0.2}$, MPa | Rm, MPa | $\sigma_d$, MPa | $\tau_d$, MPa |
|---|---|---|---|---|---|---|---|
| 32Cr-1MoV | 360 | 215 | 0.3 | 980 | 1140 | 594 | 380 |

*Fretting Analysis: Fretting Fatigue Map*

The fretting fatigue analysis has been performed for different normal forces : 40N < P < 140 N which corresponds to the following pressure domain 388 MPa< $p_0$ <726 MPa and contact widths 33$\mu$m< a < 62 $\mu$m. The sliding condition and the damage evolution are reported by the authors as a function of the external loading $\sigma_{max}$ and the normal force. The different fretting regimes are observed in separate domains where as the damage evolution is still related to the regime condition.

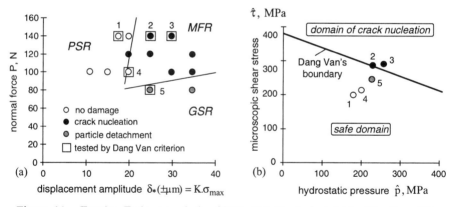

Figure 11 – *Fretting Fatigue analysis of* 32Cr-1MoV *steel : (a) Identification of the damage evolution through the MRFM (1 : modeled condition ); (b) Dang Van diagram of crack nucleation taking into account the size effect.*

Table 7 – *Cases of the study : experimental results and FEM fatigue modeling [15].*

| Case | damage | P, N | $p_0$, MPa | $\sigma_{max}$, MPa | $\hat{p}$, MPa | $\hat{\tau}$, MPa | d |
|---|---|---|---|---|---|---|---|
| 1 | No nucleation | 140 | 726 | 350 | 180 | 200 | 0.66 |
| 2 | nucleation | 140 | 726 | 500 | 229 | 286 | 1.007 |
| 3 | nucleation | 140 | 726 | 600 | 257 | 291 | 1.07 |
| 4 | No nucleation | 100 | 616 | 400 | 200 | 214 | 0.73 |
| 5 | No nucleation | 80 | 545 | 500 | 226 | 246 | 0.86 |

Considering the compliance variable K, the different sliding regimes and damage evolutions can be reported as a function of displacement and normal force (Fig. 11a). Similar conclusions can be extrapolated. Partial and mixed fretting regimes mainly favor cracking whereas wear induced by debris formation is promoted under gross slip regime. This indicates that the fretting map analysis can easily be transposed to analyze results obtained in fretting fatigue tests. To quantify the crack nucleation, the authors have performed a FEM analysis of the test and computed the stress loading path for the Dang Van crack analysis. The friction coefficient is assumed to be 0.9, according to the value measured at the transition to the gross slip regime. Analysing the different critical loading path, both the maximum shear stress amplitude and hydrostatic pressure associated with the Dang Van computation are determined and reported in Table 7. The fatigue analysis which is limited to five experimental situations, has demonstrated that the Dang Van criterion appears to be a powerful means to quantify the crack nucleation. An excellent correlation is effectively observed between the experimental results and the computations. It must be pointed out that the horizontal and vertical spacings of nodes on and below the FEM meshed surface are 4μm and 5μm respectively. The choice of these dimensions which controls the FEM stress averaging and then the fatigue quantification is not explained by the authors. It can nevertheless be observed that these dimensions are perfectly coherent with the intrinsic length parameter (l=5μm) which seems to characterize this kind of tempered steel.

## Synthesis and Conclusions

This study, conducted on similar aeronautical steel alloys has shown that crack nucleation in fretting can be quantified independently on the test if the contact stress analysis considers the size effect. By averaging the loading path on the specific volume of material representative of the microstructure it becomes possible to associate both the contact stress and the external bulk load. Then the transposition of a multiaxial fatigue approach (Dang Van criterion) permits one to quantify very precisely the crack nucleation risk condition. The determination of an intrinsic length parameter ($l=5\mu m$), specific to the studied steel alloys, permits the convergence of the fatigue prediction whatever the test situation (fretting wear, fretting wear on a pre-stress specimen and fretting fatigue).

The study has demonstrated the very strong stability of the proposed approach (size effect, multiaxial computation). A very good prediction is always obtained even if the loading condition and the contact dimensions are very different. Indeed, through this study none, static and repeated bulk loading have been superimposed, whereas sphere on plane but also cylinder on plane configurations have been investigated. More ever the studied systems evolve from 70 to 1350 $\mu m$ contact radius.

The analysis has also shown that for all the test conditions, the experimental damage analysis can be undertaken through fatigue mapping. RCFM and MRFM allow respectively the determination of the sliding condition and the identification of the crack nucleation boundary under partial and mixed regimes. It strongly clarifies the stress analysis and permits a coherent prediction of the crack nucleation.

This study also confirms a classical fatigue result, which indicates that the crack nucleation is a function of an alternating shear stress amplitude related to a given tensile state. To compare the different test situations, it appears interesting to superimpose the different damage evolutions on a single graph. Considering the Dang Van diagram, a normalized representation can be introduced by dividing the critical stress couple ($\hat{\tau}, \hat{p}$) respectively by ($\tau_d$ and $\dfrac{\tau_d}{\alpha} = \dfrac{\tau_d \cdot \sigma_d/3}{\tau_d - \sigma_d/2}$).

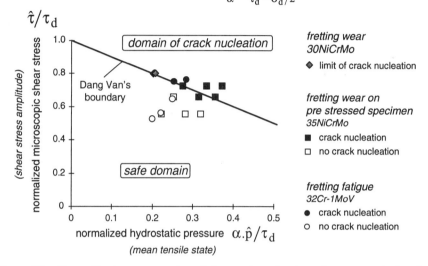

Figure 12 – *Normalized representation of the crack nucleation phenomenon in fretting. (The size effect is considered through the intrinsic length parameter : l=5μm)*

Figure 12 considers the results obtained for the three steel alloys studied through different test situations. For each contact stress analysis, the same intrinsic length parameter (l=5μm) has been considered assuming that the tempered steels present a similar microstructure. This assumption is effectively supported by similar ratio between the bending and shear fatigue limits: 30NiCrMo $\sigma_d/\tau_d = 1.61$; 35NiCrMo $\sigma_d/\tau_d = 1.56$; 32Cr-1MoV $\sigma_d/\tau_d = 1.56$. The crack nucleations are above the cracking limit whereas the undamaged situations are below. This representation indicates that the FSS test is particularly well adapted to analyse the crack nucleation through a large domain of hydrostatic pressure. Indeed the tensile stress level is directly monitored by the static stress imposed. In opposition, for the fretting fatigue test, the contact loading is directly linked to the external stressing, and the results are displayed along a diagonal line. This test configuration strongly limits the domain of investigation of the nucleation phenomenon. Thanks to this normalized representation, all the experiments could be compared whereas different results obtained by different authors could be associated. Morever the effect of palliative treatments such as shot peening or carbonitriding can be investigated. Associated with normalized pressure crack nucleation fretting map [17], it also appears as a powerful means for the designer to optimize contact geometry and surface treatments.

## Acknowledgments

The authors are grateful to Prof. Ky Dang Van for his helpful comments and suggestions.They would like also to thank C. Petiot and B. Journet for their helpful discussions and the Aubert et Duval company for the furnishing of steel alloys.

## References

[1]    Waterhouse, R.B., *Fretting Fatigue*, Applied Science publishers, 1981.

[2]    Lindley, T.C., Nix, K.J., "The Role of Fretting in the Initiation and Early Growth of Fatigue Cracks in Turbo-Generator Materials", *ASTM STP 853,* ASTM, Philadelphia, 1982, pp. 340-360.

[3]    Hoeppner, D., "Mechanisms of fretting fatigue and their impact on test methods development", *ASTM STP 1159*, M.H. Attia and R.B. Waterhouse, Eds, ASTM, Philadelphia, 1992, pp. 23-32.

[4]    Cattaneo, C., "Sul contatto di due corpi elastici: distribuzione locale degli sforzi", *Rendiconti dell'Accademia dei lincei*, 6, 27, 1938, pp. 343-348, 434-436.

[5]    Mindlin, R.D., Deresiewicz, H., "Elastic spheres in contact under varying oblique forces", *ASME Trans, Series E, Journal of Applied Mechanics*, 20, 1953, pp. 327-344.

[6]    Bill, R.C., "Fretting of AISI 9310 Steel and selected fretting-resistance surface treatments", *ASLE transactions*, Vol. 21, 3, 1977, pp.236-242.

[7]    Hills, D.A., Nowell, D., and O'Connor, J.J., "On the mechanics of fretting fatigue", *Wear*, 125, 1988, pp. 129 – 146.

[8]    Vincent, L., Berthier, Y., Godet, M., "Testing methods in fretting fatigue: a critical appraisal", *ASTM STP 1159*, M.H. Attia and R.B. Waterhouse, Eds, ASTM, Philadelphia, 1992, pp.33-48.

[9]    Zhou, Z.R., Vincent, L.," Cracking behaviour of various aluminium alloys during fretting wear", *Wear*, 155, 1992, pp. 317-330.

[10]   Johnson, K.L., *Contact Mechanics*, Cambridge Univ.Press, Cambridge, 1985.

[11]   Pellerin, V, Ph. D thesis, 1990, N° 90-01.

[12]   Vingsbo O., Soderberg S., "On fretting maps", *Wear*, 126, 198), pp. 131-147.

[13]  Blanchard, P., Colombier, C., Pellerin, V., Fayeulle, S., Vincent, L., "Material effect in fretting wear : application to iron, titanium and aluminium alloys", *Metallurgica Transaction*, volume 22A, 1991, pp.1535-1544.

[14]  Fouvry, S., Kapsa, Ph., Vincent, L., "Quantification of fretting damages", *Wear*, 200, pp. 186-205.

[15]  Petiot, C., Vincent, L., Dang Van, K., Maouche, N., Foulquier, J., Journet, B., "An analysis of fretting-fatigue failure combined with numerical calculations to predict crack nucleation", *Wear*, 185, 1995, pp. 101-111.

[16]  Fouvry, Ruiz, F.S., Kapsa, Ph., Vincent, L., "Stress and Fatigue analysis of fretting on a stressed specimen ", *Tribotest Journal 3-1*, 1996, pp. 23-44.

[17]  Fouvry, S., Kapsa, Ph., Vincent, L., "A Multiaxial Fatigue Analysis of Fretting Contact Taking Into Account the Size Effect " ASTM STP 1367, D. W. Hoeppner, V. Chandrasekaran, and C. B. Elliott, Eds., American Society for Testing and Materials, West Conshohocken, PA, 1999.

[18]  G. M. Hamilton, "Explicit equations for the stresses beneath a sliding spherical contact", *Proc. Inst. Mech. Ing. 197c*, 1983, pp. 53-59.

[19]  Dang Van, K., "Macro-Micro Approach in High-Cycle Multiaxial fatigue", *ASTM STP 1191*, 1993, pp. 120-130.

[20]  Dubar, L, PhD Thesis ENSAM (Talence), 1992, N°92-18.

T. Hansson,[1*] M. Kamaraj,[1**] Y. Mutoh[1] and B. Pettersson[2]

# High Temperature Fretting Fatigue Behavior in an XD™ γ-base TiAl

REFERENCE: Hansson T., Kamaraj M., Mutoh Y. and Pettersson B., "High Temperature Fretting Fatigue Behavior in an XD™ γ-base TiAl," *Fretting Fatigue: Current Technology and Practices, ASTM STP 1367*, D. W. Hoeppner, V. Chandrasekaran, and C. B. Elliott, Eds., American Society for Testing and Materials, West Conshohocken, PA, 2000.

ABSTRACT: The plain and fretting fatigue behavior was studied of a γ-base TiAl intermetallics at 675°C in air. he TiAl material was produced by the XD™ process and had a nominal composition of Ti-47Al-2Nb-2Mn+0.8vol%TiB$_2$. The contact material was Inconel 718. The reduction of fatigue strength by fretting was very small (around 20%) compared to metallic materials (30~60%). The plain and fretting fatigue S-N curves were flat, which resembled the behavior of ceramic materials. The flat S-N curves are corresponding to the very steep crack growth curve reported for TiAl-based materials containing TiB$_2$. A fretting fatigue crack nucleated in the very late stage of fatigue life (later than 0.75N$_f$), while it is known to nucleate in the very early stage of fatigue life (less than few percent of N$_f$) for steels. The fatigue lifetimes of this material were suggested to be determined by the cycles needed for nucleation of a crack rather than by crack propagation.

KEYWORDS: fretting fatigue, high temperature, contact pressure, intermetallics, mechanism

## Introduction

In recent years a great deal of effort has been devoted to the study of mechanical and associated phenomena of γ-based TiAl intermetallics. They are light in weight and possess excellent high-temperature properties. These attractive mechanical properties make them most suitable for applications in the aerospace turbine structures and

[1] Department of Mechanical Engineering, Nagaoka University of Technology, Nagaoka-shi, 940-2188 Japan.
[2] Materials Technology, Volvo Aero Corporation, S-46181 Trollhattan, Sweden.
* Present address: 2.
** Present address : Insitut für werkstoffe, Ruhr-Universität Bachum, D-44801 Bochum, Germany.

automobile industries[1-3]. Studies on the elevated temperature fracture, creep and fatigue behavior of various γ-TiAl based materials have been published in recent years[4-11]. Engine components are often subjected to high temperature fretting fatigue situations[12, 13] and under these conditions, the fatigue strength of the material can be reduces significantly (half to one-third in the case of metals). When these new materials are considered for high temperature applications, fretting fatigue is an important material property to establish for design considerations.

Based on the detailed observations of crack growth processes[14], a fatigue crack in TiAl intermetallics propagates by repeating nucleation, growth and coalescence of microcracks at the crack tip. These processes are significantly different from those in metallic materials with plastic deformation as nonlinear phenomenon to induce cyclic fatigue crack growth, but rather similar to those in ceramic materials[15, 16]. The microcracks nucleate along the cleavage plane, interface of lamellar structure or grain boundaries. Their nature, orientation and shape strongly depend on the microstructure[14].

On the other hand, based on the pile-up research works on fretting fatigue of metallic materials, a fretting fatigue crack nucleates near the edge of the contact region in the very early stage of fatigue life under high stress concentration superimposing applied stress and tangential force (frictional force) on the contact surface[17]. Plastic deformation also plays an important role in the fatigue crack nucleation and propagation processes in fretting fatigue similar to the case of plain fatigue. Since the plastic deformation can be rarely expected during crack nucleation and propagation processes in TiAl intermetallics, fretting fatigue mechanisms and hence characteristics may be different from those in metallic materials. However, there have been no reports available on fretting fatigue of γ-based TiAl, to the present authors' knowledge.

In the present study, basic fretting fatigue characteristics and mechanisms were discussed based on the results of fretting fatigue tests of a γ-based TiAl, which was produced by the sintering (XD™) process, at elevated temperature.

## Materials

The γ-based titanium aluminide used in the present investigation was prepared by XD™ process and investment casting into cylindrical bars. Following the casting, bars were isostatically pressed at 1260°C/175 MPa for 4h and heat treated at 1010 °C for 50h. Details of the composition and mechanical properties are listed in Table 1 and 2, respectively. Typical microstructure of the material is shown in Fig. 1. Note the $TiB_2$ particles, sizes of which are small and less than 10 μm, are randomly distributed in a lamellar matrix of the material. The contact material for the fretting fatigue tests was Inconel 718 with a composition 52.5Ni-19Cr-3.0Mo-5.1(Co+Ta)-0.90Ti-0.50Al-18Fe. The material was solution heat treated at 968°C before machining and precipitation hardened at 720°C/8h+620°C/8h after machining. Inconel 718 was chosen as a contact material to simulate a possible situation in an aircraft engine. The room temperature hardnesses of the γ-based TiAl and Inconel 718 used were Hv 418 and 513, respectively.

**Experimental Procedures**

The experimental set-up is shown in Fig. 2. The fatigue specimen had a total length of 97 mm with bottom head ends and a 20 mm long gauge section with a rectangular cross section of 4 × 6 mm. The specimens were polished before testing in the longitudinal direction using successively finer emery paper down to a grit size of 1500 and degreased with acetone. The contact pad had a wedge-like shape with flat, 2 mm high and 4 mm wide, machine-ground contact surfaces. As shown in Fig. 2, the holders for the contact pads had matching wedge-shape slots for the contact pads and were fixed to the lower grips. When a cyclic load is applied on the specimen, the contact pads will move relative to the specimen and cause a fretting action on the specimen surface. The contact load was applied to the contact pads on the 6 mm wide specimen surfaces using push rods and a calibrated proving ring outside the furnace. The frictional force was measured using high temperature strain gauges attached to the inner and outer faces of the contact pad holders as shown in Fig. 2. The signal for frictional force was calibrated at elevated temperature using a "dummy" specimen, having the same geometry as the real specimens, which was disconnected from the bottom part of the fixture to allow the whole applied force to be transmitted through the contact pads. The alignment was checked before the series of tests started and was also checked several times during the course of this study. In all the cases the amount of misalignment was below 5% at a total average strain of 0.15%.

Table 1 – *Chemical composition of the γ-based TiAl used (wt%).*

| Ti | Al | Mn | Nb | B | Cu | Si | Fe | O | C | N | H |
|----|-----|------|------|------|------|------|-------|--------|------|--------|--------|
| Bal | 31.67 | 2.77 | 4.85 | 0.34 | 0.08 | 0.02 | 0.052 | 0.0938 | 0.01 | 0.0117 | 0.0008 |

Table 2 – *Mechanical properties of the γ-base TiAl used.*

|  | RT | 675°C |
|---|-----|-------|
| Yield strength (0.2% offset) (MPa) | 414 | 346 |
| Ultimate tensile strength (MPa) | 482 | 528 |
| Elongation (%) | 0.8 | 4.4 |
| Reduction of area (%) | 0.8 | 4.2 |

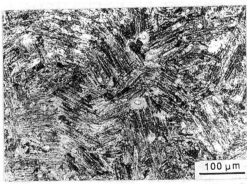

Figure 1 – *Microstructure of the material used.*

The temperature was measured and controlled using a calibrated Type K thermocouple in contact with the specimen surface at the center of the gauge length. Prior to testing, the temperature in the center of the specimen was measured by machining small holes and inserting calibrated thermocouple in contact with the specimen surface. All the tests in this study were performed at a temperature of 675°C in the center of the specimen which required a reading on the controlling thermocouple at the surface of 682°C. The temperature was kept constant within ±1°C during testing. The temperature was highest at the center of the gauge length and a temperature difference was less than 5°C within the gauge length. The specimens were kept at the test temperature 2h before starting the tests.

All the tests were performed in air at a frequency of 20 Hz and a stress ratio R of 0.1 using a conventional servohydraulic testing machine equipped with contact pressures of 50, 100 and 200 MPa. In addition, two-stage tests [18], where the contact pads were removed after a certain number of cycles in fretting fatigue and then continued in plain fatigue until failure or $10^7$ cycles, were conducted to identify the fretting fatigue crack nucleation and growth behaviors. The plain fatigue tests were also done under the same testing conditions as those of the fretting fatigue tests. The fracture surfaces and fretted surfaces were characterized using a scanning electron microscope (SEM).

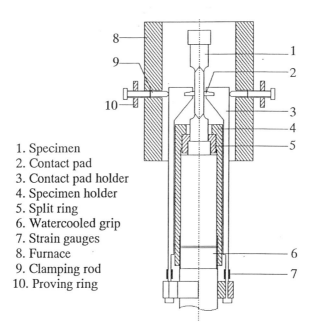

1. Specimen
2. Contact pad
3. Contact pad holder
4. Specimen holder
5. Split ring
6. Watercooled grip
7. Strain gauges
8. Furnace
9. Clamping rod
10. Proving ring

Figure 2 – *Schematic illustration of fretting faituge test setup.*

## Results and Discussion

### Fatigue Strength

The S-N curves for plain and fretting fatigue are shown in Fig. 3, where the arrows indicate that no failure occurred up to $10^7$ cycles. The plain fatigue strength ($\Delta\sigma_{wpf}$) was about 340 MPa at $10^7$ cycles, while the fatigue life was only a few thousand cycles at a $\Delta\sigma$ of 350 MPa. The resultant S-N curve for plain fatigue was very flat compared to those of steels[19] and exhibited a behavior similar to those of ceramic materials[20, 21]. The level that the maximum stress coincides with yield stress (0.2% proof stress) is also indicated in the figure. Fatigue failure in plain fatigue occurred at higher stress levels compared to yield stress. Similar high fatigue strengths in high cycle fatigue of TiAl-based materials have been reported[22]. The S-N curves for fretting fatigue became slightly steeper but they were still flat compared to the fretting fatigue S-N curves for steels[19]. Fretting fatigue S-N curves with similar flat appearance have been reported[23,24] for ceramic materials.

The reduction of fatigue strength by fretting was in the range from 18 to 26% for the present TiAl-based intermetallics. This small reduction indicates superior fretting fatigue resistance of TiAl-based intermetallics compared to metallic materials (30~60%) [19].

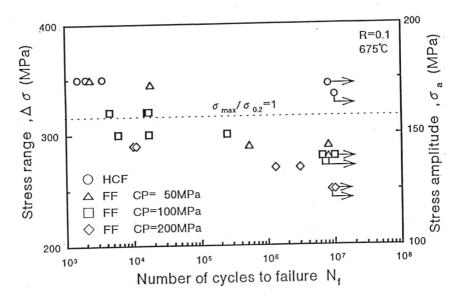

Figure 3 – *S-N curves for plain and fretting fatigue at 675 °C.*

*Tangential Force*

Figure 4 shows the variation of tangential force (frictional force) during fretting fatigue tests. The data indicated in the figure were for the run-out specimens with partial slip. The tangential force was almost constant during fretting fatigue tests. The average values of tangential force for contact pressures of 50, 100 and 200 MPa were about 62.5, 65 and 70 MPa, respectively, which indicated a slight increase in tangential force with increasing contact pressure. The tangential force coefficients, which are defined as the ratio of tangential force and contact pressure, were calculated as 1.25, 0.65 and 0.35 for contact pressures of 50, 100 and 200 MPa, respectively. These values except for the case of contact pressure 200 MPa were significantly high compared to those for steels[*19*].

*Observations of Fretted Surfaces and Fracture Surfaces*

The SEM observations of the fretted surfaces of the specimens tested under contact pressures of 50 and 200 MPa are shown in Figs. 5 and 6, respectively. Overall fretting damage looked very light except near the contact edge regions for all the specimens tested. Microcracks were found in the contact edge region, as shown in Fig. 6(b). No evidence for forming glaze oxide layer which brought about a large reduction in the coefficient of friction and the wear of the surfaces [*25, 26*] was observed on the contact region. The high tangential force coefficients in the present experiments corresponded to the glaze oxide layer which was not produced during the fretting fatigue tests.

Fractographs of the plain fatigue specimens are shown in Fig. 7. A flat facet without particular features was observed at the crack nucleation point, while mixed quasi-cleavage and interlamellar fractures were observed in fatigue crack growth region. In some cases, (a) processing defect was also found at the crack nucleation point, as shown in Fig. 7(c).

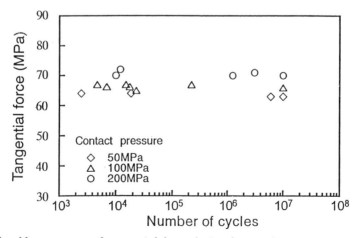

Figure 4 – *Measurements of tangential force during fretting fatigue test at the fatigue limit.*

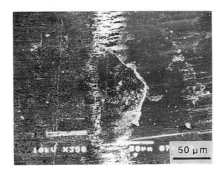

Figure 5 – *Fretted surface of the specimen tested at Δσ=290 MPa under contact pressure of 50 MPa.*

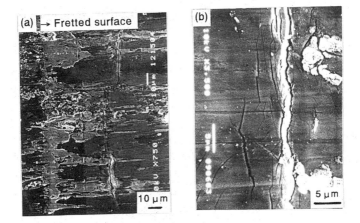

Figure 6 – *Fretted surfaces of the specimen tested at (a) Δσ=270 MPa and (b) Δσ=290 MPa under contact pressure of 200 MPa.*

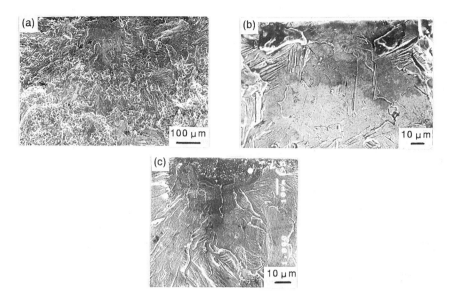

Figure 7 – *Fractographs of the plain fatigue specimens tested at Δσ=350 MPa. (a) crack nucleation point, (b) magnifying (a) and (c) defect at the crack nucleation point.*

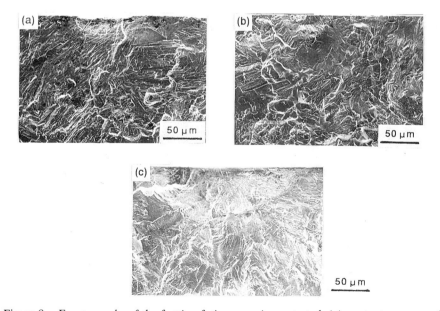

Figure 8 – *Fractographs of the fretting fatigue specimens tested. (a) contact pressure of 50 MPa, (b) contact pressure of 100 MPa and (c) contact pressure on 200 MPa.*

Fractographs of the fretting fatigue specimens are shown in Fig. 8. As can be seen from the figure, clear flat facet without particular features, which could be evident in plain fatigue, was rarely observed at the crack nucleation point, while mixed quasi-cleavage and interlamellar fracture was found in both crack nucleation and growth regions. Figure 9 shows SEM micrograph for observing both fracture surface and fretted surface near the crack nucleation point. There was no evidence for severe plastic deformation at the crack nucleation point: a fretting fatigue crack will nucleate in a brittle manner by quasi-cleavage or interlamellar fracture. Figure 10 shows SEM micrograph of near fretting crack wake region on the longitudinal cross-section of the fracture specimen. It seems from this kind of observation that a fretting crack propagates along with microcrack nucleation, growth and coalescence, as reported for $\gamma$-based TiAl intermetallics by Gnanamoorthy et al [*14*]

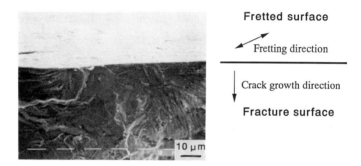

Figure 9 – *Observation of both fracture surface and fretted surface at the crack initiation point.*

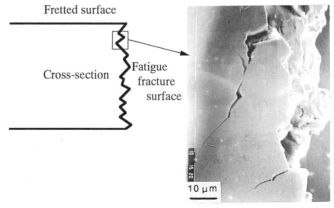

Figure 10 – *SEM micrograph of near fretting crack wake region on the longitudinal cross-section of the fractured specimen.*

*Two-Stage Tests*

It is known that a fretting fatigue crack nucleates in the very early stage of fatigue life (less than few percent of $N_f$) in metals [27]. Consequently the crack propagation life is dominant in the total fretting fatigue life. In the present study, the two-stage tests [18] were carried out to find the crack nucleates stage. The initial fretting fatigue tests were carried out up to 25, 50 and 75% of $N_f$ cycles under a stress range $\Delta\sigma$ of 270 MPa and a contact pressure of 200 MPa and the subsequent plain fatigue tests without contact pads were conducted under the same test conditions. The results are shown in Fig. 11. After initial fretting fatigue tests, the contact regions of the specimens were observed in detail. No cracks were observed in the contact region of the specimens tested up to 75% of $N_f$ cycles under fretting condition, while only light traces of fretting wear were found on the specimens. Fatigue fracture did not occur up to $10^7$ cycles in the subsequent plain fatigue tests. These two-stage test results indicate that a fretting fatigue crack nucleates in the very late stage of fatigue life in the TiAl-based intermetallics. That is, the crack nucleation life is dominant in the total fretting fatigue life of TiAl-based intermetallics compared to the crack propagation life. Additional similar two-stage tests were carried out

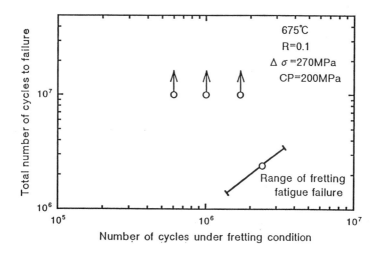

Figure 11 – *Results of two-stage tests.*

Table 3 – *Results of subsequent plain fatigue tests at room temperature.*

|   | Fretting fatigue test at 675°C | | | Subsequent plain fatigue test at RT | |
|---|---|---|---|---|---|
|   | $\Delta\sigma$ | contact pressure | cycles | $\Delta\sigma$ | $N_f$ |
| 1 | 250 | 200 | $10^7$ | 340 | 4100 |
| 2 | 250 | 100 | $10^7$ | 370 | 4200 |
| 3 | 280 | 50 | $10^7$ | 330 | $1.0 \times 10^5$ |

using the specimens run-out in the fretting fatigue tests. The results are listed in Table 3. In the subsequent plain fatigue tests at room temperature, fatigue fracture occurred out of contact region in the gauge part. Therefore, the run-out specimens would not have any cracks with critical size, which could propagate in the subsequent plain fatigue test.

To confirm the domination of crack nucleation stage in fretting fatigue life of the present material the crack propagation life has been estimated according to the fracture mechanics model proposed by Tanaka-Mutoh [13,19,28], schematics of which are shown in Fig. 12. The stress intensity factor range $\Delta K$ is given by the components due to the alternating applied stress range $\Delta\sigma$ and the alternating frictional force $\Delta F$,

$$\Delta K = 1.12\Delta\sigma\sqrt{\pi a} + \Delta K_{F1} + \Delta K_{F2} \qquad (1)$$

where

$$\Delta K_{F1} = 1.29\alpha\Delta Fl\sqrt{\frac{1}{\pi a}}$$

$$\Delta K_{F2} = 1.29(1-\alpha)\Delta F\sqrt{\pi a}\left[\frac{3}{2\pi}\ln\left(\frac{l+\sqrt{l^2+a^2}}{a}\right) - \frac{1}{2\pi\sqrt{l^2+a^2}}\right],$$

$\alpha$ is the ratio of frictional force concentration at the front edge of contact, a the crack depth and l the contact length. $\Delta K_{F1}$ is the value due to the component of frictional force concentrated at the front of the pad foot, which was obtained assuming that a point force in the y-direction $\alpha\Delta Fl$ was applied at the origin in the coordinate system as illustrated in Fig. 12. $\Delta K_{F2}$ is an approximate solution due to the uniform distribution of tangential force $(1-\alpha)\Delta Fl$ between the region from y=0 to y=l. In the calculation of crack propagation life, the following assumptions were made:

(1) The crack growth curve is given by , $\dfrac{da}{dN} = 3.7\times10^{-24}\Delta K^{18}$,which is determined based on the data indicated in Ref.[29], where the crack growth test was conducted under R=0.1 and at 600°C in air using nearly the same material.

(2)  The initial crack depth $a_0$ is 30 μm, which is determined based on SEM observation of fretting fatigue nucleation sites (Figs. 8 and 9).

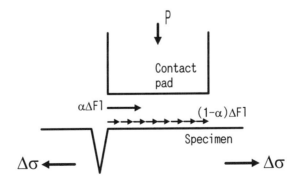

Figure 12 – *Fracture mechanical model of a fretting fatigue crack.*

(3) The final crack depth is 0.63mm, which is fracture-mechanically determined based on the applied maximum stress and the fracture toughness 15 MPa√ m[30].
(4) The value of α is 0.2, referring the previous reports[13,19,28]

The estimated crack propagation life for the fretting fatigue condition of the two stage test was very short and approximately $1.0 \times 10^3$ cycles, which was less than 1% of total fretting fatigue life. This estimation is not in contradiction to the fatigue test results of specimens run-out in fretting tests, where fatigue fracture occurred out of contact region in the gauge length, as mentioned above.

*Effect of Contact Pressure*

The relationship between fretting fatigue strength $\Delta\sigma_{wff}$ and contact pressure is shown in Fig. 13. The well-known contact pressure dependence of fretting fatigue strength for metals[27, 32], where fretting fatigue strength reduces with increasing contact pressure, was also found for TiAl-based intermetallics. It has been reported that the tangential force is the dominant factor influencing contact pressure dependence of fretting fatigue strength, while the contact pressure has an additional role [27]. As shown in Fig. 4, the tangential force for the present material slightly increased with increasing contact pressure. The result corresponded well to the contact pressure dependence of fretting fatigue strength, where the fretting fatigue strength slightly decreased with increasing contact pressure, as shown in Fig. 13.

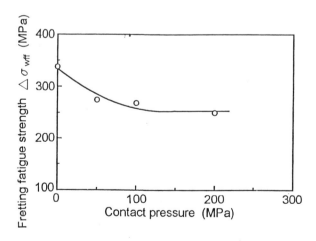

Figure 13 – *Relationship between fretting fatigue strength $\Delta\sigma_{wff}$ and contact pressure.*

*Fretting Fatigue Mechanisms in TiAl-Based Intermetallics*

From the foregoing results and discussion, it is thought that fretting fatigue failure of TiAl-based intermetallics occurs in the following manner. A fretting fatigue crack nucleates in a brittle manner by quasi-cleavage or interlamellar fracture without severe wear and plastic deformation at very late stage of the fretting fatigue life. Once a fretting crack nucleates, it propagates very rapidly to failure. The steep crack propagation curves [*29, 31*] and the low fracture toughness of about 15 MPa$\sqrt{}$ m [*30*] have been reported for γ-based TiAl-TiB$_2$ materials, which support the above process. Since the tangential force coefficient of the present material is high compared to that of metallic materials, the light fretting wear damage and the late crack nucleation may contribute to the less reduction of fatigue strength by fretting, hence the superior fretting fatigue resistance.

**Conclusions**

The reduction of fatigue strength by fretting of the γ-based TiAl intermetallics used was small compared to metallic materials, which indicates superior fretting fatigue resistance of TiAl-based intermetallics. A fretting fatigue crack nucleates in a brittle manner without severe wear damage and plastic deformation in the very late stage of fatigue life and propagated rapidly to failure. The late crack nucleation will dominantly contribute to its superior fretting fatigue resistance together with light wear damage on the contact region. An increase in contact pressure reduced fretting fatigue strength, which is commonly observed in metals.

**References**

[1]  H.A.Lipsitt (1985) *Processing and Properties of Advanced High Temperature Alloys*, Eds. S. M. Allen et al., ASM, Metal Park, OH, 157.

[2]  G.Sauthoff (1989) *Z.Metallkd.*, 80, 337.

[3]  Y. W. Kim (1989) *JOM*, July, 24-30.

[4]  W.O.Soboyejo, J.E.Deffeyes and P.B.Aswath (1991) Investigation of room- and elevated-temperature fatigue crack propagation in Ti-47Al/TiB$_2$ composites, *Scripta Met.*, 24, 1895-1900.

[5]  K.S.Chan and Y.W.Kim (1992)Influence of microstructure on crack-tip micromechanics and fracture behaviors of a two-phase TiAl alloy, *Metall. Trans.*, 23A, 1663-1677.

[6]  S.Tsuyama, S.Mitao and K.Minakawa (1992) *Mater. Sci. Eng.*, A153, 451.

[7]  A.W.James and P. Bowen (1992) Elevated temperature crack growth resistance of TiAl under monotonic and cyclic loading, *Mater. Sci. Eng.*, A253, 486-492.

[8]  D.L.Davidsson and J.B.Cammpbell (1993) Fatigue crack growth through the lamellar microstructure of an alloy based on TiAl at 25°C and 800°C, *Metall. Trans.*, 24A, 1555-1574.

[9]  R.Gnanamoorthy, Y.Mutoh, N.Masahashi and M.Matsuo (1993) High-temperature strength and fracture toughness in γ-phase titanium aluminides, *J. Mater. Sci.*, 28, 6631-6638.

[10] P.Bowen, R.A.Chave and A. W.James (1995) Cyclic crack growth in Titanium Aluminides, *Mater. Sci. Eng.*, A192/193, 443-446.

[11] R.Gnanamoorthy, Y.Mutoh, N.Masahashi and Y.Mizuhara (1995) Fracture toughness of gamma base titanium aluminides, *Metall. Mater. Trans.*, 26A, 305-313.

[12] K.J.Nix and T.C.Lindley (1988) The influence of relative slip range and contact material on the fretting fatigue properties of 3.5NiCrMoV rotor steel, *Wear*, 125, 147-162.

[13] Y.Mutoh, T.Satoh, K.Tanaka and E.Tsunoda (1989) Fretting fatigue at elelvated temperatures in two steam turbine steels, *Fatigue Fract. Engng. Mater. Struct.*, 12(5), 409-421.

[14] R.Gnanamoorthy, Y.Mutoh and Y.Mizuhara (1996) Fatigue crack growth behavior of equiaxed, duplex and lamellar microstructure $\gamma$-base titanium aluminides, *Intermetallics*, 4, 525-532.

[15] R.H.Dauskardt and R.O.Ritchie (1991) Cyclic fatigue of ceramics, *Fatigue of Advanced Materials*, Eds. R.O.Ritchie et al., MCEP, Birmingham, 133-151.

[16] Y.Mutoh, M.Takahashi and M.Takeuchi (1993) Fatigue crack growth in several ceramic materials, *Fatigue Fract. Engng. Mater. Struct.*, 16(8), 875-890.

[17] Y.Mutoh (1995) Mechanisms of fretting fatigue, *JSME Int. Journ.*, 38(4), 405-415.

[18] M.H.Wharton, D.E.Taylor and R.B.Waterhouse (1973) Metallurgical factors in the fretting-fatigue behavior of 70/30 brass and 0.7% carbon steel, Wear, 23, 251-260.

[19] Y.Mutoh and K.Tanaka (1988) Fretting fatigue of several steels and a cast iron, *Wear*, 125, 175-191.

[20] F.Guiu, M.J.Reece and D.A.J.Vaugham (1991) Cyclic fatigue in some structural ceramics, *Fatigue of Advanced Materials*, Eds. R.O.Ritchie et al., MCEP, Birmingham, 193-210.

[21] Y. Mutoh, M. Takahashi, T. Oikawa and H. Okamoto (1991) Fatigue crack growth of long and short cracks in silicon nitride, *Fatigue of Advanced Materials*, Eds. R. O. Ritchie et al., MCEP, Birmingham, 211-225.

[22] E.J.Dollay, N.E.Ashbaugh and B.D.Worth (1996) Isothermal high-cycle-fatigue of a Ti-46.6Al-3.0Nb-2.1Cr-0.2W(at%) gamma titanium aliminide, *Proc. of Fatigue '96*, Berlin, Pergamon, III, 1755-1760.

[23] M.Okane, T.Satoh, Y.Mutoh and S.Suzuki (1994) Fretting fatigue behavior of silicon nitride, *Fretting Fatigue*, Eds. R.B.Waterhouse and T.C.Lindley, Mech. Engng. Publ., London, 363-371.

[24] M.Okane, Y.Mutoh, Y.Kishi and S.Suzuki (1996) Static and cyclic fretting fatigue behaviors of silicon nitride, *Fatigue Fract. Engng. Mater. Struct.*,19(12),1493-1504

[25] F.H.Scott, D.S.Lin and G.C.Wood (1973) The structure and mechanism of formation of the glaze oxide layers on nickel-based alloys during wear at high temperatures, *Corrosion Sci.*, 13, 449-469.

[26] R.B.Waterhouse (1981) Fretting at high temperatures, *Tribology Int.*, 14(4), 203-207.

[27] T.Satoh and Y.Mutoh (1994) Effect of contact pressure on fretting fatigue crack growth behavior at elevated temperature, *Fretting Fatigue*, Eds. R.B.Waterhouse and T.C.Lindley, Mech. Engng. Publ., London, 405-416.

[28] K.Tanaka, Y.Mutoh, S.Sakoda and G.Leadbeater(1985) Fretting Fatigue in 0.55C spring steel and 0.45C carbon steel, *Fatigue Fract. Engng. Mater. Struct.*, 8(2),129-

142.

[29] A.L.Mckelvey, J.P.Campbell, K.T.Venkateswara Rao and R.O.Ritchie (1996) High temperature fatigue crack growth behavior in an XDTM γ-TiAl intermetallic alloy, *Proc. of Fatigue '96,* Berlin, Pergamon, III, 1743-1748.

[30] B.London, D.E.Larsen, D.A.Wheeler and P.R.Aimone (1993) Investment cast titanium aluminide alloys: processing properties and promise, *Structural Intermetallics,* Eds. R.Doriola et al., The Minerals, Metals and Materials Society, 151-157.

[31] P.S.Rao, A.Pattanaik, S.J.Gill, D.J.Michel, C.R.Feng and C.R.Crowe (1990) Room-temperature fatigue crack propagation in Ti-47Al/TiB$_2$ composits, *Scripta Met.,* 24, 1895-1900.

[32] R.B.Waterhouse (1992) Fretting fatigue, *Int. Mater. Reviews,* 37(2), 77-97.

Antonios E. Giannakopoulos,[1] Trevor C. Lindley[2] and Subra Suresh[1]

**Applications of Fracture Mechanics in Fretting Fatigue Life Assessment**

---

**REFERENCE:** Giannakopoulos, A. E., Lindley, T. C. and Suresh, S., **"Applications of Fracture Mechanics in Fretting Fatigue Life Assessment,"** *Fretting Fatigue: Current Technology and Practices, ASTM STP 1367*, D. W. Hoeppner, V. Chandrasekaran, and C. B. Elliott, Eds., American Society for Testing and Materials, West Conshohocken, PA, 2000.

**ABSTRACT:** Models based on fracture mechanics are developed in order to describe crack growth in fretting fatigue. Particular attention is given to a recently developed model, the 'Crack Analogue' in which certain aspects of equivalence are recognized between contact mechanics and fracture mechanics. Here, an analogy is invoked between the sharp edged contact region between two contacting surfaces and the geometry of the near-tip regions of a double edge-cracked plate. Based on this approach, a new life prediction methodology is developed for fretting fatigue.

**KEYWORDS:** fretting fatigue, crack growth, crack analogue, contact mechanics, fracture mechanics, fatigue threshold.

**Introduction**

The conventional method of establishing the important variables which can affect fretting fatigue has been to measure stress-life curves, with and without fretting, thereby providing fatigue strength reduction factors for a particular material combination of interest [1-3]. The extent of this reduction has been found to be particularly dependent on the following parameters: (a) magnitude and distribution of contact pressure (b) frictional forces and the associated near surface stresses (c) slip amplitude (d) cyclic frequency

---

[1]Department of Materials Science and Engineering, Massachusetts Institute of Technology, Cambridge, MA 02139, USA.

[2]Department of Materials, Imperial College of Science, Technology and Medicine London, SW7 2BP, England.

(e) temperature (f) environment and (g) the particular material combination and other conditions at the mating surfaces. This S-N method [4-10] is extremely useful in several respects in that it gives a ranking of various combinations of materials in terms of fretting fatigue performance. Here, an indication is given [11-16] of certain material combinations which can be expected to cause significant fretting problems or which should be avoided completely. The S-N route is used to assess the effectiveness of palliative treatments which can be devised to combat fretting such as surface coating and or peening.

Assessment procedures based on fretting fatigue strength reduction factors have been formulated by several investigators [17-21], but these have a number of limitations: (i) they are specific to the materials combinations studied, (ii) they are valid only for the experimental parameters e.g. contact pressure or slip amplitude used in each study. Although such information is available for certain important practical examples of fretting fatigue, such as the contacting materials in steam turbines [2,8], more often than not it is necessary to generate the S-N data (with and without fretting) as each new practical occurrence of fretting arises.

A number of independent research studies have shown [7,22,23] that fretting fatigue cracks can be 'initiated' at a very early stage (<5-10%) of fretting fatigue life. The cracks propagate obliquely to the specimen surface under the combined action of tangential fretting forces and the cyclic body stresses. When the crack reaches a certain depth, which depends on the applied cyclic and mean body stresses, as well as the prevailing contact conditions (contact pressure, friction coefficient, relative slip amplitude), further crack propagation occurs perpendicularly to the applied stresses which then play a more dominant role in influencing fracture than the contact stresses. It is well established that as well as accelerating crack 'initiation', fretting can also cause an increase in the rate of crack growth [22,23]. Additionally, a crack which might be dormant under pure fatigue loading might resume propagation under fretting fatigue conditions. Since under fretting, cycles to develop a propagating crack can be greatly reduced, we need to accurately predict the rate and direction of crack growth and methodologies based on fracture mechanics provide an obvious starting point.

## Assessment Procedures for Fretting Fatigue Based on Fracture Mechanics

In fracture mechanics models [14,24,25], the stress intensity factor at the tip of a crack growing beneath a fretting contact arises not only from the body stresses but also from contributions arising from the tangential and vertical forces due to the fretting contact. This composite stress intensity factor can be evaluated [26-29] by several distinct methods: (1) finite element stress analysis which might be necessary for the highly complex geometries found in practice, (2) by using stress intensity factors arising from the tangential and normal forces at the fretting position using Green's functions, and (3) by the use of distributed dislocation methods. In these fracture mechanics models of fretting fatigue, it is necessary to have a knowledge of frictional forces which arise from the specimen or component contacts [24,25] and subsequent application requires a knowledge of the fatigue crack growth properties including data for short cracks [30].

Fretting damage in the form of small cracks is often found at an early stage in the life of a specimen or component. For the case of fretting fatigue involving high numbers of cycles accumulated at low stresses (and also of constant amplitude for rotor shafts), Lindley and his co-workers [2,16,25], and independently Hattori et al [31], recognized the importance of employing the experimentally measured fatigue crack growth threshold parameter $\Delta K_{th}$. Here, the composite $\Delta K_{app}$ can be compared with the experimentally determined fatigue crack growth threshold $\Delta K_{th}$ in order to predict the balance between crack growth or arrest [2,16,25,31]. If $\Delta K_{app} > \Delta K_{th}$, then sustained growth of a fretting fatigue crack will occur). The fracture mechanics model demonstrates that the applied stress intensity factor for a small crack, $\Delta K_{app}$, will be increased under fretting conditions, possibly promoting growth, whereas the crack would remain dormant in a non-fretting situation. Such analyses will require a knowledge of both short and long crack behavior and the valid application of linear elastic fracture mechanics will need to be demonstrated.

The growth of a developing fretting crack tends to be impeded by a variety of barriers each having a corresponding threshold parameter. Both for the cases of plain fatigue [32] and fretting fatigue, three thresholds can potentially be encountered by a developing crack: (i) a microstructural threshold since the metallurgical structure and particularly grain boundaries are known to be effective barriers to growth, (ii) the crack tip driving force for the small and angled crack comprises both shear and tensile components (shear being enhanced by fretting) and for growth to be sustained, a mixed mode threshold $\Delta K_I$ / $\Delta K_{II}$ threshold must be exceeded and (iii) at the transition from shear dominated (angled crack) to tensile dominated (crack perpendicular to the stress axis), the pure mode I threshold becomes the critical threshold parameter.

A major factor impeding progress toward a better understanding of fretting fatigue is the limited direct observation of the growth of small cracks at the contact, even at room temperature and in air environment. Although it would constitute a vital step toward understanding fretting fatigue, only limited progress has been made in evaluating the growth of small cracks under mixed mode loading without fretting.

In general terms, fretting fatigue strength reduction factors are generally limited to ranking in a qualitative manner, the susceptibility of different contacting combinations to fretting fatigue as mentioned previously. Despite various complications outlined below, Lindley and Nix [2,16,25] have met with some success in modeling fretting fatigue damage in turbo-generators using fracture mechanics. However, the following simplifications were made in their analysis, each item requiring substantial further research.

(a) cracks were assumed to propagate at right angles to the surface rather than the 10-75° found in practice.

(b) only mode I stress intensity factors were used in the analysis, a consequence of the lack of experimental data for mixed mode I / II loading under fretting fatigue conditions.

(c) multiple cracking commonly occurs under fretting conditions which would introduce crack tip shielding [33,34].

(d) fretting involves multiaxial fatigue [35] whereas uniaxial models only are generally used in the fracture mechanics based analysis .

The mixed mode I / II aspect has been studied by Nowell, Hills and O'Connor [36]. As a small fretting crack grows in the region of contact, the crack tip will experience mode II as well as mode I stress intensity factors. Nowell et al. [36] have used the experimental data of Bramhall [37] to calculate $K_I$ and $K_{II}$ factors for a crack at a cylinder/flat contact and then to present the information in terms of contour maps which give values of $K_I$ and $K_{II}$ as a function of the ratio b / a where b is the crack length and a is the half contact width. The maximum value of $K_{II}$ occurs when the crack is inclined at an angle of about 45° to the free surface and from this aspect, growth under the contact is clearly favored. The calculated $K_I$ reaches a maximum for a crack approximately at right angles to the free surface and $K_I$ becomes negative for a crack inclined beneath the contact. Even in the very early stages of crack growth, a small $K_I$ component is needed in order to 'unlock' the crack faces. For an inclined crack growing under the contact under mixed mode I /II loading, the crack closure condition is correctly predicted. If the transition from mode II to mode I growth does not occur at an early stage in crack propagation life, then closure and a non-propagating crack are likely to result. The diagram can correctly predict the observed direction of crack propagation in several experiments.

Therefore fracture mechanics potentially provides an attractive assessment route to model the growth of fretting cracks but there are substantial qualifications as follows:
(1) plasticity in both the surface layers and at the crack tip must be minimal if linear elastic fracture mechanics is to be valid.
(2) although extensive research has been carried out in characterizing the growth of short cracks in plain fatigue, similar studies are required for fretting fatigue.

## A New Approach-The 'Crack Analogue'

By identifying certain aspects of equivalence between contact mechanics and fracture mechanics, the present authors have recently suggested [38] a "Crack Analogue" model for contact problems. Here, an analogy is invoked between the geometry of the near-tip regions of cracked specimens (see refs. 39-43) and that between the sharp-edged contact region between two contacting surfaces (see refs. 44-49) and for some combinations of the elastic properties of the contacting bodies which result in a square root stress singularity. It can be shown that the asymptotic elastic stress and strain fields around the periphery of the contact region (as derived from classical contact mechanics) are identical to those given by linear elastic fracture mechanics for analogous geometries. In this model, the geometry of the contact pad/substrate system naturally introduces a fictitious crack length, thereby providing a physical basis for the analysis of crack growth in contact fatigue. Based on this approach, it has been possible to develop a new life prediction methodology for fretting fatigue.

## Theoretical validation of the 'crack analogue' for contact mechanics

For a two-dimensional geometry, we now demonstrate the equivalence between the fields for sharp-edged contacts (derived from classical contact mechanics analyses) and the fields at stationary cracks (as determined from fracture mechanics). For the case of frictionless normal loading, the steps involved are as follows: 1. By considering the contact region developing under a sharp-edged punch which is normally pressed against a planar surface of the same material, we identify the cracked specimen configuration which gives geometric equivalence. 2. From classical contact analyses, we find the solutions to the stress and strain fields at the edges of sharp contacts. 3. Using linear elastic fracture mechanics analysis, we determine the corresponding stress and strain field solutions for the analogous cracked body. The assumption is invoked that the cracked body is subjected to a normal compressive load whose magnitude P is the same as that of the load pressing the punch against the planar surface. The scalar amplitude of the singular fields at the crack tip is the stress intensity factor $K_I$. 4. The different components of the stress fields determined from steps 2 and 3 are made equal and solved for $K_I$ which is obtained as a function of P, the overall geometry of the cracked body, and the characteristic punch dimension. 5. Compendia of stress intensity factor solutions are consulted e.g., [50,51,52] in order to determine the stress intensity factor $K_I$ as a function of P and the geometry of the cracked body. 6. The quantitative equivalence between the contact mechanics and fracture mechanics solutions is proven by showing that steps 4 and 5 lead to identical results. A slightly different procedure which gives an identical result can be obtained as follows. Steps 1 and 2 are performed as shown above. Next perform the above step 4, and from the stress intensity factor so determined, find the mode I crack-tip fields. Show that these fields are identical to those found in step 2.

*Two-dimensional 'crack analogue'*

The two-dimensional contact between a rectangular punch of width 2a and a flat-surfaced substrate is shown schematically in Figure 1. The most general loading case is considered where the contact interface transmits a compressive normal force P and a shear force Q per unit thickness of the contact area (i.e. P and Q have units of force per unit length). For practical cases involving sliding-contact or fretting fatigue, typical loading conditions are represented by fixed P and fluctuating Q. The uniform traction, $\sigma_{xx}$ acts along the slip direction (i.e. the x direction in Figure 1) and may include oscillatory mechanical loads or residual stresses. Noting the symmetry of the contact geometry about the y-z plane, we focus attention at the corner, (x, y) = (-a, 0) in Figure 1, without any loss of generality.

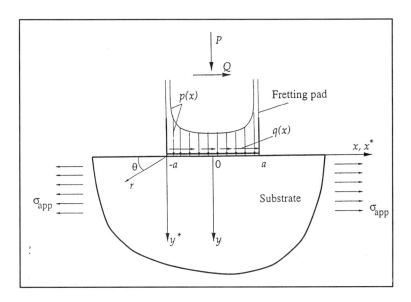

*Figure 1 Schematic representation of the contact between a two-dimensional rectangular punch and a substrate showing normal and shear loading at the contact interface and the external loading of the substrate.*

In developing the static analysis, the following assumptions are made: (1) The rectangular punch (sometimes called the contact pad) is assumed to be rigid. (2) The substrate is assumed to be semi-infinite and of depth and width of at least 6a. It is also assumed to be linear elastic and isotropic and subject to small-strain deformation at all times. (3) It is assumed that small-scale yield conditions exist such that the size of the plastic zone or damage zone, $r_p$ at the sharp edges of the contact region is small compared to the width of the contact area, i.e., $r_p \leq a/5$. (4) A requirement is that gross sliding does not occur between the punch and the substrate. (5) The asymptotic solutions are not significantly modified by any partial slip occurring at the outer edges of the contact area. This will be strictly true for rigid pads on surfaces which are incompressible (Poisson ratio $v = 0.5$) or frictionless (i.e. the coefficient of friction, $\mu = 0$). Ref. [*38*] gives a more detailed examination of pad-substrate elastic mismatch and its implications on the present analysis.

In the crack analogue analysis, the following assumptions are made:
* Linear, small strain elasticity remains valid with all material components being isotropic.
* Other pads could be present but must be separated by a distance of at least 6a away from each other.

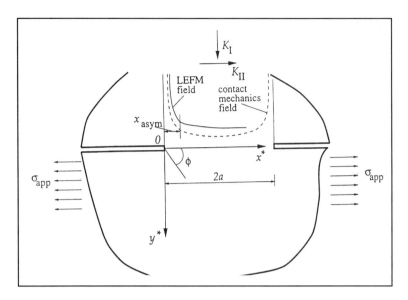

*Figure 2 Crack analogue of Figure 1 showing the double-edge cracked plate which is subject to mode I and mode II stress intensity factors $K_I$ and $K_{II}$.*

An analogy can be made between the punch/substrate contact system shown in Figure 1 and the geometry of a two-dimensional cracked specimen of Figure 2. This cracked plate is infinitely wide and contains double-edge (semi-infinite) cracks whose tips are separated by a distance 2a. The plate is remotely subjected to tensile and shear loads, P and Q, respectively, such that they induce mode I and mode II stress intensity factors, $K_I$ and $K_{II}$ respectively, at the crack-tips. Figures 1 and 2 show the x* - y* co-ordinate axes centered at the contact edge or the crack tip. The cracked plate of Figure 2 may additionally be subjected to a uniform normal stress $\sigma_{xx} = \sigma_{app}$ (which acts as the well known T stress) along the x direction to simulate conditions which represent superimposed mechanical loads during sliding or fretting contact and/or residual stresses which arise from surface treatments such as coating, shot peening etc.

In the contact situation shown in Figure 1, it is evident that a crack is likely to propagate at an angle θ below the contact surface and outside the contact area. In the crack analogue of this contact geometry shown in Figure 2, it is expected that normal and shear loading would lead to the introduction of a branched crack at the tip of one or both mode I 'cracks' (which are modeled here as non-closing under the applied compressive normal load). This initial branch angle is denoted as $\varphi_{in}$ in Figure 3. The cracking conditions in Figures 2 and 3 can therefore be made equivalent by introducing a coordinate transformation whereby θ in Figure 1 is made to equal $\pi - \theta = \varphi$ in Figure 3 for $0 \leq \theta \leq \pi$.

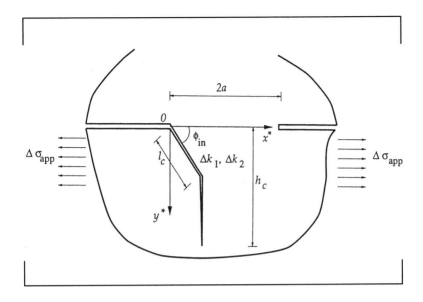

*Figure 3 Analysis of cracking in fretting fatigue using the crack analogue model.*

*Normal loading, plane strain (Mode I 'crack analogue')*

The asymptotic contact analyses of Sadowski [44] and Nadai [45] provide the singular stress fields at the sharp edges of the contact between the rectangular punch and the substrate. Using the polar coordinates $(r,\theta)$ Figure 1, the stresses at the left edge $(-a,0)$ are found to vary as:

$$\begin{pmatrix} \sigma_{rr} \\ \sigma_{\theta\theta} \\ \sigma_{r\theta} \end{pmatrix} \propto -\frac{3}{4\sqrt{r}} \begin{pmatrix} \sin\dfrac{3\theta}{2}+5\sin\dfrac{\theta}{2} \\ -\sin\dfrac{3\theta}{2}+3\sin\dfrac{\theta}{2} \\ \cos\dfrac{3\theta}{2}-\cos\dfrac{\theta}{2} \end{pmatrix}. \tag{1}$$

From the 'crack analogue' transformation given in Figure 2, we note that

$$\theta \to \pi - \theta = \varphi; \quad 0 \le \theta \le \pi. \tag{2}$$

When the trigonometric terms in Eq. (1) are expanded, the angular ($\varphi$ in Figure 2) variation of stresses {the strains ($\varepsilon_{rr}$, $\varepsilon_{\varphi\varphi}$, $\varepsilon_{r\varphi}$) can be found from the stresses and the elastic constitutive equations} are written as:

$$\begin{pmatrix} \sigma_{rr} \\ \sigma_{\phi\phi} \\ \sigma_{r\phi} \end{pmatrix} \propto \frac{1}{\sqrt{r}} \begin{pmatrix} \cos\frac{\phi}{2}\left(1+\sin^2\frac{\phi}{2}\right) \\ \cos^3\frac{\phi}{2} \\ \sin\frac{\phi}{2}\cos^2\frac{\phi}{2} \end{pmatrix} \tag{3}$$

It can be seen that the stress fields are identical to those derived by Williams [39] for a Mode I crack.

For the calculation of the Mode I stress intensity factor in the crack analogue, we note that the contact pressure gives the only non-zero component of stress at the contact surface [44]:

$$\sigma_{yy} = -\frac{P}{\pi\sqrt{a^2 - x^2}} \tag{4}$$

Taking the asymptote of Eq. (4) for the left crack tip in Figure 2, $x \to -a$, we find that

$$\sigma_{yy} \to -\frac{P}{\pi\sqrt{2ar}}, \qquad a+x = r \to 0 \tag{5}$$

The opening stress at the crack tip is related to the stress intensity factor by

$$\sigma_{yy} \to \frac{K_I}{\sqrt{2\pi r}}, \qquad r \to 0 \tag{6}$$

Comparing Eqs. (5) and (6), it is clear that

$$K_I = -\frac{P}{\sqrt{\pi a}}. \tag{7}$$

The same $K_I$ calibration result is found in standard fracture mechanics handbooks (e.g., [50]) for a double-edge-cracked infinite plate (containing two semi-infinite cracks whose tips separated by a distance of 2a) loaded remotely with a concentrated symmetric normal load P. A similar analysis can be carried out for tangential loading, plane strain which is the Mode II 'crack analogue'.

Additionally, the present authors have developed a three dimensional 'crack analogue' model, details being found in ref. [38].


**Analysis Using the 'Crack Analogue'**

The fretting crack developing at the contact is modeled as a crack branch using in particular the analysis of Cotterell and Rice [53]. As the crack branch becomes longer, it adopts a curved path eventually becoming normal to the surface, Figure 3. At that stage, the classical fracture mechanics approaches can be used to model crack growth, e.g.[25, 50, 54]. Life predictions can be made using the Paris and Erdogan [55] fatigue crack propagation law.

## 'Crack' branching

A criterion which has been found to give a good description of the onset of crack branching, Suresh [56], involves the crack attempting to achieve a crack growth direction which will give a pure mode I local condition under maximum values of $|K_I|$ and $K_{II}$. Then,

$$\max|K_I|\left(\sin\frac{\phi_{in}}{2}+\sin\frac{3\phi_{in}}{2}\right)=\max K_{II}\left|\cos\frac{\phi_{in}}{2}+3\cos\frac{3\phi_{in}}{2}\right|, \qquad (8)$$

where

$$\max|K_I|=\frac{\max P}{\pi a}, \max K_{II}=\frac{\max Q}{\pi a} \qquad (9)$$

Equation (8) can be solved for crack angle $\varphi$ but given the value of $\varphi$, it can be solved for the ratio $Q/P$ which is commonly referred to as the 'apparent friction coefficient'. Corresponding to the value of $\varphi$ for which eq. (8) is satisfied, the local fatigue crack growth amplitude will be

$$\Delta k_I=\frac{1}{4}\left(-|\Delta K_I|\left(3\cos\frac{\phi_{in}}{2}+\cos\frac{3\phi_{in}}{2}\right)+3\Delta K_{II}\left(\sin\frac{\phi_{in}}{2}+\sin\frac{3\phi_{in}}{2}\right)\right) \qquad (10)$$

where

$$\Delta K_I=\frac{\max P-\min P}{\sqrt{\pi a}} \qquad (11)$$

$$\Delta K_{II}=\frac{\max Q-(|\min Q|+\min Q)/2}{\sqrt{\pi a}} \qquad (12)$$

in order to avoid 'crack' closure. Note that $\Delta K_I=0$ for constant P, which is typically the case for fretting tests.

*Subsequent crack deflection*

The surface stresses, $\sigma_{xx}$, will compete with the local fields of the deflected or kinked crack. Then, the crack will deflect again so as to adopt a growth direction perpendicular to the contact surface (where $\sigma_{xx}$ will dominate), as indicated in Figure 3. The critical condition is then

$$\Delta K_{th} = \frac{\Delta\sigma_{app}\sqrt{\pi.l_c}}{4} \times \tag{13}$$

$$\left[ F_1\left(\phi_{in}\right)\left(3\cos\frac{\pi - 2\phi_{in}}{4} + \cos\frac{3\left(\pi - 2\phi_{in}\right)}{4}\right) + 3F_2\left(\phi_{in}\right)\left(\sin\frac{\pi - 2\phi_{in}}{2} + \frac{3\left(\pi - 2\phi_{in}\right)}{2}\right)\right]$$

where

$$\Delta\sigma_{app} = \max\sigma_{xx} - (|\min\sigma_{xx}| + \min\sigma_{xx})/2 \tag{14}$$

and $F_1(\phi_{in})$, $F_2(\phi_{in})$ are known functions that depend on the angle $\phi_{in}$ and provide the stress intensity factor calibrations. Tables of $F_1$ and $F_2$ can be found in Isida [54].

Finally, at a depth, $h_c$, the crack will become unstable, according to the relation:

$$1.12\max\sigma_{xx}\sqrt{\pi h_c} = K_{Ic} \tag{15}$$

where $K_{Ic}$ is the fracture toughness of the material. Equation (15) should be solved for $h_c$ and then inspected to show that:

$$1.12\max\sigma_{xx}\sqrt{\pi l_c} < K_{Ic} \tag{16}$$

**Fretting Fatigue Life Predictions**

Adopting the Paris law, which is judged to be appropriate in view of the initial 'long crack' operative in the 'crack analogue' and with the values of $\varphi_{in}$, $l_c$, $h_c$ and $\Delta k_l$ already established, the following fretting fatigue life estimates can be found. For the crack advancing from 0 to $l_c$, the number of cycles is

$$N_1 = \int_0^{l_c} \frac{dl}{C_1\left(\Delta k_1\right)^{m_1}} \approx \frac{l_c}{C_1\left(\Delta k_1\right)^{m_1}} \qquad (17)$$

where $C_1$ and $m_1$ are material constants specific to the surface mechanical and environmental conditions of the fretting case.

For the crack advancing from $l_c$ to $h_c$, the number of cycles is

$$N_2 = \int_{l_c \sin\phi_{in}}^{h_c} \frac{dl}{C_2\left(1.12\Delta\sigma_{app}\sqrt{\pi l}\right)^{m_2}} = \qquad (18)$$

$$\frac{1}{C_2\left(1.12\Delta\sigma_{app}\sqrt{\pi}\right)^2} \ln\frac{h_c}{l_c \sin\phi_{in}}, \quad m_2 = 2$$

$$\frac{1}{C_2\left(1.12\Delta\sigma_{app}\sqrt{\pi}\right)^{m_2}} \left(\frac{2}{m_2-2}\right)\left(\left(l_c \sin\phi_{in}\right)^{\frac{2-m_2}{2}} - h_c^{\frac{2-m_2}{2}}\right), \quad m_2 \neq 2$$

where $C_2$ and $m_2$ are material constants specific to the substrate mechanical and environmental conditions of the fretting case. The total fretting fatigue life will be:

$$N = N_1 + N_2 \qquad (19)$$

Regarding other possible extensions to the present work, one could list other effective stress intensity factors such as max $K_1$ and $K_{eff} = \sqrt{(k_1^2 + k_2^2)}$.

**Fretting Fatigue Design Using the Fatigue Crack Growth Threshold**


If the fatigue crack growth threshold, $\Delta K_{th}$ (for specific mechanical and environmental conditions) has previously been established experimentally, then a possible design criterion to avoid the formation of a crack branch is:

$$\Delta k_1 < \Delta K_{th}, \qquad (20)$$

where $\Delta k_1$ is given by eq. (10). If $\Delta k_1 \geq \Delta K_{th}$, fatigue crack propagation is predicted to commence. Experimental fretting fatigue crack growth data found in the literature can be used in order to validate the 'crack analogue' model (see ref. [38]), the necessary steps being as follows: 1) Evaluate the initial growth direction i.e. the angle $\varphi_{in}$ made by the crack with the contact surface see Figure 3). 2) Evaluate the critical length, $l_c$ at which the crack branches due to the action of $\sigma_{xx}$ (if operative). This length $l_c$ marks the transition from shear dominated (stage 1) growth to tensile dominated (stage 2) growth. 3) Evaluate the driving force for crack growth i.e. the local crack tip stress intensity factor $\Delta k_1$ which is used as an effective threshold for fatigue crack growth, here viewed as crack branching. In particular, the crack tip driving force $\Delta k_1$ is compared to the experimentally determined fatigue crack growth threshold parameter $\Delta K_{th}$ in order to determine whether a crack will arrest ($\Delta k_1 \leq \Delta K_{th}$) or continue propagating ($\Delta k_1 \geq \Delta K_{th}$). 4) Use fatigue crack growth and life assessment methods to calculate the number of cycles to failure of the fretted component. The calculated values of $\varphi_{in}$, $l_c$ and $\Delta k_1$ using equations (8), (13) and (10) can be compared to experimentally measured values for a variety of materials found in the literature.

A worked numerical example using this life prediction methodology based on the 'crack analogue' concept can be found in ref. [56] for the case of fretting fatigue in a large electrical generator. We present here an illustrative example of the crack analogue life prediction methodology for fretting fatigue with possible application to the blade/disk fixing in a jet engine. The specific details of such an analysis, as well as the relevant criteria for crack growth, should however be chosen on the basis of detailed stress analysis of the actual jet engine components. We assume contact pressures of either 160MPa or 350MPa and for a contact width of $2a = 3.6$mm, we can find the corresponding line loads (Table 1) expressed as N/m. Additionally, fatigue crack growth growth rates and the threshold $\Delta K_{th}$ are required for the material of interest Ti-6Al-4V. From the experiments of Chesnutt et. al. [57], we take the threshold at R=0.1 to be 3.5MPa$\sqrt{m}$ and $da/dN = 1 \times 10^{-14} \Delta K^{6.22}$ m/cycle. Assuming applied stress values of 195MPa and 390MPa, and taking the initial crack angle $\varphi_{in}$ to be 55° from the work of Antoniou and Radtke [58] and Davidson [59], we can use equation (13) with F($\phi_{in}$) equal to 4.51 to estimate the size of the slant crack $l_c$ and we find values of 80μm and 20μm respectively.    Using equations (8) and (9) and with $\varphi_{in} = 55°$, we find

$$Q_{max}/P = 1.136$$

Using equations (10) and (12) we find (see Table 1)

$$\Delta k_I = +/- K_{IImax} \times 0.744$$

$$\text{and } K_{IImax} = Q_{max}/\sqrt{\pi a} = 1.136 P/\sqrt{\pi a}$$

Table 1 *Values of $\Delta k_I$ and $K_{II}$*

| Nominal    Contact Pressure  (MPa) | Line Load P  (N/m) | $K_{II}$  (MPa√m) | $\Delta k_I$  (MPa√m) |
|---|---|---|---|
| 160 | $0.576 \times 10^6$ | 8.7 | 6.5 |
| 350 | $1.260 \times 10^6$ | 19.0 | 14.2 |

We use equation (17) to calculate the number of cycles $N_1$ (Table 2) to grow the crack from 0 to $l_c$.

Table 2 *Values on $N_1$*

| $l_c$ μm | $\Delta k_I = 6.5$ MPa√m | $\Delta k_I = 14.2$ MPa√m |
|---|---|---|
| 20 | $1.76 \times 10^4$ cycles | $1.36 \times 10^2$ cycles |
| 80 | $7.11 \times 10^4$ cycles | $5.51 \times 10^2$ cycles |

We now use equation (14) to calculate $N_2$ for two cases: (1) to grow the crack from $l_c$ = 20μm to $h_f$ = 50μm with $\Delta\sigma_{app}$ = 390MPa and (2) from $l_c$ = 80μm to $h_f$ = 100μm with $\Delta\sigma_{app}$ = 195MPa where $h_f$ is the final crack size, see Table 3.

Table 3 *Values of $N_2$*

| $\Delta\sigma_{app}$ (MPa) | $N_2$ Cycles |
|---|---|
| 390 | $5.63 \times 10^5$ |
| 195 | $15.1 \times 10^5$ |

For a fracture toughness $K_{IC}$ of 60MPa√m for the Ti-6Al-4V alloy, we find that the critical defect sizes for complete fracture, at the applied stress levels given in Table 3, are approximately 12mm at 390MPa and 24mm at 195MPa.

**Conclusions**

1.) Since cracks are known to be present at an early stage in fretting fatigue life in both specimens and components, models based on fracture mechanics can be developed in order to describe crack growth. 2.) The similarities in both geometries and elastic stress fields between sharp cornered rigid punches against flat surfaces and double edge cracked specimens provide the basis for the newly proposed 'crack analogue' model. 3.) Problems in contact mechanics can then be treated using classical fracture mechanics. 4.) The 'crack analogue' model can be validated by using experimental fretting fatigue data found in the literature. 5.) An important aspect of the present approach is that the sharp pad/flat surface contact introduces a 'fictitious crack' from the outset, thereby avoiding the difficulties encountered in characterising crack 'initiation' and growth of a small fretting crack.

6.) The 'crack analogue' makes it possible to unify fretting fatigue with classical fatigue crack growth models based on fracture mechanics. 7. The blade/disk fixing in an aero-engine has been used as an example in applying the 'crack analogue'.

**Acknowledgments**

This work was supported by the Multi-University Research Initiative on "High Cycle Fatigue" which is funded at MIT by the Air Force Office of Scientific Research, Grant No. F49620-96-1-0478, through a subcontract from the University of California at Berkeley. We would like to thank Dr. D.L.Davidson for valuable discussions on the application of the crack analogue.

**References**

[1] Waterhouse, R.B., *Fretting Corrosion*, Pergamon Press, Oxford,1972.

[2] Lindley, T.C. and Nix, K.J., "Fretting Fatigue in the Power Generation Industry" in *Standardisation of Fretting Fatigue Test Methods and Equipment*, *ASTM STP 1159*, Editors Attia, H.A. and Waterhouse R.B, 1992, pp. 153-169.

[3] Fenner, A.J. and Field, J.E., "Fatigue Under Fretting Conditions," *Revue Metallurgique*, Vol. 55, 1958, pp. 475-478.

[4] Nishioka, K. and Hirakawa, K., "Fundamental Investigation of Fretting Fatigue: Part 2, Fretting Fatigue Test Machines and Some Results," *Bulletin of the Japan Society of Mechanical Engineers*, Vol 12, No 50, 1969, pp. 180-187.

[5] Endo K, Goto, H and Nakamura, T., "Fretting Fatigue Strength of Several Material Combinations," *Bulletin of Japan Society of Mechanical Engineers*, Vol 17, No 92, 1973.

[6] Waterhouse, R.B., 'Fretting Fatigue', *International Metallurgical Reviews*, Vol 37, No 2, 1992, pp. 77-97.

[7] Nix, K.J. and Lindley, T.C., "The Influence of Relative Slip and Contact Materials on the Fretting Fatigue of a 3.5% NiCrMoV Rotor Steel," *Wear*, Vol 125, 1988, pp. 147-162.

[8] Lindley, T.C. and Nix, K.J., "An Appraisal of the Factors which Influence the Initiation of Cracking by Fretting Fatigue in Power Plant," *Fretting Fatigue*, MEP Publications, London, 1992, pp. 239-256.

[9] Spink, G.M., "Fretting Fatigue of a 2.5% NiCrMoV Low Pressure Turbine Shaft Steel: The Effect of Different Contact Pad Materials and of Variable Slip Amplitude," *Wear*, Vol. 136, 1990, pp. 281-297.

[10] Field, J.E. and Waters, D.M., "Fretting Fatigue Strength of En26 Steel: Effect of Mean Stress, Slip Amplitude and Clamping Conditions," National Engineering Laboratory Report UK, 1967, No 275.

[11] Duquette, D.J., *Strength of Materials and Alloys*, Publishers Pergamon Press, 1980. p 214.

[12] Sproles, F.S. and Duquette, D.J., "The Mechanism of Material Removal in Fretting," *Wear*, Vol. 49, 1978, pp. 339-352.

[13] Vingsbo, O. and Soderberg, D., "On Fretting Maps," *Wear*, Vol. 126, 1988, pp. 131-147.

[14] Rayaprolu, D.B. and Cook, R., "A Critical Review of Fretting Fatigue Investigations at the Royal Aerospace Establishment," *Standardisation of Fretting Fatigue Test Methods and Equipment, ASTM STP 1159*, Philadelphia, 1992, pp. 129-152.

[15] Nakazawa, K., Sumita, M. and Maruyama, N., "Effect of Contact Pressure on Fretting Fatigue of High Strength Steel and Titanium Alloy," *Standardisation of Fretting Fatigue Test Methods and Equipment, ASTM STP 1159*, Philadelphia, 1992, pp. 115-125.

[16] King, R.N. and Lindley, T.C., "Fretting Fatigue in a 3.5%NiCrMoV Rotor Steel," *Advances in Research*, ICF 5, Pergamon, Oxford, 1980, pp. 631-640.

[17] Nishioka, K. and Hirakawa, K., *Proceedings International Conference on Mechanical Behaviour of Materials*, Kyoto, Japan, 1972, pp. 308-318.

[18] Nishioka,K. and Hirakawa, K., "Fundamental Investigation of Fretting Fatigue: Part 5 - The Effect of Relative Slip," *Bulletin of Japan Society of Mechanical Engineers*, Vol. 12, No. 52, 1969, pp. 692-697.

[19] Nishioka, K. and Hirakawa, K., "Fundamental Investigation of Fretting Fatigue: Part 6 - Effect of Contact Pressure and Hardness of Materials," *Bulletin of Japan Society of Mechanical Engineers*, Vol. 15, No. 80, 1972, pp. 135-144.

[20] Wharton, M.H., Waterhouse, R.B., Hirakawa,K. and Nishioka, K., "The Effect of Different Contact Materials on the Fretting Fatigue Strength of an Aluminium Alloy," *Wear*, Vol. 26, 1973, pp. 253-260.

[21] Nowell, D. and Hills, D.A., "Crack initiation Criteria in Fretting Fatigue," *Wear*, Vol. 136, 1990, pp. 329-343.

[22] Endo, K. and Goto, H., "Initiation and Propagation of Fretting Fatigue Cracks," *Wear*, Vol. 38, 1976, pp. 311-320.

[23] Endo, K., "Practical Observations of Initiation and Propagation of Fretting Fatigue Cracks," in *Fretting Fatigue*, Applied Science Publishers, London, 1981, pp. 127-141.

[24] Edwards, P.R., "The Application of Fracture Mechanics to Predicting Fretting Fatigue," in *Fretting Fatigue*, Applied Science Publishers, London, 1981, pp. 67-97.

[25] Nix, K.J. and Lindley, T.C., "The Application Of Fracture Mechanics to Fretting Fatigue," *Fatigue and Fracture of Engineering Materials and Structures*, Vol. 8, No. 2, 1985, pp. 143-160.

[26] Rooke, D.P., "The Development of Stress Intensity Factors", *Fretting Fatigue*, MEP Publications, London, 1992, pp. 23-58.

[27] Dai, D.N., Hills, D.A. and Nowell, D., "Stress Intensity Factors for Three Dimensional Fretting Fatigue Cracks," *Fretting Fatigue*, MEP Publications, London, 1992, pp. 59-72.

[28] Faanes, S. and Harkegard, G., "Simplified Stress Intensity Factors in Fretting Fatigue", *Fretting Fatigue*, MEP Publications, London, 1992, pp 73-82.

[29] Sheikh, M.A., Fernando, U.S., Brown, M.W. and Miller, K.J., "Elastic Stress

Intensity Factors for Fretting Fatigue using the Finite Element Method," *Fretting Fatigue*, MEP Publications, London, 1992, pp. 83-102.

[30] Pearson, S., "Initiation of Fatigue Cracks in Commercial Aluminium Alloys and the Subsequent Propagation of Very Short Cracks," *Engineering Fracture Mechanics*, Vol. 7, 1975, pp. 235-247.

[31] Hattori, T., Nakamura, M., Sakata, H. and Watanabe, T., "Fretting Fatigue Analysis Using Fracture Mechanics," *Japan Society Mechanical Engineers*, Int. J. Ser. I, Vol. 31, 1988, pp. 100-107.

[32] Miller, K.J., "The Three Thresholds for Fatigue Crack Propagation", $27^{th}$ *Conference on Fatigue and Fracture Mechanics, ASTM STP 1296*, 1997, pp. 267-286.

[33] Dubourg, M-C and Villechaise, B., "Analysis of Multiple Fatigue Cracks: Part 1, Theory," *Journal of Tribology*, Vol. 114, 1992, pp. 455-461.

[34] Dubourg, M-C., Godet, M. and Villechaise, B., "Analysis of Multiple Fatigue Cracks: Part 2, Results," *Journal of Tribology*, Vol. 114, 1992, pp. 462-468.

[35] Fouvry, S., Kapsa, Ph., Vincent, L. and Dang Van, K., "Theoretical Analysis of Fatigue Under Dry Friction For Fretting Loading Conditions," *Wear*, Vol. 195, 1996, pp. 21-34.

[36] Hills, D.A., Nowell, D. and O'Connor, J.J., "On the Mechanics of Fretting Fatigue," *Wear*, Vol. 125, 1988, pp. 129-146.

[37] Bramhall, R., *"Studies in Fretting Fatigue,"* Ph.D. Phil. Thesis, Department of Engineering Science, Oxford University 1973.

[38] A.E.Giannakopoulos, T.C.Lindley, and S.Suresh, "Aspects of Equivalence Between Contact Mechanics and Fracture Mechanics: Theoretical Connections and a Life Prediction Methodology for Fretting Fatigue," *Acta Materialia*, 1998, Vol. 46, No. 9, pp. 2955-2968.

[39] Williams, M.L., "On the Stress Distribution at the Base of a Stationary Crack," *Journal of Applied Mechanics*, Vol. 24, 1957, pp. 109-114.

[40] Irwin, G.R., "Analysis of Stresses and Strains Near the End of a Crack Traversing a Plate," *Journal of Applied Mechanics*, Vol. 24, 1957, pp. 361-364.

[41] Westergaard, H. M., "Bearing Pressures and Cracks," *Journal Applied Mechanics*, Vol. 6, 1939, pp. 49-53.

[42] Kanninen, M.F. and Popelar, C.H., *Advanced Fracture Mechanics*, Oxford Engineering Science Series, Oxford University Press, 1985.

[43] Anderson, T.L., *Fracture Mechanics*, CRC Press, 1991.

[44] Sadowski, M., "Zweidimensionale probleme der elastizitatstheorie," *Z.angew. Math. v. Mech.*, Vol. 8, 1928, pp. 107-121.

[45] Nadai, A. I., *Theory of Flow and Fracture of Solids*, Vol. II, McGraw-Hill, N.Y., 1963.

[46a] Sneddon, I. N., "The Distribution of Stress in the Neighbourhood of a Crack in an Elastic Solid," *Proc. Royal Society of London*, Vol. A-187, 1946, pp. 229-260.

[46b] Sneddon, I. N., "Boussinesq's Problem for a Flat-ended Cylinder," *Proc. Cambridge Philosophical Society*, Vol. 42, 1946, pp. 29-39.

[46c] Sneddon, I. N., "The Distribution of Stress in the Neighbourhood of a Flat Elliptical Crack in an Elastic Solid," *Proc. Cambridge Philosophical Society*, Vol. 46, 1950 pp. 159-163.

[47] Gladwell, G.M.L., *Contact Problems in the Classical Theory of Elasticity*, Alphen aan den Rijn: Sijthoff and Noordhoff, 1980.

[48] Johnson, K. L., *Contact Mechanics*, Cambridge University Press, U.K., 1985.

[49] Hills, D.A. Nowell, D. and Sackfield, A., *Mechanics of Elastic Contacts*, Butterworth Heinemann, Oxford,1993.

[50] Tada, H., Paris, P. C. and Irwin, G. R.,*The Stress Analysis of Cracks Handbook*, Hellertown: Del Research Corporation, 1973.

[51] Rooke, D.P. and Cartwright, D.J., *Compendium of Stress Intensity Factors*, Her Majesty's Stationary Office, London, 1976.

[52] Murakami, Y. *Stress Intensity Factors Handbook*, Pergamon Press, New York, 1987.

[53] Cotterell, B. and Rice, J. R., "Analysis of Crack Branching," *International Journal of Fracture*, Vol. 16, 1980, pp. 155-162.

[54] Isida, M., "Tension of a Half Plane Containing Array Cracks, Branched Cracks and Cracks Emanating from Sharp Notches," *Transactions Japan Society Mechanical Engineers*, Vol. 45, 1979, series A, pp. 306-317.

[55] Paris, P. C. and Erdogan, F., "A Critical Analysis of Crack Propagation Laws," *Journal of Basic Engineering*, Vol. 85, 1963, pp. 528-534.

[56] Suresh, S., *Fatigue of Materials*, Second Edition, Cambridge University Press, U.K., 1998.

[57] Chesnutt, J.C., Thompson, A.W. and Williams, J.C., "Influence of Metallurgical Factors on the Fatigue Crack Growth Rate in Alpha-Beta Titanium Alloys," Report AFML-TR-78-68, 1978.

[58] Antoniou, R.A. and Radtke, T.C., "Mechanisms of Fretting Fatigue of Titanium Alloys," *Materials Science and Engineering*, A237, 1997, pp. 229-240.

[59] Davidson, D. L., To be published.

Steven E Kinyon[1] and David W. Hoeppner[2]

## Spectrum Load Effects on the Fretting Behavior of Ti-6Al-4V

**REFERENCE:** Kinyon, S. E., and Hoeppner, D. W., **"Spectrum Load Effects on the Fretting Behavior of Ti-6Al-4V,"** *Fretting Fatigue: Current Technology and Practices, ASTM STP 1367,* D. W. Hoeppner, V. Chandrasekaran, and C. B. Elliott, Eds., American Society for Testing and Materials, West Conshohocken, PA, 2000.

**ABSTRACT:** An investigation was performed to determine the effects of spectrum loading on the fretting fatigue life of titanium alloy Ti-6Al-4V. The research program consisted of a 10 Hz spectrum load cycle applied at six conditions of interest with a normal pressure of 31 MPa. Baseline tests were performed at three stress levels, 345, 310, and 286 MPa maximum load using a conventional haversine waveform. Tests were performed using a spectrum load waveform consisting of 2 cycles, one at 345 MPa, followed by another at 310 MPa, repeated to failure. Additional tests were performed using 345 and 286 MPa. The final condition of interest was investigated using a 345/310/286 MPa waveform. Miners rule was applied to each of the spectrum load cases to test the hypothesis that the titanium alloy followed Miners Rule under the loading conditions. The three conditions of interest varied from Miners Rule by 23, 47, and 62%. From these results it was concluded that Miners Rule was insufficient to predict failure lives of Ti-6Al-4V in fretting fatigue.

**KEYWORDS:** fretting, fatigue, titanium, spectrum, miners rule

Material parameters influence the decision that design engineers use to complete any engineering task. Fretting fatigue is a material response influenced by the materials in contact as well as variables such as the loading condition and the geometry near the fretting area. When two materials are in contact, fretting fatigue can occur whenever there exists a small relative motion between the two members. Typical examples where fretting fatigue failures have been studied are in dental couplings, orthopedic implants, wire ropes, electrical contacts, propeller shafts, electrical transmission lines, and dovetail joints for turbine discs [1-5] to name just a few. Knowledge of each material in use and the response to loading parameters such

---

[1]   Project Engineer, Aerospace Division, MTS Systems Corp., Eden Prairie, Minnesota, MN 55344-2290 USA.

[2]   Professor, Department of Mechanical Engineering, The University of Utah, Salt Lake City, UT 84112.

as fatigue and contact fatigue are important for the designer. Testing may be the only method to determine a material response to loading in a fretting contact condition, since analytically deriving the response to a loading condition is not possible.

As mentioned above, fretting fatigue is usually the result of small relative motion between two members. The motion may be due to vibration, or the result of loading of one or both members. A crack can prematurely nucleate as a result of fretting when compared to crack nucleation due to pure fatigue. Since the crack nucleates at an early stage in the fatigue life, the onset of crack propagation is also earlier than with pure fatigue, and can significantly reduce the time to failure for any part. In some cases, fretting fatigue lives can be reduced by a factor of greater than ten when compared to fatigue without fretting [6]. Once a crack nucleates and propagates beyond the region affected by fretting, conventional fracture mechanics can be used to analyze the crack growth to failure.

It is generally accepted that fretting fatigue failure is a phenomenon characterized by the synergistic action of wear, corrosion and fatigue. The interaction of wear, corrosion and fatigue can be influenced by parameters such as coefficient of friction, magnitude of slip, contact pressure, microstructure, environment, fatigue stress magnitude, frequency of oscillation, and others [7]. A complete understanding of the fretting fatigue phenomena is not currently possible, since so many variables can greatly effect each situation. Eden, Rose, and Cunningham first suggested that fretting is a wear process [8], and much work has been completed to identify and characterize the effects of these variables. One of the variables that has not received much attention is the role of spectrum loading on the fretting fatigue life of materials. The work outlined in this dissertation is a study to determine the effects of load spectrum on fretting fatigue life of a titanium alloy.

**Hypothesis**

The main parameter in this investigation was the load spectrum effects on the fretting fatigue behavior of titanium alloy Ti-6Al-4V. The following hypotheses were formulated for this research program:

> Load Spectrum Fretting Fatigue lives will be shorter than normal Fretting Fatigue lives for Ti-6Al-4V, based on number of cycles to failure.
> Fretting Fatigue for Ti-6Al-4V will follow Miners Rule.

To test the hypothesis that fretting fatigue lives will be shorter than normal fretting fatigue, the average number of cycles to failure for the fretting and the load spectrum tests was compared. The number of cycles to failure, on an average basis, were used to determine the tendencies of the fretting fatigue life of the titanium alloy, Ti-6Al-4V.

Miners Rule is a measurable method to determine lives of test and service components. Application of miners rule requires fatigue data without spectrum loading, and then comparing the data to spectrum tests.

**Fretting Fatigue**

In 1911 Eden, Rose and Cunningham found corrosive debris where a test specimen fit into its holders [8]. What they had found was the first record of fretting fatigue damage. They hypothesized that this corrosive product was the result of the varying stress between the specimen and holder. The answer to their query remained an untouched mystery for nearly two decades. Further technological advancement and more complex geometries in machinery led to significant fretting fatigue/corrosion problems. Tomlinson was first to investigate the phenomenon when he devised a set of experiments in 1927 to investigate the cause [9]. The experiments he created involved placing a normally loaded spherical ball onto a flat steel plate. Various normal pressures, surface finishes, and types of motion were used and resulted in damage, usually scarring and oxidation. Through the geometry of the test equipment, the relative motion was calculated to be in the range of micrometers. Tomlinson attributed the wear to molecular attrition.

In the 1930s, premature fatigue failures of railway car axles forced the railroad companies to investigate the causes. Investigations were initiated to determine the reasons behind press fit fatigue failures [10,11]. Discontinuities, inhomogeneities and stress concentrations from manufacturing and design were given most of the credit. Fretting fatigue and/or corrosion were not the terms used to discuss these failures. Most of the materials in use at that time were steels, and it was noticed that most of the areas where failures were occurring were pitted and covered with a rust-like deposit, strong indicators that fretting was a possibility.

In 1939 Thomlinson, with the help of Thorpe and Gough, performed another early investigation into the effects of fretting corrosion [12]. What they observed in experimentation and in-field experience convinced them that the corrosion products were mechanical rather than chemical in nature. It was later recognized that fretting fatigue, although referred to as fretting corrosion at the time, was mechanical and environmental in nature.

In the early 1940s Warlow-Davies conceived an experiment where steel specimens were first fretted and then fatigued to failure [13]. The fretting and fatigue portions of the fretting tests were kept separate to investigate the effect of fretting on fatigue life. Pure fatigue tests were conducted without prefretting to examine the change in fatigue life when compared to the pre-fretted tests. A 20% decrease in life was observed from the results of these tests. At this time, the effect of fretting fatigue was attributed to the relative motion between the surfaces in contact. Mindlin showed this to be the case by using a sphere on a flat surface [14].

With the development of the relative motion theory, the stage was set for researchers to embark on the search to determine the mechanisms and parameters of fretting. Halliday and Hirst [15] proposed...

1.    Metal is removed by grinding or tearing of adhered asperities. The metal particles evolved from the wear/tear action oxidize but cause no further damage.

2.    The metal particles oxidize and are then ground into a powder. The powder in turn increases the wear rate.

3.    Oxide or oxidized particles are removed from the surface. The newly exposed bare metal is oxidized. The process of oxidation and removal continues.

They concluded that the first and second were dominant mechanisms were dominant when they first used electron microscopy to investigate fretting fatigue.

In modern days, investigations into a possible relationship between particle production and the subsequent ability of the debris bed to kinematically support the normal load has been investigated [16,17]. Much of the recent work has been performed to determine ways to prevent fretting, such as surface treatments, protective coatings and lubrication [18-20].

**Spectrum Loading**

Fatigue research has been conducted to determine the effects of stress magnitudes, mean stress magnitudes, frequency, environment, and many other variables on a material's fatigue life. Due to the influence of so many variables on the fatigue process, many researchers have attempted to characterize each by focusing on variations of one variable, while holding others constant.

*Early Spectrum Fatigue* - Early fatigue work usually consisted of applying sinusoidal loads through the use of specifically designed equipment. Stress range-life ($\Delta\sigma$-N) curves for fatigue were derived from such tests and were used by engineers to estimate material response characteristics in design. It was found in the early fatigue research that varying the stress ratio as well as the stress magnitude gave different responses in fatigue. In practice, few components see constant amplitude fatigue stresses. They may be subjected to consistent stress magnitudes for some known (or usually unknown) number of cycles, and then other stress magnitudes for a different number of cycles. Miner [21] suggested what later came to be known as Miners Rule to approximate fatigue life or fatigue damage for aluminum using fatigue curves generated from conventional constant amplitude fatigue tests. The life of an article subjected to varying stresses could be approximated using the life at each stress level. Failure was defined by the summation over all relevant stress levels of the ratios of the number of cycles at a stress level to the number of cycles for failure at that stress level. When the sum of the ratios equals unity, expected failure life could be determined. Miners Rule for fatigue is written

$$\frac{n_1}{N_1} + \frac{n_2}{N_2} + \frac{n_3}{N_3} + ... + \frac{n_f}{N_f} = 1 \qquad (1)$$

where,

$n_x$ = Number of cycles at $\sigma_x$

$N_x$ = Number of cycles for failure at $\sigma_x$ from a constant amplitude $\sigma$-N curve.

The tests performed by Miner to test equation 1 gave reasonable results. It should be noted that many undergraduate and graduate level materials courses introduce Miners Rule with little attention to the boundary conditions outlined by Miner. Miners Rule may approximate failure in fatigue accurately for one material, but may not represent test data for another. For this reason Miners Rule should be used with caution, and test data should be examined whenever Miners Rule is used to determine fatigue lives.

*Recent Spectrum Fatigue* - The work that Miner reported in his original paper was significant because it gave engineers a new way to use the stress range-life data to approximate component lives. The equipment used to perform fatigue tests in Miners day did not allow for more complex load sequences to be applied. It has been well accepted that the effects of constant stress fatigue do not quantify the life of a component for varying load applications.

The use of Miners Rule is limited in that the actual load case may not be adequately mimicked by Miners Rule. With the development of more complex testing equipment came the ability for researchers to apply ever more complex load sequences. Several load spectrum have been developed in an attempt to test fatigue lives using realistic load environments.

Much of the work was completed to support aircraft fatigue studies. The 1970s marked a transition for the development of spectrum sequences for the use in fatigue research. A number of international working groups formed, which led to the definition of loading standards for fighter aircraft, transport aircraft wing skins, helicopter components, and engine disks [22].

Some work has been performed in Fretting Fatigue research using spectrum fatigue loading. Mutoh, Tanaka, and Kundoh found agreement using Miners Rule when performed fretting fatigue tests on two steels using a two-step and a random loading sequence [23-24]. Edwards and Ryman found less damage when performing fretting fatigue tests on L65 aluminum alloy using a Gaussian random sequence and a gust spectrum loading sequence [25].

## Titanium

The discovery of titanium was reported in 1791 by William Gregor [*38*]. He discovered a black magnetic sand in the parish of Menaccan in Cornwall, England that contained an oxide of an unknown element. Gregor proposed the name menaccanite. At about the same time a German chemist named M. H. Klaproth had made the same discovery while examining rocks and minerals. Klaproth opted to name the oxide titanium after the Titans in Greek mythology. The new element posed several obstacles during its development. The first was the extraction of titanium from its sources.

The first successful extraction was performed by Nilson and Petersson in 1887 [*39*]. The process outlined by Nilson and Petersson was refined by Hunter to the level that 98% pure titanium could be extracted from titanium dioxide, commonly known as rutile. One of the most common methods for titanium extraction became known as the Hunter process.

The aerospace industry began to enjoy the benefits of titanium's high strength to weight ratio during the 1960s when titanium began to be used in gas turbine engines. As with any high strength alloys, titanium has undergone some significant research. As titanium gained popularity as an engine disk material, what later became known as the dwell effect became one of several challenges to be overcome.

Early research of the dwell effect showed that the processing methods of the alloys were at fault. The most significant change at that time was the move to a triple vacuum melt. The titanium alloys used in engine disks at the time were melted and cast in a vacuum twice to remove impurities such as oxygen. The low oxygen environment was necessary to block the oxidation of the alloy. The new triple melt processing reduced uncontained disk bursts [40].

Room temperature creep and the dwell effect have been studied extensively in titanium alloys. Much of the work was driven by the aerospace industry as the result of engine disk failures [38,39].

Titanium alloys display a significant response to spectrum loading. The research performed to investigate the dwell effects were using a simple (dwell) spectrum. This investigation was performed to investigate the material with a load spectrum under fretting fatigue conditions.

## Fretting Fatigue of Titanium

1970 marked a new decade and a new awareness of fretting fatigue and titanium alloys. The first work in the area of fretting fatigue on titanium alloys was in 1970 by Milestone [40]. In 1971 NACE held a conference on Corrosion Fatigue. Several papers were presented that outlined new research in the area of fretting fatigue of titanium and its alloys [20,38-40].

Most of the early work on fretting fatigue was done to examine material responses to various environmental and loading conditions. Waterhouse and Dutta reported that a 60% life reduction was found when fretting titanium alloys in a 1% NaCl corrosive environment [38]. Similar results were found by Waterhouse along with Wharton using different alloys tested in the same environment [39]. Goss and Hoeppner found that the life dependency of titanium was sensitive to variable normal pressures in fretting fatigue [40].

Other researchers have investigated the variation in fretting fatigue lives of titanium alloys. Surface conditions have been found to play a significant role in the life of Ti-6Al-4V [38]. Shot peened specimens had longer lives than did shot peened and polished specimens or polished only specimens. Temperature effects have been investigated as well [39].

## Test Loads

The main parameter investigated in this research was the effect of variable amplitude loading on the fretting fatigue life of Ti-6Al-4V alloy. In an attempt to determine if there in fact exists an effect, six loading sequences were selected. The normal load applied remained constant for each test and test condition. The six test conditions evaluated were:  1) large amplitude haversine;  2) medium amplitude haversine; 3) small amplitude haversine; 4) combined large and medium amplitude haversine; 5) combined large and small amplitude haversine; 6) combined large, medium and small amplitude haversine. Each of the load conditions was applied until specimen failure or 10 million cycles (run-out).

In order to investigate the effects of spectrum loads using Miners Rule, the magnitude of the large, medium and small amplitude waveforms had to be determined. Since the application of Miners Rule required the failure lives at each level without spectrum loading, baseline tests were necessary to determine the average life at each stress level. Fatigue loads were determined by trial and error, although the initial tests were performed using data from Hoeppner and Goss [40]. The normal loads for all conditions were the same, 31 MPa, citing the same article.

The initial test performed found that below a maximum stress level around 283 MPa, the fretting fatigue lives were several orders of magnitude higher than at slightly higher stresses. The first baseline tests were performed at 286 MPa. Additional baseline tests were performed at maximum stresses of 310 and 345 MPa. The data used generated the average life to failure for each stress level and a stress range-life curve for the alloy in fretting fatigue. Minimum stress levels were found using a stress ratio (R) of 0.1. The load spectrums used for each condition are briefly explained below.

*345, 310, and 286 MPa Loads* - The three conditions labeled 345, 310, and 286 MPa were conventional fretting fatigue tests with the maximum loads equal to 345, 310, and 286 MPa. These tests form the baseline test data used to determine the average number of cycles to failure for the application of Miners Rule.

*345/310 and 345/286 MPa Spectrum Loads* - To determine the effects of the combined large and small amplitude loads, a large amplitude fatigue load was applied (345 MPa), followed by one cycle of a smaller magnitude (310 MPa). The spectrum will be continued, one cycle of each stress level, to failure. The same test was performed using 345 MPa and 286 MPa.

*345/310/286 MPa Spectrum Loads* - To further test the hypothesis that Miners Rule will aid in determining the fretting fatigue life of the titanium alloy, similar tests to the above were performed. A 345 MPa maximum fatigue load, followed by a 310 MPa fatigue load, and finally a 286 MPa maximum fatigue load were consecutively applied to the specimens. The three-stress cycles continued until specimen failure.

**Material**

Ti-6Al-4V is a commonly used material in the aircraft industry because of its high strength to weight ratio, and corrosion resistance. Ti-6Al-4V was selected for this investigation because of its use as turbine disk material. The connection between the disk blades and the disk is known to be a likely location of fretting fatigue [6].

The material selected for this investigation was Mill Annealed Ti-6Al-4V. The material composition for this alloy meets mil standard MIL-T-9046H Type 3 Comp C. The Ultimate Tensile Strength is 965 MPa and the 0.2% offset yield strength is 924 MPa.

The fretting pads and the dogbone specimens, shown in Figure 1, were machined from the same 4.318-mm thick sheet. The fretting specimens and fretting pads were numbered by applied loads and test number for identification. The specimens were numbered at the time of testing to remove any experimental bias. Each test condition had up to four tests. Each specimen was prepared for the test by polishing with incremental sandpaper to 800 grit. The final polishing step was performed in the direction of the fatigue loading, along the length of the specimen.

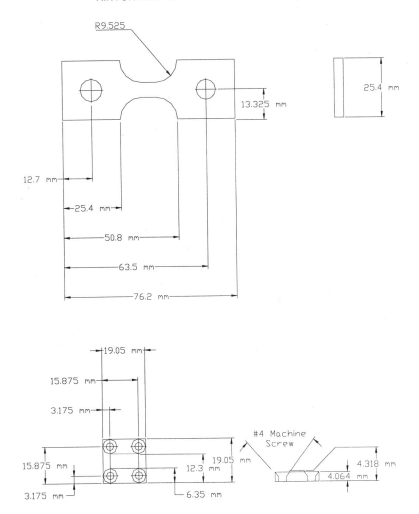

Figure 1 - Dogbone specimen and fretting pad.

Since this research was exploratory in nature and will guide future work regarding material responses in spectrum fretting fatigue, four specimens were tested at each condition of interest.

The lower support for the test specimen was a phenolic material pad to react the normal load through and beneath the specimen, yet not cause fretting on the lower face of the specimen. All the tests that failed did so at the upper pad interface.

## Test Equipment

The load frame used for the tests was built at the Structural Integrity Laboratory at the University of Utah. The frame consists of dual actuators, one used for the fatigue load, and one for the normal load. The fatigue actuator was a 24.5 kN capacity, and the normal actuator was 8.9 kN capacity. Sensotec load cells provide the control channel feedback for both actuators. The normal load actuator was not original. Modifications were made prior to this program using the original frame, and an MTS Servo-Valve donated by MTS for this program.

The fatigue and normal loads were applied through a closed loop servo-hydraulic test system controlled by two MTS 442 controllers, one for each channel. The load spectrum was provided by an MTS 481 Mircroprofiler. Hydraulic pump control was performed using an MTS 413 control panel, and an MTS 410 data display was used to indicate the loads. A cycle counter was installed to monitor the number of cycles applied to the specimen. A data acquisition system was used to monitor the tests. The data acquisition system consisted of a Macintosh Centris desktop computer utilizing Labview. The computer system recorded maximum and minimum loads, at 1000 cycle increments. Little deviation from target loads was recorded during the system operation.

## Results

### Baseline Fretting Fatigue Tests

Several tests were performed using a conventional fatigue test waveform form. These tests were conducted to determine the average number of cycles to failure at each stress level, and in effect create a stress-life curve. Tests were conducted at 345, 310, 296, 286 and 276 MPa maximum stresses and a stress ratio of 0.1. The normal load for all fretting fatigue tests was 31 MPa. The results from these tests are included in (Table 1). The data from (Table 1) are plotted in Figure 2. The results from the 345, 310, and 286 MPa tests were averaged to develop the baseline data used for the spectrum tests. (Table 1) contains the average number of cycles for failure of the baseline tests. Specimen 286-2 failed in the grip section at nearly 33 million cycles and was not included in the average life calculation. Posttest inspection of the specimen showed that there was no evidence of fretting at the pad interface, the test failed near the edge of the grip region, slightly outside of the gage section. One result of this test was that a decision was made that run-out for the remaining tests would be 10 million cycles. It was assumed that failure due to fretting would occur before this life, or the fretting damage was not significant enough to cause failure.

*Spectrum Tests*

Load spectrum tests were conducted to investigate the effects of spectrum loading on the fretting fatigue behavior of Ti-6Al-4V. Results from the spectrum tests are included in (Table 2). A stress-life plot for the spectrum fretting fatigue tests is included in Figure 3.

Table 1 - *Results from baseline tests.*

| Specimen I.D. | Cycles to Failure | Average Life |
|:---:|:---:|:---:|
| 286-1 | 251,432 | |
| 286-2 | 32,663,827[1] | |
| 286-3 | 762,023 | |
| 286-4 | 860,742 | 625,732 |
| 310-1 | 292,101 | |
| 310-2 | 590,395 | |
| 310-3 | 836,083 | |
| 310-4 | 730,688 | 612,317 |
| 345-1 | 434,394 | |
| 345-2 | 190,062 | |
| 345-3 | 612,676 | 412,377 |

[1]These test data were included in this table, but were not used further in the analysis

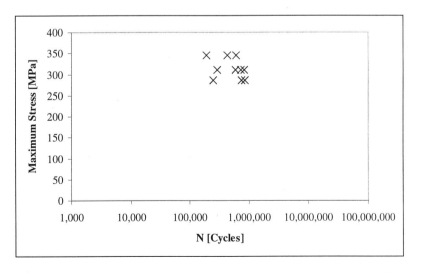

Figure 2 - *Baseline fretting fatigue results.*

Table 2 - *Results from load spectrum tests.*

| Specimen I.D. | Cycles to Failure | Average Life |
|---|---|---|
| 345/310-1 | 342,706 | |
| 345/310-2 | 529,256 | |
| 345/310-3 | 10,000,000[1] | |
| 345/310-4 | 267,795 | 379937 |
| 345/286-1 | 1,777,378 | |
| 345/286-2 | 242,394 | |
| 345/286-3 | 931,208 | |
| 345/286-4 | 926,459 | 928834 |
| 345/310/286-1 | 1,754,421 | |
| 345/310/286-2 | 1,591,005 | |
| 345/310/286-3 | 870,297 | |
| 345/310/286-4 | 10,357,544 | 1405241 |

[1]These test data were included in this table, but were not used further in the analysis

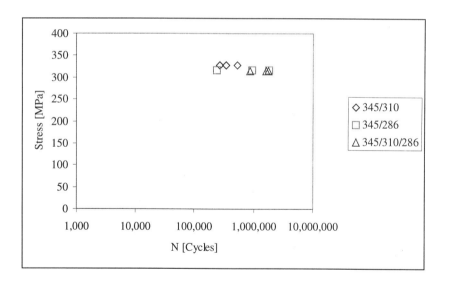

Figure 3 - *Spectrum fretting fatigue results.*

## Comparison Using Miners Rule

These tests were performed to investigate the hypothesis that load spectrum tests using a sinusoidal waveform would follow Miners Rule. In order to compare the tests, Miners Rule was used following baseline tests fatigued using a conventional sinusoidal waveform. The results from the baseline tests are compared to the spectrum tests in (Table 3). The data in (Table 3) show that Miners Rule better approximated experimental values for the first condition, 345/310. The difference using Miners Rule and experimental values was 23% difference. The other two test conditions, 345/286 and 345/310/286 did not compare as well, with a 47 and 62% difference, respectively.

Table 3 - *Miners Rule compared to experimental values.*

| Test condition | Miners Rule Approximation | Experimental average life | Percent Difference |
|---|---|---|---|
| 345/310 | 492841 | 379937 | 23 |
| 345/286 | 496814 | 928834 | 47 |
| 345/310/286 | 530149 | 1405241 | 62 |

## Hypothesis Results

The results from the tests were examined to investigate the response validity to the hypotheses outlined earlier. From the experimental results, the average life using conventional loading at 345, 310, and 286 MPa maximum stress were 412,377, 612,317, and 624,732 cycles, respectively. The average lives of the 345/310, 345/286, 345/310/286 MPa spectrum tests were 379,937, 928,834, 1,406,241 cycles. The spectrum fretting fatigue tests lasted longer than the conventional fretting fatigue tests with the exception of the 345/310 MPa tests. The average lives of the 345/310 MPa tests were the shortest of the conditions investigated.

Based on these results, the fretting fatigue tests with spectrum loading are longer than the conventional loading tests. The hypothesis is false from the data retrieved experimentally.

The application of Miners Rule to the experimental data showed that the difference between experimental results and Miners Rule were 23, 47, and 62%. Miners Rule underestimated the fretting fatigue lives in all but the first case. The data show that Ti-6Al-4V does not follow Miners Rule for fretting fatigue in the conditions tested.

## Conclusions

A program has been concluded to test the hypotheses that load spectrum fretting fatigue lives will be shorter than normal fretting fatigue lives for Ti-6AL-4V, based on number of cycles to failure, and fretting fatigue for Ti-6Al-4V will follow Miners Rule.

### Spectrum Fretting Lives

From the experimental results, the average life using conventional loading at 345, 310, and 286 MPa maximum stress were 412,377, 612,317, and 624,732 cycles, respectively. The average life of the 345/310, 345/286, 345/310/286 MPa spectrum tests were 379,937, 928,834, 1,406,241 cycles. The spectrum fretting fatigue tests lasted longer than the conventional fretting fatigue tests with the exception of the

345/310 MPa tests. The average life of the 345/310 MPa tests were the shortest of the conditions investigated.

Based on these results, the fretting fatigue tests with spectrum loading are longer than the conventional loading tests.

*Miners Rule in Fretting Fatigue*

The application of Miners Rule to the experimental data showed that the difference between experimental results and Miners Rule were 23%, 47%, and 62%. The data show that Ti-6Al-4V does not follow Miners Rule for fretting fatigue in the conditions tested. The sensitivity of the titanium alloy to microstructural variability may be the reason for the discrepancy between the test results and Miners Rule.

**References**

[1]    Hoeppner, D., and Chandrasekaran,V., "Fretting in Orthopedic Implants: A Review," *Wear*, Vol. 173, 1994, pp. 187-189.

[2]    Zhou, Z. R., Cardou, A., Fiset, M., and Goudreau, S., "Fretting Fatigue in Electrical Transmission Lines," *Wear*, Vol. 173, 1994, pp. 179-188.

[3]    Harris, S. J., Waterhouse, R. B., and McColl, I. R., "Fretting Damage in Locked Coil Steel Ropes," *Wear*, Vol. 170, 1993, pp. 63-70.

[4]    Waterhouse, R.B., "Fretting Fatigue in Aqueous Electrolytes," *Fretting Fatigue*, Pergamon Press, 1981, p. 159.

[5]    Boddington, P. H. B., Chen, K. C., Ruiz, C., "An Investigation of Fatigue and Fretting in a Dovetail Joint," *Experimental Mechanics*, September, 1984, pp. 208-217.

[6]    Hoeppner, D.W., and Poon, C., "The Effect of Environment on the Mechanism of Fretting Fatigue," *Wear*, Vol. 52, 197, pp. 175-191.

[7]    Hoeppner, D. W., "Mechanisms of Fretting Fatigue," *International Conference on Fretting Fatigue*, Keynote Presentation, Sheffield, England, 1993.

[8]    Eden, E. M., Rose, W. N., and Cunningham, F. L., "The Endurance of Metals," *Proceedings of the Institute of Mechanical Engineers*, 1911, p. 875.

[9]    Tomlinson, G. A., "The Rusting of Steel Surfaces in Contact," *Proceedings of the Royal Society*, Series A, Vol. 115, 1927, pp. 472-483.

[10]   Buckwalter, T. V., and Horger, O. J., "Investigation of Fatigue Strength of Axles, Press-fits, Surface Rolling, and Effect of Size," *Transactions of American Society for Metals*, Vol. 25, 1937, p. 229.

[11]    Peterson, R. E., and Wahel, A. M., "Fatigue of Shafts at Fitted Members, with Related Photo-elastic Analysis," *Transactions of American Society for Mechanical Engineers, Journal of Applied Mechanics*, Vol. 57, 1935, p. A-1.

[12]    Tomlinson, B. A., Thorpe, P. L., and Gough, H. J., "An Investigation of the Fretting Corrosion of Closely Fitting Surfaces," 1939.

[13]    Warlow-Davies, E. J., "Fretting Corrosion and Fatigue Strength:  Brief Results of Preliminary Experiments," *Proceedings of the Institute of Mechanical Engineers*, Vol. 146, 1941, pp. 32-38

[14]    Mindlin, R. D., 1949, "Compliance of Elastic Bodies in Contact," *Transactions of American Society for Mechanical Engineers, Journal of Applied Mechanics*, Vol. 71, pp. 259-268.

[15]    Halliday, J. S., and Hirst, W., "The Fretting Corrosion of Mild Steel," *Proceedings of the Royal Society*, Vol. 236, March 1956, pp. 411-425.

[16]    Godet, M., 1984, "The Third-Body Approach: A Mechanical View of Wear, " *Wear*, Vol. 100, pp. 437-452.

[17]    Colombie, Ch., Berthier, Y., Floquet, A., Vincent, L., and Godet, M., "Fretting: Load Carrying capacity of Wear Debris," *Transaction of the ASME, Journal of Tribology,* Vol. 106, April, 1984, pp. 194-201.

[18]    Hoeppner, D. W., Gates, F. L., "Fretting Fatigue Consideration in Engineering Design," *Wear*, Vol. 70, 1981, pp. 155-164.

[19]    Beard, J., "The Avoidance of Fretting," *Materials and Design*, Vol. 9, No. 4 July, 1988.

[20]    Lum, D. W., and Crosby, J., J., "Fretting Resistant Coatings for Titanium Alloys," *NACE International Conference on Corrosion Fatigue:  Chemistry, Mechanics and Microstructure*, 1971.

[21]    Miner, M. A., "Cumulative Damage in Fatigue," *Journal of Applied Mechanics*, 1945, pp. 159-164.

[22]    Ten Have, A. A., "European Approaches in Standard Spectrum Development," *Development of Fatigue Loading Spectra, ASTM STP 1006*, 1989, pp. 17-35.

[23]    Mutoh, Y., Tanaka, K.,  and Kondoh, M, "Fretting Fatigue in JIS S45C Steel Under Two-Step Block Loading," *Japan Society of Mechanical Engineers International Journa,* Vol. 30, 1987, pp. 386-393.

[24]    Mutoh, Y., Tanaka, K.,  and Kondoh, M, "Fretting Fatigue in SUP9 Spring Steel Under Random Loading," *Japan Society of Mechanical Engineers International Journal,*Vol. 32, 1989, pp. 274-281.

[25]    Edwards, P., Ryman, R., "Studies in Fretting Fatigue Under Service Loading Conditions," Eigth International Committee on Aeronautical Fatigue, Lausanne, Switzerland, 1975.

[23]    McQuillan, A. D., McQuillan, M. K., *Metallurgy of the Rare Metals-4: Titanium,* Academic Press, London, 1956, pp. 1-50.

[27]    Hunter, M. A., "Metallic Titanium," *Journal of the American Chemical Society,* Vol. 32, Part 1, 1910, pp. 330-338.

[28]    United States of America Federal Aviation Administration Aircraft Certification Service Engine and Propeller Directorate, "Titanium Rotating Components Review Team Report," Dec. 1990.

[29]    Ryder, J. T., Petit, D. E., Krupp, W. E., Hoeppner, D. W., *Final Report: Evaluation of Mechanical Property Characteristics of IMI 685,* Lockheed-California Company, Rye Canyon Laboratory, October 1973.

[30]    Hatch, A. J., Partridge, J. M., and Broadwell, R. G., "Room-Temperature Creep and Fatigue Properties of Titanium Alloys," *Journal of Materials,* Vol. 2, No. 2, March 1967, pp. 111-119.

[31]    Milestone, W. D., "A new Apparatus for Investigating Friction and Metal-to-Metal Contact in Fretting Joints," *Effects of Environment and Complex Load History on Fatigue Life, ASTM STP 462,* 1970, pp. 318-328.

[32]    Hoeppner, D. W., and Goss, G. L., "Research on the Mechanism of Fretting Fatigue," *NACE International Conference on Corrosion Fatigue: Chemistry, Mechanics and Microstructure,* 1971.

[33]    Salkind, M. J., Lucas, J., J., "Fretting Fatigue in Titanium Helicopter Components," *NACE International Conference on Corrosion Fatigue: Chemistry, Mechanics and Microstructure,* 1971.

[34]    Waterhouse, R., B., "The Effect of Fretting Corrosion in Fatigue Crack Initiation," *NACE International Conference on Corrosion Fatigue: Chemistry, Mechanics and Microstructure,* 1971.

[35]    Waterhouse, R. B., and Dutta, M. K. "The Fretting Fatigue of Titanium and Some Titanium Alloys in a Corrosive Environment," *Wear,* Vol. 25, 1973, pp. 171-175.

[36]    Waterhouse, R. B. and Wharton, M., H., "The Behavior of Three High-Strength Titanium Alloys in Fretting Fatigue in a Corrosive Environment," *Lubrication Engineering,* Vol. 32, 1976, pp. 294-298.

[37]    Goss, G. L., and Hoeppner, D. W., "Normal Load Effects in Fretting Fatigue of Titanium and Aluminum Alloys," *Wear,* Vol. 27, 1974, pp. 153-159.

[35-38]     Lutynski, C., Simansky, G., and McEvily, A., J., "Fretting Fatigue of Ti-6Al-4V Alloy," *Materials Evaluation Under Fretting Conditions, ASTM STP 780,* 1982, pp. 150-164.

[39]     Waterhouse, R. B., and Iwabuchi, A., "High Temperature Fretting Wear of Four Titanium Alloys," *Wear,* Vol. 106, 1985, pp. 303-313.

[40]     Wulpi, D. J., *Understanding How Copmponents Fail,* American Society of Metals, April 1987, pp.183-204.

# Fretting Fatigue Parameter Effects

*D.L. Anton,[1] M. J. Lutian,[2] L. H. Favrow,[3] D. Logan,[4] and B. Annigeri[5]*

**The Effects of Contact Stress and Slip Distance on Fretting Fatigue Damage in Ti-6Al-4V/17-4PH Contacts**

**REFERENCE:** Anton, D. L., Lutian, M. J., Favrow, L. H., Logan, D., and Annigeri, B., **"The Effects of Contact Stress and Slip Distance on Fretting Fatigue Damage in Ti-6Al-4V/17-4PH Contacts,"** *Fretting Fatigue: Current Technology and Practices, ASTM STP 1367*, D. W. Hoeppner, V. Chandrasekaran, and C. B. Elliott, Eds., American Society for Testing and Materials, West Conshohocken, PA, 2000.

**ABSTRACT:** A comprehensive evaluation of fretting fatigue variables was conducted on shot-peened Ti-6Al-4V forging material in the β-STOA condition in contact with 17-4PH pins, a material couple representative of helicopter dynamic component interfaces. Utilizing test equipment incorporating independent fatigue stress and fretting slip displacement control (as described elsewhere in this symposium), a test matrix spanning slip distances, δ, of $25 \leq \delta \leq 75$ μm and contact stresses, $\sigma_f$, of $70 \leq \sigma_f \leq 200$ MPa. Fatigue stresses were used which resulted in cycle lives ranging from run out, $>10^7$ to $10^3$. A flat against flat contact geometry was used with the contact area covering ~10 mm$^2$. Representative fretting scars through out the test matrix were examined via serial section analysis and the crack number, density, location, length and flank angle noted through the scar volume. A fretting fatigue endurance map of contact stress vs. slip distance shows that slip amplitude dominates mean fretting fatigue strength at $10^7$ cycles under the tested conditions with contact stress playing only a moderate role. The fretting surface could be characterized as moderately pitted with dense third body debris. The debris was determined to be $TiO_2$. Critical cracks formed through a linking of smaller cracks across the fretting scar as evidenced by a number of nucleation sites on the fracture surface. The coefficient of friction, COF, was observed to increase from its initial value of 0.15 to a stable ~0.75 through the first $10^3$-$10^4$ cycles.

**KEYWORDS:** fretting fatigue, titanium, Ti-6Al-4V, life prediction, mechanisms, coefficient of friction, energy of fretting

[1] Principal Scientist, United Technologies Research Ctr., 411 Silver Lane, E. Hartford, CT 06108
[2] Sr. Materials Engineer, Sikorsky Aircraft, 6900 Main Street, Stratford, CT 06601
[3] Research Engineer, United Technologies Research Ctr., 411 Silver Lane, E. Hartford, CT 06108
[4] Assistant Research Engineer, United Technologies Research Ctr., 411 Silver Lane, E. Hartford, CT 06108
[5] Principal Scientist, United Technologies Research Ctr., E. Hartford, CT, USA.

**Introduction**

Titanium alloys and especially Ti-6Al-4V are used extensively in aerospace structural applications where maximum temperature requirements are below 400°C, including rotorcraft dynamic component applications. The majority of these applications require mechanical joining to other structural members, also typically Ti-6-4. Where these joint interfaces are subjected to vibratory loading, they are potentially susceptible to fretting fatigue mechanisms. Many studies have shown previously that titanium alloys are particularly susceptible to fretting fatigue. Design factors need to be taken into account when such jointed structures are used. The typical high-cycle fatigue strength reduction for fretting fatigue is 50% [1]. That is, the stresses must be reduced by 50% in the region susceptible to fretting fatigue, as compared to open section areas, in order to insure that the component will not fail prematurely due to fretting fatigue. This is usually accomplished through enlarging part dimensions and thus reducing, somewhat, the high strength to density titanium alloys possess. This reduction in fatigue allowable strength is typically known as the fretting fatigue Knock Down Factor, KDF.

The mechanisms of fretting fatigue surface degradation under spherical contact in titanium alloys, as deduced from friction log plots, are (i) surface oxide elimination, (ii) metal to metal contact, (iii) debris generation and (iv) stable debris layer [2]. Microscopic investigations of the fretted surfaces revealed the debris to consist largely of titanium oxide (rutile $TiO_2$) with fragments of unoxidized metal [3]. The accumulation of wear debris partially separates the metal surfaces causing wear to intensify in the remaining areas of metal contact, leading to the formation of pits. Failure is associated with the formation of ridges on the specimen surface, which developed into piles of platelets as the fretting action continued. These ridges also tended to trap the wear debris [4]. Crack nucleation occurred at the boundary of the fretting wear damage area. This is the site of the fretting stick on each cycle, and thus the region of maximum surface shear.

A fretting fatigue map under spherical contact conditions has been defined for Ti-6Al-4V having regions of (i) no degradation, (ii) cracking and (iii) particle detachment [4], roughly correlating to the stick/slip and gross slip defined by Vingsbo and Sölderberg [5]. In the region of zero to 50μm slip, slip amplitude was found to have a much greater influence on fretting mechanism than contact stress.

In order to achieve higher efficiency, more reliable and lighter weight aircraft, methodologies must be developed to reduce the KDF. The two most universally accepted methodologies are through shot peening and contact materials substitution. Shot peening is used in most instances where fretting fatigue is possible. This returns nearly half of the 50% KDF noted above. By introducing alternate materials into the joint, such as can be achieved with coatings, sleeves or liners, additional credit towards the KDF is obtained.

Utilizing a new fretting fatigue testing apparatus [6] which incorporates independent control of applied stress and slip amplitude, this study is intended to produce base line of fretting fatigue data for a representative component interface under a matrix of contact stress and slip distance. The materials couple of interest is Ti-6-4 in the β-STOA condition with a typical shot-peened surface in contact with 17-4PH stainless steel. The β-STOA heat treatment is given in order to gain more uniform mechanical properties in large forgings. This material specification and combination has been applied to many forged rotorcraft dynamic component applications. Toward the goal of reducing the fretting fatigue KDF in future rotorcraft designs via application of improved materials and processes, understanding of the mechanisms of fretting fatigue damage accumulation for the baseline test conditions is an important step. To initiate this effort, an intensive metallographic examination of the subsurface damage state was conducted utilizing a serial section analysis [7].

**Experimental Details**

*Materials*

Fretting fatigue specimens were fabricated from Ti-6Al-4V cylinders 150 mm in diameter and 300 mm long, received in the β-STOA condition from Sikorsky Aircraft. These cylinders are cores extracted from and heat treated along with a production forging lot. The forgings were manufactured from premium quality, triple-melt Ti-6Al-4V meeting AMS 2380, Grade 2 requirements. The heat treatment condition of the as-received cylinders was a solution heat treatment from above the β transus, water quench, and overage at approximately 700°C for 4 hours. This heat treatment yields a fully transformed β microstructure comprised of α in the lath morphology with a prior β grain size approximately 1.6mm in diameter. Occasional prior β grain boundaries were observed to have a thin layer of α resident. The typical titanium microstructure resulting from this processing is given in Fig. 1.

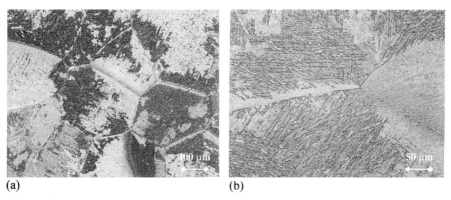

(a)                                (b)

Figure 1 – *Typical microstructure of Ti-6Al-4V in β-STOA condition showing large grains of approximately 1.5mm with a semi-continuous film of α titanium in the grain boundaries*

Fretting fatigue test specimens were machined having a rectangular gage section measuring 7.1 x 12.2 mm and 60 mm long. The gage section was shot-peened to 8-12A resulting in an average surface roughness, $R_a$, of 1.8μm. A drawing of the fatigue specimen is given in [6] elsewhere in these proceedings.

Fretting pins were fabricated 17-4PH steel rod, heat treated to the H1025 condition (155 ksi minimum tensile strength). These pins, for which a drawing is shown elsewhere [6], maintained a circular flat contact area with the fretting fatigue specimens having an area of 9.9 mm². The contact area was finished to a 600-grit finish with the final polishing scratches aligned at 45° to the slip direction.

*Open Section Fatigue*

Fatigue testing in the absence of fretting contact was performed on nine specimens so as to create a base line from which to compare the fretting fatigue test results. The mean stress was kept constant for all tests at 150MPa. It is thus important to realize that the R ratio changes with the stress amplitude. All testing was conducted at ambient laboratory conditions with temperature and humidity recorded continuously during the test. The test frequency was maintained at 20Hz utilizing a load controlled to a sine wave in a closed loop servo-hydraulic test frame.

*Testing & Matrix*

The testing methodology details are covered elsewhere in this proceedings [6]. The independent fretting fatigue test variables studied here are slip distance, $\delta$, and fretting stress, $\sigma_f$, or the contact stress. All references to slip amplitude in this paper refer to command slip amplitude and not the actual sliding amplitude. The command slip distance utilized were 25, 50 and 75 $\mu$m while the fretting stress used were 70, 100 and 200 MPa. The full matrix of these test parameters was investigated by the determination of fatigue S-N (stress vs. number of cycles) curves under all nine of the above stated conditions. This was accomplished by testing minimally nine specimens at three stress amplitudes, $\sigma_a$, ranging from 150 through 500 MPa. From this data, an analytical curve fit could be made and a mean fatigue stress for $10^7$ cycles established for each condition. Specimen failure is defined as the separation of the fretting fatigue specimen into two pieces.

*Analysis*

*Fractography & Serial Section* - After testing, all specimens were documented by photographing the fretting scars, notation of the visual location of the fretting fatigue failure (i.e. left or right fretting scar), and SEM documentation of the fracture surface. One specimen from each of the independent $\delta$ and $\sigma_f$ test conditions cycled at $\sigma_a = 200$ MPa was analyzed via a serial section analysis [7] and laser profilometry. Briefly, serial section analysis entailed successive cutting at 0.2mm intervals and metallographic preparation followed by optical and SEM examination of the fretting fatigue specimen cross section just below the fretting scar to observe incipient cracking and fretting debris build-up. The serial section analysis yields a quasi three-dimensional view of the fretting specimen subsurface structure and damage state. Incipient crack locations were recorded as well as their angle, depth and location. The angle of the initial crack growth or Crack

Take Off Angle, CTA, was measured from the surface of the specimen parallel to the direction of slip during the compression portion of the applied stress cycle.

COF - Load measurements of both at the upper and lower specimen grips, were continuously monitored and recorded at a rate of 160 measurements per cycle during the course of the test. These load data were later reduced to coefficient of friction (COF) and fretting energy data as reported elsewhere in this proceedings [6].

**Results and Discussion**

*Open Section Fatigue*

The open section fatigue test results are given in Table 1. Figure 2 illustrates this data in graphical form. Gage section failures were obtained in all but one specimen where a thread failure occurred. The scatter in the data is minimal with respect to fatigue data in general.

Table 1 - *Open Section Fatigue Results*

| Specimen ID | Stress Amplitude (Mpa) | Cycles to Failure | Failure Location |
|---|---|---|---|
| SATIOS-1 | 350 | 631,096 | gage section |
| SATIOS-2 | 350 | 1,127,000 | threads |
| SATIOS-3 | 300 | 2,649,734 | gage section |
| SATIOS-4 | 400 | 587,340 | gage section |
| SATIOS-5 | 350 | 787,542 | gage section |
| SATIOS-6 | 300 | 1,895,579 | gage section |
| SATIOS-7 | 400 | 280,898 | gage section |
| SATIOS-8 | 275 | 3,651,855 | gage section |
| SATIOS-9 | 275 | 5,596,685 | gage section |

*Fretting Fatigue*

The fretting fatigue results are given in Table 2. Additionally, typical fretting fatigue curves for the $\delta=25\mu m$, $\sigma_f=100MPa$ and $\delta=75\mu m$, $\sigma_f=100MPa$ condi-tions are given in Fig. 2 for comparison with the open section fatigue results. The fretting fatigue failures are seen to occur randomly between both the left and right fretting pads, with nearly all of the failures occurring under one of the fretting pads. Significant fatigue resistance debt is observed for both of the fretting fatigue conditions depicted in Fig. 2. Here fretting fatigue knockdown factor is defined as:

$$KDF^{N_f} = 1 - \left( \sigma_{ff}^{N_f} \Big/ \sigma_o^{N_f} \right)$$

(1)

where:    $KDF^{N_f}$ = the knockdown factor at $N_f$ cycles

$\sigma_{ff}^{N_f}$ = the mean fretting fatigue strength at $N_f$ cycles

$\sigma_o^{N_f}$ = the mean open section fatigue strength at $N_f$ cycles.

Referring to Fig. 2, the $KDF^{10^6}$ is much greater for the $\delta=75\mu m$, $\sigma_f=100MPa$ fretting conditions, ~0.57, than for the $\delta=25\mu m$, $\sigma_f= 100MPa$ conditions, ~0.29.

Table 2 - *Fretting Fatigue Test Results*

| Specimen ID | Stress Amplitude (MPa) | Fretting Stress (MPa) | Slip Amplitude (mm) | Cycles to Failure | Failure Location |
|---|---|---|---|---|---|
| SATI-19 | 200 | 70 | 25 | 3,697,107 | right pin |
| SATI-21 | 250 | 70 | 25 | 258,942 | right pin |
| SATI-22 | 300 | 70 | 25 | 599,958 | left pin |
| SATI-23 | 300 | 70 | 25 | 822,772 | under lft. pin |
| SATI-24 | 250 | 70 | 25 | 10,000,000 + | |
| SATI-24-2 | 250 | 70 | 25 | 10,000,000 + | |
| SATI-25 | 300 | 70 | 25 | 644,556 | left pin |
| SATI-26 | 200 | 70 | 25 | 673,037 | right pin |
| SATI-27 | 200 | 70 | 25 | 197,763 | left pin |
| SATI-56-2 | 500 | 70 | 25 | 29,820 | outside frett |
| SATI-59 | 150 | 70 | 50 | 970,948 | left pin |
| SATI-65 | 200 | 70 | 50 | 257,250 | left pin |
| SATI-66 | 200 | 70 | 50 | 2,193,597 | right pin |
| SATI-75 | 300 | 70 | 50 | 94,503 | right pin |
| SATI-76 | 300 | 70 | 50 | 61,804 | right pin |
| SATI-83 | 150 | 70 | 50 | 269,222 | right pin |
| SATI-88 | 150 | 70 | 50 | 10,000,000 + | |
| SATI-91 | 300 | 70 | 50 | 93,744 | left pin |
| SATI-92 | 200 | 70 | 50 | 1,280,066 | right pin |
| SATI-93 | 150 | 70 | 50 | 2,010,922 | left pin |
| SATI-60 | 150 | 70 | 75 | 2,215,082 | left pin |
| SATI-69 | 200 | 70 | 75 | 986,574 | right pin |
| SATI-70-2 | 200 | 70 | 75 | 1,094,808 | left pin |
| SATI-73 | 300 | 70 | 75 | 159,735 | right pin |
| SATI-74 | 300 | 70 | 75 | 235,447 | left pin |
| SATI-82 | 150 | 70 | 75 | 447,072 | right pin |
| SATI-84 | 150 | 70 | 75 | 2,670,365 | right pin |
| SATI-87 | 300 | 70 | 75 | 135,395 | left pin |

Table 2 - *Fretting Fatigue Test Results (Cont'd)*

| Specimen ID | Stress Amplitude (MPa) | Fretting Stress (MPa) | Slip Amplitude (mm) | Cycles to Failure | Failure Location |
|---|---|---|---|---|---|
| SATI-14 | 300 | 100 | 25 | 517,550 | right pin |
| SATI-15 | 300 | 100 | 25 | 145,417 | right pin |
| SATI-16 | 250 | 100 | 25 | 343,292 | left pin |
| SATI-17-2 | 250 | 100 | 25 | 1,805,183 | right pin |
| SATI-18-2 | 250 | 100 | 25 | 9,333,041 | right pin |
| SATI-20 | 250 | 100 | 25 | 10,000,000 + | |
| SATI-20-2 | 500 | 100 | 25 | 13,240 | right pin |
| SATI-24-3 | 500 | 100 | 25 | 10,865 | bellow pin |
| SATI-55 | 500 | 100 | 25 | 30,433 | right pin |
| SATI-6-2 | 300 | 100 | 25 | 104,735 | right pin |
| SATI-7 | 200 | 100 | 25 | 10,000,000 + | |
| SATI-8 | 200 | 100 | 25 | 10,000,000 + | |
| SATI-35 | 300 | 100 | 50 | 227,101 | left pin |
| SATI-36 | 200 | 100 | 50 | 176,746 | right pin |
| SATI-37 | 250 | 100 | 50 | 111,765 | right pin |
| SATI-38 | 300 | 100 | 50 | 80,248 | right pin |
| SATI-39 | 200 | 100 | 50 | 215,445 | left pin |
| SATI-40 | 250 | 100 | 50 | 121,308 | left pin |
| SATI-41 | 300 | 100 | 50 | 97,864 | left pin |
| SATI-42 | 300 | 100 | 50 | 106,088 | right pin |
| SATI-43 | 300 | 100 | 50 | 97,117 | right pin |
| SATI-44 | 200 | 100 | 50 | 287,655 | right pin |
| SATI-54-2 | 500 | 100 | 50 | 16,607 | right pin |
| SATI-58 | 150 | 100 | 50 | 501,038 | right pin |
| SATI-45 | 200 | 100 | 75 | 156,100 | right pin |
| SATI-46-2 | 200 | 100 | 75 | 271,729 | right pin |
| SATI-47 | 200 | 100 | 75 | 346,348 | right pin |
| SATI-48 | 250 | 100 | 75 | 120,148 | right pin |
| SATI-49 | 250 | 100 | 75 | 186,613 | left pin |
| SATI-50 | 250 | 100 | 75 | 150,584 | right pin |
| SATI-51 | 150 | 100 | 75 | 871,390 | right pin |
| SATI-52 | 150 | 100 | 75 | 1,094,854 | right pin |
| SATI-53 | 500 | 100 | 75 | 20,665 | right pin |

Table 2 - *Fretting Fatigue Test Results (Cont'd)*

| Specimen ID | Stress Amplitude (MPa) | Fretting Stress (MPa) | Slip Amplitude (mm) | Cycles to Failure | Failure Location |
|---|---|---|---|---|---|
| SATI-28-2 | 250 | 200 | 25 | 10,000,000 + | |
| SATI-29 | 200 | 200 | 25 | 10,000,000 + | |
| SATI-30 | 250 | 200 | 25 | 6,836,250 | runout |
| SATI-31 | 300 | 200 | 25 | 245,884 | left pin |
| SATI-32 | 300 | 200 | 25 | 89,390 | left pin |
| SATI-33 | 300 | 200 | 25 | 3,279,055 | left pin |
| SATI-34 | 300 | 200 | 25 | 72,136 | right pin |
| SATI-57 | 500 | 200 | 25 | 38,793 | right pin |
| SATI-61 | 150 | 200 | 50 | 2,571,901 | right pin |
| SATI-63 | 200 | 200 | 50 | 1,026,277 | left pin |
| SATI-64 | 200 | 200 | 50 | 329,637 | right pin |
| SATI-77 | 300 | 200 | 50 | 199,817 | right pin |
| SATI-78 | 300 | 200 | 50 | 239,405 | left pin |
| SATI-81 | 150 | 200 | 50 | 10,000,000 + | |
| SATI-89 | 200 | 200 | 50 | 10,000,000 + | |
| SATI-90 | 300 | 200 | 50 | 157,007 | right pin |
| SATI-62 | 200 | 200 | 75 | 590,130 | right pin |
| SATI-67 | 200 | 200 | 75 | 1,491,598 | left pin |
| SATI-68 | 200 | 200 | 75 | 307,780 | right pin |
| SATI-71 | 300 | 200 | 75 | 131,543 | left pin |
| SATI-72 | 300 | 200 | 75 | 121,688 | left pin |
| SATI-79 | 150 | 200 | 75 | 1,142,114 | left pin |
| SATI-80 | 150 | 200 | 75 | 248,296 | right pin |
| SATI-85 | 150 | 200 | 75 | 551,565 | left pin |
| SATI-86 | 300 | 200 | 75 | 1,102,375 | left pin |

The results of this analysis of the fretting fatigue data is given in Table 3. The data can be best assessed by graphical means in plotting the contours of knockdown factor against the independent variables $\sigma_f$ vs. $\delta$, given in Fig. 3.

The $KDF^{10^7}$ is seen to be a very strong function of fretting fatigue slip amplitude, $\delta$, and only weakly related to contact stress, $\sigma_f$. At large slip amplitudes, one notes that the $KDF^{10^7}$ is 0.7. This indicates that in a hypothetical joint where controlled contact slippage of 75μm occurs under all loading conditions, only 30% of the open section fatigue stress can be designed for. Alternatively, if contact slippage can be reduced to 30μm, the $KDF^{10^7}$ is 0.2 and 80% of the open section run out stress can be designed to.

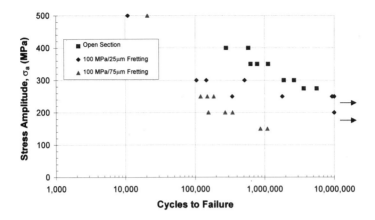

Figure 2 – *S-N curve for open section fatigue and fretting fatigue of Ti-6Al-4V in contact with 17-4PH stainless steel showing significant loss in fatigue resistance with fretting.*

Table 3 - *Run-Out Stress and KDF Deduced From Fretting Fatigue Curves*

| $\sigma_f$ (MPa) | $\delta$ (mm) | $\sigma_{ff}^{10^7}$ (MPa) | $KDF^{10^7}$ |
|---|---|---|---|
| 70 | 25 | 170 | 0.7 |
| 70 | 50 | 130 | 0.5 |
| 70 | 75 | 100 | 0.4 |
| 100 | 25 | 225 | 0.9 |
| 100 | 50 | 100 | 0.4 |
| 100 | 75 | 75 | 0.3 |
| 200 | 25 | 210 | 0.8 |
| 200 | 50 | 150 | 0.6 |
| 200 | 75 | 50 | 0.2 |

*COF & $E_{ff}$* - The difference between the upper and lower load cell measurements, $\Delta P = P_{upper} - P_{lower}$, is plotted as a function of the instantaneous slip displacement in Fig. 4 for a typical fretting sample (i.e. SATI40 tested at $\sigma_a$=250 MPa, $\sigma_f$=100MPa and $\delta$=50μm) for cycles 30, 300, 3,000, 30,000, and 90,000. One notes that typical hysteric response is observed as previously reported by Vincent et. al. [2]. The dynamic coefficient of friction, $\mu_d$, is taken as the constant $\Delta P$ at either end of the slip trace where full pin sliding occurs divided by the $2\sigma_f$ (where the factor of two takes into consideration that we employ two fretting pins). In addition, the area inscribed by the hysteresis loop is the nonconservative energy dissipated during each fretting cycle. This energy goes into local contact heating, local plastic deformation in both the pin and specimen and chemical modifications of the pin and specimen alloys into their respective oxides/nitrides.

The $\mu_d$ as a function of cycles is given in Fig. 5 for at $\sigma_a$=200MPa and $\sigma_f$=100MPa for the three slip amplitudes used in this study. The starting $\mu_d$ is seen to be on the order of 0.175 and increases logarithmically over the first $10^3$ cycles at which time reaches a saturation level. In the case of the larger slip amplitudes, this saturation $\mu_d$ is ~0.8, while at $\delta$=25μm, $\mu_d$ reaches a saturation of only 0.65 to 0.7.

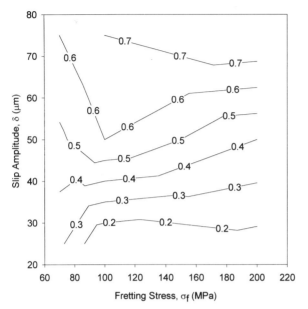

Figure 3 –*Fretting fatigue map showing knock down factors, KDF as a function of the independent fretting variables contact stress, $\sigma_f$ and slip displacement, $\delta$.*

The energy of fretting, $E_f$, is plotted as a function of cycle count in Fig. 6 for the same specimens denoted in the preceding paragraph. The response form is similar in that initial low values increase logarithmically to approximately $10^3$ cycles after which saturation is reached. Very little change in fretting energy with cycles is observed at 25μm slip, with energy dissipation of ~0.125 joules per cycle. As slip amplitude increases, saturation energy dissipation $E_f$=0, one obtains the minimum slip amplitude, $\delta_m$=10μm, as the slip amplitude at which no energy will dissipate during fretting. As was shown in [6], this is approximately the slip amplitude at which full sliding is anticipated to occur. *Fractography/Profilometry*

*Fretting Wear Characterization* - Laser profilometry of the fretting scars as well as SEM documentation was conducted on the fretting scar leading to failure for each specimen tested. The specimens tested at low slip displacements, 25μm, showed very light fretting damage while those specimens tested under higher slip amplitudes, 50 & 75μm, displayed a much greater appearance of fretting damage. The relative severity of damage is illustrated in Fig. 7 where a comparison is made between samples tested at (a) 25μm and (b) 75μm slip displacement. At 25μm slip, the shot-peened surface roughness is still visible with little occurrence of fretting debris adhering to the surface. Localized surface cracking is visible due to this lack of debris under these conditions. A crack is clearly visible in the lower portion of the fretting scar depicted in Fig. 7a. Under larger slip displacements, as shown in Fig. 7b, the fretting scar is much more clearly defined with all of the prior shot-peened surface features worn away by the fretting action.

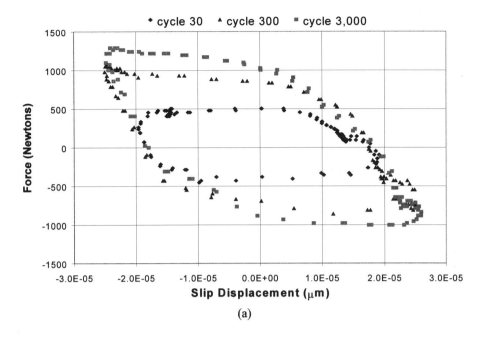

(a)

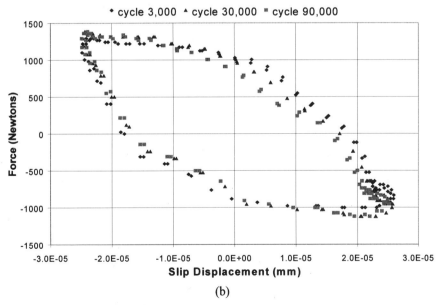

(b)

Figure 4 - *Plots of fretting load cell difference, ΔP, vs. slip displacement, δ, for specimens SATI-40 tested at σf=100MPa, σa=250MPa and δ=50μm for cycles
(a) 30, 300 and 3,000 and (b) cycles 3,000, 30,000 and 90,000*

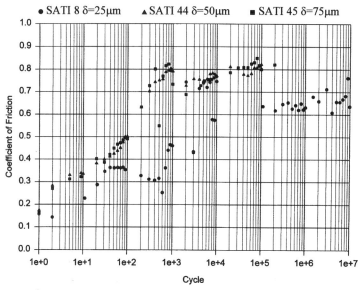

Figure 5 - *Average coefficient of friction as a function of cycle count for fretting at* $\sigma_a = 200MPa$, $\sigma_f = 100MPa$, $\delta = 25$, 50 and 75$\mu$m

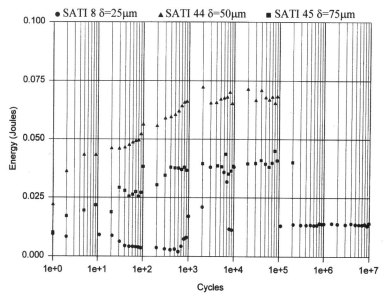

Figure 6 - *Energy expended in fretting as a function of cycle count at* $\sigma_a = 200MPa$, $\sigma_f = 100MPa$ and $\delta = 25$, 50 and 150$\mu$m.

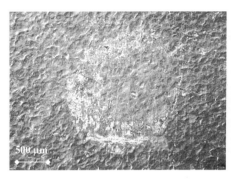 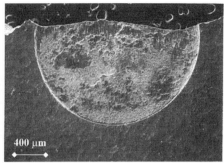

Figure 7 - *Fretting scar from specimens (a) SATI-29, $\delta = 25\mu m$, $\sigma_f = 200MPa$ and (b) SATI-47, $\delta = 75\mu m$, $\sigma_f = 100MPa$, showing very light fretting surface wear and heavy fretting surface wear.*

Accumulation of fretting debris obscures the surface, essentially no information on fretting damage can be gained through surface observation.

A laser profilometry examination was made of each fretting scar to quantitatively characterize the degree of fretting wear. A typical laser profilometry scan and a line scan are given in Fig. 8. By this technique, both maximum depths of wear and average fretting scar depth can be accurately measured. The average fretting scar depth (in μm) can be plotted as function of the two independent variables, $\delta$ and $\sigma_f$, and is given in Fig. 9. Fretting stress, $\sigma_f$, primarily impacts the average fretting scar depth. Fig. 9 shows that fretting wear increases dramatically at fretting stresses of 100MPa and below. Increasing slip amplitude also increases fretting wear, but to a significantly lower degree.

*Initiation Characterization*

*Open Section Fatigue* - Fatigue crack growth could easily be identified as intergranular, nucleating at near surface intercolony α noted in Fig. 1. Fig. 10 gives a typical fracture for an open section fatigue site is clearly visible as lying less than 100-200μm from the specimen surface in Fig. 10a with a magnified view of the nucleation site in Fig. 10b. The initiation site is inferred to be in intergranular α which resides in the tensile region adjacent to the shot-peened compressive residual stress surface.

*Fretting Fatigue* - An analysis of critical crack location within the fretting contact region was performed by measuring the average distance of the final fracture surface from the bottom edge of the fretting scar. This is graphically depicted in Fig. 11 along with a histogram of the results. The crack locations are clearly bimodal in location, with peaks at both the 1/3 and 2/3 scar diameter locations and apparently Gaussian in distribution. By far, the predominant failure location occurred at the upper 2/3 location on the fretting surface. Surprisingly, absolutely no failures were observed either at the center of the fretting scar or at the top or bottom edge where edge of contact stresses would be expected to be highest. No correlation of crack location with $\sigma_f$ or $\delta$ could be identified.

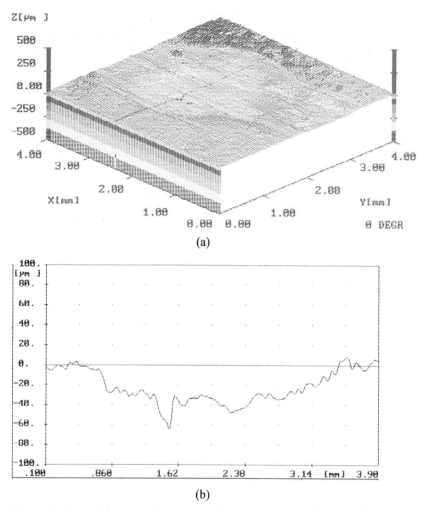

Figure 8 – *Typical laser profilometry results showing (a) total scan and (b) horizontal line scan at location indicated in (a).*

The fracture surfaces of the fretting fatigue specimens can be typified those given in Figs. 12-14 for specimens tested at $\sigma_f = 100$MPa and $\delta = 25$, 50 and 75μm respectively. Crack nucleation in Fig. 12 occurred at a prior β grain boundary denoted by the arrow. It is unclear whether this grain boundary intersected the surface under the fretting contact. The nucleation of fatigue failures at grain boundary α, adjacent to the fretting contact, occurred occasion-ally in those specimens tested at $\sigma_f = 70$ or 100 MPa. Such sub-surface nucleation events were observed only in 25μm slip tests.

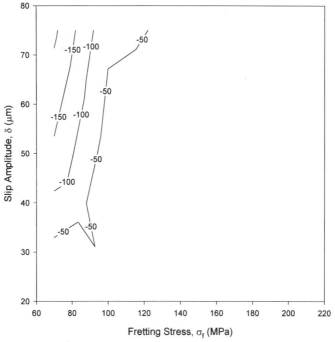

Figure 9 – *Average fretting wear scar depth (in μm) as a function of δ and σ_f.*

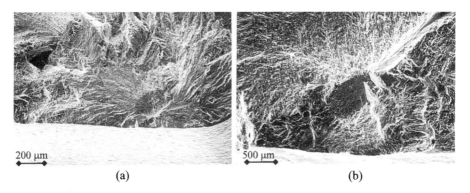

(a)                                                    (b)

Figure 10 – *SEM fractograph of open section fatigue of β-STOA heat treated Ti-6Al-4V showing nucleation sites at near surface prior β grain boundaries.*

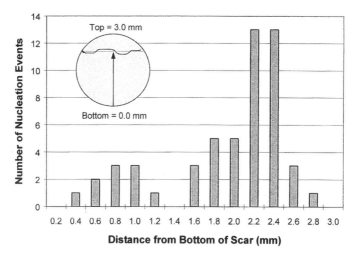

Figure 11 – *Fretting fatigue crack location as measured from the bottom edge of the fretting scar.*

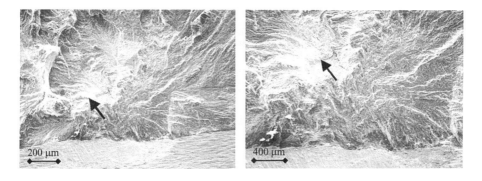

Figure 12 – *Fracture surface of SATI-18-2 tested at δ = 25mm, σ_f = 100 MPa showing subsurface initiation at inter-colony α denoted by arrows.*

More typical of the fracture surfaces are those given in Figs. 13 & 14, which show a number of fretting fatigue nucleation sites indicated by the arrows. These nucleation sites are surface connected with the initial embryonic crack propagating at an approximate 45° angle to the specimen surface. These embryonic cracks, which nucleated independently, appear to have coalesced into a large crack which eventually penetrated the specimen perpendicularly to the stress axis, and grew as a typical long crack as has been well documented and understood by use of linear elastic fracture mechanics. For the purposes of understanding the useful life of a structure subjected to fretting fatigue, it will be necessary to understand the nucleation and growth of these embryonic fretting fatigue cracks.

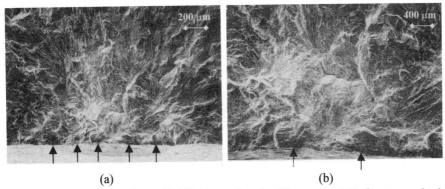

(a)                                          (b)

Figure 13 – *Fracture surface of SATI-44 tested at δ = 50 μm, σ$_f$ = 100 showing multiple surface initiated cracks denoted by arrows, which linked together to form one crack which propagated to failure.*

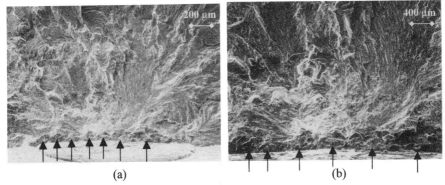

(a)                                          (b)

Figure 14 – *Fracture surface of STI-47 tested at δ = 75 μm, σ$_f$ = 100 MPa showing multiple surface initiated cracks denoted by arrows, which linked together to form one crack which propagated to failure.*

*Serial Section*

Subcritical cracks, surface pitting and third body debris morphology originating from the fretting action were clearly observed and denoted by **A**, **B** and **C** respectively in Fig. 15, a typical section from specimen SATI-8 tested at δ = 25μm, σ$_f$ = 100MPa & σ$_a$ = 200MPa. In this and the following photomicrographs, both the slip direction and applied stress axis are horizontal. This photomicrograph represents approximately half of the fretting scar from this section. It is, however, representative of the general surface damage typical for all of the specimens characterized in this study. It is noteworthy that cracks did not nucleate necessarily from pits and crack nucleation was not identified with pits. Cracking and pitting was observed, but their simultaneous occurrence was considered incidental.

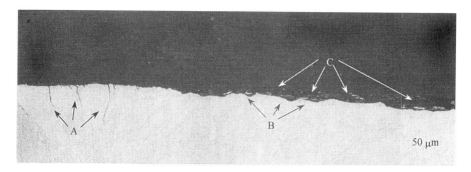

Figure 15 – *Serial section at 0.8mm from SATI-8 showing cracks (A), pits (B) and third body debris.*

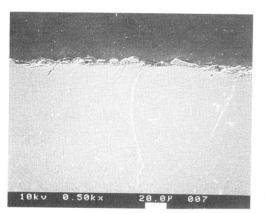

Figure 16 – *SEM image of fretting fatigue cracks shown in Fig. 15 as A.*

A higher magnification SEM image of the subcritical fretting fatigue cracks is given in Fig. 16. The surface cracks denoted by A in this figure typically nucleated at the surface at an angle to the surface of approximately 60°. After penetration to ~20-50 μm, the cracks transformed into penetrating cracks which propagated perpendicularly to the applied stress. Debris was observed pushed deep into the crack yielding the bright contrast. Energy dispersive spectroscopy (EDS) quantitative analysis indicated this debris to be essentially $TiO_2$, with no evidence of Fe or N playing a significant role. This packing of the crack with third body debris is expected to lead to a high degree of crack closure.

Fig. 17 shows a SEM view of the third body debris and pitting which typically accompanied it. This debris is seen to be quite layered in appearance. Adjacent to the titanium alloy surface, flakes of titanium are clearly visible. These flakes also showed up brightly in the optical micrograph of Fig. 15. The layered structure is composed entirely of $TiO_2$. Extensive EDS investigations found no evidence of metal flakes within this layered structure.

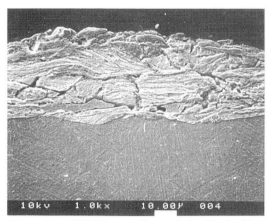

Figure 17 – *SEM image of layered third body debris consisting primarily of $TiO_2$ with multiple internal cracks and entrapped Ti-6-4 alloy flakes as shown in Fig. 16 as B.*

*Crack Length vs. Initial Angle*

Figure 18a shows a correlation between $\sigma_f$ and the Crack Length, as measured from the surface to the crack tip, vs. Crack Take Off Angle, CTA, the initial crack propagation direction measured from the specimen surface. At the two lower fretting stresses, a greater number and longer cracks emanated from CTA's ranging from 45 to 90°. At a contact stress of 200 MPa, this peak moved to 90° to 135°. Penetrating cracks having initial CTA of 180° were observed only in the 100 and 200 MPa fretting stress tests. No cracks with a CTA greater than 90° were observed for tests conducted at $\sigma_f$ = 70 MPa. These results are in Table 4 where average CTA, $\overline{CTA}$, maximum crack length, $L_{max}$, average crack length, $\overline{L}$, and number of cracks observed, N, are given along with fretting stress, $\sigma_f$. Both $\overline{CTA}$ and N increase with increasing $\sigma_f$ while $L_{max}$ and $\overline{L}$ decreases. Increasing the fretting stress increases the number of cracks nucleated, but their maximum and average lengths are smaller.

The length of the fretting cracks weakly correlated with the angle at which the crack initially propagated. This is shown in Fig. 18b which plots crack length vs. CTA for $\sigma_f$ = 100MPa and $\delta$ = 25, 50 and 75µm. The highest number and longest cracks initially grew at an angle ranging from 40° to 80° from the fatigue specimen surface. Within this range, a large number of cracks having a length greater than 0.1mm were measured. At 50µm slip, two cracks were observed which were significantly larger than the others, approaching 0.3 mm in length, which originally grew at an inclination of 100-110° to the specimen surface. While these cracks are of concern since they are long in length, They do not reside in the normal crack length distribution. Another large distribution of cracks lies at 180°. The specimen run at 25µm slip contained a large number of cracks in the region of CTA=70°. At $\delta$=50µm, the number of cracks was well distributed along CTA while the $\delta$=75µm specimen contained a large distribution of cracks having a CTA=180°. These are penetrating cracks which did not grow at the usual inclined angle to the applied stress direction. No correlation between slip amplitude and $\overline{CTA}$, $\overline{L}$ or $L_{max}$ could be determined.

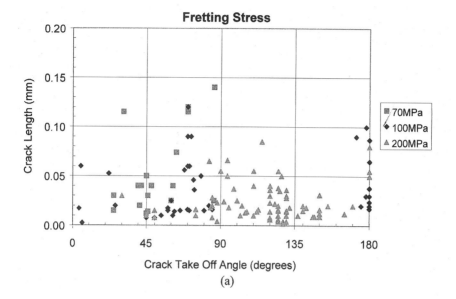

(a)

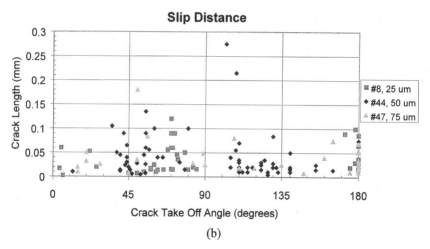

(b)

Figure 18 – *Crack length vs. nucleation take-off angle for (a) $\delta = 25\,\mu m$*
*showing increasing maximum depth angle with fretting*
*stress and (b) $\sigma_f = 100$ MPa showing longer cracks at*
*~45-65° with another large population at 180°.*

Table 4 - *Serial Section Summary for 25 μm Slip Fretting Fatigue*

| σf (MPa) | *CTA* (degrees) | $L_{max}$ (mm) | *L* (mm) | N |
|---|---|---|---|---|
| 70 | 51.7 | 0.140 | 0.053 | 17 |
| 100 | 93.0 | 0.120 | 0.037 | 41 |
| 200 | 117.5 | 0.085 | 0.023 | 76 |

## Conclusions

The slip characterized in this study can be described as gross slip since the entire pin was in slip. No area of the contact was found to be in a stick situation. Third body debris was distributed uniformly across the contact area. The fretting KDF, as defined in the text, was found to be as high as 0.7 for $10^7$ cycles. This occurred at slip distances of 75μm. Smaller slip displacements lead to lower KDF's. Contact stress played little role in KDF in this test procedure which separated these two independent parameters. The coefficient of friction increased logarithmically from initial values of 0.15 to saturation values of 0.7-0.8, depending on the slip amplitude. Larger slip amplitudes resulted in larger COF's. The increasing COF is due to the build up of the third body debris layer and roughening of the contact surface. The energy absorbed per fretting cycle also increased logarithmically with cycles reaching saturation levels ranging from 0.010 to 0.070 joules per cycle. This fretting energy was greater for larger slip amplitudes. This energy is the energy consumed in plastic deformation, heat generation and chemical reactions at the fretting surface. All of the fretting specimens failed under the contact zone, in a bimodal distribution residing at the 1/3 and 2/3 regions of the contact. The mechanism leading to this bimodal distribution is unknown and did not correlate with FEM calculations of maximum stress. Flat pin rounding during the fretting action may be responsible for the observed nucleation locations. Fretting wear scar depth was primarily influenced by contact stress, with lower contact stresses leading to greater metal removal. Fractography revealed subsurface crack nucleation sites in open section fretting fatigue at residual α prior β grain boundaries. Fretting conducted at 25μm slip also yielded some nucleation sites at subsurface α. All of the failures at 50 and 75μm slip resulted in multiple surface nucleation sites which linked across the fretting contact. The fretting debris was found to be composed of $TiO_2$ in a layered morphology with embedded fragments of unoxidized titanium alloy. Embryonic fretting cracks nucleated at angles ranging from 40° to 135° from the specimen surface with those cracks growing at ~60° having longer lengths.

## Acknowledgments

This work was jointly funded by the Rotorcraft Industry Technology Association, the National Rotorcraft Technology Center, Sikorsky Aircraft and United Technologies Corp. The very diligent support of Messrs. R. Brown, E. Roman and R. Holland in conducting the experimental and microstructural portions of this study is greatly appreciated.

**References**

[1]   K. Nakazawa, M. Sumita and N. Maruyama *in Standardization of Fretting Fatigue Test Methods and Equipment*, ASTM STP 1159, M. Attia and R. Waterhouse eds., ASTM, USA, pp.115-125, (1992).

[2]   C. Chamont, Y. Honnorat, Y. Berthier, M. Godet and L. Vincent, in the proceedings of the *Sixth World Conference on Titanium*, pp. 1883-1887, (1988).

[3]   R.K. Betts, Air force Materials Laboratory Technical Report, AFML-TR-71-212, Wright Patterson Air Force Base, Dayton, OH, (1971).

[4]   P. Blanchard, C. Colombie, V. Pellerin, S. Fayeulle and L. Vincent, Met. Trans. A, 22A, pp. 1535-1544, (1991).

[5]   O. Vingsbo and D. Solderberg, Wear, 126, pp. 131-147, (1988).

[6]   L.H. Favrow, D. Werner, D. Pearson, K. Brown, M. Lutian and D. Anton, in *Fretting Fatigue: Current Technology and Practices, ASTM STP 1367*, D.W. Hoeppner, V. Chandrasekaran and C.B. Elliot, Eds., American Society for Testing and Materials, 1999.

[7]   D.L. Anton, R. Guillemette, J. Reynolds and M. Lutian, *in Surface Performance of Titanium*, J. Gregory, H. Rack and D. Eylon eds., TMS, Warrendale, PA, USA, pp.187-198, (1997).

D. Nowell[1], D.A. Hills[1] and R. Moobola[1]

**Length Scale Considerations in Fretting Fatigue**

---

**REFERENCE:** Nowell, D., Hills, D. A., and Moobola, R., **"Length Scale Considerations in Fretting Fatigue,"** *Fretting Fatigue: Current Technology and Practices, ASTM STP 1367*, D. W. Hoeppner, V. Chandrasekaran, and C. B. Elliott, Eds., American Society for Testing and Materials, West Conshohocken, PA, 2000.

**ABSTRACT:** The presence of fretting can be shown to decrease the fatigue life of a component by accelerating crack initiation and subsequent propagation. Behavior in the "long crack" regime may be adequately explained using linear-elastic fracture mechanics. In contrast crack initiation is complex and much more difficult to understand. A size effect has been recorded in which the fretting fatigue life for contacts loaded to the same magnitude of stress varies with contact size. This phenomenon is used to explore the various length scales present in the crack initiation problem. It is shown that it is important to consider the relative magnitudes of these dimensions when attempting to analyse any particular initiation problem.

**KEYWORDS:** fretting fatigue, crack initiation, length scales

**Nomenclature**

| | |
|---|---|
| $a$ | Contact half-width |
| $b$ | Crack length |
| $b_0$ | Threshold crack size for long crack propagation |
| f | Coefficient of friction |
| $p_0$ | Peak contact pressure |
| $P$ | Normal force applied to contact |
| $Q_{max}$ | Maximum tangential force applied during loading cycle |
| $x$ | Co-ordinate along contact surface |
| $\Delta K$ | Stress intensity factor range |
| $\Delta K_{TH}$ | Threshold stress intensity factor range |
| $\Delta\sigma_{fl}$ | Fatigue limit in plain fatigue |
| $\Delta\tau_{max}$ | Maximum shear stress amplitude |

---

[1] University Lecturer, Professor, and research student, respectively, Department of Engineering Science, University of Oxford, Parks Road, OXFORD, OX1 3PJ, U.K.

141

**Introduction**

Mechanical engineers are used to designing components on the basis of stress alone; the tendency to yield is dependent only on the stress state, and the size or extent of zones of severe stress is unimportant. The first indication that scales may also be important in certain types of strength calculation arose in the development of the analysis of brittle fracture, and subsequently fatigue, where the absolute size of defects was shown to be critical in determining the conditions for failure. Fretting fatigue may be thought of as a form of plain fatigue, but where the crack nucleation process is controlled by micro-displacement of a contacting body. As soon as the crack grows to a significant size, and its growth rate may be correlated with the crack tip stress intensity factor, there is no significant difference between plain fatigue and fretting fatigue; it is simply that in the latter case the contact stress field contributes to the stress state experienced by the crack tip. In the case of fretting fatigue a size effect has been observed over and above that which may be correlated by the stress intensity factor. Experiments by Bramhall [1], subsequently refined by Nowell [2] used the Hertzian contact geometry of a cylindrical pad on a flat specimen. By varying the normal force and the pad radius it is possible to vary the peak contact pressure (and thus the magnitude of the stresses) and the contact size (and therefore the extent of the contact stress field) independently. Both sets of experiments showed clearly and repeatably that for contacting pairs subject to stress states of identical magnitude, but varying extent, large contacts would initiate a fatigue crack and lead to a finite life, while small contacts would exhibit an infinite life. Figure 1 shows a typical set of results obtained by Nowell [2].

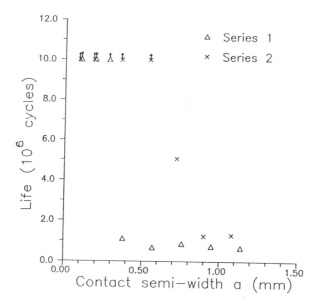

Figure 1 - *Variation of fatigue life with contact size (from [2])*.

The effect of contact size on crack propagation can be accounted for using linear elastic fracture mechanics and the Paris Law [3]. It is found that this does not explain the experimental effect observed. Thus, the transition from short to 'infinite' lives must be attributable to the initiation or short crack phases of crack development. The initiation process of both plain and fretting fatigue cracks is not well understood, and it is difficult to provide a comprehensive answer to the question 'Under what conditions will a crack initiate, and if initiation occurs, how long does the process take?' using continuum mechanics concepts alone. However, the observation of the size effect in fretting problems does, at least, provide one clue to the nature of the process, and prompts the question of what the salient length scales are which arise in the problem. Most of the fretting fatigue tests which have been carried out at Oxford have employed a Hertzian contact, simply because this affords a simple, closed-form solution to the state of stress induced, surface displacements and other salient variables [4, 5, 6], but a convincing explanation would have to be equally applicable to *any* well-defined contact geometry.

In what follows we will use the term 'initiation' in its most general sense i.e. referring not to a single event, but to a gradual evolution from the smallest embryo fissure existing across only a relatively small number of atoms, to a moderately short crack, where LEFM concepts are just beginning to dominate the problem. Each phase of this initial development may be thought of either in terms of macroscopic parameters, such as the fatigue crack growth threshold (not, in this context, a true material property) or in terms of the micromechanics of the local destruction of the crystal lattice by accumulating bond failures. We believe that there are several processes at work, and each is controlled, at least in part, by a characteristic length dimension. In analyzing crack initiation in any fretting fatigue problem it is important to consider the various length scales present in order that the correct processes may be taken into account.

**The Early Stages of Crack Initiation**

If the material contains macroscopic defects, such as inclusions, or significant pre-existing surface defects, such as score marks, it may well be that nucleation in the form of ratchetting of dislocations does not occur. However, in other cases, the first step in the formation of an embryo crack is often the formation of persistent slip bands (PSBs). Mura and Nakasone [7] present a model of the initiation process in which dislocations move on adjacent glide planes forming a PSB. At some point it becomes energetically favourable for the accumulated slip to be released, transforming the PSB into an embryo crack. In the long term a quantitative explanation of initiation along these lines looks very promising. In the shorter term, however, it is very difficult to obtain all the underlying physical properties needed to obtain a quantitative assessment of how long an embryo crack will take to form: what is clear is that early initiation is controlled by the *range* of shear stress experienced along a particular potential slip band. Clearly, such bands are confined to potential dislocation glide planes, so that if the grains are large there may not be an ideally oriented slip plane. However, if we discount these cases and examine only those where the grain size is sufficiently small for there always to be an ideally oriented slip plane, it is clear that, exterior to the contact, initiation will be controlled by the range of shearing stress found on a 45 degree plane, as the principal directions lie normal to and parallel with the free surface. Most fretting fatigue cracks arise not exterior to the contact, but within a partial slip zone beneath the contact. This means that the planes of maxmimum shear stress range are no longer at 45° to the surface. Further, determining the *range* of shearing stress experienced during a single reversing pass of the contact load is not simple, as the local orientation of the principal directions changes continuously during the load cycle, and all possible directions must be investigated to determine the maximum value. Detailed calculations for a Hertzian contact were carried out by Fellows et al. [8], and the results confirm that, for many combinations of applied load and load reversal, the maximum shearing stress range does occur within the slip zone, although the variation of the maximum range with position was too weak to enable an accurate pin-pointing of the favoured initiation site to be predicted for correlation with experiment. Sample results are shown in Figure 2.

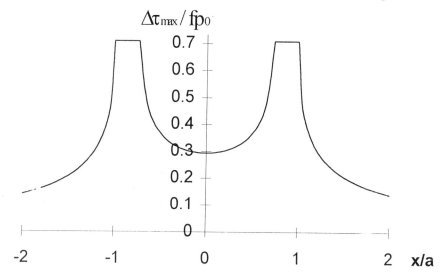

Figure 2 - Variation of maximum shear stress amplitude with position for a Mindlin-Cattaneo contact with $Q_{max}/fP = 0.5$

Returning to the question of length scales, it is apparent that this model itself raises several length scale questions. We have already mentioned that in our idealised calculations we assumed that there was always a perfectly oriented grain available within which dislocations would glide in the externally determined maximum shear stress direction, and clearly, in the case of a very large grain material or small contacts, this may not be so. Thus, the chances of there being an ideally oriented grain (in the sense of having a close-packed potential glide plane close to the orientation of the plane containing the maximum shear stress range), are maximised for a material containing very small grains. On the other hand, in Mura and Nakasone's paper they suggest that the PSB may well pile up against an obstacle such as a grain boundary and they consider that the the PSB length may be regarded as being of the order of the grain size; this is used in their model to explain the dependence of the fatigue limit on grain size. From a materials science perspective this is clearly an over-simplification, and the PSB length should properly be regarded as an independent material parameter. It is clear that two important length scales emerge from our consideration of the mechanics of fatigue crack initiation: (i) the length of the PSB, (ii) the characteristic material grain size, and (iii) the overall size of the contact. It is the relative magnitude of (ii) and (iii) which governs the number of grains subjected to high stress levels.

**Surface Finish Effects**

The arguments advanced in the previous section implicitly assumed that the contact can be thought of as smooth. However, the stress state at the near surface is profoundly influenced by the exact nature of the surface finish of the contact. No real contact is perfectly smooth, and it follows that the smooth curves of contact pressure derived from a conventional calculation do not satisfactorily represent what happens in reality very close to the contact. The contact pressure itself will be localised at the asperity, and a detailed knowledge of how this is distributed may become important in determining the surface stress state needed for the slip band calculation. It is not appropriate here to provide a comprehensive review of surface roughness effects. An appropriate introduction is given by Williams [9]. There are, in practice, three strands of analysis commonly used in quantifying this effect; the first is to obtain a family of profilometer traces from the surface, and then to re-construct its true form, from which the contact pressure may be deduced. This is the course followed extensively by Sayles et al. [10, 11]. It provides realistic information, but each problem must be analyzed and measured separately, which is a long and expensive procedure. The second approach is to use classical ideas of quantifying surface topography using statistical methods, and thereby reducing the number of independent variables significantly [12, 13]. This provides useful information, but it is notoriously difficult to obtain reliable estimates of the local pressure peaks as there is inevitably some very localised plasticity, and it is also difficult, even in principle, to quantify the asperity tip radius. The third approach is to make a gross idealization of the surface as an array of asperities of similar curvature and spacing. Clearly, this is indeed gross, and lacks realism, although it may be quite realistic in the case of strongly periodic surfaces, such as those produced by surface grinding. Also, it does provide some physical insight into the mechanics of initiation.

Figure 3 shows a typical idealized contact in the form of a cylinder, on whose surface there is an array of uniformly spaced asperities the shape of which is also idealized as a cylinder. Each individual asperity contact may therefore by thought of as a Hertzian contact, and the influence of each contact on the others found by superposition. The underlying assumption made is that the effective displacement at a particular asperity site caused by the other asperities is constant across the asperity in question. This is appropriate providing that the asperities are not too closely spaced. The load carried by each asperity is determined by the overall form of the contacting bodies, whilst the distribution is controlled by the shape of individual asperities. In many problems there will be a distinct gap between each asperity contact, so that the problem becomes multiply connected, and the localization of the overall contact pressure at each asperity has several effects [14]; the first is that the maximum contact pressure sustained, albeit over a very small area, is much greater than that implied by the bulk contact, so that the local shear stresses able to drive dislocations on persistent slip bands are correspondingly high. This, in turn, implies that the chances of nucleating a crack are greater than for a 'smooth' contact, although it should be borne in mind that the macroscopic manifestation of dislocation glide is plastic flow, so that this may, in many cases limit the asperity contact pressure.

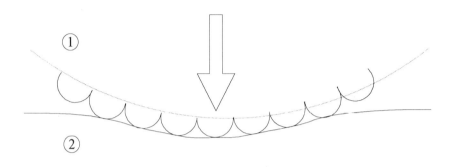

Figure 3 - *Idealized rough contact with uniformly distributed cylindrical asperities.*

A second effect of the localization of the contact pressure into discrete zones is that, for asperities near the extreme edges of the contact, where the gross relative displacement of the contacting bodies is large, a surface particle may undergo several cycles of local surface loading for each bulk cycle of shear load. It may be seen, with reference to Figure 4, that if we assume that one body is perfectly smooth whilst the other carries an idealized rough surface of the kind described, it is the smooth body which sustains the multiplied number of cycles of loading, while the rough body sustains cyclic loading at the 'bulk' rate. This multiplication of the number of load cycles in the vicinity of the contact can occur only when the surface displacement is greater than the characteristic asperity spacing. This provides a possible explanation for the size effect noted in the introduction and for the widely reported effect of relative slip displacement on fretting fatigue life. For a particular slip-slip regime the slip amplitude is given by the integral of the relative strain in the two bodies. This scales with the contact size. Hence, for larger contacts there are larger slip amplitudes and the number of asperity load passes experienced during a single 'bulk' loading cycle will increase. These arguments indicate the existence of a further three salient length scales: (iii) the typical asperity contact size, (iv) the characteristic asperity spacing, and (v) the slip amplitude.

A possible way of exploiting the above phenomenon as a means of understanding the relevant length scales in a fretting fatigue test is by carrying out two sets of tests, in one of which the asperities are placed on the specimen carrying the bulk load, and in the second the asperities are placed on the pad. An alternative way of attaining the same effect is to produce an artificial rough surface in which the asperities are deliberately formed as long 'furrows' on the pads; if these are oriented across the specimen the number of local stress cycles will be enhanced, whilst if they are oriented longitudinally this will not occur, although the macroscopic stress state is identical. However, in trial experiments where we have attempted to use these ideas we have found only a modest effect of 'furrow' orientation, and the influence of interchanging roughnesses has not been successfully demonstrated. A development of the 'idealized asperity' model is to examine the effect of applying a shearing force. This will now create a characteristic Cattaneo-Mindlin contact at each individual asperity with what is normally the macroscopic stick zone, while asperities which are in the macroscopic slip band will slide [15]. This model illustrates why, in Johnson's well-known ball on flat fretting tests [16,17], the damage within the bulk slip annulus does not end abruptly at the notional stick/slip interface.

## Slip amplitude

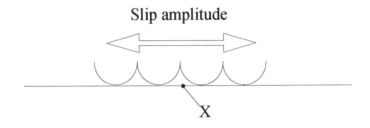

X

*Figure 4 - The effect of slip amplitude on crack initiation: multiple asperity passes experienced by a point X on the contacting body when the slip amplitude is larger than the asperity spacing.*

### Short Crack Effects

The very earliest stages of crack initiation have been described, where a de ɔct forms within a single grain. Such defects are considerably smaller than those which may be treated by linear-elastic fracture mechanics since the assumptions of material isotropy and of small-scale yielding do not apply. A considerable amount of effort has been applied in recent years to the problem of very early crack growth ('short cracks'). A review of this work is given by Miller [18]. Many short crack concepts are clearly applicable to the fretting problem, although as yet there have been few reported attempts to implement them. Of particular interest is the concept of a 'threshold stress intensity factor range', $\Delta K_{TH}$ Early work on slow crack growth rates sought to draw an analogy between the 'fatigue limit' exhibited by many bcc metals with the notion that there was a characteristic stress intensity range, the threshold stress intensity, below which a crack would not grow. Providing that this is interpreted as relating to a relatively long crack and low applied load, the concept works very well, and $\Delta K_{TH}$ may be regarded as a material property, $\Delta K_0$. However, the approach does *not* work for high load, low crack length problems; short cracks are found to propagate at $\Delta K$ values below $\Delta K_{TH}$ if the stress amplitude is sufficiently high. For un-notched plain fatigue specimens the phrase 'sufficiently high' may be interpreted as meaning above the fatigue limit $\Delta \sigma_{fl}$. The threshold for crack propagation may be conveniently presented on a Kitagawa-Takahashi diagram [19] (Fig. 5.). For long cracks, propagation occurs when $\Delta K$ exceeds $\Delta K_0$. For short cracks in plain fatigue, propagation occurs when $\Delta\sigma$ exceeds $\Delta\sigma_{fl}$. This second condition may alternatively be expressed in terms of the stress intensity factor to give a threshold $\Delta K$ for short cracks which is a function of crack length:

$$\Delta K_{TH} = C\,\Delta\sigma_{fl}\sqrt{\pi b} \qquad (1)$$

where $C$ is the usual geometric factor, approximately equal to 1.1215 for a surface breaking crack in plane strain. Thus, plotted on a $\Delta K$ vs $b$ diagram as in Figure 5, the short crack threshold is seen to be proportional to $b^{0.5}$. In plain fatigue transition to long crack growth occurs at a crack size $b_0$, given by

$$\Delta K_0 = C\,\Delta\sigma_{fl}\sqrt{\pi b_0} \qquad (2)$$

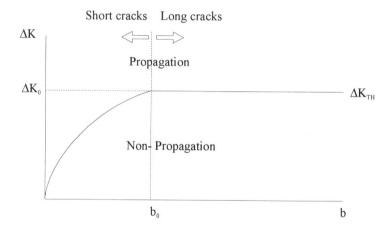

Figure 5 - *Kitagawa-Takahashi diagram showing the threshold stress intensity as a function of crack size.*

As $b_0$ depends only on $\Delta K_0$ and $\Delta\sigma_{fl}$ it is sometimes also considered as a material constant. For the case of fretting fatigue, cracks growing near to the contact experience a rapidly changing stress field and the stress intensity factor range is a more complex function of crack length. The transition crack size may still be found from the requirement that $\Delta K = \Delta K_0$, but a different transition size, $b_{0f}$, results because of the different stress intensity factor variation. This transition size may be regarded as a further length scale in the initiation problem since it represents the limit beyond which LEFM may be taken to apply.

**Summary**

The arguments outlined above demonstrate that there are at least seven salient length scales associated with the crack initiation problem in fretting fatigue. These are:

(i)     The material grain size
(ii)    The characteristic persistent slip band length
(iii)   The overall size of the contact
(iv)    The typical asperity contact size
(v)     The characteristic asperity spacing
(vi)    The slip amplitude
(vii)   The transition crack size for long crack growth

It should be noted that this list is by no means exhaustive, but it is believed that the most important dimensions have been captured.

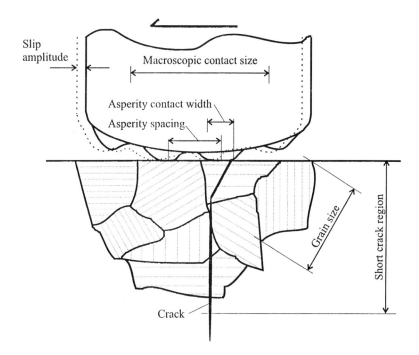

Figure 6 - *Diagrammatic representation of some of the important length scales in fretting fatigue crack initiation.*

In our experience and in our own experiments (e.g. [2]) many fretting contacts operate in the partial slip regime where wear and consequent third body generation are not important. However, in some cases the generation of 'third body' wear debris and its expulsion of the contact may be significant. In this case appropriate length scales might be the overall contact size (iii) and the characteristic third body size. This would introduce an additional length scale. The ratio of the overall contact size to the characteristic third body size would also have a pronounced influence on the behaviour of the contact, and in particular the pressure distribution.

Table 1 - *Interaction of length scales in fretting fatigue*

|  | Grain size | PSB size | Contact size | Asperity contact size | Asperity spacing | Slip amplitude | LEFM crack size |
|---|---|---|---|---|---|---|---|
| Grain size | - |  | [21] | [21] |  |  | [18] |
| PSB size |  | - |  |  |  | [8] |  |
| Contact size | Isotropy of macro contact |  | - | [14] | [14] | [20] | [6] |
| Asperity contact size | Isotropy of micro contact |  | Pressure concentr-ation | - |  |  |  |
| Asperity spacing |  |  | Number of asperity contacts |  | - | [6] |  |
| Slip amplitude |  | PSB control | Wear rate |  | Multiple asperity passing | - |  |
| LEFM crack size | Degree of isotropy |  | Closure |  |  |  | - |

In different circumstances the relative magnitudes of the length scales discussed may be very different. It is only by first evaluating the relevant dimensions that an appropriate analytical approach may be selected. For example, if the slip amplitude is much less than the characteristic asperity spacing, it will not be necessary to consider multiple asperity passes. Similarly, comparison of the grain size and the asperity and overall contact sizes enables one to determine whether a smooth or rough contact analysis should be used in analyzing the initiation of an embryo defect. The length scales are illustrated diagrammatically in Figure 6.

The existence of so many independent important dimensions is one reason why the fretting fatigue crack initiation problem has proved so difficult to analyse. Table 1 summarizes the interactions between the length scales we have determined and gives references to known investigations of the resulting size effects. The table is, inevitably, incomplete, and gives no indication of the strength of an effect on the overall fatigue life. Further, it should be borne in mind that, as stated at the outset, long cracks behave similarly in fretting fatigue and in plain fatigue, where their growth rate is satisfactorily described by the Paris equation or similar experimental 'laws'. The total life of the specimen is clearly the sum of the nucleation and propagation phases, and the majority of the effects described here influence only the former; their effect on the overall life will therefore depend on the relative proportions of fatigue life spent in initiation and propagation.

**References**

[1]    Bramall, R., "Studies in Fretting Fatigue", D.Phil. thesis, University of Oxford, 1973.

[2]    Nowell, D., "An Analysis of Fretting Fatigue", D. Phil. thesis, University of Oxford, 1988.

[3]    Paris, P.C., Gomez, M.P., and Anderson, W.P., "A Rational Analytic Theory of Fatigue", *The Trend in Engineering*, 13, 1961, 9-14.

[4]    Nowell, D., and Hills, D.A., "Mechanics of Fretting Fatigue Tests", *International Journal of Mechanical Sciences*, 29, 1987, 355-365.

[5]    Nowell, D., and Hills, D.A., "Contact Problems Incorporating Elastic Layers", *International Journal of Solids and Structures*, 24, 1, 1988, 105-115.

[6]    Hills, D.A., Nowell, D., and O'Connor, J.J., "On the Mechanics of Fretting Fatigue" *Wear*, 125, 1988, 129-156.

[7]    Mura, T., "A Theory of Crack Initiation in Solids", *Journal of Applied Mechanics*, 57, 1990, 1-6.

[8]    Fellows, L.J., Nowell, D., and Hills, D. A., "On the Initiation of Fretting Fatigue Cracks", *Wear,* 205, 1997, 120-129.

[9]    Williams, J.A., "Engineering Tribology", Oxford University Press, Oxford, 1994.

[10]   Webster, M.N., and Sayles, R.S., "A Numerical Model for the Elastic Frictionless Contact of Real Rough Surfaces", *Journal of Tribology,* 108, 1986, 314-320.

[11]   West, M.A., and Sayles, R.S., "A 3-dimensional Method of Studying 3-body Contact Geometry and Stress on Real Rough Surfaces", *Interface Dynamics: Proc 15th Leeds-Lyon Symposium on Tribology,* D. Dowson, C.M. Taylor, M. Godet, D. Berthe, Eds, Elsevier, Amsterdam, 1988.

[12]   Onions, R.A., and Archard, J.F., "The Contact of Surfaces having a Random Structure", *Journal of Physics D: Applied Physics,* 6, 1973, 289-304.

[13]   Greenwood, J.A., and Williamson, J.P.B., "Contact of Nominally Flat Surfaces", *Proceedings of the Royal Society, Series A,* 295, 1966, 300-319.

[14]   Nowell, D., and Hills, D.A., "Hertzian Contact of Ground Surfaces", *Journal of Tribology,* 111, 1, 1989, 175-179.

[15]   Ciavarella, M., Hills, D.A., and Moobola, R., "Analysis of Plane, Rough Contacts subject to Shearing Forces", *International Journal of Mechanical Sciences,* 41, 1999, 107-120.

[16]   Johnson, K.L., "Surface Interaction between Elastically Loaded Spheres under Tangential Forces", *Proceedings of the Royal Society, Series A,* 230, 1955, 531-548.

[17]   O'Connor, J.J., and Johnson, K.L., "The Role of Asperities in Transmitting Tangential Forces between Metals", *Wear,* 6, 1963, 118-139.

[18]   Miller, K.J., "Materials Science Perspective of Metal Fatigue Resistance", *Materials Science and Technology,* 9, 1993, 453-462.

[19]   Kitagawa, H., and Takahashi, S., "Applicability of Fracture Mechanics to Very Small Cracks or Cracks in the Early Stage", Proceedings of the Second International Conference on Mechanical Behavior of Materialss, pp 627-631, ASTM, 1976.

[20] Nishioka, K., and Hirakawa, K.," Fundamental Investigations of Fretting Fatigue Part 2: Fretting Fatigue Testing Machine and some Test Results", *Bulletin of the Japan Society of Mechanical Engineers*, 12, 1969, 180-187.

[21] Willis, J.R., "Hertzian Contact of Anisotropic Bodies", *Journal of the Mechanics and Physics of Solids*, 14, 1966, 163-176.

Wieslaw Switek[1]

## An Investigation of Friction Force in Fretting Fatigue

REFERENCE: Switek, W., "An Investigation of Friction Force in Fretting Fatigue", *Fretting Fatigue: Current Technology and Practices, ASTM STP 1367*, D. W. Hoeppner, V. Chandrasekaran, and C. B. Elliott, Eds., American Society for Testing and Materials, West Conshohocken, PA, 2000.

ABSTRACT: Fretting fatigue limit of mechanical joints is effected by various parameters but most important of them are friction force, relative slip, and clamping pressure which appear in the area of contacting surfaces. Unfortunately these three parameters are mutually influencing each other and their individual effect is not easy to be established. In this paper a stress model of fretting fatigue is presented according to which the fretting fatigue limit depends only on the stress distribution around the fatigue cracks developing in fretting fatigue process. That hypothesis has been proved by investigations carried out on carbon steel in ambient conditions and in an oxygen-free environment. For the last research, a special chamber has been designed and built in which the fretting junction has been investigated in an argon environment. The friction force and the relative slip between the specimen and the fretting pads have been measured by means of strain gauges. The results of the investigation are in good agreement with the theoretical formula developed by the author in previous works and prove predominant influence of friction force on the fretting fatigue limit of a mechanical joint.

KEYWORDS: fatigue of materials, fretting fatigue, crack propagation, friction force investigation, environmental testing

Fretting fatigue is most commonly found in practice when one part of a press-fitted assembly is subjected to an alternating loading. A small relative movement appears as a result of strain differences between the components of the mechanical connection. In the area of contacting surfaces a specific damage process develops which is called fretting.

[1] Professor, Universidad de las Américas-Puebla, Departamento de Ingeniería Mecánica, A.P.418, Santa Catarina Mártir, 72820 Puebla, México.

As a result of simultaneous action of fretting and fatigue, the fatigue strength of the assembly is seriously reduced. There are several factors influencing fretting fatigue limit, $Z_F$, reported in the literature [1-3] but the most important of them seem to be clamping pressure, relative slip between the elements of junction, and the coefficient of the frictional shear stress. The relative slip, s, and the coefficient of friction, ø, are strictly dependent on each other, therefore the only independent factors that determine the fretting fatigue limit, $Z_F$, are the clamping pressure and the coefficient of friction, ø, between contacting surfaces. The influence of the other factors like hardness of contacting materials, surface finishing, frequency of load or conditions of test, can be easily explained by considering the change of the basic factors they produce. The above observations caused the author establish "the stress theory of fretting fatigue damage process" [4-7], according to which the fatigue failure of the fretting junction can be explained by considering only the stress conditions of the initiation and propagation of fretting fatigue cracks. Microscope examinations of the initiation and growth of the fatigue cracks in the area of fretting [8,9] show that the fretting fatigue process from the point of view of fatigue damage development can be divided into three different stages.

- In the first, relatively short stage, no fatigue cracks can be observed, but damage of the contact surface is due to abrasive wear effects. The intensity of the wear process and the loss of weight of the contacting components are not affecting the fatigue strength of the assembly.

- In the second stage of the damage process, development of the fatigue cracks appear initially from the edge of the fretting contact (where the level of stress is the highest) and later on also at larger distance from the edge. The cracks propagate at an angle of 45° to the surface but usually when their length does not exceed about 50 μm, the cracks change direction of growth and emerge on the surface creating scars and caverns characteristic for fretting. The consideration of the state of the stress in the area of cracks initiation and development [5] leads to the conclusion that the crack growth satisfies the condition that the strain energy released from the stress field, as a result of cracks development, attains its maximum. There are also some cracks that appear in the highly stressed areas, growing initially at the 45° to the surface, but the inclination increases to 90° as the crack grows. When the fretting junction is loaded below the fretting fatigue limit these cracks develop to a certain length (in the case of mild steel about 200 μm) and arrest, creating a non-propagating fatigue cracks. The non-propagating cracks can be seen in the specimen not only when the fretting assembly is loaded over the fretting fatigue limit, $Z_F$, but also below $Z_F$.

- The third stage of the damage process takes place only when the fretting assembly is loaded over the fretting fatigue limit, $Z_F$. In that case the non-propagating cracks resume propagation until fracture of the assembly.

Applying fracture mechanics theory, the author [5] described the maximum external cyclic load which does not cause the propagation of the cracks created in the second stage of the fretting fatigue damage process.

That level of the load, which is equal to the fretting fatigue limit, $Z_F$, of the fretting assembly, can by expressed by the formula:

$$Z_F = Z_S \left[ 1 - \frac{(\Delta K_t)_l + (\Delta K_n)_l}{\Delta K_{TH}} \right] \qquad (1)$$

where:

$Z_S$ - the fatigue limit of the smooth elements

$(\Delta K_t)_l$ , $(\Delta K_n)_l$ - the range of stress intensity factors calculated from the main crack of the non-propagating length $l$ corresponding to tangential force T and normal force P, respectively,

$K_{TH}$ - the threshold stress intensity factor.

**Fatigue Investigation**

*Experimental Procedure*

The fatigue investigations were carried out in cyclic tension test with the use of a Fatigue Dynamics Machine LFE-150, Model DS-2000 with 8000 N maximum dynamic load capacity at a frequency equal to 20 Hz. A sinusoidal type of cyclic load was applied with the stress ratio, R = $\sigma_{min.}/\sigma_{max.}$ = 0. The specimen used for the investigation, shown in Fig. 1, has the testing portion flattened in order to make it possible to create a flat fretting joint together with the fretting pads fixed in a specially designed fretting attachment.

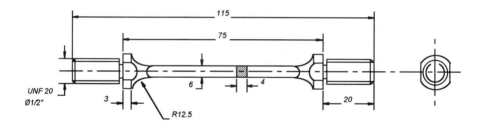

FIG. 1–*The geometry of the specimen for the fretting fatigue investigation. All dimensions are in mm.*

The fretting attachment used to produce the fretting process on the flat surface of the specimen, described in details in the paper [5], is shown in Fig. 2. In this attachment the relative slip between the fretting pad and the specimen is constant in the whole area of contact.

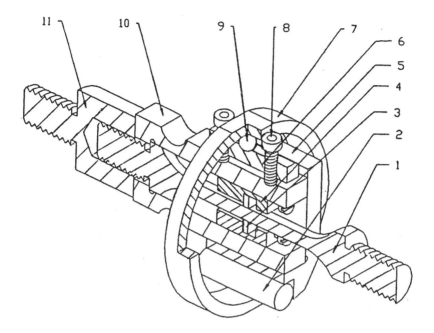

FIG. 2–*The general view of the fretting assembly: (1) specimen, (2) roller, (3) fretting pad, (4) holder, (5) body, (6) loading beam, (7) loading ring, (8) bolts, (9) ball, (10) connector, (11) adapter.*

For the fatigue investigation in an environment with controlled contents of oxygen a special chamber, shown in Fig. 3, has been designed and constructed.

The chamber is made of a glass tube (6) with strictly parallel and polished ends, clamped against the two steel plates (4 and 7) by means of six bolts (8) and nuts (3). In the upper plate (4) there are two valves (2) fixed, one for connection to the vacuum pump to remove air from the chamber and the second to fill the chamber with argon. The upper plate (4) is connected directly to the upper grip (1) of the specimen (5) but the lower plate (7) is connected to the lower grip (9) of the specimen by means of an elastic bellow (10) in order to allow small extension of the specimen. The chamber is fully hermetized by the use of seals and gaskets or by glue joints where possible.

The procedure of fretting fatigue testing in argon environment was the following:

- the fretting attachment, shown in Fig.2, is assembled to the specimen (1) and the required clamping pressure between the specimen (1) and the fretting pads (4) is set by screwing in the bolts (8) and deflecting the calibrated loading ring (7),

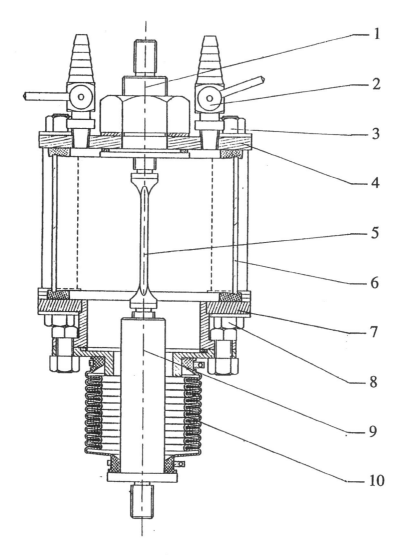

FIG. 3–*The chamber for fatigue investigations in argon environment.*

    - the grips (1) and (9) shown in Fig. 3, are connected together with the fretting attachment to the specimen, and the whole chamber is assembled with the fretting attachment inside,
    - the grips are fixed in the fatigue testing machine,
    - one valve (2) is connected to the vacuum pump and the other to the cylinder with argon through a pressure reducer. After the pump produces vacuum in the chamber, the pump valve is closed and then the chamber is refilled with argon through the argon

valve. The pressure of argon during the fatigue tests was always kept 20% above the atmospheric pressure in order to prevent leakage of air into the chamber.

The friction force between the specimen and the fretting pads was measured by means of strain gauges. The strain gauges were fixed directly to the flat surface of the specimen at both sides of the fretting pads, as shown in Fig. 4, and were connected to the Wheatstone bridge and calibrated with the use of the Tensile Testing Machine in the units of force.

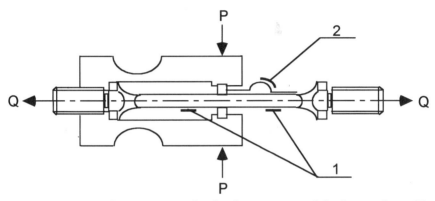

FIG. 4–*The position of strain gauges for the determination of the friction force (1) in fretting joint and the relative slip (2) between the contacting surfaces.*

The location of the strain gauges means that the difference in readings of force from the two strain gauges was exactly equal to the friction force which appears in the area of fretting contact. The relative slip between the specimen and the fretting pad was monitored by means of a special indicator, shown in Fig. 4. It was made of a strip of a sheet of metal with a strain gauge fixed to its surface. The indicator was then calibrated with the use of a dial micrometer in a special holder.

The static properties of the material were measured with the use of the Universal Tensile Testing Machine. A Brinell Hardness Machine was used to determine the hardness of the material and the optical microscope was used for the observation of the fatigue cracks.

*The Results of the Investigations*

The specimens for fatigue investigations and the fretting pads were made of the carbon steel SAE 1045 with the chemical composition: 0.45% C, 0.25% Si, 0.65% Mn, 0.035% P, and 0.035% S. Before machining, the material was stress relieved by annealing at 700° C for 2 hours and slow cooling inside the furnace. The Brinell hardness

of the material after annealing was 82±2 HRB. Static properties of the material, specified below, were determined using a standard specimen of 5 mm diameter. The mean results of 5 tests are as follows:
  - Ultimate strength -  608.7 MPa
  - Yield point        -  420  MPa
  - Elongation         - 31.2%
  - Reduction of area  - 53.4%.

For all the fatigue investigations, the uniaxial tensile cyclic load was applied at a frequency of  20 Hz. The results of fatigue tests are shown in Fig. 5.

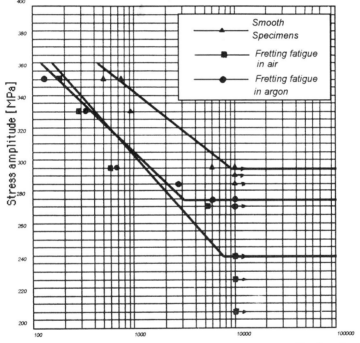

FIG. 5–*The results of the fatigue tests of the steel SAE1040 investigated in air and in argon.*

As can be seen from Fig. 5, the fatigue limit of the smooth specimen (without fretting) is equal to $Z_S = 300$ MPa , the fretting fatigue limit  in  air is equal to $Z_F = 245$ MPa and the fretting fatigue limit in argon is equal to $Z_F = 280$ MPa.

The friction force between contacting surfaces has been investigated at a constant clamping pressure of the fretting assembly equal to p = 90 MPa. According to the previous investigation made by the author [4,5], at this level of contact stress, the fretting fatigue limit, $Z_F$, of the investigated flat fretting assembly, reaches its minimum for a mild carbon steel. The coefficient of friction, ø, was measured at the beginning of the fretting fatigue process, and after 20.000, 50.000 and 100.000 cycles of the axial load. For the measurement of the friction force the machine was always stopped, the specimen was loaded statically to the required level and the strain gauges readings were taken. The results of that investigation are shown in Fig. 6.

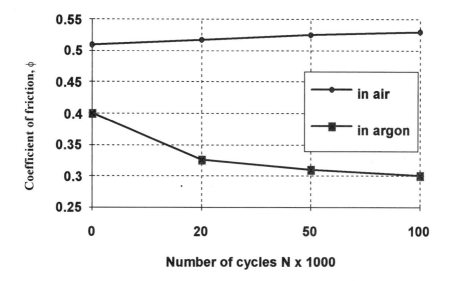

FIG. 6–*The influence of number of cyclic loads for the coefficient of friction ø in fretting. Carbon steel SAE1040, investigated in air and in argon.*

As seen from Fig. 6, the coefficient of friction, ø, in air increases slightly (less than 5%) when the process of fretting advances up to 100.000 cycles of load. In case of the investigation in non-corrosive environment ( in argon) the coefficient of friction, ø, drops considerably in the first 20.000 cycles of load, but later on remains approximately constant. The decrease of the coefficient of friction, ø, in 100.000 cycles of alternating loads was more than 30 %.

The most important influence on the coefficient of friction, ø, of the fretting junction has the relative slip, s, between contacting surfaces. The results of the investigations carried out in air and in argon are presented in Fig. 7.

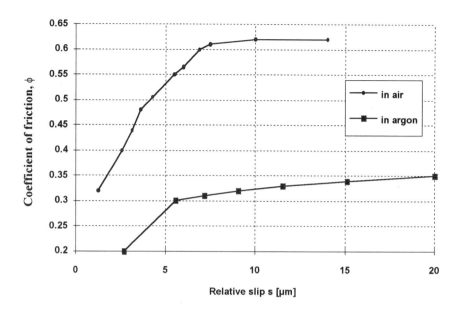

FIG. 7–*The relationship between coefficient of friction, ø, and relative slip, s, for the investigated fretting assembly.*

As can be seen from Fig. 7, the coefficient of friction, ø, increases with the increase of the relative slip up to s = 15–20 μm and remains constant for the higher relative displacements. In the argon environment the dependence between the coefficient of friction and the relative slip is basically the same but the coefficients of friction, ø, are about 40% lower than these in the air. The relation in Fig. 7 was obtained for the fretting joint after 100,000 cycles of load when the fretting process between the contacting surfaces is fully developed [8]. The fatigue investigations were carried out at the level of stress amplitude equal to the  fretting fatigue limit of the assembly and at constant clamping pressure equal to p = 90 MPa. For the measurement of the coefficient of friction, the fatigue process was arrested and the readings of forces and displacement have been taken from the Tensile Testing Machine. Different relative slip of the fretting assembly was obtained by the application of different levels of axial load, Q.

The results shown in Fig.7 are valid only for the constant clamping pressure used throughout these experiments equal to p = 90 MPa. However, it should be remembered that the clamping pressure also influences the relative slip of the fretting assembly. For the flat fretting joint investigated, the dependence between relative slip, clamping pressure and axial load is presented in Fig. 8.

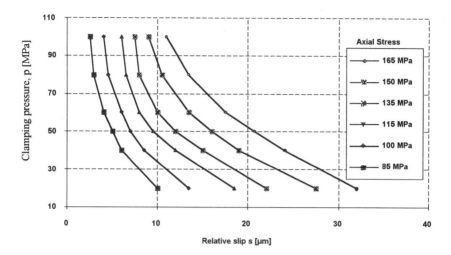

FIG. 8–*The relationship between relative slip, s, clamping pressure, p, and axial load for the investigated fretting assembly.*

In order to establish the influence of clamping pressure on coefficient of friction of the investigated fretting junction, the above described test for determination of the coefficient of friction was also performed at lower levels of clamping pressure. It was discovered that at a constant level of the relative slip, with the lower values of clamping pressure the higher results of the coefficient of friction, ø, are obtained. The relation is approximately proportional but less than a 15% increase of the coefficient of friction, ø, is observed when the clamping pressure decreases from 90 MPa to 20 MPa.

*Analysis of the Results*

The results of fatigue investigations show that the fretting fatigue limit, $Z_F$, of the fretting joint working in the non-corrosive environment (in argon) in the case of the carbon steel SAE 1040, is 12.5% higher than the fretting fatigue limit, $Z_F$, of the same joint tested in air. Microscope observations of the contacting surfaces show that in the case of the investigation in argon, not only the products of corrosion (in the form of a red oxide powder) not exist, but also the contacting surfaces are smoother than in the case of the investigation in air. However, it was observed that the process of nucleation and propagation of the fretting crack is basically the same in air and in argon and all three stages of fatigue damage development, described earlier, can be detected.

The most important difference between the fretting process in air and in argon is in the level of the coefficient of friction, ø. The coefficient of friction, ø, in the case of the fretting process in argon environment is much smaller than the coefficient of friction in the case of air environment. This observation seems to imply that the process of corrosion has negligible influence on the fretting fatigue limit, but the increase of the fretting fatigue limit in a non-corrosive environment can be explained on the basis of the stress theory of the fretting phenomenon presented by the author [5].

According to that theory, the fretting fatigue limit, $Z_F$, of a fretting assembly can be calculated by the formula (1) which takes into consideration only the stress conditions necessary for the development of the main fatigue crack. For the fretting assembly presented in this paper, both the components of the stress intensity factors due to the frictional force, $(\Delta K_t)_l$, and the thrust (clamping) force, $(\Delta K_n)_l$, can be determined according to Fig.9 [5].

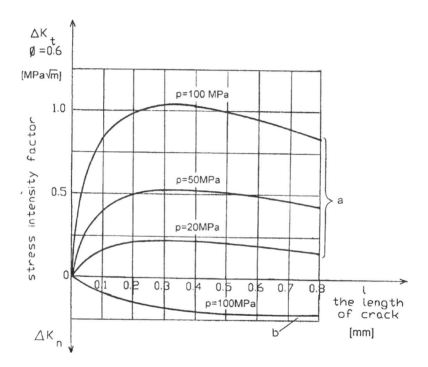

FIG. 9–*Stress intensity factor components due to: (a) the frictional force and (b) the thrust force[5].*

The range of the stress intensity factor due to friction force, $(\Delta K_t)_l$ in Fig. 8 is directly proportional to the coefficient of friction, ø . The graphs in Fig.8 are made for the coefficient of friction ø = 0.6, but for different coefficient of friction, the results obtained from Fig. 9 should be multiplied by the factor $k = \dfrac{\phi_{act}}{0.6}$ where: $\phi_{act}$ – the actual coefficient of friction.

In order to calculate according to the formula (1), the fretting fatigue limit, $Z_F$, of the investigated fretting joint, the fatigue limit of the smooth element, $Z_S = 300$ MPa, (see Fig. 2) and the threshold stress intensity factor, $K_{TH} = 6$ MN/m$^{3/2}$ [10] must be inserted into equation (1). The components of the stress intensity factors due to friction force, $\left(\Delta K_t\right)_l$, as well as to the clamping force, $\left(\Delta K_n\right)_l$, can be determined from Fig. 9. for the length of the main crack equal to 190 µm. This length of the main crack has been determined in microscopic observations made by the author in the earlier investigation [8]. As a result of these calculations, the fretting fatigue limit of the fretting assembly in the air is equal to $Z_F = 256$ MPa , but the fretting fatigue limit in argon is equal to $Z_F = 278$ MPa. As it is seen, the results of these calculations based on equation (1) and the results of investigations shown in Fig. 2 are very close to each other. The difference between the theoretical result and the result of the fatigue tests in the case of the investigations in air is about 4%, but in argon the exactness of the calculations is even higher since the difference is equal to less than 1%. The above results may justify the conclusion that the stress approach can be successfully applied to the calculation of a fretting fatigue limit, $Z_F$, of a fretting assembly. According to this approach, the main factors that have an influence on the fretting fatigue limit, $Z_F$, of the fretting assembly are those which influence the state of stress in the cyclically loaded element in the area of the contacting surfaces. The clamping pressure and the friction force are the major factors that determine the fatigue life of any fretting assembly.

It is worthy to notice, that the clamping force always produces a negative value of the stress intensity factor, $\left(\Delta K_n\right)_l$, (see Fig. 9) and from that point of view it has a positive effect on the fretting fatigue limit of the assembly. However, it is the clamping pressure that causes the appearance of the friction force which has a very detrimental influence on the fatigue properties of a fretting assembly. In order to increase the fatigue properties of a fretting assembly the friction force between contacting surfaces should be minimized. This can be done either by decreasing the clamping pressure or by minimizing the coefficient of friction, ø. Also interesting is the possibility of increasing the fretting fatigue limit, $Z_F$, when the clamping pressure is higher than 90 MPa. This happens because with the increase of the clamping pressure, p, the relative slip, s, and the coefficient of friction, ø, decrease (see Fig. 8 and Fig. 7), and as a result, the stress intensity factor, $\left(\Delta K_t\right)_l$ is also decreasing and the fretting fatigue limit, $Z_F$, of the assembly can be increased [5,6].

**Conclusions**

1. The fretting fatigue limit, $Z_F$, of the carbon steel SAE 1040 investigated in an argon environment is 12.5% higher than the fretting fatigue limit, $Z_F$, of the same steel investigated in air.

2. The coefficient of friction, ø, between the contacting surfaces in fretting is about 40% lower in the case of the investigation in argon than when the fretting assembly was investigated in air.

3. Although in the case of investigation in argon there is no formation of the products of oxidation in form of a red powder, the process of the initiation of cracks and their further propagation is similar to that in the case of the investigation in air.

4. The stress approach for the calculation of the fretting fatigue limit, $Z_F$, can be successfully applied to the case of the investigations in air and in argon. The difference between the results of the calculations and the investigations is less than 5%.

5. The friction force that appears between contacting surfaces in fretting connection has the most important influence on the fretting fatigue limit, $Z_F$, and should be minimized in order to increase the fatigue properties of any fretting assembly.

## References

[1] Collings, J.A. and Marco, S.M., "The Effect of Stress Direction During Fretting on Subsequent Fatigue Life," *Proceedings*, American Society for Testing and Materials, Philadelphia, Vol.64, 1964, pp.547-560.

[2] Waterhouse, R.B., *Fretting Corrosion*, Pergamon Press, Oxford, 1972.

[3] Dobromirski, J., "Variables of Fretting Process: Are There 50 of THEM?" *Standardization of Fretting Fatigue Test Materials and Equipment, ASTM STP 1159*, Attia and Waterhouse, Eds., Philadelphia, 1992.

[4] Switek, W., "Fretting Fatigue Strength of Mechanical Joints," *Theoretical and Applied Fracture Mechanics 4*, North-Holland, Amsterdam, 1985, pp.59–63.

[5] Switek, W., "Fretting Fatigue Strength of Specimens Subjected to Combined Axial and Transversal Loading," *Multiaxial Fatigue and Deformation Testing Techniques, ASTM STP 1280*, S. Kulluri and P.J. Bonacuse, Eds., American Society for Testing and Materials, 1997, pp. 208–223.

[6] Switek.W., "Fatigue Properties of Mechanical Joints under Fretting Conditions," *Scientific Report of Technical University, Mechanics*, Vol. 60, 1980, pp.3–86 (in Polish).

[7] Switek, W., "The Investigation of Fretting Fatigue Strength of Steel Elements," *I Congreso International de Materiales, XVI Encuentro de Investigation Metalurgia*, 1994, pp.647–690.

[8] Switek, W., "Early Stage Crack Propagation in Fretting Fatigue," *Mechanics of Materials 3*, North–Holland, Amsterdam, 1984, pp.257–267.

[9] Switek, W. and Gomez, C., "Aspectos metalograficos del daño de fatiga por rozamiento," *Rev.XV Encuentro de Investigacion Metalurgica*, Saltillo, Mexico, 1993, pp. 270–285.

[10] Nix, K.J. and Lindley, T.C., "The Application of Fracture Mechanics to Fretting Fatigue," *Fatigue Fract. Enging. Mater. Struct.*, Vol.8, No.2, 1985, pp.143–160.

Siegfried Fouvry,[1] Philippe Kapsa,[1] and Léo Vincent [2]

## A Multiaxial Fatigue Analysis of Fretting Contact Taking Into Account the Size Effect

**REFERENCE:** Fouvry, S., Kapsa, P., and Vincent, L., **"A Multiaxial Fatigue Analysis of Fretting Contact Taking Into Account the Size Effect,"** *Fretting Fatigue: Current Technology and Practices, ASTM STP 1367,* D. W. Hoeppner, V. Chandrasekaran, and C. B. Elliott, Eds., American Society for Testing and Materials, West Conshohocken, PA, 2000.

**ABSTRACT:** Fretting damage that consists of cracking or wear generated by debris formation is induced by very small alternated displacements between contacting surfaces. This paper focuses on the quantification of the fretting crack nucleation. Fretting experimental results obtained with a well defined quenched aeronautical steel are analyzed by means of a multiaxial fatigue approach. The plane/sphere configuration was studied under partial slip situations. Validated for classical fatigue conditions, the Dang Van's fatigue prediction is compared to fretting cracking mechanisms. The correlation is achieved according to some conditions :
- The friction coefficient operating in the annular partial slip contact has to be identified.
It permits more accurate estimation of the stress loading path.
- The loading states which are computed to determine the crack nucleation risk must be averaged on an elementary volume representative of the microstructure of the steel. It allows a convenient size effect consideration regarding the very small material volume stressed below the contact.

**KEYWORDS:** fretting, crack nucleation, high-cycle fatigue, size effect, fretting map.

### Introduction

Fretting commonly refers to the degradation of the fatigue properties of a material due to repeated sliding of two contacting surfaces. The small relative displacement amplitudes range typically between 10-100 μm [1]. Examples of practical situations

---

[1] UMR CNRS 5513, Ecole Centrale de Lyon, BP 163, 69131, Ecully, France.
[2] UMR CNRS 5621, Ecole Centrale de Lyon, BP 163, 69131, Ecully, France.

where cracking or wear induced by fretting influence mechanical integrity include such various applications as bolted and riveted joints, key-way-shaft coupling and shrink-fitted couplings [2]. To prevent fretting damage, fretting wear and fretting fatigue have been widely studied for more than thirty years [1-4].

The principal objective of this work consists of the prediction of the fretting cracking behavior by means of classical high cycle fatigue approaches [5]. To achieve this purpose, a well known aeronautical steel (30NiCrMo) [6] verifying the Dang Van multiaxial fatigue approach is studied. The experimental contact conditions are optimized using a sphere/plane configuration under partial slip regime. Under such sliding conditions, the contact displays a central stick domain surrounded by a quasi unworn sliding annulus. The wear of the surface is very limited. Therefore the contact geometry is unaltered. Using a very large ball radius (R=50 mm), macroscopic elastic conditions are imposed whereas very smooth surfaces limit the roughness effects.

**Experiments**

*Fretting test*

Fretting tests were conducted using a tension compression hydraulic machine already described elsewhere (Figure 1) [7]. During a test, the normal force (P), the tangential force (Q) and the displacement ($\delta$) are recorded. The magnitude of the tangential force versus the displacement is recorded for each cycle. This permits the plotting of a 3D fretting log [7], where the (Q-$\delta$) fretting cycles are reported versus the number of cycles along a log scale (Figure 2).

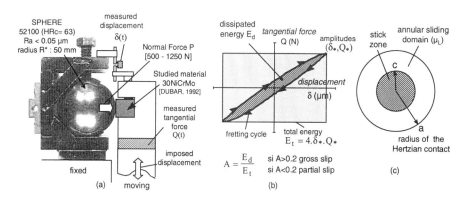

Figure 1 - *(a) Fretting test contact; (b) Analysis of partial slip fretting cycle and determination of the energy sliding criterion A to quantify the sliding condition [3]; (c) Partial slip contact.*

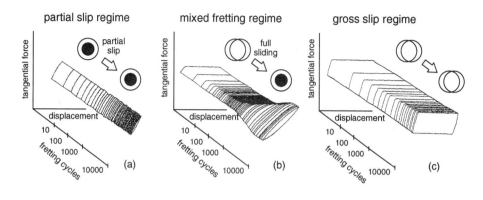

partial slip regime          mixed fretting regime          gross slip regime

Figure 2 - *Fretting Log : (a) partial slip regime; (b) mixed fretting regime; (c) gross slip regime.*

*Material*

A 50 mm radius 52100 chromium steel ball (Table 1) with a very low surface roughness (Ra<0.05 μm) is applied on an aeronautical steel plane which after a mirror polishing also displays a very low surface roughness (Ra<0.01). This 30NiCrMo low alloyed steel alloy is supplied by Aubert & Duval and presents the following weight composition 0.31% C, 4.08% Ni, 1.39% Cr, 0.25% Si, 0.49% Mn, 0.5% Mo. It was heat treated at 850°C, oil quenched and tempered at 585°C during 1 hour to obtain the mechanical properties given in Table 1. The studied steel is coming from a set of fatigue specimens previously studied under multiaxial conditions [6]. All the fatigue properties have been completely identified (Table 1) [6].

Table 1 - *Mechanical properties of steels [6].*

| Materials | Hardness | E, GPa | ν | $\sigma_{Y0.2}$, MPa | Rm, MPa | $\sigma_d$, MPa | $\tau_d$, MPa |
|-----------|----------|--------|------|----------|---------|----------|----------|
| 30NiCrMo | 400 Hv | 200 | 0.3 | 1080 | 1200 | 690 | 428 |
| 52100 | 62 HRc | 210 | 0.3 | 1700 | - | - | - |

$\sigma_d$ (MPa) : Alternated bending fatigue limit ($10^6$ cycles).
$\tau_d$ (MPa) : Alternated shear fatigue limit ($10^6$ cycles).

*Experimental conditions*

The partial slip condition is maintained during all the test duration (Figure 2a). It corresponds to a partial slip regime, where the contact never supports a full sliding between the surfaces ($Q_* < \mu.P$). The contact displays two parts, a central stick domain and an annular sliding domain (Figure 3a). Verifying the Mindlin's assumption of pressure and shear distributions, the stick and contact radii are measured at the end of the test which permits the computation of a stabilized local friction coefficient ($\mu_L$).

The experimental objective of this study was to determined as finest as possible the limit tangential force amplitude associated to a given normal force which induces the crack nucleation after $10^6$ cycles. For each normal force several fretting tests have been performed maintaining a constant tangential force amplitude ($Q_*$) by monitoring the displacement amplitude ($\delta_*$). Under partial slip regime the contact is very stable and the chosen tangential force amplitude is progressively reached before 1000 cycles. The critical value ($Q_{c*}$) is extrapolated by increasing or decreasing the applied tangential value if respectively no crack or a crack is observed. At least more than 10 fretting tests are required to bracket the critical tangential force ($Q_{c*}$) with a precision inferior to 5 Newton.

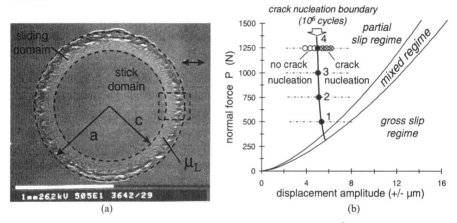

Figure 3 - *(a) SEM observation of a Partial slip fretting scar ($10^6$ cycles, P=1250N, $\overline{Q}_* = 440$N ); (b) Identification of the crack nucleation boundary through the Material Response Fretting Map (cf. Table 2).*

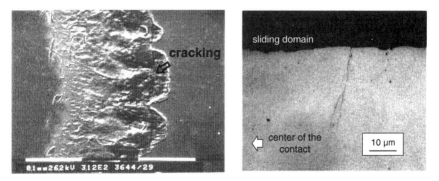

Figure 4 - *First crack nucleation observed on the contact border (Figure 3a, $10^6$ cycles, P=1250N, $Q_* = 440$N ).*

The cracking identification is performed by surface SEM observations and optical cross sections (Figure 4). The resolution of the microcrack observation is around 10-20 μm. Therefore, the crack nucleation condition is here associated to the observation of a crack length superior to 10 μm. It corresponds approximately to the austenitic grain size. The first crack nucleation is observed symmetrically on the surface at the contact borders along the sliding axis (Figure 4). The four critical conditions of first crack nucleation defined through the Material Response Fretting Map are summarized in Table 2.

Table 2 - *Fretting crack nucleation conditions of the aeronautical steel ($10^6$ cycles) (cf. Figure 3b) ( $p_0$ : Maximum Hertzian pressure; $\sigma_{e\,max}$ : Maximum Von Mises stress).*

| Conditions | (1) | (2) | (3) | (4) |
|---|---|---|---|---|
| P, N | 500 | 750 | 1000 | 1250 |
| $p_0$, MPa | 788 | 902 | 993 | 1070 |
| a, μm | 550 | 630 | 693 | 747 |
| $Q_{c*}$, N | 260±5 | 325±5 | 385±5 | 435±5 |
| k=c/a | 0.6 | 0.65 | 0.68 | 0.7 |
| $\mu_L = (Q_{c*}/P)/(1-(c/a)^3)$ | 0.66 | 0.6 | 0.56 | 0.53 |
| $\sigma_{e\,max}$, MPa | 848±15 | 861±15 | 880±15 | 890±15 |

## Mechanical Analysis of the Crack Nucleation

*Elastic analysis of the partial slip contact*

An elastic stress description of the partial slip contact is assumed ( $\sigma_{e\,max}$ <$Re_{0.2\%}$). The surface loading can be described as the sum of a constant loading induced by the normal force and an alternated shear stress field induced by the cyclic tangential force as described by the Mindlin's formalism [8, 9].

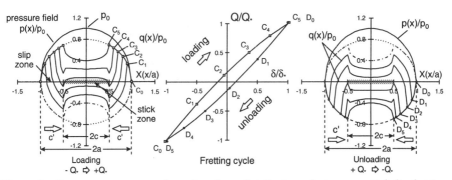

Figure 5 - *Constant pressure and varying shear distributions during a partial slip fretting cycle (c/a=0.5, μ=0.8, c' : radius of the sliding front ).*

The surface loading during the fretting cycle is then expressed as the superposition of a constant Hertzian pressure distribution p(x) and an alternated shear field distribution q(x) (Figure 5). The loading path is deduced by combining the Mindlin description with the Hamilton formulation expressed for a full sliding sphere/plan contact [10].

*Dang Van multiaxial fatigue criterion*

The partial slip loading path is complex, therefore a multiaxial fatigue approach is required (Figure 6).

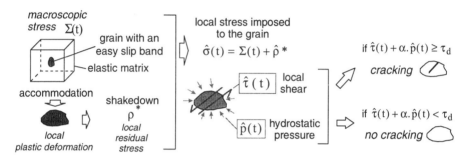

Figure 6 - *Illustration of the high cycle Dang Van's fatigue approach.*
$$\alpha = \left(\tau_d - \sigma_d/2\right)/\left(\sigma_d/3\right)$$

The Dang Van fatigue approach was shown to describe the fatigue behavior of the studied steel [6]. This fatigue criterion considers that the first crack nucleation is observed on a grain which presents an easy slip band direction with regard to the macroscopic loading direction (Figure 6) [5]. Surrounded by an elastic matrix, the grain will first support a plastic deformation before reaching an elastic shakedown state. The elastic stabilization and the initial plastic hardening which is associated corresponds to the introduction of a local residual stress $\rho^*$. The stabilized local stress $\hat{\sigma}(t)$ loading imposed to the grain on which the fatigue analysis must be performed is consequently defined as the sum of the macroscopic $\Sigma(t)$ loading and the stabilized residual stress $\rho^*$:

$$\hat{\sigma}(t) = \Sigma(t) + \rho^* \qquad (1)$$

From the local stress $\hat{\sigma}(t)$, the two microscopic stresses: shear amplitude stress applied on the plane defined by the normal $\bar{n}$, $\hat{\tau}(\bar{n},t)$, and hydrostatic pressure, $\hat{p}(t)$, are determined. The cracking risk is postulated to be a function of the local shear amplitude $\hat{\tau}$ and the tensile state $\hat{p}$. If during the loading path the combined influence of these two variables presents a maximum superior to a limit value, then there is a risk of crack nucleation (Figure 6). The cracking risk is more easily quantified through a double maximization of time evolution (t) and plane orientation ($\bar{n}$).

$$d = \max_{\bar{n},t} \left\{ \frac{\|\hat{\tau}(\bar{n},t)\|}{\tau_d - \alpha \cdot \hat{p}(t)} \right\} \quad \text{with} \quad \alpha = \frac{\tau_d - \sigma_d/2}{\sigma_d/3} \tag{2}$$

*If "d" is superior or equal to 1 there is a cracking risk;*
*if "d" remains smaller than 1 there is no cracking risk.*

Note that the cracking risk is calculated by combining alternated bending and shear fatigue limits ($\sigma_d, \tau_d$). The local residual stress is estimated considering an isotropic kinematic hardening law [5, 6]. We also consider the author's postulate which assumes that a sufficient number of micrograins is included in the elastic macroscopic domain such that at least one will present its easy slip band direction collinear to the maximum shear stress [5] :

$$\hat{\tau}(t) = \frac{1}{2} \text{tresca} \left( \hat{\sigma}(t) \right) \;,\;\; \hat{p}(t) = \frac{1}{3} \text{trace}(\hat{\sigma}(t)) = \frac{1}{3} \text{trace}(\Sigma(t)) \tag{3}$$

The crack nucleation risk is calculated on each point of the top surface and below the contact (Figure 7) [11]. It is shown that for this steel configuration with large friction coefficient, the maximum risk is observed on the surface at the contact border along the sliding axis. This first investigation confirms the experimental observation of the first crack nucleation under high cycle conditions.

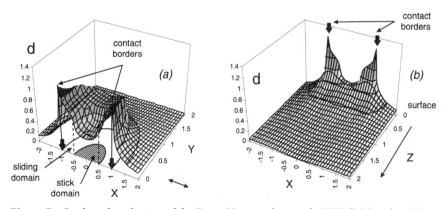

Figure 7 - *Surface distribution of the Dang Van cracking risk (30NiCrMo, c/a =0.5,*
*μ=0.8); (a) surface distribution (X,Y); (b) subsurface distribution (X,Z).*

On this particular point the stress loading is simplified and the Dang Van formulation can be expressed analytically [11].

The critical hydrostatic pressure at the contact border is given by

$$\hat{p} = \frac{1}{3} q_0 \left( f_{11}^{Q*} + f_{22}^{Q*} \right) \tag{4}$$

Whereas the corresponding critical local shear stress is expressed by:

$$\hat{\tau} = \frac{1}{2} q_0 \, f_{11}^{Q*} \tag{5}$$

Where,

$$f_{11}^{Q*} = \frac{1}{4} \left[ (4+v) \left( \frac{\pi}{2} - \Phi \right) + k\left(1-k^2\right)^{1/2} \left(4+v-2v\,k^2\right) \right], \tag{6}$$

$$f_{22}^{Q*} = \frac{1}{4} \left[ 3v \left( \frac{\pi}{2} - \Phi \right) + vk\left(1-k^2\right)^{1/2} \left(3+2k^2\right) \right], \tag{7}$$

$$\Phi = \tan^{-1} \left( k \Big/ \left(1 - k^2\right)^{1/2} \right) \tag{8}$$

with $\mu = \mu_L$, $k = c/a = \left(1 - Q*/(\mu.P)\right)^{1/3}$ and $q_0 = \mu.p_0$.

$p_0$ : Hertzian pressure at the center of the contact,

$q_0$ : Hertzian shear stress at the center of the contact.

The critical value of the Dang Van parameter is then expressed through the simplified relationship:

$$d = \frac{\dfrac{q_0}{2} f_{11}^{Q*}}{\tau_d - \dfrac{\alpha}{3} \left( f_{11}^{Q*} + f_{22}^{Q*} \right)} \tag{9}$$

Note that the cracking risk is mainly controlled by the alternated tangential loading rather than the static normal loading. Moreover the Dang Van formulation is very close to the formalisms developed by Socie D. and other multiaxial approaches which assume that the crack nucleation phenomenon is a function of a shear stress amplitude associated with a tensile state [12]. Considering that at the contact borders the stress loading is very closed to an uniaxial condition [13], it can then be understood that most of multiaxial fatigue analysis of fretting contacts tend to have similar results and permit a good estimation of cracking failures [14].

*Comparison with experiments*

The Dang Van crack nucleation risk is determined for each of the critical conditions leading to the nucleation of the very first crack (Table 3).

Table 3 - *Dang Van predictions extracted from the analysis of limit crack nucleation conditions (Table 2); X(d) : relative position of the maximum value.*

| Conditions | (1) | (2) | (3) | (4) |
|---|---|---|---|---|
| Dang Van d | 1.24 | 1.23 | 1.24 | 1.23 |
| X(d)= x(d)/a | 1 | 1 | 1 | 1 |
| $\|\rho *\|$, MPa | 150 | 170 | 190 | 200 |

The corresponding norm of the stabilized residual stress $\|\rho *\|$ is also calculated. The obtained values, comprised between 100 and 200 MPa, appear to be equivalent to the residual stress evolution measured by X-ray diffraction on a similar shot peened steel submitted to gross slip fretting loading [15]. The calculated values of the cracking risk are nevertheless systematically higher than the predicted threshold (d=1). The fatigue predictions obtained from the conventional fatigue properties appear too pessimistic compared to the fretting cracking analysis. It predicts cracking, when no failures are observed through the contact. It can lead to over-dimensioning of components which is expensive and unjustified. This difference cannot be induced by the surface asperities, first, because the surface roughness is very low, and second, the surface asperities by inducing local over-stressing should accelerate the crack nucleation.

**Size Effect**

In opposition to classical fatigue analysis, where the maximum stress field concerns a rather large volume of matter, contact loading is characterized by a very sharp stress gradient on and below the surface (Figure 7). Indeed, the maximum stress field concerns volume dimensions which can be inferior to the grain size. To estimate the contact failure from macroscopic fatigue approach a size effect must be taken into account. The local fatigue approach based on a point stress analysis (point M) must be replaced by a non local fatigue description which considers a mean loading state $\overline{\Sigma}(M, t)$ averaged on a micro volume V(M) surrounding the point on which the fatigue analysis is performed. Such a micro volume is obviously highly dependent on the microstructure. We consider a cubic volume, whose edges are assimilated to the physical length scale "l" (Figure 8). To simplify the computation, this cubic edge can be normalized versus the contact radius inducing L=l/a.

As shown in Figure 8, which takes into account the size effect influences, the maximum cracking discontinuity on contact border is strongly smoothened whereas the maximum value is progressively translated towards the sliding area. The physical reality of the stress loading endured by the material is included between the discontinuous point analysis and the whole contact averaging: L=1. The "fatigue" fretting analysis consists of extrapolating the dimension of the elementary volume which permits the linking of classical fatigue variables with the cracking process involved in the contact.

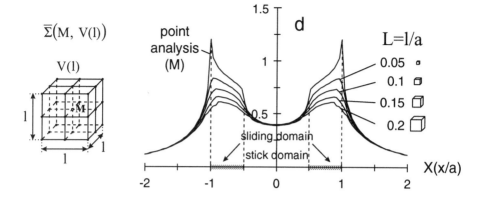

Figure 8 - *Cracking risk distributions considering the size effect ( c/a=0.5, μ=0.8 ).*

Both the normalized value "L" and the physical parameter "l" are considered in order to estimate the influence of the contact dimension. The four critical conditions are investigated. The characteristic dimensions are determined by successive iterations on "L" , averaging the stress and calculating the corresponding Dang Van "d" value until it corresponds to the threshold condition: d=1 (Table 4). It can be noted that the physical variable "l" is much more stable than the normalized parameter "L". The microvolume appears therefore more related to a physical constant dimension characteristic of the microstructure rather than to the contact size.

Table 4 - *Determination of the elementary volume associated to the experimental limit crack nucleation conditions (X(d)/a : relative position where d is maximum).*

| Conditions | (1) | (2) | (3) | (4) |
|---|---|---|---|---|
| $L(l/a)$ $(10^{-3})$ | 12.1 | 9.4 | 8.79 | 7.03 |
| l, μm | 6.6 | 5.9 | 6.1 | 5.3 |
| X (d)/a | 0.99 | 0.99 | 0.99 | 1 |
| $\|\rho *\|$, MPa | 135 | 155 | 170 | 180 |

The calculated value corresponds to 6-5 μm cubic edge with a cubic diagonal (8-10.5) more or less equivalent to the prior austenitic grain size ($\varnothing \approx 10$ -15 μm). To interpret such a small dimension two hypothesis can be considered.

Based on the Dang Van description of the crack nucleation phenomena, the microvolume can be related to the probability of activation of an easy slip band through the tempered microstructure (Figure 9a). Indeed it is fundamental to consider the microstructure of the steel. After quenching, the steel displays a very fine martensitic structure. Besides, the high temperature of tempering (585°C) allows the diffusion of carbons atoms and the crystallization of very small micro domains of ferrite and cementite. The calculated microvolume can then be related to the initial austenitic grain size which after tempering contains a large enough density of micro domains of ferrite where the first crack nucleation will occur.

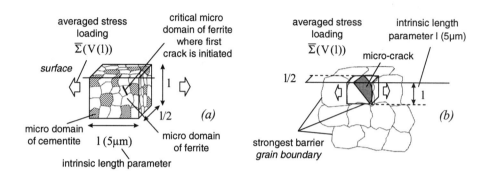

Figure 9 – *Schematic view of a two phase alloy including ferrite, physical interpretation of the fatigue limit : (a) Based on the Dang van's description the fatigue limit is related to the probability to initiate a crack in a micro domain of ferrite; (b) The Miller's description associates the fatigue limit to the blocking of an inherent micro-crack at the strongest barrier which are the grain boundaries.*

A second explanation of this characteristic length scale considers the K.J. Miller approach: "the fatigue limit can be seen to be a function of the maximum non-propagating crack length associated with a particular stress level" [16] (Figure 9b). Therefore, the calculated length dimension (l=5 μm), defined through an average stress approach, can be associated with the maximum dimension that an incipient crack nucleating through the grain can reach before being stopped by the strongest barrier: the grain boundary. Both the length parameter (l=5μm) and the corresponding mean stress level can then be associated with a non propagating condition.

It is shown that the two hypothesis can explain the dimension of the microvolume which was determined through the present "Fatigue" approach of fretting. To enlighten the fatigue interpretation of this phenomenon, which is still a major question in the field of research on fatigue, various approaches can be considered. For instance, the fatigue analysis could be studied on different microstructures whereas different contact geometries could be investigated.

From this high cycle fatigue analysis a specific fretting contact crack nucleation diagram can be plotted transposing the Dang Van formalism (Figure 10). It represents the distribution of the cracking domain as a function of the maximum local shear and the hydrostatic pressure.

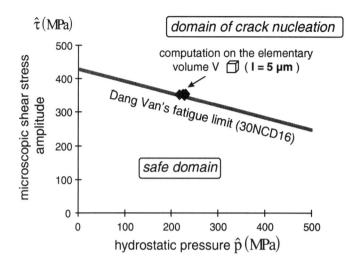

Figure 10 - *Dang Van diagram of crack nucleation taking into account the size effect.*

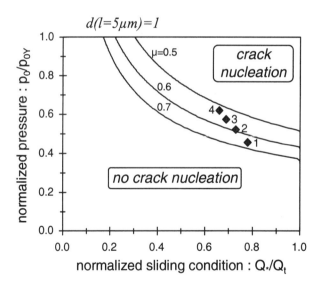

Figure 11 - Normalized pressure crack nucleation fretting map : ___ Dang Van crack nucleation limits ($d_{(l=5\mu m)}=1$) as a function of the friction coefficient ◆ : Experimental fretting crack nucleation conditions (cf. Figure 3b; Table 2); $p_{0Y} = 1.6 \cdot \sigma_{Y0.2}$ : Yielding Maximal Hertzian pressure; $Q_* / Q_t = Q_* / \mu_L . P = (1 - k^3)$

It is also possible to develop a normalized fretting map (Figure 11) where the crack nucleation risk is expressed through the five fundamental parameters which are :

- the Hertzian pressure $p_0$ (normalized by the Hertzian yield pressure under pure indentation $p_{0y} = 1.6\,\sigma_{y0.2}$),
- the fretting sliding condition here expressed by the ratio between the tangential force amplitude versus the amplitude at the transition ( $Q_t = \mu.P$ ),
- the friction coefficient operating through the sliding domain under partial slip condition,
- the conventional fatigue properties associated to the shear and bending fatigue limits,
- the intrinsic length scale parameter associated to the studied material.

By defining the admitted Hertzian pressure as a function of these parameters, the predictive approach allows an instantaneous estimation of the safe domain where no crack are initiated.

This normalized representation can be extended by displaying the crack nucleation boundary as a function of pressure condition, the sliding condition and the characteristic length scale (Figure 12). In spite of strong variations induced by the friction coefficient fluctuation under partial slip situations, a good correlation is obtained between the experimental crack nucleation conditions and the predictive curves (Figure 12b). Moreover from this representation it can be deduced that for very small contacts no propagating crack are nucleated. This result is in full accordance with the conclusions exposed by Hills and Nowell concerning a critical contact size [17, 18].

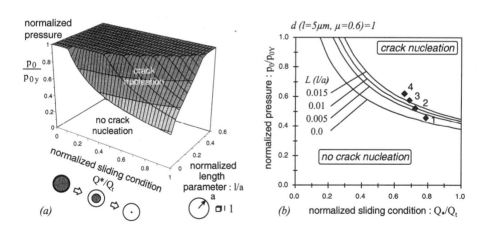

Figure 12 - *Normalized fretting chart displaying the crack nucleation boundary as a function of the pressure, the sliding condition and the structural length parameter ($\mu=0.6$) : (a) 3D representation ; (b) Comparison with experiments (cf. Table 4).*

## Conclusion

Fatigue analysis of fretting cracking phenomena implies to precisely identify the surface loading path through the determination of the local coefficient of friction ($\mu_L$). Moreover, contact problems are characterized by a very sharp stress gradient and very small stressed material volumes. This implies to consider the size effect introducing a non-local fatigue description of the cracking phenomenon. The study of an aeronautical steel (30NiCrMo) indicates that:
- a direct prediction of crack nucleation from classical fatigue shear and bending fatigue variables is possible, if the loading state is averaged over a critical elementary volume defined by a 5-6 μm cubic edge,
- this length scale value appears as a constant physical variable equivalent to the original austenitic grain size. It appears as a characteristic of the steel microstructure and remains independent of the contact dimension.

This critical material volume can be considered as an intrinsic microstructural variable which can be analyzed in terms of probability to activate a micro defect trough the tempered microstructure. Moreover, when defining the fatigue limit as a non propagation condition, the length variable can also be associated with the distance between the micro structural barriers (grain boundary) which stop the propagation of an incipient crack through the grain. Future investigations will focus on enlightening the fatigue interpretation of fretting crack nucleation. Taking into account the different parameters which control the crack nucleation under high cycle fatigue conditions, a normalized crack nucleation fretting mapping is proposed.

## Acknowledgments

The authors would like to thank the Commission of the European Community for partial financial support (BRE-CT92-0224), Prof. K. Dang Van, F. Sidoroff, and A.B. Vannes for helpful discussions and, A. Barrier, R. Lambrech and J.M Vernet for technical assistance.

## References

[1]    Waterhouse, R.B., *Fretting Fatigue*, Applied Science publishers, 1981.

[2]    Lindley, T.C., Nix, K.J., "The role of fretting in the initiation and early growth of fatigue cracks in turbo-generator materials", *ASTM STP 853,* ASTM, Philadelphia, 1982, pp. 340-360.

[3]    Hoeppner, D., "Mechanisms of fretting fatigue and their impact on test methods development", ASTM STP 1159, M.H. Attia and R.B. Waterhouse, Eds, ASTM, Philadelphia, 1992, pp. 23-32.

[4]   Vincent, L., Berthier, Y., Godet, M., "Testing methods in fretting fatigue: a critical appraisal", ASTM STP 1159, M.H. Attia and R.B. Waterhouse, Eds, ASTM, Philadelphia, 1992, pp.33-48.

[5]   Dang Van, K., "Macro-micro approach in high-cycle multiaxial fatigue", *Advances in Multiaxial Fatigue, ASTM STP 1191*, D.L. McDowell and R.Ellis, Eds., ASTM, Philadelphia, 1993, pp. 120-130.

[6]   Dubar, L., "Fatigue multiaxiale des aciers, passage de l'endurance à l'endurance limitée, prise en compte des accidents géométriques", PhD Thesis ENSAM (Talence), 1992, N°92-18.

[7]   Blanchard P., Colombie C., Pellerin V., Fayeulle S., and Vincent L., "Material Effects in Fretting Wear: Application to Iron, Titanium, and Aluminium Alloys", Metallurgical Transaction A, Volume 22A, July 1991, pp. 1535-1544.

[8]   Mindlin, R.D., Deresiewicz, H., "Elastic spheres in contact under varying oblique forces", *ASME Trans, Serie E, Journal of Applied Mechanics*, 20, 1953, pp. 327-344.

[9]   Johnson K.L., "Surface interaction between elastically loaded bodies under tangential forces", *Proc. Roy. Soc., Ser. A230*, 1955, pp. 531-548.

[10]  G. M. Hamilton, Explicit equations for the stresses beneath a sliding spherical contact, Proc. Inst. Mech. Ing. 197c, (1983), pp. 53-59.

[11]  Fouvry, S., Kapsa, Ph., Vincent, L., Dang Van, K., "Theoretical analysis of fatigue cracking under dry friction for fretting loading conditions", *Wear, 195*, 1996, pp. 21-34.

[12]  Socie, D., "Critical Plane Approaches for Multiaxial Fatigue Damage Assessment", *Advances in Multiaxial Fatigue, ASTM STP 1191*, D.L. McDowell and R. Ellis, Eds., ASTM, 1993, pp.7-36.

[13]  Fouvry, S., Kapsa, Ph., Vincent, L., "Stress and Fatigue analysis of fretting on a stressed specimen under partial slip conditions", *Tribotest Journal 3-1*, 1996, pp. 23-44.

[14]  Szolwinski, M.P., Farris, T.N., "Mechanics of fretting crack formation", *Wear, 198*, 1996, pp. 93-107.

[15]  Benrabah, A., Vannes, A.B, "Influence des contraintes résiduelles sur le comportement en fretting des matériaux", SF2M, *Edit. Revue de Metallurgy* N°9, 1995, pp. 262-269.

[16]   Miller, K.J., "Material science perspective of metal fatigue resitance", *Materials Science and Technology*, June 1993, Vol. 9, pp. 453.

[17]   Hills, D.A., Nowell, D., and O'Connor, J.J., "On the mechanics of fretting fatigue", *Wear*, *125*, 1988, pp. 129 – 146.

[18]   Nowell,D. and Hills, D.A., "Crack Initiation in Fretting Fatigue" , *Wear*, *136*, 1990 pp. 329-343.

Rebecca Cortez,[1] Shankar Mall,[2] and Jeffrey R. Calcaterra[3]

## Interaction of High-Cycle and Low-Cycle Fatigue on Fretting Behavior of Ti-6-4

**REFERENCE:** Cortez, R., Mall, S., and Calcaterra, J. R., "Interaction of High-Cycle and Low-Cycle Fatigue on Fretting Behavior of Ti-6-4," *Fretting Fatigue: Current Technology and Practices, ASTM STP 1367*, D. W. Hoeppner, V. Chandrasekaran, and C. B. Elliott, Eds., American Society for Testing and Materials, West Conshohocken, PA, 2000.

**ABSTRACT:** The fretting behavior of Ti-6Al-4V when subjected to combined high-cycle fatigue/low-cycle fatigue (HCF/LCF) was investigated. Constant amplitude fretting fatigue tests were first conducted at room temperature at both 1 Hz and 200 Hz frequencies. The fretting fatigue life was smaller for the higher frequency condition than for the lower one. Then, fretting tests combining the interaction of both high-frequency and low-frequency waveforms were conducted which yielded similar lives to the constant amplitude fretting fatigue tests. Tests consisting of HCF/LCF blocks of 50:1, 500:1, and 5000:1 were examined where the ratio represented the number of HCF cycles to LCF cycles. Overall, Miner's linear damage summation rule provided a reasonable estimate of combined HCF/LCF fretting lives for the HCF/LCF tests considered.

**KEYWORDS:** fretting fatigue, high-cycle fatigue, low-cycle fatigue, titanium base alloy

### Introduction

The detrimental impact of fretting on the overall fatigue resistance of numerous materials is well documented in the literature. Fretting fatigue occurs when the relative oscillatory motion of two bodies in contact is combined with a cyclically applied axial load. As pointed out by Hoeppner and Goss [1], fretting acts as a flaw generator which leads to premature crack nucleation when compared to fatigue situations void of fretting contact. One might expect that for applications subject to high-cycle fatigue, in which the majority of life is spent in crack nucleation, that the influence of fretting damage could be significant.

[1] Materials Research Engineer, Air Force Research Laboratory, Sensors Directorate, Wright-Patterson AFB, OH 45433-7322.
[2] Principal Materials Research Engineer, Air Force Research Laboratory, Materials & Manufacturing Directorate, Wright-Patterson AFB, OH 45433-7817.
[3] Aerospace Engineer, Air Force Research Laboratory, Air Vehicles Directorate, Wright-Patterson AFB, OH 45433-7605.

The purpose of the present study was to examine the fretting fatigue behavior of Ti-6Al-4V when subjected to both constant amplitude and superimposed high-frequency fatigue loading on a low-frequency waveform. This alloy is one typically used in aerospace applications, such as for blades and disks in turbine engines. In general, turbine disk failures result from a complex interaction of both low-cycle fatigue (LCF) and high-cycle fatigue (HCF) in which fretting and variable amplitude loading are two key components. The effects of frequency and loading interaction on the fretting fatigue resistance of this alloy are presented based on experimental results and microscopic examination of the fretting fatigue surfaces. Within the context of the present research, LCF refers to conditions of low load ratios, R ($\approx$0.1), where R represents the ratio of minimum load to maximum load, while HCF conditions refer to fatigue lives typical of higher R ratios.

**Experimental**

The fretting fatigue specimens used in the present study originated from two sources of stock Ti-6Al-4V (weight percent): one was from bar stock while the other was from a forged plate. The bar material was processed according to Aerospace Material Specifications Titanium Alloy, Bars, Wire, Forging, and Rings, 6Al4V, Annealed (AMS 4928). The pre-machined material was heat treated at 705°C in vacuum for two hours; cooled below 149°C in an argon environment; vacuum annealed at 549°C for two hours; then argon cooled. By comparison, the Ti-6Al-4V plate was solution heat treated at 935°C for one hour; air fan cooled; vacuum annealed at 705°C for two hours; then slow cooled. The resulting microstructures for the bar and plate reveal that the bar is more textured than the plate and is characterized by wide, plate-like $\alpha$ with intergranular $\beta$ while the plate contains acicular $\alpha$ (Figure 1).

The measured yield stress ($\sigma_y$), ultimate stress ($\sigma_u$), and elastic modulus (E) for the Ti-6Al-4V bar were 990 MPa, 1035 MPa, and 109 GPa, respectively; while measured $\sigma_y$, $\sigma_u$, and E values for the plate material were found to be 930 MPa, 978 MPa, and 118 GPa, respectively [2]. Although the mechanical properties of this material are highly dependent on thermal processing and the resultant microstructure, no apparent difference in the fretting fatigue performance of the bar and plate stock were noticed as will be discussed in the next section. As the pure tensile fatigue response of the plate stock [3] was found to be lower than the bar stock [4], it was necessary to consider any potential differences between the fretting fatigue response of the two source materials in the present study. No pure fatigue tests were conducted in the present study as uniaxial tensile fatigue tests have been conducted on the plate or bar Ti-6Al-4V source material at 60 Hz [3], 70 Hz [4,5], and 400 and 1800 Hz [5].

Both fatigue specimens and fretting pads were machined and ground to precise specifications. The specimen gage section was nominally 6.35 mm wide and 1.93 mm thick while the ends of the fretting pads (9.53 mm by 9.53 mm cross section) were ground to a 50.8 mm radius curvature. The fatigue specimens were polished to a #8 surface finish while the pads were ground to an 8 RMS surface finish. All fretting fatigue tests were conducted using a servo-hydraulic uniaxial load frame equipped with an attached

## Ti-6Al-4V  Bar

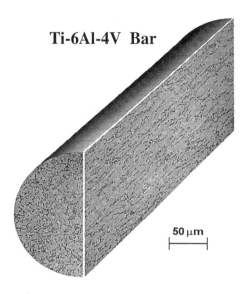

50 μm

## Ti-6Al-4V Plate

Longitudinal

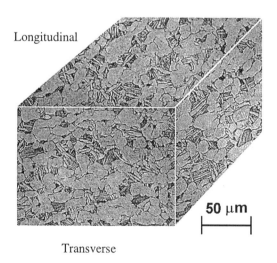

50 μm

Transverse

Figure 1 - *Microstructure for bar and plate Ti-6Al-4V.*

fretting fixture. Details of the fretting fixture have been presented elsewhere [6]. Fretting contact to the specimens is provided via Ti-6Al-4V pads (Figure 2). A bulk normal load of 1334 N was chosen for the present experimental program. Tests were run at ambient room temperature during which time the laboratory temperature and humidity were not controlled.

All tests were run in displacement control. The axial load ratio, R, ranged from 0.1 to 0.8. Peak contact pressure for the round pads based on Hertzian contact solution [7] was determined to be 277 MPa. The constant amplitude and superimposed HCF/LCF tests were conducted using a sinusoidal waveform. The HCF/LCF waveforms were performed such that 200 Hz small-amplitude cycles (in cycle groups of 50, 500, or 5000) were applied at the peak stress of a comparatively larger sine wave cycling at 1 Hz (Figure 3). Essentially, the HCF/LCF waveform can be considered as the summation of two separate waveforms. The HCF portion cycles between two loads representing a large R value (the R value for one test may be any one value ranging from 0.4 to 0.7). Similarly, the LCF portion cycles between a nominal R = 0.1 minimal load and the same maximum load as the HCF portion. The HCF portion occurs at the peak of the LCF cycle and introduces a 200 Hz cycling "dwell" for a time period of either 0.25, 2.5, or 25 seconds.

Two hundred hertz was chosen as the upper limit for the fretting fatigue tests as this frequency was the maximum reliable test frequency for the mechanical test frame system used. Initial shakedown of the fretting fatigue set-up included an examination of the system resonance at various frequencies ranging from 1 Hz to 200 Hz in increments of 10 Hz. The cyclic load response above the fretting contact region on the specimen (measured by the acceleration compensated system load cell) as well as below the contact area (measured via strain gages) at 200 Hz were in-phase and in conformity with the requested cyclic command shape. A more detailed discussion of the experimental procedures is provided elsewhere [6].

## Results and Discussion

The fretting fatigue lives for the constant amplitude test series are summarized (Table 1) and are graphically presented as a function of axial stress range (Figure 4), where the axial stress range represents the difference of the maximum and minimum stress. Throughout this manuscript, $\sigma_{max}$ represents the actual maximum axial stress, $\Delta\sigma$ represents the actual axial stress range, $Q_{max}$ and $Q_{min}$ represent the tangential load at maximum and minimum stress, respectively, and R represents the ratio of the minimum to maximum axial load. Note that the measured tangential loads (Q) were calculated by taking half of the difference between the loads measured below and above the fretting contact zone. The tabulated data for the constant amplitude tests has been reported previously [6]. The specimens numbered 97-Jxx originated from the bar material while the remaining specimens were machined from plate stock.

The 200 Hz fretting fatigue resistance of Ti-6Al-4V appears insensitive to the microstructural differences between the plate and the bar as all of the high-frequency data fall within the limits of scatter (Figure 4). The authors believe that the additional fretting contact stress component dominates the overall fatigue response of both the bar and plate

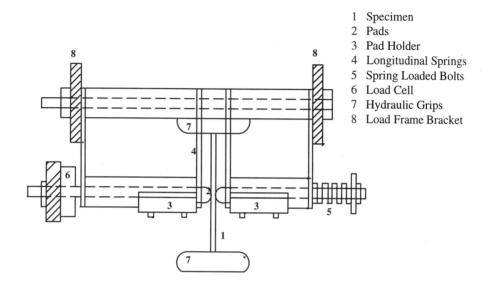

1  Specimen
2  Pads
3  Pad Holder
4  Longitudinal Springs
5  Spring Loaded Bolts
6  Load Cell
7  Hydraulic Grips
8  Load Frame Bracket

Figure 2 – *Fretting fatigue fixture schematic.  Not to scale.*

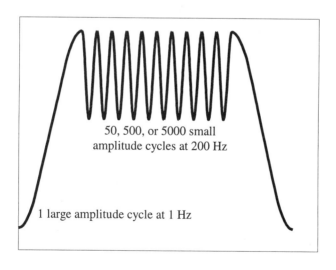

50, 500, or 5000 small
amplitude cycles at 200 Hz

1 large amplitude cycle at 1 Hz

Figure 3 - *HCF/LCF waveform schematic.*

Ti-6Al-4V. Therefore, for the axial fatigue stress conditions chosen, the influence of microstructure is secondary to the fretting effects. Caution should be exercised in generalizing this observation to all fretting fatigue conditions. A decrease in the normal load applied to the specimen has been found to increase the fretting fatigue resistance of mill annealed Ti-6Al-4V [8]. Therefore, it may be possible that for test conditions involving lower contact or axial cyclic stresses that the resultant microstructure of the fatigue specimen may influence cycles to failure. Note that the data points represent actual fretting fatigue lives for a given axial stress range while the curves represent the fit through the data. These constant amplitude curves will be used for future comparison to the HCF/LCF experimental data.

Table 1 - *Constant Amplitude Fretting Fatigue Test Summary*

| Specimen | f, Hz | $\Delta\sigma$ (MPa) | Cycles to Failure | $\sigma_{max}$ (MPa) | R | $Q_{max}$ (N) | $Q_{min}$ (N) |
|---|---|---|---|---|---|---|---|
| 98-550 | 200 | 698 | 31 570 | 716 | 0.02 | 694 | -427 |
| 97-J27 | 200 | 586 | 13 145 | 595 | 0.02 | 685 | -245 |
| 98-546 | 200 | 562 | 53 445 | 584 | 0.04 | 596 | -298 |
| 97-J25 | 200 | 508 | 57 573 | 519 | 0.04 | 494 | -133 |
| 98-551 | 200 | 396 | 123 940 | 698 | 0.43 | 454 | -231 |
| 97-J30 | 200 | 339 | 100 850 | 625 | 0.46 | 1019 | -214 |
| 98-549 | 200 | 309 | 117 797 | 558 | 0.45 | 342 | -218 |
| 98-547 | 200 | 305 | 262 460 | 554 | 0.45 | 285 | -200 |
| 98-553 | 200 | 251 | 371 112 | 716 | 0.65 | 654 | 156 |
| 97-J28 | 200 | 233 | 416 450 | 659 | 0.65 | 925 | 574 |
| 97-J42 | 200 | 217 | 1 159 063 | 606 | 0.64 | 783 | 338 |
| 97-J37 | 200 | 190 | 10 000 000+ | 534 | 0.64 | 578 | 236 |
| 98-554 | 200 | 169 | 4 141 431 | 549 | 0.69 | 587 | 414 |
| 98-N80 | 200 | 151 | 6 852 776 | 681 | 0.78 | 558 | 259 |
| 98-574 | 1 | 681 | 37 214 | 684 | 0.00 | 693 | -786 |
| 98-573 | 1 | 536 | 150 147 | 538 | 0.00 | 587 | -603 |
| 98-577 | 1 | 533 | 127 070 | 540 | 0.01 | 559 | -578 |
| 98-575 | 1 | 377 | 210 806 | 664 | 0.43 | 493 | -307 |
| 98-578 | 1 | 293 | 2 250 000+ | 536 | 0.46 | 530 | -80 |

The choice to use stress range as the driving parameter for the fretting fatigue tests was based on the presumption that stress range is directly related to the relative displacement at the contact interface between the fatigue specimen and the Ti-6-4 fretting pads. Although no mean stresses were taken into consideration, the use of stress range versus cycles to failure resulted in a single curve for the fretting fatigue results of the present study. Note that as the stress range decreased, the difference between the tangential loading at maximum and minimum axial load also decreased.

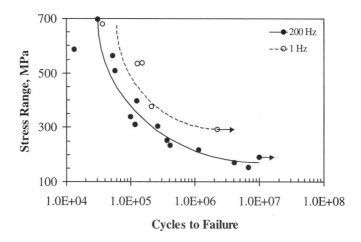

Figure 4 - *Constant amplitude fretting fatigue test results. The arrows represent run-out where the tests did not fail.*

The fretting fatigue lives are longer for the 1 Hz test condition when compared to the 200 Hz test condition as graphically shown (Figure 4). Differences in fatigue lives are most prevalent for lower axial stress ranges. The effect of frequency on the fretting fatigue behavior of other materials has been documented [9,10]. In a fretting fatigue study of carbon steels under reversed twisting and bending, it was found that increases in the test frequency led to increased fretting fatigue strength [9]. As fretting fatigue involves the rubbing of surfaces, localized increases in temperature may influence the failure mechanism. Test frequency may also influence the coefficient of friction [10]. The impact of temperature and the coefficient of friction for fretting fatigue tests conducted with the current experimental design will be examined in future studies.

Microscopic examination of the fretting surfaces revealed qualitatively distinct fretting fatigue damage features for the 1 Hz and 200 Hz test frequencies. Comparisons of the fretting surfaces using a scanning electron microscope (SEM) (Figure 5a and 5b) revealed rougher surface damage for the 200 Hz test (Figure 5a) when compared to the 1 Hz test (Figure 5b). As shown, the area near the fracture surface for the 200 Hz test is laden with accumulated Ti-6Al-4V debris, pits and gouged regions (Figure 5a). The shorter fatigue life exhibited by the 200 Hz data most probably resulted from the additional wear caused by the high-frequency vibratory rubbing between the titanium pad and the specimen. Note that the test conducted at 1 Hz (Figure 5b) also contains areas marred by rubbing, however, the damage is less severe than that of the 200 Hz test. As mentioned, localized heating of the fretting surface may have occurred at 200 Hz.

Quantification of the differences in the fretting surfaces as a function of frequency as well as in cross-sectional regions is ongoing and will be presented elsewhere.

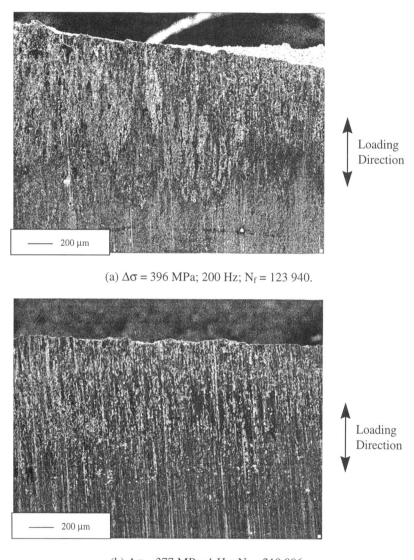

(a) $\Delta\sigma = 396$ MPa; 200 Hz; $N_f = 123\,940$.

(b) $\Delta\sigma = 377$ MPa; 1 Hz; $N_f = 210\,806$.

Figure 5 - *SEM micrographs of fretting surfaces for constant amplitude fatigue tests. Frequency comparison.*

Tabulations of the experimental combined HCF/LCF fretting fatigue tests (Table 2) are provided for comparison to the constant amplitude fatigue data (Table 1). In this study a HCF/LCF test was designated by the ratio of the small-amplitude, high-frequency cycles to the large-amplitude, low-frequency cycle. For example, a variable amplitude loading of 50:1 represented a loading combination of fifty 200 Hz cycles for every 1 Hz cycle and is presented as $N_S:N_L$ (Table 2). The other symbols tabulated (Table 2) hold the same meaning as those previously used (Table 1) with the subscripts S and L representing the small-amplitude and large-amplitude cycles, respectively.

The effect of superimposing high-frequency cycles at the maximum stress of a single, large-amplitude cycle is plotted as cycles to failure as a function of stress range (Figure 6). In general, fatigue lives for the HCF/LCF tests fell within the scatter band of the constant amplitude data, however, a couple tests run at $N_S:N_L$ of 50:1 were biased toward the 1 Hz frequency data where $N_f$ was found to be greater than the 200 Hz data under constant amplitude loading. Specifically, the HCF/LCF data fall within the band of ±2 times the failure cycles indicated by the 200 Hz constant amplitude trend curve. The bias toward the 1 Hz constant amplitude test curve for a couple of the 50:1 tests points is not surprising when one considers the dwell time at the maximum stress for the three HCF/LCF conditions. For instance, 50:1 represents a 0.25 second dwell at 200 Hz, while 500:1 and 5000:1 represent dwells at 200 Hz for 2.5 and 25 seconds, respectively. As the majority of life for the 50:1 HCF/LCF loading is spent at 1 Hz, it is reasonable to expect that the damage mechanisms prevalent under the 1 Hz frequency dominate the 50:1 HCF/LCF series. By comparison, different damage parameters, such as localized temperature rises may dominant the 500:1 and 5000:1 HCF/LCF loading blocks. The deviation between the lower stress range HCF/LCF fretting fatigue lives for the $N_s:N_L$ equal to 500:1 and the 200 Hz constant amplitude trend line is worth noting (Figure 6).

Miner's linear damage summation rule was used to predict the combined HCF/LCF fretting fatigue lives (Figure 7). The predictive curve (Figure 7) was generated by incorporating the baseline 200 Hz and 1 Hz constant amplitude data with Miner's rule. The curve represents the condition of $N_S:N_L$ of 500:1, however, the curves for $N_S:N_L$ equal to 50:1 or 5000:1 could easily have been used as all combinations yield similar lives for the range of axial stress conditions examined. Differences in the predictive lives are only apparent at the low axial stress range ($\Delta\sigma$ near 180 MPa), however, as the curve is already flat in this region, the difference in curve shape between the three conditions is negligible. A comparison of the actual fatigue lives with Miner's predictive curve (Figure 7) shows that, overall, Miner's rule provided a reasonable estimate of the expected HCF/LCF fretting fatigue lives. The deviation of a couple of the 50:1 HCF/LCF points is a carry-over from the previous observation made regarding the bias of the 50:1 tests toward the 1 Hz constant amplitude fatigue data.

Microscopic examination of the fretting fatigue surfaces was completed to characterize the fretting surface damage occurring during both constant amplitude and HCF/LCF fretting fatigue. An SEM image of a constant amplitude test (Figure 8a) showed that major cracking occurred near the location of the trailing edge of contact (contact interface on the loading side, i.e., the actuator end of the specimen) which is the location of the maximum contact stress based on theoretical analysis of cylindrical

Table 2 – HCF/LCF Waveform Fretting Fatigue Test Summary

| Specimen | Cycles to Failure | $N_S:N_L$ | $\sigma_{max}$ (MPa) | $\Delta\sigma_S$ (MPa) | $Q_{max}$ (N) | $Q_{min,S}$ (N) | $R_S$ | $\Delta\sigma_L$ (MPa) | $Q_{min,L}$ (N) | $R_L$ |
|---|---|---|---|---|---|---|---|---|---|---|
| 98-N83 | 251 200 | 50:1 | 715 | 404 | 570 | -282 | 0.43 | 702 | -628 | 0.02 |
| 98-571 | 667 182 | 50:1 | 553 | 290 | 524 | 88 | 0.48 | 507 | -394 | 0.08 |
| 98-N75 | 462 060 | 50:1 | 695 | 246 | 548 | 10 | 0.65 | 662 | -630 | 0.05 |
| 98-572 | 3 432 045 | 50:1 | 572 | 191 | 705 | 317 | 0.67 | 520 | -390 | 0.09 |
| 97-J54 | 78 155 | 500:1 | 653 | 402 | 494 | -196 | 0.38 | 653 | -1085 | 0.00 |
| 98-563 | 78 155 | 500:1 | 696 | 375 | 516 | -89 | 0.46 | 655 | -764 | 0.06 |
| 97-J29 | 143 286 | 500:1 | 517 | 297 | 536 | 125 | 0.43 | 507 | -681 | 0.02 |
| 98-588 | 211 422 | 500:1 | 544 | 290 | 443 | 31 | 0.47 | 514 | -596 | 0.06 |
| 98-560 | 232 908 | 500:1 | 695 | 265 | 810 | 329 | 0.62 | 695 | -985 | 0.00 |
| 98-565 | 275 048 | 500:1 | 696 | 243 | 483 | 28 | 0.65 | 659 | -769 | 0.05 |
| 97-J26 | 453 906 | 500:1 | 539 | 172 | 981 | 145 | 0.68 | 523 | -709 | 0.03 |
| 98-566 | 1 811 115 | 500:1 | 580 | 193 | 687 | 299 | 0.67 | 528 | -475 | 0.09 |
| 98-N84 | 87 628 | 5000:1 | 709 | 408 | 635 | -358 | 0.42 | 673 | -697 | 0.05 |
| 98-N82 | 98 221 | 5000:1 | 695 | 401 | 549 | -370 | 0.42 | 673 | -654 | 0.03 |
| 98-570 | 156 073 | 5000:1 | 562 | 291 | 605 | 145 | 0.48 | 524 | -495 | 0.07 |
| 98-N69 | 665 133 | 5000:1 | 706 | 253 | 613 | -8 | 0.64 | 671 | -699 | 0.05 |
| 98-567 | 1 825 014 | 5000:1 | 573 | 191 | 724 | 334 | 0.67 | 520 | -433 | 0.09 |

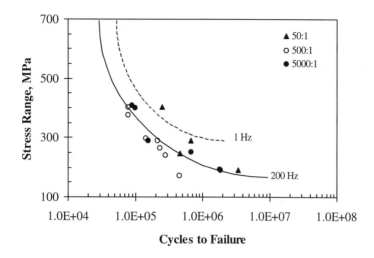

Figure 6 – *Combined HCF/LCF fretting fatigue results.*

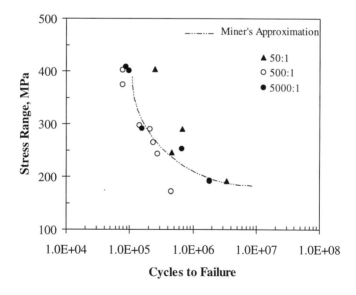

Figure 7 – *Miner's rule prediction for HCF/LCF fretting fatigue lives.*

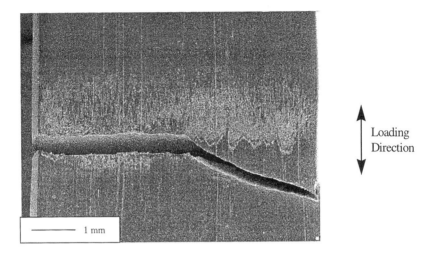

Loading
Direction

(a)  Constant amplitude loading. $\Delta\sigma = 681$ MPa; R = 0.1; f = 1 Hz, $N_f = 37\ 214$.

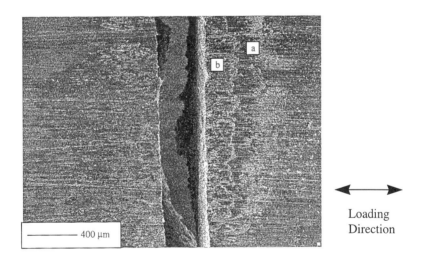

Loading
Direction

(b)  HCF/LCF loading. $\Delta\sigma_S = 291$ MPa; $N_S:N_L = 5000:1$; $R_S = 0.5$; $N_f = 156\ 073$.

Figure 8 - *SEM micrographs detailing differences in fretting surface damage as a function of axial loading mode.*

contact [7]. Fatal cracks were also noted near this location for constant amplitude fretting of an aluminum alloy [11].

For HCF/LCF loadings, evidence of two distinct bands of fretting damage was noted on the specimen surface (Figure 8b). The propagation of the fatal crack for the HCF/LCF loading occurred near an interface where the fretting wear patterns were distinct. The availability of additional interfaces suggests the possibility of alternative paths from which the cracks may nucleate. The line separating two regions of damage in a HCF/LCF test (Figure 8b) most likely represents the demarcation between the 1 Hz large-amplitude loading (labeled area a), and the 200 Hz small-amplitude fretting region (labeled area b). Recall that for the HCF/LCF tests, 200 Hz cycles in cycle blocks of 50, 500, or 5000 were superimposed at the peak of a single 1 Hz sine wave. Therefore, area a (Figure 8b) most probably contains damage features characteristic of 1 Hz testing (similar to Figure 5a), while area b contains damage typical of both 200 and 1 Hz fretting tests. In a separate study characterizing fretting surface damage on an aluminum alloy, it was also concluded that fatigue cracks formed at or near boundaries of fretted regions [12]. As summarized by Waterhouse [13], fretting fatigue crack initiation sites are points where a stress concentration is present, i.e., at the slip-non-slip boundary or near the outer contact boundary.

Damage characteristic of fretting fatigue was evident on all of the constant amplitude loading specimens. An SEM image (Figure 9a) provides evidence of transverse cracking of the fretting debris. Notice the presence of both large debris platelets and ground-up debris. Evidence of transverse cracking and large platelet formation for mill-annealed Ti-6Al-4V was also found for flat on flat fretting fatigue at 30 Hz [14]. An SEM image for a test conducted at a higher load ratio (Figure 9b) provides contrast to the rough surface features noted on the surface for a lower load ratio test (Figure 9a). For the test conducted at a load ratio of 0.7 (Figure 9b) the surface damage appeared flatter than the one conducted at a load ratio value of 0.5 (Figure 9a). This flatness most probably resulted due to the constant rubbing generated by the relatively small amplitude loading along the specimen surface. This observation is in agreement with the fatigue lives found at higher stress ratio values. The longer lives imply a delayed influence by fretting damage as cracks that may be forming on the surface appeared to be continuously rubbed away.

## Conclusions

Both constant amplitude and combined HCF/LCF fretting fatigue tests were conducted to study the interaction of low and high-cycle fatigue. Axial cyclic loads at both 1 Hz and 200 Hz were combined with a normal applied load to flat Ti-6Al-4V specimens with Ti-6Al-4V round-ended pads. Fretting fatigue resistance was greater for the 1 Hz constant amplitude loading condition when compared to similar loadings at the higher frequency (200 Hz). Microscopic examination of the fretting fatigue surfaces revealed qualitative differences in the damage features between the two frequencies with the higher frequency exhibiting more surface damage. Combined HCF/LCF fretting fatigue tests yielded fatigue lives similar to the 200 Hz constant amplitude fretting fatigue

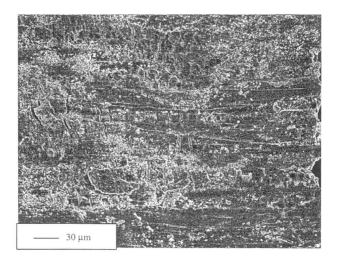

(a) $\Delta\sigma = 552$ MPa; R = 0.5; $N_f = 117\ 797$.

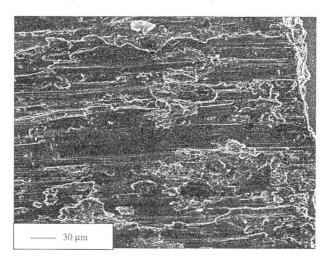

(b) $\Delta\sigma = 552$ MPa; R = 0.7; $N_f = 4\ 141\ 431$.

Loading
Direction

Figure 9 - *SEM micrographs of fretting surfaces for 200 Hz constant amplitude fatigue tests.*

test series for high proportions of high-cycle fatigue blocks (500:1 and 5000:1). However, for the condition of fifty 200 Hz cycles to one 1 Hz cycle, the fretting fatigue lives were similar to the 1 Hz constant amplitude data series. Miner's rule provided reasonable approximation of the combined HCF/LCF fretting fatigue lives. The development of numerous wear interfaces on the fretting surface under HCF/LCF loading conditions provides additional regions of stress concentrators for the nucleation of fatigue cracks.

## Acknowledgments

The authors thank Dr. Ted Nicholas (AFRL/MLLN) for supporting this work. The efforts of John Porter of the Materials & Manufacturing Directorate at Wright-Patterson AFB and Eric Shell of the University of Dayton in putting together the microstructure photomontage are appreciated. The assistance of Mr. Vince Prinzo in conducting several of the fretting fatigue tests is also appreciated.

## References

[1]    Hoeppner, D.W. and Goss, G.L., "A Fretting-Fatigue Threshold Concept," *Wear*, 27, 1974, pp. 61-70.

[2]    Lanning, D., Haritos, G. K., and Nicholas, T., "Notch Size Effects in HCF Behavior of Ti-6Al-4V," submitted to the *International Journal of Fatigue*, 1998.

[3]    Bellows, R. S., Muju, S., and Nicholas, T., "Validation of the Step Test Method for Generating Haigh Diagrams for Ti-6Al-4V," submitted to the *International Journal of Fatigue*, June 1998.

[4]    Maxwell, D. C. and Nicholas, T., "A Rapid Method for Generation of a Haigh Diagram for High Cycle Fatigue," submitted to *Fatigue and Fracture Mechanics: 29th Volume, ASTM STP 1321*, T. L. Panontin and S. D. Sheppard, Eds. American Society of Testing and Materials, Philadelphia, 1998.

[5]    Morrissey, R. J., McDowell, D. L., and Nicholas, T., "Frequency and Stress Ratio Effects in High Cycle Fatigue of Ti-6Al-4V," submitted to the *International Journal of Fatigue*, June 1998.

[6]    Cortez, R., Mall, S., and Calcaterra, J., "Investigation of Variable Amplitude Loading on Fretting Fatigue Behavior of Ti-6Al-4V," submitted to the *International Journal of Fatigue*, July 1998.

[7]    Hills, D.A. and Nowell, D., *Mechanics of Fretting Fatigue*. Kluwer Academic Publishers, Dordrecht, The Netherlands, 1994.

[8]    Goss, G.L. and Hoeppner, D.W., "Normal Load Effects in Fretting Fatigue of Ti and Al Alloys," *Wear*, 27, 1974, pp. 153-159.

[9]    Endo, K., Goto, H., and Nakamura, T., "Effects of Cycle Frequency on Fretting Fatigue Life of Carbon Steel," *Bulletin of the Japanese Society of Mechanical Engineers*, 12 (54), 1969, pp. 1300-1308.

[10]    Dobromirski, J.M., "Variables of Fretting Process: Are There 50 of Them?," *Standardization of Fretting Fatigue Test Methods and Equipment, ASTM STP 1159*, M. Helmi Attia and R.B. Waterhouse, Eds., American Society for Testing and Materials, Philadelphia, PA, 1992, pp. 60-66.

[11]    Szolwinski, M.P., Harish, G., and Farris, T.N., "Comparison of Fretting Fatigue Crack Nucleation Experiments on Multiaxial Fatigue Theory Life Predictions," *Proceedings of the American Society of Mechanical Engineers Symposium on High Cycle Fatigue*, Volume 55, 1997, pp. 449-457.

[12]    Alic, J.A., Hawley, A.L., and Urey, J.M., "Formation of Fretting Fatigue Cracks in 7075-T7351 Aluminum Alloy," *Wear*, 56, 1079, pp.351-361.

[13]    Waterhouse, R.B. "Theories of Fretting Processes," *Fretting Fatigue*, R.B. Waterhouse, Ed., Applied Science Publishers, London, 1981, pp. 203-219.

[14]    Goss, G.L. and Hoeppner, D.W., "Characterization of Fretting Fatigue Damage by SEM Analysis," *Wear*, 24, 1973, pp. 77-95.

Sung-Keun Lee,[1] Kozo Nakazawa,[2] Masae Sumita,[2] and Norio Maruyama[2]

## Effects of Contact Load and Contact Curvature Radius of Cylinder Pad on Fretting Fatigue in High Strength Steel

**REFERENCE:** Lee, S.-K., Nakazawa, K. Sumita, M., and Maruyama, N., **"Effects of Contact Load and Contact Curvature Radius of Cylinder Pad on Fretting Fatigue in High Strength Steel,"** *Fretting Fatigue: Current Technology and Practices, ASTM STP 1367,* D. W. Hoeppner, V. Chandrasekaran, and C. B. Elliott, Eds., American Society for Testing and Materials, West Conshohocken, PA, 2000.

**ABSTRACT:** Fretting fatigue of high strength steel was studied at various contact loads using bridge-type fretting cylindrical pads with various contact curvature radii. The fretting fatigue life showed a minimum at a certain contact load irrespective of the pad radius even though the frictional force increased monotonously with contact load. At a high contact load, the fretting fatigue life remained almost unchanged with a change in pad radius. At a low contact load, however, the fretting fatigue life increased with an increase in pad radius even though the frictional force was independent of the radius.

**KEYWORDS:** fretting fatigue, contact load, cylindrical pad, fretting pad radius, frictional force, high strength steel

### Introduction

Fretting fatigue is controlled by many factors. Contact pressure is one of the most important factors, and hence, the effect of contact pressure on fretting fatigue has been studied by many researchers. However, fretting fatigue test has not yet been standardized, so that the results obtained so far show different phenomena depending on the test methods used by the researchers. Under the test geometry where fatigue specimens had gauge section of flat faces and fretting pads had flat faces (hereafter, this is termed the test geometry of flat/flat), most of the researchers found that the fretting fatigue life or strength decreased monotonously with an increase in contact pressure [1-6]. A few researchers, however, found that the fretting fatigue life showed a minimum at a certain contact pressure [7,8]. The present authors found a singular phenomenon that under a certain test condition, the fretting fatigue life exhibited a minimum and a maximum with an increase in contact pressure [9,10]. Under the test geometry where cylinders were used instead of flat

---

[1]Professor, Department of Metallurgical Engineering, Dong-A University, 840 Hadan-dong, Saha-gu, Pusan, 604-714 Korea.
[2]Senior Researcher, Research Team Leader (retired), and Senior Researcher, respectively, National Research Institute for Metals, 1-2-1 Sengen, Tsukuba, 305-0047 Japan.

fretting pads (hereafter, this is termed the test geometry of flat/cylinder), similar phenomena that the fretting fatigue life or strength decreased with the increase in contact pressure [11-13] or showed a minimum at a certain contact pressure [14,15] were found. It is necessary to obtain a fundamental and consistent understanding of the phenomenon as to the effect of contact pressure on fretting fatigue.

This paper aims to make clear experimentally the effects of contact pressure and radius of cylindrical pad on fretting fatigue in high strength steel. Obtained results are discussed with various mechanisms proposed.

## Experimental Procedure

The material used was a quenched and tempered SNCM439 steel. Table 1 shows the chemical composition of the steel. The steel was quenched and tempered in the following sequence: heated to 1123 K for 1 h then oil quenched: tempered at 873 K for 1 h then air cooled. Mechanical properties of the steel are shown in Table 2. Fatigue data for the present steel are shown elsewhere [16]. The fretting fatigue device is shown in Fig. 1. Dimensions of the fatigue specimen and the fretting pad are shown in Fig. 2. Bridge-type fretting pads of the same material as the fatigue specimen were used. Pads having a span length of 8 mm and contact radii of 15, 30 and 60 mm and of infinite length (flat) were used along with pads with a span length of 12 mm and a contact radii of 30 mm and infinite length (flat) to investigate the effect of span length of the pad on fretting fatigue. The thickness of all the pads was 12 mm, however, contact length was 6 mm. The gauge parts of the fatigue specimens and the fretting pads were polished with 0-grade emery paper and then degreased with acetone.

The fretting fatigue facility was the same one used in the previous paper [10]. The fretting fatigue tests were performed on a 100 kN capacity closed loop electro-hydraulic fatigue testing machine. A constant normal pad load was applied by a small hydraulic actuator. The contact load was maintained below 2 kN. The tests were carried out using a sinusoidal wave form at a frequency of 20 Hz, at a stress ratio of 0.1 in laboratory air of 40 to 70% relative humidity at room temperature. An axial stress amplitude of 220 MPa was used throughout the present study. Frictional force between the fatigue specimen and the pad was measured using strain gauges bonded to the side of the central part of the pad.

Table 1- *Chemical composition of SNCM439 steel (mass%).*

| C | Si | Mn | P | S | Cu | Ni | Cr | Mo |
|------|------|------|-------|-------|------|------|------|------|
| 0.40 | 0.29 | 0.77 | 0.020 | 0.022 | 0.06 | 1.64 | 0.81 | 0.16 |

Table 2- *Mechanical properties of SNCM439 steel.*

| 0.2% P.S. ( MPa ) | U.T.S. ( MPa ) | El. ( % ) | R.A. ( % ) | Vickers H. |
|-------------------|----------------|-----------|------------|------------|
| 910 | 1000 | 20 | 66 | 330 |

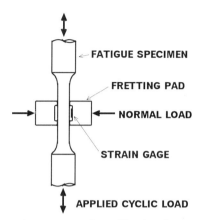

Fig. 1- *Schematic representation of fretting fatigue test.*

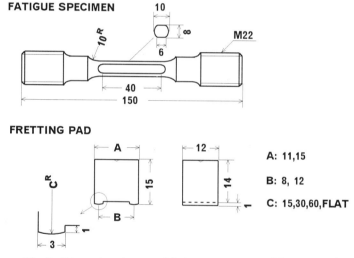

Fig. 2- *Dimensions in mm of fatigue specimen and fretting pad.*

The relative slip amplitude between the specimen and the outer edge of the pad foot was measured using calibrated extensometers. The frictional force and the relative slip amplitude were continuously recorded by a personal computer system during the test, and the data were analyzed after the tests.

## Results

*Contact pressure*

Under the test geometry of flat/flat, a contact pressure means nominal contact pressure that is calculated by dividing contact load by apparent contact area. It is assumed that the distribution of the contact pressure is uniform over the whole contact area. This nominal contact pressure is unchanged with repeated fretting cycles as long as a constant contact load is applied even though cyclic wear occurs at the fretted area since the apparent contact area is usually unchanged. Whereas, under the test geometry of flat/cylinder, the contact pressure is maximum at the central portion and zero at the edges of the contact area. The contact area is sometimes increased by cyclic wear with repeated fretting cycles. As described in the section of discussion, actual contact breadth observed in the present study is larger than that calculated by the Hertz equation [17], probably due to plastic deformation and wear at the fretted area. The increase in contact breadth has already been reported by a few researchers [11,13]. Thus, the meaning of the calculated maximum contact pressure is obscure. The change of contact load, however, corresponds qualitatively to that of contact pressure, although their physical meaning is different. Therefore, as a measure of normal force component in fretting fatigue, the contact load is used instead of the calculated contact pressure.

*Fretting fatigue life*

The effect of contact load on fretting fatigue life for the pad span length of 8 mm with various radii is shown in Fig. 3. The number of cycles to failure at a contact load of zero corresponds to the plain fatigue life without fretting, and is beyond $10^7$ cycles. With an increase in contact load, the fretting fatigue life decreases sharply for all the pads. At a

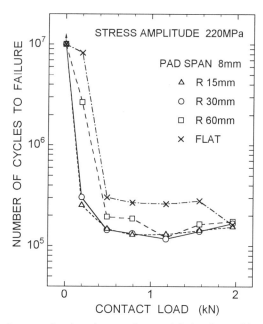

Fig. 3-*Effect of contact load on fretting fatigue life for the pad length of 8 mm.*

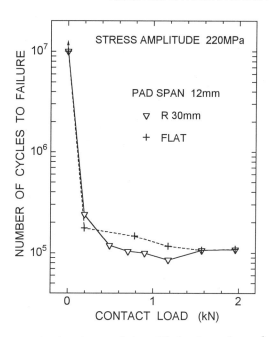

Fig. 4-*Effect of contact load on fretting fatigue life for the pad span length of 12 mm.*

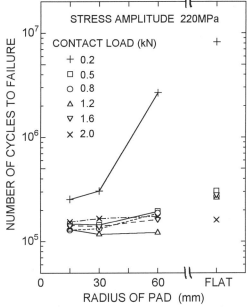

Fig. 5-*Effect of pad radius on fretting fatigue life for the pad span length of 8 mm.*

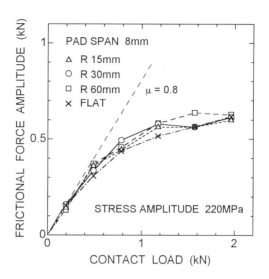

Fig. 6-*Relations between contact load and frictional force amplitude for the pad span length of 8 mm.*

contact load of more than 0.4 kN, the life changes gradually. When the pad radius is 15 to 60 mm, the fretting fatigue life shows a minimum at a contact load of 0.8 to 1.2 kN. When the pad radius is infinite, that is, the contact surface of the pad is flat, the fretting fatigue life shows a slight minimum at a contact load of 1.2 kN and a maximum at a contact load of 1.6 kN. The effect of contact load on fretting fatigue life for the pad span length of 12 mm is shown in Fig. 4. When the pad radius is 30 mm, the fretting fatigue life shows a minimum at a contact load of 1.2 kN. However, when the contact surface of the pad is flat, the fretting fatigue life decreases monotonously without showing a minimum or a maximum. It is also noted that the fretting fatigue life for the pad span length of 12 mm is shorter than that for the pad span length of 8 mm at the higher contact load. The effect of pad radius on fretting fatigue life under a given contact load is shown in Fig. 5 for the pad span length of 8 mm. The data for the flat pad are also included. When the contact loads are 0.5 to 2 kN, the fretting fatigue lives remain almost unchanged with an increase in the pad radius. When the contact load is 0.2 kN, the fretting fatigue life increases remarkably with the pad radius.

*Frictional force*

Frictional force between the fatigue specimen and the pad varied with the number of cycles. In fretting fatigue, the crack initiation and the acceleration of crack growth usually occurs after $10^4$ to $10^5$ cycles [3,12,18-20]. Hence, a frictional force amplitude determined around $10^4$ cycles was used. Relations between contact load and frictional force amplitude are shown in Figs. 6 and 7 for the pad span lengths of 8 and 12 mm, respectively. For the

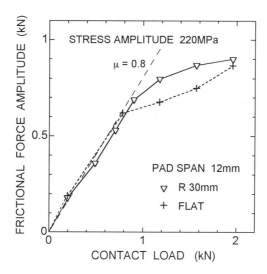

Fig. 7-*Relations between contact load and frictional force amplitude for the pad span length of 12 mm.*

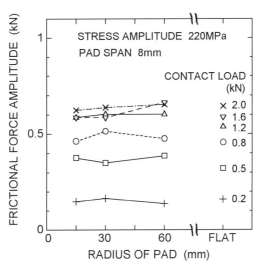

Fig. 8-*Relations between pad radii and frictional force amplitude for the pad span length of 8 mm.*

pad span length of 8 mm in Fig. 6, the frictional force amplitudes increase almost linearly at contact loads below 0.5 kN. At contact loads of more than 0.8 kN, the rate of increase drops and the frictional force amplitudes appear saturated irrespective of the pad radius. For the pad span length of 12 mm in Fig. 7, the increase in frictional force amplitudes with contact load is linear up to a contact load of 1 kN, and the rate of increase drops at contact loads of more than 1.2 kN. Concerning the effect of pad span length on frictional force, the frictional force amplitude is almost independent of the pad span length at low contact loads. At high contact loads, however, the frictional force amplitude is higher for the pad span length of 12 mm than for the pad span length of 8 mm. The coefficient of friction $\mu$ is defined by the relation, $\mu = F/P$, where $F$ and $P$ are frictional force amplitude and contact load, respectively. Dashed lines in Figs. 6 and 7 represent the relation of $\mu = 0.8$. At the lower contact load where gross slip occurs at the contact area, the coefficient of friction is approximately 0.8. At the higher contact load where both of the stick and slip regions exist at the contact area, the coefficient of friction is smaller than 0.8, and the higher the contact load, the smaller the coefficient of friction. Relations between pad radii and frictional force amplitude under a given contact load are shown in Fig. 8 for the pad span length of 8 mm. The frictional force amplitude is independent of the pad radii for all the contact loads.

*Relative slip amplitude*

Relations between contact load and relative slip amplitude for the span lengths of 8 and 12 mm are shown in Fig. 9. The relative slip amplitude varied with the number of cycles under a given contact load since the frictional force also varied during the test. The relative slip amplitude measured around $10^4$ cycles was used. The relative slip amplitude at a contact load of zero represents the value calculated by assuming that the pad is rigid. With an increase in contact load, the relative slip amplitude tends to decrease although each set of data does not show a smooth change. The relative slip amplitude for the pad span length of 12 mm also has a tendency to be larger than that for the pad span length of 8 mm.

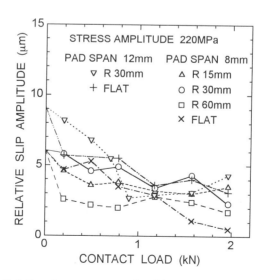

Fig. *9-Relations between contact load and relative slip amplitude.*

## Discussion

*Effect of contact load on fretting fatigue*

Under the test geometry of flat/cylinder, Waterhouse [14] reported the effect of maximum contact pressure on fretting fatigue life under a fixed breadth of contact using alpha brass and aluminum alloy. The radii of cylindrical pads were varied in order that the contact area might have a fixed breadth. From a knowledge of the Hertz equation for the breadth of contact, it was possible to calculate the radii of cylindrical pads and the loading conditions to give a maximum contact pressure in the contact region with a fixed breadth of contact. A minimum in life was observed at a certain maximum contact pressure in aluminum alloy, but not in alpha brass. He obtained a fretting fatigue limit for the aluminum alloy by the two stage type of test and showed that the lower values of maximum contact pressure were more efficient in initiating a fatigue crack although he did not mention the reason for this. Nishioka and Hirakawa [11] reported the effect of maximum contact pressure on fretting fatigue in a carbon steel using various radii of cylindrical pads under a constant clamping load. The maximum contact pressure decreased with an increase in pad radius under a constant clamping load as shown from the Hertz equation. They showed that fatigue strength based on fracture decreased at first with contact pressure and became nearly constant at high contact pressure and that the fatigue strength based on crack initiation decreased with contact pressure. They thought that the crack initiation was controlled by a tangential friction force at the contact area and the decrease in fretting fatigue strength was caused by the increase in tangential force. They also reported that the width of contact area increased with the number of cycles due to fretting wear. Del Puglia et al [15] reported fretting fatigue strength in steel at contact load range of 0.4 to 1 kN and showed that the fretting fatigue strength was lowest at a contact load of 0.6 kN. Satoh et al [13] reported the effect of contact pressure on fretting fatigue in turbine rotor steels using test geometries of flat/flat and flat/cylinder. They showed that the configuration of the cylindrical pad was changed by wear and the contact situation became similar to the situation under the test geometry of flat/flat and the actual contact width was larger than that calculated by the Hertz equation. The mean contact pressure calculated using the actual contact width was more useful than the maximum contact pressure for the evaluation of fretting fatigue properties under the flat/cylinder geometry. Fretting fatigue strength or life decreased with an increase in mean contact pressure and attained a constant value at mean contact pressures of more than 100 MPa. As a reason for this result, they thought that the frictional stress became saturated with an increase in contact pressure. In some cases, fretting fatigue strength tended to increase with a further increase in mean contact pressure. For this phenomenon, they thought that the contact pressure itself gave a static compressive stress near the crack front and caused suppression of crack growth [4,5]. Fernando et al [8] reported the influence of contact load on crack growth and fretting fatigue life in 4 percent copper-aluminum alloy using the test geometry of flat/flat. They showed that a critical value of contact load existed where the fretting fatigue life attained a minimum value, and postulated that the increase in life at high contact pressures was caused by the retardation of crack growth produced by crack closure due to the high compressive contact load.

In the present study, the fretting fatigue life under the test geometry of flat/cylinder shows a minimum at a contact load of 0.8 to 1.2 kN irrespective of the pad radius as shown in Figs. 3 and 4. The frictional force amplitude which is generally thought to control crack initiation and growth increases monotonously with an increase in contact load as shown in Figs. 6 and 7. Hence, the minimum fretting fatigue life phenomenon cannot be explained directly by the change in frictional force amplitude. A possible mechanism to explain the phenomenon is the concept of crack growth retardation due to a high static compressive stress near the crack front or due to crack closure effects at high contact loads as described above [4,5,8,13]. Another possible mechanism is a concept of local stress concentration at the contact area in which the situation of fretting damage is considered in terms of contact pressure and relative slip amplitude [9,10]. The effect of contact load involves the relative slip amplitude effect since the relative slip amplitude decreases with an increase in contact load. A fretting map in terms of two variables, such as these two factors proposed by Vingsbo and Sörderberg [21] contributes to an understanding of fretting damage. Zhou et al [22] reported a fretting map which included crack initiation. Their map was based on fretting wear tests of aluminum alloys using curved samples on flat contacts. They showed that longer cracks were initiated for a given number of cycles at an intermediate or mixed fretting regime between the stick regime and the slip regime. They pointed out that the cracks were nucleated mainly due to overstressing in the mixed fretting regime, although they did not show the mechanism to cause overstressing. Nakazawa et al [9,10] reported the effects of contact pressure and relative slip amplitude on fretting fatigue in a high strength steel using the test geometry of flat/flat. They made a fretting fatigue life map in terms of these two factors. They found a singular phenomenon that under a certain test condition, the fretting fatigue life exhibited a minimum and a maximum with an increase in contact pressure. They also found that the fretting fatigue life under a given contact pressure exhibited a minimum at a certain relative slip amplitude. The fretting fatigue life map indicated that these two minimum life conditions in terms of relative slip and contact pressure were the same and consequently the underlying mechanisms were the same. On the basis of experimental observations on the situation of fretting damage, they thought that the minimum life was probably caused by stress concentration at a narrow stick region of the fretted area, and that the maximum life observed at the higher contact pressure was a transitional phenomenon caused by the occurrence of minimum life. The situation of partial slip with narrow stick region corresponded probably to the mixed fretting regime described above.

Figure 10 shows a fretted surface near the initiation site of fracture in a specimen fractured using a pad with the span length of 8 mm and radius of 15 mm. Contact breadth $2a$ is given by the equation, $2a=[4RN(1-v^2)/\pi E]^{1/2}$, where $E$, $v$, $R$ and $N$ are the Young's modulus and the Poisson's ratio of the material, the radius of cylindrical pad and the contact load per unit length, respectively [17]. The contact breadth calculated using material constants, $E$=208 GPa and $v$=0.30, is 0.24 mm at a contact load of 0.5 kN and 0.48 mm at a contact load of 2 kN. Actual contact breadth observed in Fig. 10 is approximately 0.6 mm at 0.5 kN and 0.9 mm at 2 kN, that is, two or three times as large as the calculated one. Similar increase in contact breadth has already been reported [11,13]. At a contact load of 2 kN, a relatively wide stick region develops at the middle portion of the fretted area. The stick region remains as-polished. At a contact load of 0.5 kN, gross

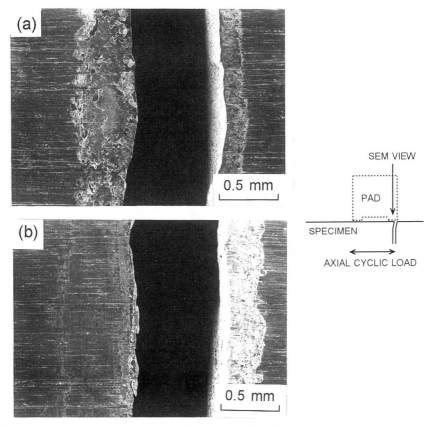

Fig.10-*Scanning electron micrographs of fretted surface near the initiation site of fracture in a specimen fractured using a pad with the span length of 8 mm and radius of 15 mm. The contact loads are (a) 0.5 kN and (b) 2 kN.*

slip occurs, and the as-polished surface almost disappears. The occurrence of gross slip can also be judged from the observation of the wave form of frictional force. A critical contact load below which gross slip occurs is approximately 0.5 kN for the pad span length of 8 mm and 0.9 kN for the pad span length of 12 mm irrespective of the pad radius. These critical values correspond well to the deflection points from the linear portion at low contact loads in the curves shown in Figs. 6 and 7. In Figs. 3 and 4, the minimum fretting fatigue life is observed at a contact load slightly higher than these critical loads where partial slip with narrow stick region probably exists. This suggests that the minimum life is closely related to the latter mechanism of local stress concentration at contact area.

Under the test geometry of flat/flat in the present study, the fretting fatigue life for the pad span length of 8 mm shows a slight minimum and a maximum as shown in Fig. 3. Whereas, the fretting fatigue life for the pad span length of 12 mm decreases monotonously with contact load as shown in Fig. 4. The relative slip amplitude for the span length of 12 mm is larger than for the span length of 8 mm. The above results correspond to those obtained in the previous paper [10] in which the fretting fatigue life exhibits a minimum and a maximum when the relative slip amplitude is small, while it decreases monotonously with an increase in contact pressure when the relative slip amplitude is large. It is impossible to explain the result in Fig. 3 by the mechanism of crack growth retardation at high contact load. To explain this result, the mechanism of local stress concentration at the fretted area described above should be considered. Whether a significant influence of local stress concentration exerts itself on fretting fatigue life or not depends probably on test conditions.

*Effect of cylindrical pad radius on fretting fatigue*

Waterhouse [14], Hills et al [23], Satoh et al [13], and Szolwinski and Farris [24] reported fretting fatigue behavior using various cylindrical pad radii. However, the effect of the pad radius on fretting fatigue life was not shown directly. Nishioka and Hirakawa showed that fretting fatigue strength of a carbon steel under a contact load of 0.72 kN was nearly constant in the range of pad radii from 15 to 60 mm [11]. In the present study, the fretting fatigue lives under contact loads of 0.5 to 2 kN remain almost unchanged with the increase in pad radius as shown in Fig. 5. This result corresponds well to their result. However, when the contact load is 0.2 kN, the fretting fatigue life increases remarkably with the pad radius . The frictional force amplitude remains nearly constant irrespective of the pad radius as shown in Fig. 8. According to the Hertz equation, the contact width of pad is increased and the maximum contact pressure is decreased with the increase in pad radius. The increase in fretting fatigue life with an increase in pad radius may be caused by the decrease in contact pressure. However, a more in depth study is needed to explain the reason why the increase in life occurs only at a contact load of 0.2 kN.

**Conclusions**

The following conclusions can be drawn from the present study.
(1) The fretting fatigue life shows a minimum at a certain contact load irrespective of the pad radius even though the frictional force increases monotonously with the increase in contact load.
(2) At a high contact load, the fretting fatigue life remains almost unchanged with a change in pad radius. At a low contact load, however, the fretting fatigue life increases with an increase in pad radius even though the frictional force is independent of the radius.

**References**

[1] Gaul D. J., and Duquette D. J., "The Effect of Fretting and Environment on Fatigue Crack Initiation and Early Propagation in a Quenched and Tempered 4130 Steel", *Met. Trans.*, Vol. 11A, 1980, pp. 1555-1561.

[2] Satoh T., Mutoh Y., Yada T., Takano A., and Tsunoda E., "Effect of Contact Pressure

on High Temperature Fretting Fatigue", *J. Soc. Mater. Sci., Japan*, Vol. 42, 1993, pp. 78-84 (in Japanese).

[*3*] Sato K., and Fujii H., "Fretting-Fatigue Strength and Fracture Morphology of Carbon Steel S45C", *J. Japan Soc. Strength Frac. Mater.*, Vol. 18, 1984, pp. 98-113 (in Japanese).

[*4*] Nagata K., Matsuda T., and Kashiwaya H., "Effect of Contact Pressure on Fretting Fatigue Strength", *Trans. Japan Soc. Mech. Engnrs*, Vol. 53, 1987, pp. 196-199 (in Japanese).

[*5*] Mutoh Y., Nishida T., and Sakamoto I., "Effect of Relative Slip Amplitude and Contact Pressure", *J. Soc. Mater. Sci., Japan*, Vol. 37, 1988, pp. 649-655 (in Japanese).

[*6*] Adibnazari S., and Hoeppner D. W., "A Fretting Fatigue Normal Pressure Threshold Concept", *Wear*, Vol. 160, 1993, pp. 33-35.

[*7*] Switek W., "Fretting Fatigue Strength of Mechanical Joints", *Theoret. Appl. Frac. Mech.*, Vol. 4, 1985, pp. 59-63.

[*8*] Fernando U. S., Farrahi G. H., and Brown M. W., "Fretting Fatigue Crack Growth Behaviour of BS L65 4 percent Copper Aluminum Alloy under Constant Normal Load", *Fretting Fatigue, ESIS 18* (Edited by R. B. Waterhouse and T. C. Lindley), 1994, Mechanical Engineering Publications, London, pp. 183-195.

[*9*] Nakazawa K., Sumita M., and Maruyama N., "Effect of Contact Pressure on Fretting Fatigue of High Strength Steel and Titanium Alloy", *Standardization of Fretting Fatigue Test Methods and Equipment, ASTM STP 1159*, 1992, pp. 115-125.

[*10*] Nakazawa K., Sumita M., and Maruyama N., "Effect of Relative Slip Amplitude on Fretting Fatigue of High Strength Steel", *Fatigue Fract. Engng Mater. Struct.*, Vol. 17, 1994, pp. 751-759.

[*11*] Nishioka K., and Hirakawa K., "Fundamental Investigation of Fretting Fatigue (Part 6, Effects of Contact Pressure and Hardness of Materials)", *Bull. of the JSME*, Vol. 15, 1972, pp. 135-144.

[*12*] Endo K., and Goto H., "Initiation and Propagation of Fretting Fatigue Cracks", *Wear* Vol. 38, 1976, pp. 311-324.

[*13*] Satoh T., Mutoh Y., Nishida T., and Nagata K., "Effect of Contact Pad Geometry on Fretting Fatigue Behavior", *Trans. Japan Soc. Mech. Engnrs*, Vol. 61, 1995, pp. 1492-1499 (in Japanese).

[*14*] Waterhouse R. B., "The Effect of Clamping Stress Distribution on the Fretting Fatigue of Alpha Brass and Al-Mg-Zn Alloy", *Trans. Am. Soc. Lubric. Engnrs*, Vol. 11, 1968, pp. 1-5.

[*15*] Del Puglia A., Pratesi F., and Zonfrillo G., "Experimental Procedure and Parameters Involved in Fretting Fatigue Tests", *Fretting Fatigue, ESIS 18* (Edited by R. B. Waterhouse and T. C. Lindley), 1994, Mechanical Engineering Publications, London, pp. 219-238.

[*16*] Nakazawa K., Takei A., Kasahara K., Ishida A., and Sumita M., "Effect of Ni-TiC Composite Film Coating on Fretting Fatigue of High Strength Steel", *J. Japan Inst. Metals*, Vol. 59, 1995, pp. 1118-1123 (in Japanese).

[*17*] O'Connor J. J., "The Role of Elastic Stress Analysis in the Interpretation of Fretting Fatigue Failures", *Fretting Fatigue*, Edited by R. B. Waterhouse, 1981, Applied Science Publishers, London, pp. 23-66.

[*18*] Fenner A. J. and Field J. E., "La Fatigue dans les Conditions de Frottement", *Revue de Metallergie*, Vol. 55, 1958, pp. 475-485.

[*19*] Wharton M. H., Taylor D. E., and Waterhouse R. B., "Metallurgical Factors in the Fretting-Fatigue Behaviour of 70/30 Brass and 0.7% Carbon Steel", *Wear*, Vol. 23, 1973, pp. 251-260.

[*20*] Nakazawa K., Sumita M., Maruyama N., and Kawabe Y., "Fretting Fatigue of High Strength Steels for Chain Cables in Sea Water", *ISIJ International*, Vol. 29, 1989, pp. 781-787.

[*21*] Vingsbo O., and Sörderberg S.,"On Fretting Map", *Wear*, Vol. 126, 1988, pp. 131-147.

[*22*] Zhou Z. R., Fayeulle S., and Vincent L., "Cracking Behavior of Various Aluminum Alloys during Fretting Wear", *Wear*, Vol. 155, 1992, pp. 317-330.

[*23*] Hills D. A., Nowell D., and O'Connor J. J., "On the Mechanics of Fretting Fatigue", *Wear*, Vol. 125, 1988, pp. 129-146.

[*24*] Szolwinski M. P., and Farris T. N., "Observation, Analysis and Prediction of Fretting Fatigue in 2024-T351 Aluminum Alloy", *Wear*, Vol. 221, 1998, pp. 24-36.

Bettina U. Wittkowsky,[1] Paul R. Birch,[2] Jaime Domínguez,[3] and Subra Suresh[2]

An Experimental Investigation of Fretting Fatigue with Spherical Contact in 7075-T6 Aluminum Alloy

REFERENCE: Wittkowsky, B. U., Birch, P. R., Dominguez, J., and Suresh, S., "An Experimental Investigation of Fretting Fatigue with Spherical Contact in 7075-T6 Aluminum Alloy", *Fretting Fatigue: Current Technology and Practices, ASTM STP 1367*, D. W. Hoeppner, V. Chandrasekaran, and C. B. Elliott, Eds., American Society for Testing and Materials, West Conshohocken, PA, 2000.

ABSTRACT: A new fretting fatigue loading system which facilitates continuous monitoring and control of such parameters as normal and tangential contact loads, and cyclic displacements in addition to the specimen fatigue loading parameters has been designed and built. Systematic and controlled experiments have been carried out with a spherical-tip contact pad fretting against a flat surface. The pad and specimen materials were both 7075-T6 aluminum alloy. For this pad geometry, an analysis of the mechanics of the fretting test has been carried out. In addition, the influence of a number of fretting parameters on the fatigue lifetime has been experimentally investigated. The geometry of fretting scars and the stick-slip annuli has been established for the different contact conditions employed. An approach has been developed to determine the eccentricity produced by the axial load on the stick zone, which correlates very well with the experimental results. Metallographic and fractographic examinations have also been performed on the fretted specimens.

KEYWORDS: fretting fatigue, fatigue testing, fretting device, contact stresses

## Introduction

Fretting fatigue denotes the detrimental effect on a material's fatigue properties arising from the cyclic sliding of two contacting surfaces with small relative displacements between the surfaces. In addition to local stress fields in the vicinity of contact, one or both of the components may be subject to bulk stresses caused by cyclic loads. Its result is the nucleation of many cracks in the process zone. These cracks may produce the deterioration of the surface by spallation. They also may grow until final fracture occurs across one of the elements under contact. Locations where fretting fatigue is observed in service include

[1]Institute for Materials Research, GKSS Research Center, D-21502 Geesthacht, Germany.
[2]Graduate student and Professor, respectively, Department of Materials Science and Engineering, Massachusetts Institute of Technology, Cambridge, MA 02139, USA.
[3]Professor, Departamento de Ingeniería Mecánica, ESI, Universidad de Sevilla, Camino de los Descubrimientos s/n, 41092 Sevilla, Spain.

213

riveted and bolted joints, shrink-fitted couplings, metal ropes and cables, coil wedges in generator rotors, and the blade-dovetail contact sections in gas turbine engines [1]. Various investigators have undertaken experimental, analytical, and computational studies of fretting fatigue since it was reported for the first time in 1911 [2], but much about the phenomenon remains unknown.

Fretting fatigue crack nucleation and growth is a complicated phenomenon. A number of factors influence fretting fatigue behavior, including the contact pressure, the amplitude of relative slip, tangential tractions at the surface, contact materials, friction coefficient, surface conditions and environment. In the past, different geometries, producing very different stress fields, have been used to experimentally investigate the process [3-10]. Tests simulating real working conditions [6, 7] or using fretting bridges [8-10] are the most frequently used.

In tests with fretting bridges, the pads used have usually been planar, cylindrical or spherical. Different loading conditions have been considered. In some cases, only a constant tensile axial load was applied to the specimen simultaneously with a constant normal contact pressure and alternating tangential tractions applied to the fretting pad. In other cases, the axial load applied to the specimen has also been an alternating load, simulating more accurately the usual real working conditions. This variety of testing conditions and the complexity of the process have impeded a better understanding of this subject.

There is general agreement that fretting fatigue process must be studied considering at least two different phases [11, 12]: (i) the fatigue crack nucleation and early propagation through a zone close to the contact surfaces, where the stress field is multiaxial, dominated by the contact stresses, and with the stress components varying out of phase [13]; and (ii) long crack propagation through the bulk of the specimen, where the local contact conditions have a negligible effect. This second phase can be analyzed using the common procedures applied for long cracks under far-field applied loads (in the absence of fretting). The first phase, similar to the situation involving the fatigue of notched specimens, is important in high cycle fatigue. When the number of cycles to final failure is small, the importance of this phase decreases and the total life is only marginally affected by fretting. However, in high cycle fatigue, fretting has a predominant effect [5]. As a limiting case of high cycle fatigue, fretting may reduce the fatigue limit by a factor of two or three [14, 15].

In order to study the fretting fatigue mechanisms, crack nucleation and growth through the contact zone stress field can also be divided into two further phases. The first one includes the initial process until the first crack appears or until it advances through the first several grains. The second phase commences when the crack growth problem can be considered as a continuum mechanics problem, but highly influenced by the complex contact zone stress field variation.

The aim of this paper is to analyze some aspects of the nucleation and early growth of fretting fatigue cracks in 7075-T6 aluminum alloy and to determine the effect of different fretting parameters on the fatigue life. This is accomplished by analyzing the results of fretting fatigue tests carried out on specimens of this material under spherical-plane contact. These highly idealized conditions enable an analytical 3D-analysis of the

Table 1 - *Mechanical properties of the Al 7075-T6 extruded bars*

| UTS | $\sigma_{y0.2}$ | $E$ | Fatigue strength (Axial load; $R= -1$) MPa | | | $\Delta K_{th}$ ($R= -1$) |
|---|---|---|---|---|---|---|
| MPa | MPa | GPa | $10^6$ cycles | $10^7$ cycles | $10^8$ cycles | MPa m$^{-1/2}$ |
| 572 | 503 | 72 | 214 | 175 | 140 | 2.2 |

complex stress fields under the contact area during the fretting cycle, without the need to incorporate stress singularities at the edge of contact.

**Material and Experimental Procedure**

The two contacting bodies were made of Al 7075-T6. The spherical surfaces and plane specimens were machined from extruded bars with a diameter of 25.4 and 12.7 mm, respectively. Specimens were flat-sided plates with reduced rectangular gage sections. The radius of the spherical pads was 25.4 mm. The surfaces of the pads and the specimens were polished with a 0.5 μm alumina powder and then ultrasonically cleaned in pure alcohol. The main mechanical properties of pads and specimens are listed in Table 1. The microstructure of the material was quite irregular. The grains were elongated due to extrusion. The average grain sizes in the extrusion direction and perpendicular to it were 100 and 60 μm, respectively.

The fretting apparatus used was designed and constructed in conjunction with a servo-hydraulic uniaxial test system (Figure 1). The specimen is subjected to a variable axial load, $P$, through the actuator of the servohydraulic testing machine. The supports of the spherical pads are mounted on a compliant system and attached to the structure of the testing machine. Its compliance can be changed to allow the variation of the tangential load, $Q$, or the relative displacement between the specimen and pads, $\delta$, maintaining all other test parameters constant. The system also allows one to apply and measure independently the normal contact load, $N$, which is transmitted to the specimen through the spherical pads. A detailed description of the fretting fatigue apparatus is reported elsewhere [16].

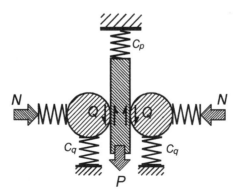

Figure 1 - *Principle sketch of the test set up. All springs are adjustable.*

The bulk stresses applied to the specimen were all well below the fatigue strength. The maximum contact pressure under the spheres never exceeded the yield strength of the material. The average friction coefficient, $\mu$, was found to be $\mu = 1.2$. The determination of the friction coefficient is described in detail in [16]. The value of $\mu$ is in good agreement with the values reported by other researchers [4].

In all tests, the frequency was f = 10 Hz, and the axial load ratio was R = -1. Three series of tests (labeled A to C) were conducted. (1) In series A, the influence of the ratio $Q/N$ on the fatigue life has been investigated. The amplitude of the tangential load and the bulk stress, $\sigma_{ax}$, were constant in all tests, $Q = \pm 15$ N and $\sigma_{ax} = \pm 83$ MPa. Normal contact load, $N$, was varied from test to test. (2) In test series B, the influence of decreasing contact stresses ($N$ and $Q$) on the fatigue life was investigated. The axial stresses and the ratio $Q/N$ were constant in all tests of this series. (3) In test series C, the influence of the axial bulk stress on the fatigue strength was investigated. In this series, the contact load conditions were constant for all tests: $N = 20$ N and $Q = \pm 15$ N. The axial stress was decreased from test to test until no failure of the specimens could be detected. Table 2 shows the test parameters for these three series.

## Contact Conditions under a Sphere-Plane Configuration: A Brief Review

In a sphere-plane configuration under a normal contact load, $N$, the contact area is circular, with a radius, $a$:

Table 2 - *Test parameters for series A to C.*

| Test Number | Normal Contact Load $N$ N | Tangential Contact Load $Q$ N | Axial Bulk Stress $\sigma_{ax}=P/A$ MPa | Number of Cycles to Failure[1] | Ratio $Q/N$ |
|---|---|---|---|---|---|
| A-1 | 30 | ± 15 | ± 85 | 480,000 | 0.5 |
| A-2 | 20.8 | ± 15 | ± 83 | 449,500 | 0.72 |
| A-3 | 15.6 | ± 15 | ± 85 | 395,000 | 0.96 |
| A-4 | 12.5 | ± 15 | ± 83 | 361,000 | 1.2 |
| B-1 | 20.8 | ± 15 | ± 83 | 449,500 | 0.72 |
| B-2 | 16 | ± 11.7 | ± 83 | 530,000 | 0.73 |
| B-3 | 13.9 | ± 10 | ± 83 | 803,000 | 0.72 |
| B-4 | 10.3 | ± 7.5 | ± 83 | 2,940,000 | 0.73 |
| B-5 | 12 | ± 9 | ± 83 | 2,680,000 nf | 0.75 |
| C-1 | 20 | ± 15 | ± 83 | 549,000 | 0.75 |
| C-2 | 20 | ± 15 | ± 70 | 516,000 | 0.75 |
| C-3 | 20 | ± 15 | ± 56 | 1,540,000 nf (600)[2] | 0.75 |
| C-4 | 20 | ± 15 | ± 63 | 2,940,000 nf | 0.75 |
| C-5 | 20 | ± 15 | ± 59 | 1,777,000 nf (500)[2] | 0.75 |

[1] nf means no failure after the indicated number of cycles.
[2] The length of the longest crack detected at the surface after testing is written in parenthesis (μm).

$$a = \left(\frac{3NR}{4E^*}\right)^{\frac{1}{3}},$$

(1)

where $R$ is the radius of the sphere, and

$$\frac{1}{E^*} = \frac{1-v_1^2}{E_1} + \frac{1-v_2^2}{E_2}.$$

(2)

Here $E_i$ and $v_i$ are Young's modulus and Poisson's ratio of the specimen ($i = 1$) and sphere ($i = 2$), respectively. The contact pressure distribution is Hertzian, with

$$p(r) = \frac{3N}{2\pi a^2}\sqrt{1 - \frac{r^2}{a^2}} \qquad\qquad 0 \le r \le a$$

(3)

When a tangential load, $Q$, is applied, no global sliding takes place if the maximum admissible friction force, $F_R = \mu N$, is larger than $Q$ ($F_R > Q$). If the specimen and pad materials have the same mechanical properties, the normal pressure distribution does not change and tangential tractions appear at the interface. The contact area is then divided into two zones (Figure 2): a ring shaped zone (B-A) where micro-slip occurs, and a central, circular zone (B-B), the stick-zone [17-19] which deforms with no relative motion. The radius $b$ for that circular area can be obtained from

$$\frac{b}{a} = \left(1 - \frac{Q}{\mu N}\right)^{\frac{1}{3}}$$

(4)

The tangential tractions produced at the contact zone can be represented by the equations

$$\tau_{xz} = \frac{3\mu N}{2\pi a^3}\left(a^2 - r^2\right)^{\frac{1}{2}} = \mu p(r) \qquad\qquad b \le r \le a$$

(5)

$$\tau_{xz} = \frac{3\mu N}{2\pi a^3}\left[\left(a^2 - r^2\right)^{\frac{1}{2}} - \left(b^2 - r^2\right)^{\frac{1}{2}}\right] \qquad\qquad r \le b$$

(6)

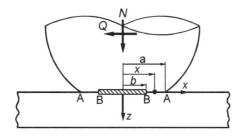

Figure 2 – *Scheme of the contact zone in plane y=0.*

When the tangential load, $Q$, reaches the value $\mu N$ ($Q \geq F_r$), the stick zone disappears and the entire contact area begins to slide.

Once the tangential and normal traction distributions are known as functions of $Q$ and $\mu N$, the elastic stresses produced at any point of the elements in contact can be obtained by using the solution of Hamilton and Goodman [20]. The explicit expressions for the stresses were subsequently obtained by Hamilton [21] and by Sackfield and Hills [22], which are easier to process.

Once $Q$ ceases to increase, whether before or after reaching a global sliding situation, sliding ceases at every point and the tangential traction distribution is that given by eqs. (5) and (6). If, from that point, $Q$ decreases to any new value, $Q^*$, a new sliding process begins in the reverse sense, and a new sliding ring is generated. Any point having slid previously will have to change the tangential traction from $\mu p(r)$ to $-\mu p(r)$ before reverse sliding. That is, the increase of tangential traction will have to be $-2\mu p(r)$. Thus, the size of the sliding zone and the increase of the tangential tractions are defined by

$$\frac{c}{a} = \left(1 - \frac{Q - Q^*}{2\mu N}\right)^{\frac{1}{3}},$$
(7)

$$\Delta\tau_{xz} = -\frac{3\mu N}{\pi a^3}\left(a^2 - r^2\right)^{\frac{1}{2}} = -2\mu p(r) \qquad c \leq r \leq a,$$
(8)

$$\Delta\tau_{xz} = -\frac{3\mu N}{\pi a^3}\left[\left(a^2 - r^2\right)^{\frac{1}{2}} - \left(c^2 - r^2\right)^{\frac{1}{2}}\right] \qquad r \leq c,$$
(9)

where $c$ is the radius of the new stick zone.

When a load $P$ is applied to the specimen, a tangential force $Q$ is induced (Figure 1). The value of $Q$ at any moment can be obtained from:

$$Q = \frac{C_p}{C_q}(P - 2Q),$$
(10)

where $C_p$ and $C_q$ are the compliance of the test machine system and the fretting pads support, respectively. If $Q < \mu N$, a mixed stick-slip situation is produced, but the effect of the bulk load applied to the specimen would be to modify the stick and slip zones, and the stress distribution under the contact zone.

To analyze the effect of the bulk load in cylindrical contacts, Hills and Nowell [10] considered the problem as a perturbation of the Mindlin problem (Figure 3). They obtained that the stick zone displaces an amount $e$, and that the tangential tractions distribution, $q(x)$, in that zone changes from

$$q(x) = q'(x) + q''(x) = -\mu p_0\sqrt{1 - \left(\frac{x}{a}\right)^2} + \mu p_0\left(\frac{c}{a}\right)\sqrt{1 - \left(\frac{x}{c}\right)^2}$$
(11)

when $P = 0$, to

$$q(x) = -\mu p_0 \sqrt{1 - \left(\frac{x}{a}\right)^2} + \mu p_0 \left(\frac{c}{a}\right)\sqrt{1 - \left(\frac{e-x}{c}\right)^2} \qquad (12)$$

when $P \neq 0$, where $p_0$ is the maximum pressure at the contact zone. That is, the second term, $q''(x)$, maintains its shape but moves with the stick zone by an amount $e$, which has the value

$$e = \frac{\sigma_{ax}a}{4\mu p_0} \qquad (13)$$

For spherical contact, a similar approach can be undertaken assuming that the bulk stress produces a perturbation of the Mindlin solution. We assume that the bulk stress produces a displacement, $e$, of the stick zone in the $x$ direction and that the tangential traction distribution changes from eqs. (5) and (6) to the modified distribution

$$\tau_{xz} = \frac{3\mu N}{2\pi a^3}\left(a^2 - r^2\right)^{\frac{1}{2}} = \mu p(r), \qquad r_e > b;\ r \leq a, \qquad (14)$$

$$\tau_{xz} = \frac{3\mu N}{2\pi a^3}\left[\left(a^2 - r^2\right)^{\frac{1}{2}} - \left(b^2 - r_e^2\right)^{\frac{1}{2}}\right], \qquad r_e \leq b, \qquad (15)$$

where

$$r_e = \sqrt{(x - e)^2 + y^2}. \qquad (16)$$

Within the stick zone we ensure that the surface particles do not slide (Figure 3). Initially, this condition can be relaxed to avoid sliding only in the $x$ direction. Then

$$\frac{\partial u_{x1}}{\partial x} - \frac{\partial u_{x2}}{\partial x} = 0 \qquad (17)$$

$$|q(r)| \leq \mu|p(r)| \qquad (18)$$

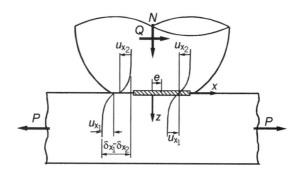

Figure 3 – *Scheme of the displacements at the contact zone.*

Where $u_{x1}$ and $u_{x2}$ represent the displacement of any point at the surface of the specimen and pad, respectively (Figure 3). The expressions of the displacements produced at the contact surfaces by the tangential tractions represented by eqs. (14) and (15) can be found elsewhere [23]. The displacement $u_{x1}$ is obtained adding the displacement produced by the bulk stress in the specimen (plane strain is assumed). Substituting the final values of $u_{x1}$ and $u_{x2}$ in expression (17), the expression for the displacement, $e$, of the stick zone can be obtained

$$e = \frac{1-v}{4-3v} \frac{4\sigma_a a}{\pi\mu p_0} \qquad (19)$$

Within the stick zone we must check whether the sliding in the $y$ direction produced by the stress distribution and eccentricity of the stick zone is zero or not. From the expression of $u_{y1}$ and $u_{y2}$ produced by the new displaced distribution of $\tau_{zx}$ and by the bulk stress [23], it is seen that:

$$u_{y1} - u_{y2} = \frac{v\sigma_{ax}}{2G} \frac{1-v}{4-3v} y \qquad (20)$$

and that

$$\frac{\partial u_{y1}}{\partial y} - \frac{\partial u_{y2}}{\partial y} \neq 0 \qquad (21)$$

This means that the surfaces will tend to slide perpendicularly to the $x$ axis. This causes a tangential traction distribution, $\tau_{yz}$, at the contact surface. This new component of the tractions makes some points close to the border of the stick zone reach the value $q(r)=\mu p(\mathrm{r})$, and slide, reducing the size of the stick zone. Due to the symmetry, $\tau_{yz}$ will tend to zero at $y=0$ (axis $x$), and then the effect of it on the stick zone size along this axis will be negligible. Nevertheless, it can be checked that off the axis $y=0$, but within the assumed stick zone, the relative displacements $u_y$ are at least one order of magnitude smaller than the relative displacements $u_x$. Thus, the tangential tractions $\tau_{yz}$ needed to eliminate this difference in displacement are also at least one order of magnitude smaller than $\tau_{zx}$, tending to zero close to the axis $y=0$.

Figure 4 shows a fretting scar produced in test number C-5. The white circles represent the results produced by equations (1), (4), and (19). The roughness of the surfaces and the high number of cycles made the stick and contact zone not perfectly circular. However, we can see that the approach fits the test results. Considering this result, in conjunction with the fact that the maximum stresses are produced at $y=0$, that cracks are usually nucleated at that axis, and also that $y=0$ is the axis where the error produced by this approach is the smallest, one may adopt this approach for the stress analysis of the tests performed experimentally.

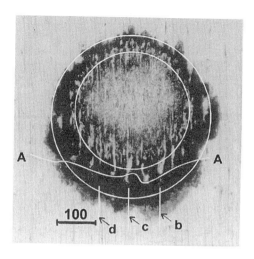

Figure 4 - *Fretting scar produced in test number C-5.*

## Test Results

Figure 5 shows the lives obtained in the three test series, A, B and C. Figure 5a shows the results for test series A. For the same value of Q, the variation of lives produced when $N$ changes is small. An increase of $N$ from 12.5 to 30 N, which makes $Q/N$ to decrease from 1.2 to 0.5, produces an increase of fatigue life of only 30%. For these test conditions, a straight line fit to the relationship between $Q/N$ and the fatigue life appears adequate.

Figure 5b represents the results of series B. In this case, as $Q$ changes its value in the same proportion to $N$, the effect of a reduction of $N$ will be an increment in fatigue life. But the variation in fatigue life with $N$ is not linear. In this case a reduction of $N$ from 20.8 to 16 (23%) enhances the number of cycles to failure by a factor of 1.18. But, a reduction from 13.9 to 10.3 N (26%) enhances the fatigue life by a factor of 3.7. From these two test series, it is inferred that the tangential contact tractions have a much higher effect on

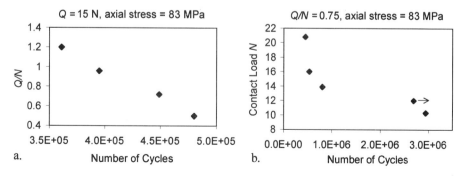

Figure 5 – *Tests result. Test series A (a). Test series B (b).* → *means no failure.*

fatigue life than $N$. Also, it is seen that these two contact tractions, normal and tangential, have opposite effects. In series A, the higher the $N$, i.e., smaller the $Q/N$ and the slip zone, the longer the fatigue life of the specimen. But this effect is small. In series B, the higher the $Q$ the shorter the fatigue life, with a very different effect depending on the level of $Q$.

Regarding the effect of the bulk stress, some data can be obtained from the test series C, although the analysis in this case is incomplete. Three tests (C3, C4 and C5) were stopped before failure, but some cracks were found when the surface was analyzed with the aid of microscopy. The number of cycles before stopping the test and the length of the cracks detected (half length at the surface between 100 and 300 μm) make it difficult to conclude whether the crack was stopped because the stress intensity factor increment was below the threshold or it was still growing. Thus, to obtain a good estimation of the bulk stress effect, it will be necessary to run at least three other tests under the same loading conditions, and some other new tests with bulk stresses between 83 and 63 MPa. In any case, test series C shows a strong effect of the bulk stress on the fatigue life: a 24% reduction in the bulk stress increases the fatigue life by a factor of 5.4.

Figure 6 shows some micrographs of nucleated cracks. Figure 6a shows two secondary cracks found in specimen A-3, at a distance between 50 and 100 μm from the fracture surface. The crack that eventually produced fracture was nucleated close to the border of the stick zone. The two cracks in the photograph were nucleated within the slip ring of the contact zone. These two cracks are between 30 and 60 μm long. Thus, considering the average grain size, which is 60 μm, they can be considered microstructurally short cracks. They are nucleated at an angle of between 20° and 30° from the surface and rotate gradually to an angle close to 70° during the first 20 to 30 μm of growth. In the first 10 μm of one of the cracks a small particle has been detached. This may be a consequence of two or more very closely nucleated cracks that merged after growing a few microns.

Figures 6b to 6d show three sections of the main crack produced in test C-5. The three sections are close to the symmetry axis of the contact zone, and so, close to the zone where the stresses are higher. The position of the planes of these sections relative to the fretting scar is represented by three lines (b to d) in Figure 4. In the same figure, a line (A-A) crossing the contact zone represents the intersection of the main crack with the specimen surface. The intersections of this line with the planes represent the local zone of the photographs in Figures 6b to 6d. The crack length at the surface is about 500 μm. The maximum crack depth is 150 μm at the center, and its shape is close to semielliptical. Many nucleated cracks close to the main crack can be found in these three sections. Most of them begin to grow at a small angle to the surface. Later, some of them rotate and grow in a direction close to perpendicular to the previous growth direction (Figures 6b and 6d). Some others rotate and grow in a direction close to 90° from the surface. Some cracks also intersect with other nucleated cracks producing the detachment of small particles. Finally, the main crack, which in this case is the only one longer than 10 to 15 μm, after growing about 20 μm, rotates and grows in a direction close to 70°-80° from the surface. The zone where cracks nucleated is about 60 to 80 μm long, which is close to the length of the slip ring along the axis $y = z = 0$.

To analyze this nucleation and growth process, an analysis of the stresses produced close to the contact zone during the fretting test has been carried out, and some multiaxial

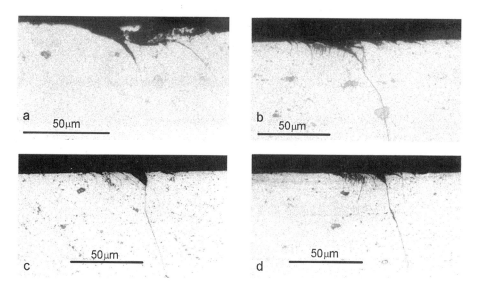

Figure 6 – *Cracks close to the nucleation site.*

fatigue criteria have been applied for the case of test C-5. Four multiaxial fatigue criteria have been applied: maximum amplitude of the tangential stress, $\Delta \tau_{max}/2$, where $\Delta \tau_{max}$ is the value of $\Delta \tau$ on the plane where it is maximum, the McDiarmid approach [24], the Crossland approach [25], and the Smith, Watson and Topper parameter [26]. The first three are related to the crack nucleation dominated by shear strains; the fourth is related to an nucleation process controlled by the maximum amplitude of the normal stresses. For each criterion, the appropriate characteristic parameter has been calculated. To check the possibility of fatigue crack nucleation, each of these parameters is compared to the value of the same parameter obtained for the stress conditions produced at the fatigue limit.

The expressions of the characteristic parameters for these criteria are as follows. For the McDiarmid criterion, the critical parameter is

$$\frac{\Delta \tau_{max}}{2} + \frac{t_1}{2\sigma_{TS}} \, \sigma_{max} \, , \tag{22}$$

where $\sigma_{max}$ is the maximum value of the normal stress on the $\Delta \tau_{max}$ plane and $\sigma_{TS}$ is the tensile strength. Its value at the fatigue limit is $t_1$. The Crossland parameter is written:

$$\sqrt{J_{2,a}} + \frac{J_{1max}}{3} \left( \frac{3t_1}{b_1} - \sqrt{3} \right) , \tag{23}$$

where $J_{2,a}$ and $J_{1max}$ are the amplitude of the second invariant of the deviatoric stress tensor and the maximum value of the first invariant, respectively, and $b_1$ is the bending fatigue limit. The value of the parameter at the fatigue limit for the previous criteria is the torsion fatigue limit, $t_1$. The values used for $b_1$ and $t_1$ in this multiaxial fatigue analysis, have been: $b_1 = 175$ MPa and $t_1 = 101$ Mpa. The Smith, Watson and Topper parameter (SWT) is:

$$SWT = \frac{\Delta\sigma_{max}}{2E}\,\sigma_{max1}\,, \tag{24}$$

where $\Delta\sigma_{max}$ is the maximum increment of normal stress and $\sigma_{max1}$ the maximum value of the normal stress on the plane where $\Delta\sigma_{max}$ is produced. At the fatigue limit, the value of this parameter is

$$SWT = \frac{b_1^2}{2E} \tag{25}$$

Figure 7 shows the equivalent stress parameters normalized by their values at the fatigue limit at plane $y=0$, for the two depths: $z=0$ (Figure 7a); and $z=10$ μm (Figure 7b), depth at which most secondary cracks have already rotated. Figure 8 shows, for the same plane and depths, the angles of the planes where the maximum amplitude of the shear stress ($\varphi \to \Delta\tau_{max}$) and normal stress ($\varphi \to \Delta\sigma_{max}$) are obtained. It is seen that the first three criteria give very similar results, predicting crack nucleation (parameter $\cong 1.6\text{-}1.8 > 1$) in the slip zone. Although the SWT criterion gives very different values, it also predicts crack nucleation in the same zone. At $z=10$ μm, the maximum values of all criteria are still higher than 1 and are produced below the central zone of the slip ring.

Looking at the main crack nucleation sites, it seems that for this material, the three criteria dominated by the shear strains are better than the SWT criterion. All those three criteria give very similar results. It would be interesting to compare these three criteria with experimental results obtained with other materials with a ratio $t_1/b_1$ different to $1/\sqrt{3}$. According to the equations defining the critical parameter of these three criteria, for ratios different from $1/\sqrt{3}$, the three criteria are expected to produce different results.

It can be seen in Figure 6 that all cracks are nucleated at angles close to the angles where the amplitude of the shear stress is maximum ($\varphi \to \Delta\tau_{max}$ in Figure 8). Many of them bend close to the surface ($z<10$ μm). Some of them rotate and grow perpendicular to the previous direction, which is also a direction of maximum $\Delta\tau$. Others, after rotation grow in a direction close to perpendicular to the contact surface. The rotation of the main crack (after growth of about 20 μm) is toward the direction which maximize the amplitude

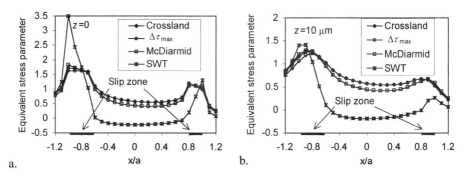

Figure 7 – *Multiaxial fatigue criteria applied to test C-5. Plane y=0. a.) At the surface. b.) At z=10 μm.*

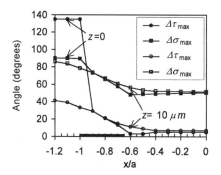

Figure 8 – *Angles of maximum increment of normal and tangential stresses.*

of the normal stress ($\varphi \rightarrow \Delta\sigma_{max}$ in Figure 8). This change in the direction of crack growth does not seem to have any relationship with the microstructure, which has an average grain size in that direction of close to 60 µm. An explanation of this rotation is that cracks cannot continue growing by shear strain because the spherical pad does not allow the motion of the crack faces outside the surface of the specimen [3]. Thus, growth continues in a direction perpendicular to the maximum normal stress amplitude because in this direction, the surface of the spherical pad will not produce any interference with the crack faces motion.

We note that the analysis presented in Figs. 7 and 8 deal exclusively with total life approaches to fatigue. It should be pointed out here that the information reported here, such as the crack trajectory in Fig. 6, could also be incorporated into models which deal with subcritical crack growth under fretting fatigue (e.g. [27, 28]. Specific examples of such approaches and a case study involving fretting fatigue in a turbogenerator can be found in [29].

**Conclusions**

Three series of fretting fatigue tests under spherical contact have been carried out to analyze the effects of different parameters on fatigue life of 7075-T6 aluminum alloy. The fretting fatigue process has been correlated with the variable bulk axial and contact loads applied to specimens. An approach which considers the effect of the eccentricity produced by the bulk axial stress on the stick zone has been developed. This approach is seen to predict the displacement of the stick zone reasonably well. Tests results gave evidence of the strong effect produced by the tangential load on fatigue life. These results also shown that if all other load parameters are maintained constant, the effect of the normal contact load is very small. A decrease of normal contact load produces a small increase in the fretting fatigue life.

From the analysis of the evolution of cracks during early stages, it can be seen that most of them nucleate at angles close to the plane of the maximum increment of the tangential stresses. Later on, some of them rotate by 90° to grow along the other plane of maximum increment of the tangential stresses. Very close to the surface, at a distance smaller than the characteristic microstructural size, the main crack rotates and continues growing along planes close to those where the maximum increment of the normal stress is produced.

## Acknowledgements

This work was supported by the Multi-University Research Initiative on "High Cycle Fatigue" which is founded at MIT by the Air Force Office of Scientific Research, Grant No. F49620-96-1-0478, through a subcontract from the University of California at Berkeley. BUW thanks GKSS Research Center for the opportunity to pursue this investigation at MIT. PRB thanks the National Science Foundation for a graduate fellowship in support of his studies at MIT. JD thanks the DGICYT and the Universidad de Sevilla for facilitating completion of this work at MIT.

## References

[1] Waterhouse, R. B. and Lindley, T. C., *Fretting Fatigue*, ESIS Publication No 18, European Structural Integrity Society. Mechanical Engineering Publications Limited, London, 1994.

[2] Eden, E. M., Rose, W. N. and Cunningham, F. L., "The endurance of metals," *Proc. of the Institution for Mechanical Engineers*, Vol. 4, 1911, pp. 839-974.

[3] Yamashita, N. and Mura, T., "Contact Fatigue Crack Initiation under Repeated Oblique Force," Wear, Vol. 91, 1983, pp. 235-250.

[4] Lamac, V., Dubourg, M. C. and Vincent, L., "A Theoretical Model for the Prediction of Initial Growth Angles and Sites of Fretting Fatigue Cracks," *Tribology International*, Vol. 30, 1997, pp. 391-400.

[5] Antoniou, R. A. and Radtke, T. C., "Mechanisms of Fretting-Fatigue of Titanium Alloys," *Materials Science and Engineering*, Vol. A237, 1997, pp. 229-240.

[6] Ruiz, C., Boddington, P.H.B. and Chen, K.C., "An Investigation of Fatigue and Fretting in a Dovetail Joint", *Experimental Mechanics*, Vol. 24, 1984, pp. 208-217.

[7] Zhou, Z. R., Goudreau, S., Fiset, M. and Cardou, A., "Single Wire Fretting Fatigue Tests for Electrical Conductor Bending Fatigue Evaluation", *Wear*, Vol. 181-183, 1995, pp. 537-543.

[8] Sato, K., "Damage Formation During Fretting Fatigue," *Wear*, Vol. 125, 1988, pp. 163-174.

[9] Nix, K. J. and Lindley, T. C., "The Application of Fracture Mechanics to Fretting Fatigue," *Fatigue and Fracture of Engineering Materials and Structures*, Vol 8, 1985, pp. 143-160.

[10] Hills, D. A. and Nowell, D., "The Development of a Fretting Fatigue Experiment with Well-Defined Characteristics," *Standardization of Fretting Fatigue Tests Methods and Equipments*, ASTM STP 1159, H. M. Attia and R. B. Waterhouse, Eds., American Society for Testing and Materials, 1992, pp. 69-84.

[11] Endo, K. and Goto, H., "Initiation and Propagation of Fretting Fatigue Cracks," *Wear*, Vol. 38, 1976, pp. 311-324.

[12] Sato, K. and Fujii, H., "Crack Propagation Behaviour in Fretting Fatigue," *Wear*, Vol. 107, 1986, pp. 245-262.

[13] Domínguez, J., "Cyclic Variation in Friction Forces and Contact Stresses During Fretting Fatigue," *Wear*, vol. 218, 1998, pp.43-53.

[14] Sato, K., Fujii, H. and Kodama, S., "Crack Propagation Behaviour in Fretting Fatigue of S45C Carbon Steel," *Bulletin of JSME*, Vol. 29, 1986, pp. 3253-3258.

[15] Hattori, T., Nakamura, M., Sakata, H. and Watanabe, T., "Fretting Fatigue Analysis Using Fracture Mechanics", *JSME International Journal*, Series 1, Vol. 31, 1988, pp. 100-107.

[16] Wittkowsky, B. U., Birch, P. R., Domínguez, J. and Suresh, S., "An Apparatus for Quantitative Fretting-Fatigue Testing", *Fatigue and Fracture of Engineering Materials and Structures*, Vol. 22, 1999, pp. 307-320.

[17] Cattaneo, C., "Sul Contatto di Due Corpi Elastici: Distribuzion Locale degli Sforzi," *Reconditi dell Accademia Nazionale dei Lincei*, Vol. 27, 1938, pp. 342–348, 434–436, 474–478.

[18] Mindlin, R. D., "Compliance of Elastic Bodies in Contact, *Journal of Applied Mechanics*, Vol. 16, 1949, pp. 259–268.

[19] Mindlin , R. D. and Deresiewicz, H., "Elastic Spheres in Contact Under Varying Oblique Forces," *Journal of Applied Mechanics*, Vol. 20, 1953, pp. 327–344.

[20] Hamilton, G. M. and Goodman, L. E., "The Stress Field Created by a Circular Sliding Contact," *Journal of Applied Mechanics*, Vol. 33, 1966, pp. 371–376.

[21] Hamilton, G. M., "Explicit Equations for the Stresses Beneath a Sliding Spherical Contact," *Proc. of the Institution of Mechanical Engineering*, Vol. 197C, 1983, pp. 53–59.

[22] Sackfield, A. and Hills, D., "A Note on the Hertz Contact Problem: a Correlation of Standard Formulae," *Journal of Strain Analysis*, Vol. 18, 1983, pp. 195–197.

[23] Johnson K.L., *Contact Mechanics*, Cambridge University Press, 1985.

[24] McDiarmid, D.L., "Mean Stress Effect in Biaxial Fatigue where the Stresses Are Out-of-phase and at Different frequencies," in *Fatigue under Biaxial and Multiaxial Loading*, K. Kussmaul, D. McDiarmid and D. Socie, Eds., ESIS 10, Mechanical Engineering Publications, 1991, pp. 321-335.

[25] Crossland, B., "Effect of Large Hydrostatic Pressures on the Torsional Fatigue Strength of an Alloy Steel," *Proc. of the International Conference on Fatigue of Metals*, Institution of Mechanical Engineering, London, 1956, pp.138-149.

[26] Smith, K. N., Watson, P. and Topper, T. H., "A Stress-Strain Function for Fatigue of Metals", *Journal of Materials, JMLSA*, Vol. 5, 1970, pp. 767-778.

[27] Giannakopoulos, A. E., Lindley, T. C. and Suresh, S., "Aspects of Equivalence Between Contact Mechanics and Fracture Mechanics: Theoretical Connections and a Life Prediction Methodology for Fretting Fatigue," *Acta Materialia*, Vol. 46, 1998, pp. 2955-2968.

[28] Giannakopoulos, A. E., Lindley, T. C. and Suresh, S., "Application of Fracture Mechanics in the Assessment of Fretting Fatigue," *Fretting Fatigue: Current Technology and Practices, ASTM STP 1367*, D. W. Hoeppner, V. Chandrasekaran and C. B. Elliott, Eds., American Society for Testing and Materials,1999.

[29] Suresh, S., *Fatigue of Materials*, Second Edition, Cambridge University Press, 1998.

# Environmental Effects

M. Helmi Attia[1]

# Fretting Fatigue of Some Nickel-Based Alloys in Steam Environment at 265 °C

**REFERENCE:** Attia, M. H., **"Fretting Fatigue of Some Nickel-Based Alloys in Steam Environment at 265°C,"** *Fretting Fatigue: Current Technology and Practices, ASTM STP 1367,* D. W. Hoeppner, V. Chandrasekaran, and C. B. Elliott, Eds., American Society for Testing and Materials, West Conshohocken, PA, 2000.

**ABSTRACT:** In comparison with plane fatigue, fretting damage may lead to a substantial decrease in the fatigue strength of the material due to the addition of alternating frictional stresses, and the continuous destruction of the surface oxide. Fretting fatigue and plain fatigue tests have been conducted to establish the effect of fretting on the fatigue strength of two nickel-based alloys, Inconel 600 and Incoloy 800 against carbon steel and 410S stainless steel, respectively. The tests were conducted in steam environment at 265°C. The main test variables were the contact pressure $0 < p_c < 138$ MPa and the mean stress $0 < \sigma_m < 138$ MPa. The results of the S-N curves for plain and fretting fatigue tests indicated that the reduction factor in the fatigue strength of Inconel I-600 alloy due to fretting is in the range of 2.2 to 3.5, depending on the loading conditions. The results also showed that Incoloy I-800 is more susceptible to fretting fatigue than I-600 by approximately a factor of 1.5.

**KEYWORDS:** fretting fatigue, Inconel 600, Incoloy 800, high temperature, steam environment

## Introduction

Fretting may be defined as the damage caused by low amplitude oscillatory motion between two contacting surfaces. Within this broad definition, fretting wear and fretting fatigue mechanisms are included. In fretting wear, where the damage is measured by the volumetric material losses, there is no lower limit on the contact pressure, $p_c$, or amplitude of oscillation, $\delta$, below which wear damage will not occur [1], provided that $p_c$, and $\delta > 0$. It also occurs whether the mode of motion is micro- or macro-slip, and in the presence or absence of alternating bulk stresses $\sigma_a$ in one of the contacting bodies. Fretting fatigue, on the other hand, is defined in terms of the reduction in fatigue strength or fatigue life, regardless of the severity of the associated fretting wear damage. This effect tends to be more pronounced for high strength materials. Unlike fretting wear, fretting fatigue damage requires the presence of body stresses $\sigma_a$, and it has a lower limit, which is defined by the fretting fatigue strength $\sigma_{ff}$. If the applied body stress $\sigma_a < \sigma_{ff}$, the fatigue life is indefinite and the cracks nucleated by the fretting action will not propagate to failure. It is well established that

---

[1] Principal Research Engineer, Ontario Power Technologies, Materials Technology Dept., 800 Kipling Ave, Toronto, Ontario, Canada M8Z 6C4.

fretting fatigue affects both the crack nucleation and propagation processes, mainly due to the addition of alternating frictional stresses, and the continuous destruction of the surface oxide film. For a given material combination and environmental conditions, the most critical factors which affect the fretting fatigue strength are the normal pressure $p_c$, the mean stress $\sigma_m$, slip amplitude $\delta$, surface treatment, and environmental conditions. Comprehensive coverage and analysis of the fretting fatigue mechanisms, and the effect of various process variables can be found in references [1-4].

Due to its good oxidation resistance at high temperatures and resistance to chloride ion stress-corrosion cracking, Inconel I-600 alloy has a number of applications in nuclear and chemical industries. Incoloy I-800 is widely used in heat exchangers, and process piping for its good strength and excellent corrosion resistance in aqueous environments. Extensive literature review indicated that the available data base for the fretting fatigue strength of I-600 and I-800 is very limited. Following the fretting fatigue failure of the I-600 steam generator tube at Mihama-2, in Japan in 1991 [5,6], fretting fatigue crack nucleation data and S-N results of Inconel-600 against carbon steel were reported by the Japanese Agency of Natural Resources and Energy for contact pressures, $0 < p_c < 40$ MPa ($< 5.7$ ksi), and zero mean stress [6,7]. These tests were, however, conducted in air at room temperature. Hamdy and Waterhouse [8] have investigated the fretting fatigue behaviour of Inconel 718 against Nimonic 90 in air at temperatures in the range 20 ºC $< T < 540$ ºC.

The main objective of the present work is to examine the effect of fretting on the reduction in the fatigue strength of Inconel 600 in contact with carbon steel and Incoloy 800 in contact with 410S stainless steel, in steam environment at 265°C. To establish the individual and combined effects of the variables affecting the fretting fatigue process, different levels of contact pressure, and mean stress were tested. Plain fatigue and tensile tests were also conducted to establish the mechanical properties of the test specimens.

## Fretting Fatigue Testing Set-up

### Test Strategy and Experimental Set-up

The test strategy is based on the following:
a. Test type: one-stage, to ensure the conjoint action of fretting and fatigue.
b. Test equipment: pull-push type.
c. Controlled test conditions: contact pressure (dead weight system), specified surface finish and machining marks orientation, as well as controlled environment (steam at 265 ºC). The control system of the test equipment maintains the frequency and the amplitude of the applied body stress constant during the test duration.

The fretting fatigue tests were carried out using a uniaxial fatigue testing machine. The schematic diagram of the test set-up (Figure 1) shows the fatigue specimen and the contact pad placed inside an autoclave to control the surrounding medium (steam) and its temperature. The figure also shows the dead weight loading system used to apply a constant contact pressure. The steam generating system shown in Figure 2 was used. The oxygen content in the demineralized feed-water was maintained at a low level by continuously purging nitrogen bubbles into the water reservoir.

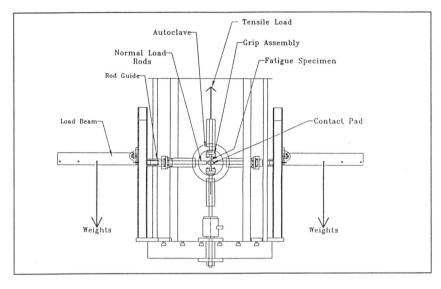

*Figure 1 A schematic of the fretting fatigue test.*

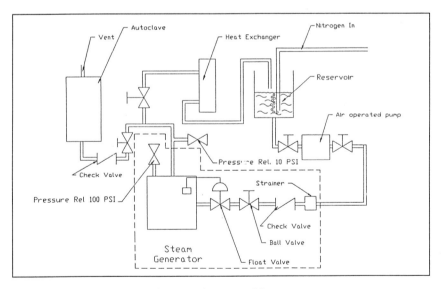

*Figure 2 A schematic diagram of the steam generator
system used to supply steam to the autoclave.*

*Test Specimens*

The fretting fatigue specimen used in these tests is shown in Figure 3a. The specimen is 133.35 mm long, with two parallel flats machined on opposite sides of its central portion to form a gauge length of 38.1 mm. The ends of the flats were given a radius, followed by a straight section and another radius to avoid a stress concentration at the change in sections. The specimens were made of Inconel 600 and Incoloy 800 bars of the following composition:

| I-600 : | C | Mn | Fe | Ni | Cr | Si | Cu | Al | Ti | Co | Cb+Ta |
|---|---|---|---|---|---|---|---|---|---|---|---|
|  | 0.07 | 0.31 | 8.85 | 73.88 | 15.73 | 0.28 | 0.24 | 0.18 | 0.26 | 0.06 | 0.14 |

| I-800 : | C | Mn | Fe | Ni | Cr | Si | Cu | Al | Ti |
|---|---|---|---|---|---|---|---|---|---|
|  | 0.08 | 0.95 | 43.08 | 34.14 | 20.27 | 0.11 | 0.22 | 0.55 | 0.60 |

Bridge type fretting contact pads, where the central portion of one side was recessed to form a bridge with a pair of rectangular feet, were used. The span L = 28.45 mm, and the cross sectional area of each foot is 2.03 mm × 8.89 mm (Figure 3b). The pads were machined from ASME SA-516-70 carbon steel and ASME SA-240-410S stainless steel (hot rolled, annealed and descaled) of the following composition:

| Carbon steel : | C | Mn | P | S | Si |
|---|---|---|---|---|---|
|  | 0.30 | 1.1 | 0.03 | 0.034 | 0.20 |

| Stainless steel-410S: | C | Mn | P | S | Si | Cr | Ni | Co | Cu | Mo | N |
|---|---|---|---|---|---|---|---|---|---|---|---|
|  | 0.07 | 0.08 | 0.026 | 0.003 | 0.25 | 12.35 | 0.16 | 0.01 | 0.16 | 0.03 | 0.06 |

The wear pad materials were tested against I-600 and I-800 fatigue specimens, respectively. The fatigue specimen flats and the contact surfaces of the fretting pads were ground to a surface roughness better than 32 $\mu$-in CLA in the direction of the fatigue loading.

*Test Procedure*

In this investigation, three types of tests were conducted:
1. Tensile tests of I-600 and I-800 alloys to establish the mechanical properties of the materials, namely, the modulus of elasticity E, the yield stress (0.2% offset proof stress), and the ultimate tensile strength. These tests were conducted in air at room temperature.
2. Base-line tests to confirm the ASME S-N curve of I-600 by conducting plain fatigue tests in steam at 265 °C.
3. Fretting fatigue tests of I-600 against ASME SA-516-70 carbon steel wear pads and I-800 against ASME SA-240-410S stainless steel pads. These tests were also conducted in steam at 265 °C.

To ensure proper alignment of the fretting fatigue specimen over the full working range of axial load ($0 \leq P \leq 11.565$ kN), two strain gauges were attached to the flat surface of each specimen. The average measured strain $\bar{\epsilon}_m$ for all tests was found repeatable and to agree with the calculated strain $\epsilon_c$, with average percentage deviation in the range 10.5-14.5 %. A temperature sensing element was placed near the reduced section of the specimen to control the temperature setting of the autoclave. During the test, the applied alternating load, frequency and the temperature are continuously monitored and automatically controlled.

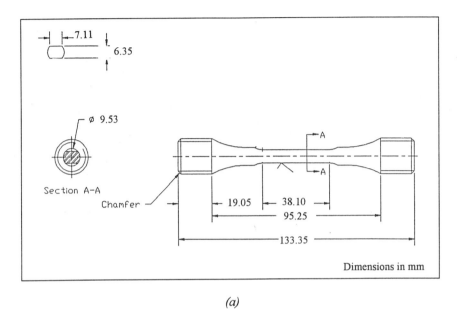

*(a)*

Note **1**: grind to sharp edge and then 0.125 to 0.250 mm radius.     Dimensions in mm

*(b)*

*Figure 3(a) Fretting fatigue test specimen, and (b) fretting pad dimensions.*

*Design of Experiments*

A critical literature review showed that the increase in the contact pressure $p_c$ up to a certain critical level $p_{cr}$ results in a reduction in the fretting fatigue strength $\sigma_{ff}$. Above $p_{cr}$, the fretting fatigue strength remains nearly unchanged. For most ferrous and nonferrous materials, this critical pressure is found to be ~ 70 MPa (10 ksi) *[1]*. The effect of the relative slip amplitude $\delta$ on the fretting fatigue strength $\sigma_{ff}$ is coupled with the magnitude of the mean body stress $\sigma_m$. For $\sigma_m = 0$, the fretting fatigue strength $\sigma_{ff}$ decreases as the slip amplitude $\delta$ increases up to a certain value $\delta_{cr}$. For all values of $\delta > \delta_{cr}$, the value of $\sigma_{ff}$ remains nearly constant or increases slightly as long as the regime of gross slip is not reached ($\delta_{gross}$ > 30-40 $\mu$m). Available data indicates that $\delta_{cr}$ is approximately 7-10 $\mu$m for most metals *[1,9,10]*. For $\sigma_m > 0$, as the mean stress increases, the fretting fatigue strength decreases first and then remains nearly constant for mean stresses above a critical level $\sigma_{cr}$. Some reported data indicated that the mean stress $\sigma_m$ does not affect the fatigue strength $\sigma_{ff}$ at relatively small slip amplitudes, $\delta_m$ < 5-7 $\mu$m. From this analysis, it was concluded that the test matrix given in Table 1 would allow us to interpolate the test results to predict the reduction in fretting fatigue strength for other combinations of variables in the following range: $0 < p_c < 138$ MPa, and $0 < \sigma_m < 138$ MPa. The validity of the interpolation scheme will be tested by using the first three S-N curve to predict the fretting fatigue strength obtained from the $4^{th}$ S-N curve.

Table 1: *Test parameters to be investigated*

| | Parameter | Number of levels | Values |
|---|---|---|---|
| Variables | Material combination | Two levels | I-600 vs. carbon steel ASME SA-516-70. I-800 vs. stainless steel ASME SA-240-410S. |
| | Contact pressure $p_c$ | Three levels | 0 (base line), 69 MPa (10 ksi) and 138 MPa ( 20 ksi). |
| | Mean stress $\sigma_m$ | Two levels | 0 and 138 MPa (20 ksi). |
| | Stress amplitude $\sigma_a$ | Five to six levels | 69 MPa (10 ksi) < $\sigma_a$< 276 MPa (40 ksi). |
| Fixed parameters | Environment | Temperature and medium | 265 °C, in steam. |
| | Frequency f | one level | 30 Hz. |

*Measurement of the Relative Slip at the Fretting Fatigue Specimen/Wear Pad Interface*

In order to interpolate the test results to other untested conditions, the relative slip amplitude at the fretting contact interface has to be estimated. The relative slip amplitude $\delta$ between the fatigue specimen and the wear pad is related to the stress amplitude $\sigma_a$, the span L between the feets of the pad, and the modulus of elasticity E of the fatigue specimen by the following relation [10,11]:

$$\delta = \frac{\sigma_a L}{2 E} \tag{1}$$

Attempts were made for direct measurement of the relative slip amplitude, using two high-temperature non-contacting proximeters attached to a special fixture, which is directly mounted onto the inside surface of the autoclave. Two targets were secured to the fatigue specimen and the wear pad. The difference in the readings obtained from the two proximeters is the relative movement $\delta$. Although a special effort was made in the design of the targets and the fixture to ensure satisfactory dynamic characteristics, the measurement signals showed a large amount of dynamic and electronic noise. In addition, the very low signal levels resulted in erratic readings. An alterative approach was followed to eliminate the dynamics problem by calibrating the proximeters under quasi-static conditions and comparing their readings with the strain gauge measurements and the analytical predictions (Eq. 1).

The percentage deviation between the measured and calculated slip amplitudes and strains was found to be < 12%. It was, therefore, concluded that the slip amplitude $\delta$ can satisfactorily be estimated from Eq. 1. A similar conclusion has been reached by other researchers [1-3]. Equation 1 was also validated by comparing the measured and calculated strain over the gauge length of the fatigue specimen.

**Test Results and Discussion**

*Tensile Tests of I-600 and I-800 Test Materials at Room Temperature*

The results of the room temperature tensile tests were as follows:
a- For Inconel I-600: E= $2.20\times10^5$ MPa ($31.9\times10^6$ psi), $\sigma_y$ = 296 MPa ($43 \times 10^3$ psi), and the ultimate tensile stress $\sigma_{UTS}$ = 709 MPa ($102.8\times10^3$ psi).
b- For Incoloy I-800: E = $1.95\times10^5$ MPa ($28.3\times10^6$ psi), $\sigma_y$ = 221 MPa (32.1 ksi), and $\sigma_{UTS}$ = 582 MPa ($84.4\times10^3$ psi).
At 265 °C, E and $\sigma_{UTS}$ of I-800 are reduced by 12.5 % and 10.5 %, respectively. The same ratios were assumed to be applicable to I-600.

*Plain Fatigue Tests of I-600 at 265°C in Steam Environment*

The I-600 plain fatigue test data obtained for mean stress $\sigma_m$ = 0 at 265°C are shown in Figure 4 (identified as OPT series). The test samples that did not fail after N = $5\times10^7$ cycles are identified by small horizontal arrows. The figure also shows the I-600 and I-800 plain fatigue data compiled by the ASME [12], the Electric Power Research Institute EPRI [13], and the Japanese Agency of Natural Resources and Energy [6,7] in air at room temperature, for different levels of mean stress. When the EPRI's data are corrected to zero mean stress using modified Goodman relationship [14], it predicts a fatigue strength of 207 MPa (30 ksi) at N= $5\times10^7$ cycles. Comparing the test results with available fatigue data indicates that the

ASME correlation can be used in subsequent sections to define the base-line fatigue strength of I-600 and I-800, $\sigma_{pf}$ = 234 MPa (34 ksi). For the purpose of comparison, the plain fatigue data of Inconel 718 against Nimonic 90 in air at 280 °C, and compressive mean stress of 552 MPa (80 ksi) is also shown [8].

*Fretting Fatigue S-N Curves of I-600 and I-800 at 265 °C in Steam Environment*

The fretting fatigue results for I-600 against carbon steel wear pads in steam at 265 °C are presented in Figs. 5 to 7. Figure 5 shows the effect of the contact pressure, $p_c$ = 0, 69 MPa, and 138 MPa, for a mean stress $\sigma_m$ = 0. As predicted, the contact pressure has a significant effect on the fretting fatigue strength $\sigma_{ff}$ only for 0<$p_c$<69 MPa ($1 \times 10^4$ psi). Above a critical value of $p_c$ = 69 MPa, the effect of the contact pressure is insignificant. While an increase in the contact pressure from 0 to 69 MPa results in 55% reduction in the fatigue strength (from 237 to 107 MPa), a further increase in the contact pressure by an equal amount (from 69 to 138 MPa) results in only an additional 10% reduction in $\sigma_{ff}$.

The effect of the mean stress $\sigma_m$ on the fretting fatigue strength is shown in Fig. 6 for $p_c$ = 138 MPa. The figure indicates that the increase in the mean stress from 0 to 138 MPa results in only 20 % reduction in the fatigue strength from 96.5 to 75.8 MPa. The combined effects of contact pressure 0 < $p_c$ <138 MPa and mean stress 0 <$\sigma_m$<138 MPa are summarized in Fig. 7 along with the fretting fatigue data of Inconel 718 at 250 °C in air [8], and I-600 at room temperature in air [6,7].

Figure 8 shows the fretting fatigue characteristics of Incoloy I-800 (against 410S stainless steel) for contact pressure, $p_c$= 69 MPa, and mean stress $\sigma_m$ = 0. At N= $5 \times 10^7$ cycles, the fretting fatigue strength $\sigma_{ff}$ is 69 MPa. For the purpose of comparison, the results of Inconel I-600 (against carbon steel), which were produced under similar conditions are also shown in the figure. It can be concluded that I-800 is more susceptible to fretting fatigue than I-600 by approximately a factor of 1.5. This may be attributed to the fact that Ni-based alloys and stainless steels are very compatible for adhesive wear and localized micro-welding. This in turn results in higher friction stresses which have a decisive effect on fretting fatigue [15].

In these tests, the fretting wear damage was found to be mild. A typical 3-D map of the topography of the fretted surface is shown in Fig. 9, for Inconel I-600 tested under these conditions: contact pressure, $p_c$= 138 MPa , mean stress, $\sigma_m$ = 0, alternating stress (defined as half the peak-to-peak stress), $\sigma_a$ = ± 124 MPa, and number of cycles, N= $4.26 \times 10^6$ cycles. The wear volume in this case was 0.02 mm$^3$, and the maximum wear depth was only 10 μm. It is evident from the figure that adhesion between the contacting surfaces was taking place, resulting in transfer of material to the I-600 specimen. The material build up was as high as 24 μm above the undamaged surface.

A simple interpolation scheme is proposed to estimate the fretting fatigue strength reduction factor (FFS-RF) for untested conditions. In this investigation, the reduction factor FFS-RF is defined as the ratio between the plain fatigue strength without fretting $\sigma_{pf}$ and the fatigue strength with fretting $\sigma_{ff}$, at a given life of N=$5 \times 10^7$ cycles: FFS-RF = $\sigma_{pf} / \sigma_{ff}$. Although the use of limited number of fretting fatigue strength FFS data is justifiable for the following reasons, its validation requires conducting more tests. First, the model is based on extensive literature review and analysis of available test results, regarding the effect of contact pressure, mean stress, and slip amplitude on fretting fatigue strength. Second, it is also based

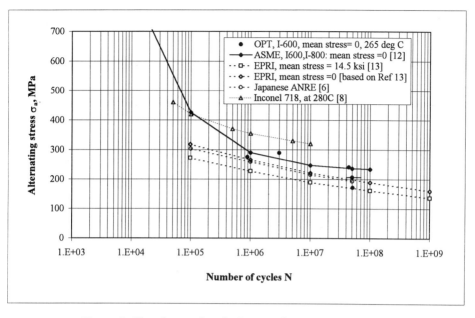

*Figure 4   Plain fatigue data for I-600 and I-800*

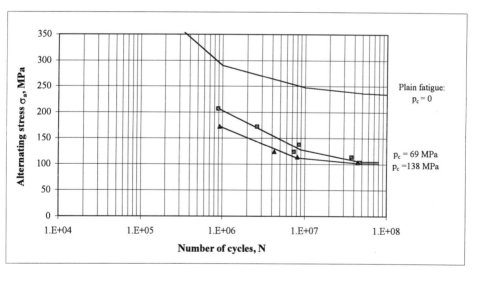

*Figure 5   Effect of contact pressure on the fretting fatigue strength of I-600*
*for mean stress $\sigma_m = 0$*

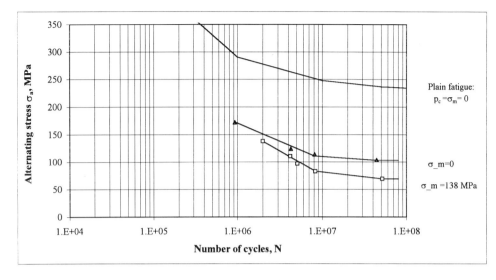

*Figure 6   Effect of the mean stress on the fretting fatigue strength of I-600 for contact pressure $p_c$ = 138 MPa*

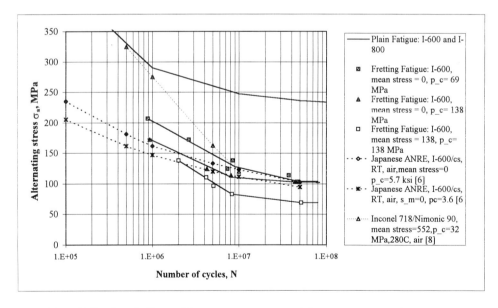

*Figure 7   Combined effects of the contact pressure and the mean stress on the fretting fatigue strength of I-600 at 265 °C in steam environment*

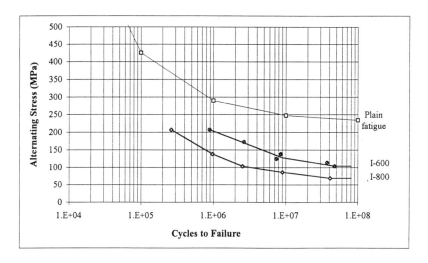

*Figure 8    Fretting Fatigue strength of I-600 and I-800 for the for the following*
*conditions: $p_c$ = 69 MPa, and $\sigma_m$ = 0 (at 265 °C in steam environment)*

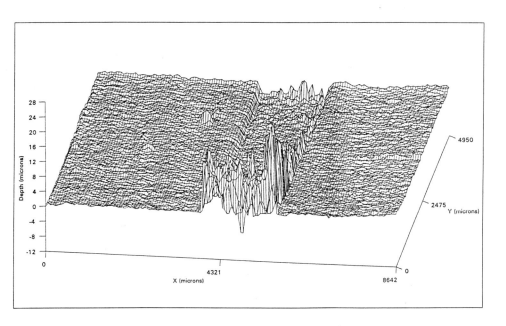

*Figure 9    A typical three-dimensional topography of fretted surface*

on using Goodman's method to predict the effect of mean stress. This approach has been verified in [16], and in the present work for the case of fretting fatigue. The proposed scheme is based on the following set of rules:

1. For a given mean stress, the fretting fatigue strength-contact pressure relation is bi-linear:

$$\left[\sigma_{ff,p}\right]_{\sigma_{m_0}} = \left[\sigma_{ff_{p_0,m_0}}\right] - k_{p_1}\, p_c\,, \qquad\qquad 0 < p_c < p_{cr}$$

$$\left[\sigma_{ff,p}\right]_{\sigma_{m_0}} = \left[\sigma_{ff_{p_0,m_0}}\right] - k_{p_1}\, p_{cr} - k_{p_2}\,(p_{cr} - p_c)\,, \qquad p_c > p_{cr} \tag{2}$$

where, $k_{p1}$ and $k_{p2}$ are the slopes of the line for $p_c < p_{cr}$, and $p_c > p_{cr}$, respectively, and $k_{p1} > k_{p2}$. The symbol $\sigma_{ff_{p_0,m_0}}$ stands for the plain fatigue strength at $p_c = 0$ and $\sigma_m = 0$.

2. For a given contact pressure p, the relationship between the mean stress $\sigma_m$ and $\sigma_{ff}$ is linear and obeys Goodman's equation, which relates the fatigue strength at zero mean stress $\sigma_{ff,mo}$ to the fatigue strength at a given mean stress $\sigma_{ff,m}$ and the ultimate tensile stress of the material $\sigma_{UTS}$ [14]:

$$\sigma_{ff,m} = \sigma_{ff,m_0}\left(1 - \bar{\sigma}_m\right) \tag{3}$$

where, $\bar{\sigma}_m$ is the ratio between the mean stress and the ultimate tensile stress, $\bar{\sigma}_m = \sigma_m/\sigma_{UTS}$.

3. For slip amplitudes $\delta < \delta_{cr}$, and $\delta_{cr} = 5$ μm, the mean stress has no effect on the fretting fatigue strength: $\sigma_{ff} \neq f\{\sigma_m\}$.

From Eqs. 2 and 3, the fretting fatigue strength $[\sigma_{ff}]_{p,m}$ at any given contact pressure p, and mean stress $\sigma_m$ is described by the following relation:

$$\left[\sigma_{ff}\right]_{p,m} = \left[\sigma_{ff_{p_0,m_0}} - (p/p_{cr})\left(\sigma_{ff_{p_0,m_0}} - \sigma_{ff_{p_{cr},m_0}}\right)\right]\left[1 - \bar{\sigma}_m\right], \qquad \text{for } p < p_{cr} \tag{4}$$

$$\left[\sigma_{ff}\right]_{p,m} = \left[\left[\sigma_{ff}\right]_{p_{cr},m_0} - k_{p_2}\left(p - p_{cr}\right)\right]\left[1 - \bar{\sigma}_m\right], \qquad \text{for } p > p_{cr} \tag{5}$$

From the test results of I-600, the following parameters are estimated: $p_{cr} = 69$ MPa, the plain fatigue strength at zero mean stress $\sigma_{ff_{p_0,m_0}} = 238$ MPa, and the fretting fatigue strength at the critical contact pressure and zero mean stress $\sigma_{ff_{p_{cr},m_0}} = 100$ MPa. The rate of change in the fretting fatigue strength with contact pressure, when $p > p_{cr}$, $k_{p2} = 0.15$.

Ignoring the effect of relative slip amplitude $\delta$ first, the measured fretting fatigue strength $\sigma_{ff}$ for $\sigma_m = 0$ (shown as **bold** in Table 2) are used to estimate the fretting fatigue strength $\sigma_{ff}$ for other conditions, $\sigma_m \neq 0$, and $0 < p_c < 138$ MPa. Through an iterative procedure, the values of $\delta$ are estimated to adjust the predicted stress amplitudes $\sigma_{ff}$.

To test the validity of this interpolation scheme, the fretting fatigue strength $\sigma_{ff}$ values measured for $\sigma_m = 0$, and $p_c = 0$, 69 and 138 MPa were used to predict $\sigma_{ff}$ for $\sigma_m = 138$ MPa, and $p_c = 138$ MPa. The predicted value of 75 MPa is in a good agreement with the test result of 69 MPa. Predicted and measured values are shown in Fig.10 as points C and D, respectively. The figure confirms that Goodman's equation can be used to predict the effect of mean stress in fretting fatigue. It should be noted that points A and E in Fig. 10 represent the ultimate tensile strength of I-600 at 265 °C and room temperature, respectively. The

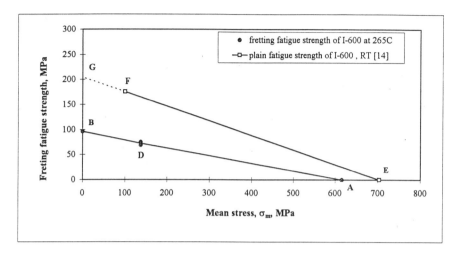

*Figure 10   Goodman's Correlation for prediction the effect of mean stress
on fretting fatigue stength for slip amplitudes  δ > 5-7  μm*

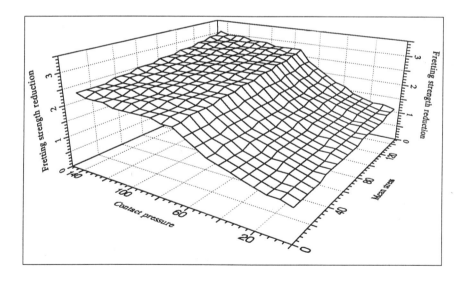

*Figure 11   Predicted fretting fatigue reduction factor of I-600, as a function
of contact and mean stress (in MPa)*

results obtained following the proposed scheme are presented in Table 2.

Normalization of the data given in Table 2 with respect to the plain fatigue strength $\sigma_{pf}$ at $p_c = \sigma_m = 0$, produces the fretting fatigue strength reduction factor SRF as a function of the contact and mean stresses; $FFS - RF = \sigma_{pf} / \sigma_{ff}$.

Table 2  Estimated fretting fatigue strength of I-600 at 265 °C in steam for different combinations of contact and mean stresses $\sigma_{ff}$ {$p_c$, $\sigma_m$}, $0 < p_c$, $\sigma_m < 138$ MPa

| Contact pressure, MPa ⇒ Mean stress, MPa ⇓ | 0 | 34.5 | 69.0 | 103.5 | 138 |
|---|---|---|---|---|---|
| 0 | **236.5** | (171.7) | **106.9** | (102.0) | **96.5** |
| 34.5 | (223.4) | (162.0) | (100.7) | (95.8) | (91.0) |
| 69.0 | (210.3) | (152.4) | (95.2) | (90.3) | (85.5) |
| 103.5 | (197.0) | (142.7) | (88.9) | (84.8) | (80.0) |
| 138.0 | (183.4) | (133.0) | (82.7) | (78.6) | **69** (75.2) |

Notes:  **bold**: measured value, (....) : estimated value

The response surface representing the effect of contact and mean stress on SRF of I-600 is shown in Figure 11.  It can be seen from the figure that the fretting fatigue SRF can be as high as ~ 3.5.  In the fretting fatigue tests conducted in this investigation, the slip amplitude at the fretting fatigue strength (corresponding to $N = 5 \times 10^7$ cycles) varies between 5.5 and 12.5 μm.

**Conclusions**

The following conclusions can be drawn from this experimental investigation.  First, the fretting action in steam environment at 265°C can accelerate crack nucleation and propagation processes and cause a significant reduction in the fatigue strength of I-600 against carbon steel, and I-800 against 410 stainless steel.  For the test conditions reported in this investigation, the fatigue strength reduction factor was found to be in the following range: 2.2 to 3.5.  Among other factors, the reduction in the fatigue strength can be attributed to high friction stresses, which are localized at the contact asperities due to adhesion and transfer of material.  Second, the effect of the main process parameters, namely, contact and mean stresses on fretting fatigue strength has been established.  A scheme for interpolation of the test results to other conditions that have not been tested is  presented.  Based on the limited test data obtained in this work, one can conclude that I-800 alloys (against 410S stainless steel) are more susceptible to fretting fatigue damage than I-600 alloy (against carbon steel) by approximately a factor of 1.5.

*Acknowledgment*

This work was funded by CANDU Owners Group COG, which the author greatly appreciates.

**References**

1.  Waterhouse, R. B., *Fretting Corrosion*, Pergamon Press Ltd., New York, 1972.

2.  Waterhouse, R. B., ed., *Fretting Fatigue*, Applied Science Publishers, London, UK, 1981.

3.  Attia, M. H. and Waterhouse, R. B., eds., "Standardization of Fretting Fatigue Test Methods and Equipment", *ASTM STP 1159*, American Society for Testing and Materials, West Conshohocken, PA, 1992.

4.  Waterhouse, R. B., and Lindley, T. C., eds., "Fretting Fatigue", *Proceedings*, International Symposium, sponsored by the European Structural Integrity Society ESIS, Sheffield, UK, published by Mechanical Engineering Publications Ltd, London, 1994.

5.  "What Happened at Mihama: Final Report Issued", Nuclear Engineering International, January, 1992, pp.36-38.

6.  Final report on the Steam Generator Tube Break at Mihama Unit No.2 of Kansai Electric Power Co., Inc. Occurred on February 9, 1991, Agency of Natural Resources and energy Ministry of International Trade and Industry, Japanese Government, Nov. 1991.

7.  Idia, K., "Recent Issues and Researches on Nuclear Pressure Vessel and Piping", *Proceedings*, International Conference on Pressure Vessel Technology, Dusseldorf, Germany, May 1992, pp. 45-62.

8.  Hamdy, M. M. and Waterhouse, R. B., "The Fretting Fatigue Behaviour of a Nickel-Based Alloy (Inconel 718) at Elevated Temperatures", *Proceedings*, International Conference on Wear of Materials, American Society of Mechanical Engineers, NY, 1979, pp.351-355.

9.  Waterhouse, R. B., "The Problems of Fretting Fatigue Testing", *ASTM STP 1159* on 'Standardization of Fretting Fatigue Test Methods and Equipment', American Society for Testing and Materials, West Conshohocken, PA, 1992, pp. 13-19.

10. Fenner, A. J. and Field, J. E., "A Study of the Onset of Fatigue Damage Due To Fretting", *Transactions of the North east Coast Institute of Engineers and Shipbuilders*, Vol. 76, 1995-1960, pp. 183-228.

11. Lindley, T. C. and Nix, K. J., "Fretting Fatigue in Power Generation industry:

Experiments, Analysis and Integrity Assessment", *ASTM STP 1159* on "Standardization of Fretting Fatigue Test Methods and Equipment", M. H. Attia and R. B., Waterhouse, eds., 1992, pp.153-169.

12. ASME (Section III) draft report cited in: G. I., Ogundele, "Corrosion Fatigue of UNS Alloy N06600 Steam Generator Tubing", OHT Report No. A-NSG-95-74-P/COG Report No. 95-299, Ontario Hydro Technologies, 1995.

13. Fatigue Performance of Ni-Cr-Fe Alloy 600 Under Typical PWR Steam Generator Conditions, EPRI Report NP-2957, prepared by Westinghouse Electric Corporation R&D Center, March, 1983.

14. Buch, A., *Fatigue Strength Calculation*, Materials Science Surveys No. 6, Trans Tech Publications, Switzerland, 1988, p. 137.

15. Collins, A.J., "Fretting Fatigue Phenomena With Emphasis on Stress Field", PhD thesis, Ohio State University, 1963.

16. Mutoh, Y. and Satoh, T., "High Temperature Fretting Fatigue", *Proceedings*, International Symposium on Fretting Fatigue, sponsored by the European Structural Integrity Society ESIS, Sheffield, UK, Mechanical Engineering Publications Ltd, London, 1994, pp. 389-404.

Charles B. Elliott III[1] and Allen M. Georgeson[2]

# Fretting Fatigue of 8090-T7 and 7075-T651 Aluminum Alloys in Vacuum and Air Environments

**REFERENCE:** Elliott, C. B., III and Georgeson, A. M., **"Fretting Fatigue of 8090-T7 and 7075-T651 Aluminum Alloys in Vacuum and Air Environments,"** *Fretting Fatigue: Current Technology and Practices, ASTM STP 1367,* D. W. Hoeppner, V. Chandrasekaran, and C. B. Elliott, Eds., American Society for Testing and Materials, West Conshohocken, PA, 2000.

**ABSTRACT:** The fretting fatigue characteristics of 8090-T7 and 7075-T651 aluminum alloys were investigated. The purpose of this investigation was to determine the relative effects of wear and environmental degradation from laboratory air on the fretting fatigue process.

The fretting fatigue test system used in this experimentation enables testing either in air or a scanning electron microscope vacuum environment. Fretting fatigue damage in vacuum tests was considered as resulting predominantly from wear, whereas damage from tests in air results from the concurrent wear and oxidation mechanisms.

Based on this experimentation, it was concluded that:
- Degradation of fatigue life in air due to fretting is caused primarily by wear mechanisms for 8090-T7 aluminum.
- Fretting fatigue life in air was greater than fretting fatigue life in a vacuum for the 8090-T7 aluminum under the conditions of this experimentation.
- Environmental mechanisms significantly decrease fatigue life due to fretting in air for 7075-T651 aluminum.
- Fretting fatigue behavior is material dependent.

**KEYWORDS:** fretting, fretting fatigue, 8090 aluminum, 7075 aluminum

## Background

The fretting process occurs when two surfaces are in contact under conditions where there is relative movement of small amplitude so that the resulting wear debris remains between the surfaces. The simultaneous action of fretting and fatigue tends to

---

[1]Research Assistant Professor, Mechanical Engineering Department, University of Utah, 50 S. Central Campus Dr. Rm. 2202, Salt Lake City, Utah 84112-9208.
[2]Senior Engineer, Structures Division, Boeing Commercial Airplane Group, P.O. Box 3707, MC 67-FF, Seattle, Washington 98124-2207.

cause material degradation and structural failure of components at a more rapid rate than would be expected based on either fatigue or fretting acting alone. An important aspect of fretting fatigue that is not well understood is the relative importance of the wear and the environmental mechanisms in the process. Wear is inherently part of the process because of the contact loading, therefore it can be argued that the environmental effects are secondary.[1] The mechanical aspects of the process have been shown to be dominant under some conditions for some materials,[2] while environmental effects have been shown to cause significant damage and degradation of fatigue lives in other situations.[3]

This work is a continuation of the fretting fatigue investigation performed by Elliott.[4,5] The basic purpose of his research (hereafter referred to as previous research) was to "provide insights into the effects of corrosion, specifically oxidation in air, on the fretting fatigue process." He developed a test system[6] that allows fretting fatigue testing in an air environment or in the vacuum environment in the chamber of a scanning electron microscope (SEM). Fretting fatigue damage in vacuum tests can be considered as resulting predominantly from wear, whereas damage from tests in air results from the concurrent wear and oxidation mechanisms. It was found that the degradation in fretting fatigue life in air was primarily due to wear for the 8090 alloy, but was primarily due to corrosion for the 7075 alloy. Also, in all cases where comparisons were possible, the average fretting fatigue lives of 8090 specimens were less in a vacuum, but greater in air, than those of 7075 alloy specimens. Table 1 summarizes these findings. In the previous research it was concluded that "general statements of the relative effects of wear and corrosion on the fretting fatigue process are meaningless unless tied to a specific material," and that "fretting fatigue life in air compared to a vacuum is material dependent." Due to the number of experimental conditions investigated, at most three specimens for each condition were tested, giving limited statistical confidence to the results. Also, some of the fatigue tests with an applied normal load, but without fretting, did not fail but cycled until runout at three million cycles. Therefore it was felt that some of the experimental variables should be further investigated with greater replication, and an increased fatigue load magnitude so that these tests might be more likely to fail instead of cycling until runout.

Table 1 - *Summary of Previous Research Findings*[5]

| Test Type | Average cycles to failure | |
| --- | --- | --- |
| | 8090-T7 | 7075-T7351 |
| Fatigue without fretting | 1,759,527 | 2,537,695 |
| Fretting fatigue in an SEM vacuum (assumed predominantly wear) | 403,223 | 2,377,757 |
| Fretting fatigue in air (oxidation and wear) | 341,980 | 106,706 |

The objective of this program is to investigate the effect of environmental degradation, specifically corrosion in ambient laboratory air, on the fretting fatigue response of 8090 and 7075 aluminum alloys, and to obtain greater statistical confidence in the results than was obtained in the previous research. Additionally, this program may help provide insights into the standardization of fretting fatigue test methods, especially with regards to the environment used for base comparisons. This program investigates the use of a vacuum environment as a standard comparison test environment, instead of 'laboratory air,' which has been the traditional base environment.

A coincident purpose of this program is to compare the fretting fatigue behavior of the relatively untested 8090 alloy to the 7075 alloy, which has been used in industry. Material comparison was needed because of the lack of standardization of fretting fatigue test methods. The 7075 alloy was chosen as the comparison material because it is readily available for testing by others, as well as being a material used where 8090 may be used.

For more detailed results of this experimentation see [7].

## Experimentation

The fretting fatigue test system used for this research (Figure 1) includes these features:
- In order to provide consistency of results, a single system is used for both air and vacuum testing.
- Both fatigue and fretting fatigue tests can be performed with the system. Fatigue tests can be run with or without a normal load from the internal fretting loading subsystem. The fatigue tests performed in this program were run with a normal load.
- To provide a vacuum environment, the system uses the vacuum chamber of an International Scientific Instruments ISI SS-40 scanning electron microscope.
- Also the system allows observation by SEM of the fretted surface of the specimen without any environment other than the vacuum being introduced to the specimen. This capability was not used in this program.

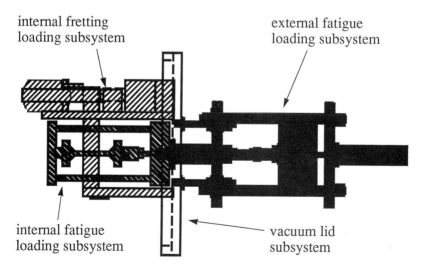

Figure 1 - *Top View of the Fretting Fatigue Load Frame[6]*

During both the fretting fatigue and the fatigue tests a normal load was applied to the specimen by the internal fretting loading subsystem through the two fretting pads (Figure 2). For fretting fatigue tests the top pad was made of the same aluminum alloy as the specimen, and the bottom pad had a low-friction phenolic material for the surface in contact with the specimen, so that fretting occurred only between the top surface of the specimen and the top fretting pad. The fatigue tests were performed using phenolic surfaces on both the top and the bottom so that fretting would not occur. The purpose of applying a normal load during fatigue tests was to approximate the stress conditions experienced by the specimens during fretting fatigue tests.

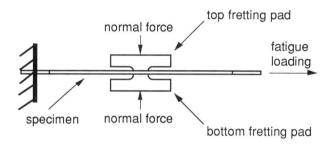

Figure 2 - *Side View Showing the Relationship between the Specimen and Fretting Pads*

The experiment consisted of the following tests:
- 3 normal load fatigue tests of 8090 in air,
- 3 normal load fatigue tests of 8090 in vacuum,
- 3 normal load fatigue tests of 7075 in air,
- 3 normal load fatigue tests of 7075 in vacuum,
- 8 fretting fatigue tests of 8090 in air,
- 8 fretting fatigue tests of 8090 in vacuum,
- 8 fretting fatigue tests of 7075 in air, and
- 8 fretting fatigue tests of 7075 in vacuum.

Of the 44 specimens, four fatigue specimens were tested during the pilot test which was necessary to establish the fatigue loading levels. The order of testing of the 40 remaining specimens was randomized and blocked so that the results would not reflect historical bias. The results were compared for statistical significance by a test for differences in means, the 't-test.'

The parameters that were not under investigation were constrained to be as consistent as possible. A 13 Hz sinusoidal fatigue loading signal with maximum stress of 37 ksi and stress ratio (R) of +0.1 were used for all testing. A nominal normal load of 3 ksi was used for both the fatigue and the fretting fatigue tests.

The 8090 material, provided by Alcoa Co., was a block of hot rolled 3.2 inch thick plate heat treated to T7 designation. The 7075 material was commercially procured, and was a block of 3.1 inch thick plate heat treated to T651 designation. Specimen blocks were machined to final dimensions, except the individual specimens were not separated. Each resulting dog-bone shaped block was sliced into individual specimens by a Buehler Isomet low speed saw with a 5 inch blade. The edges were dry-sanded with 600 grit silicon carbide abrasive paper. Specimen faces were wet sanded with 600 grit paper using a Buehler Handimet II roll grinder. All specimens were fabricated so that their length dimension was along the length (L) direction of the block of material. Fretting pads were machined, and the fretting surfaces were dry-sanded with 600 grit paper. This sanding included removal of the stress singularities that would otherwise exist due to the pad geometry.

## Results and Discussion

Results of testing and average values for groups of similar tests are in Table 2. Temperatures for the tests were from 22C to 27C, and relative humidities were from 19% to 35%. Failure occurred in a grip section for two of the fatigue specimens, but

Table 2 - *Results of Fatigue and Fretting Fatigue Tests*

| Test Type and Environment | 7075-T651 Cycles | 8090-T7 Cycles |
|---|---|---|
| Fatigue with Normal Load in Air | 426 400 | *3 122 400 |
| | #376 200 | *3 037 000 |
| | 164 000 | *3 287 400 |
| Averages | **322 200** | **3 148 933** |
| Fatigue with Normal Load in a Vacuum | 884 600 | *3 231 300 |
| | 510 100 | 1 543 600 |
| | 479 700 | #2 233 000 |
| Averages | **624 800** | **2 335 967** |
| Fretting Fatigue in Air | 32 400 | 157 200 |
| | 78 400 | 85 400 |
| | 89 400 | 74 100 |
| | 60 500 | 53 000 |
| | 67 400 | 207 800 |
| | 70 400 | 82 100 |
| | 72 500 | 68 900 |
| | 98 500 | 122 800 |
| Averages | **71 187** | **106 412** |
| Fretting Fatigue in a Vacuum | 256 300 | 47 600 |
| | 370 800 | 73 200 |
| | 495 600 | 61 000 |
| | 289 700 | 32 700 |
| | 929 100 | 82 600 |
| | 382 900 | 40 800 |
| | 354 200 | 76 300 |
| | 310 300 | 95 500 |
| Averages | **423 612** | **63 712** |

*Specimen did not fail. Test was stopped at the indicated number of cycles.
#Specimen failed in a grip section.

these were considered usable tests because the number of cycles at failure was statistically similar to other fatigue results when compared with fretting fatigue results.

The relative displacement and effective coefficient of sliding friction at the point of fretting contact were not measured in this program. However, analysis of the surfaces of fretted specimens indicated that global sliding occurred. Wear due to the low-friction phenolic pads was not an issue.

Test results were compared to determine the confidence level that means are significantly different, using the 't-test' for statistical significance. Using 95% as the critical confidence level, the results indicate:

- Fatigue lives of the test materials were reduced by the effects of fretting except for the 7075 alloy tested in vacuum where the fatigue lives were reduced by the effects of fretting, but not with 95% confidence.
- Degradation in fatigue life due to fretting is caused primarily by wear mechanisms for the 8090 alloy.
- Environmental effects significantly decrease fatigue life due to fretting in air for the 7075 alloy. The average life in air was 17% of the average life in vacuum.
- The average fretting fatigue lives of 8090 specimens were less in a vacuum, but greater in air, than those of 7075 alloy specimens.
- An unexpected result was that the fretting fatigue lives of 8090 specimens in the air environment were significantly greater than the fretting fatigue lives of 8090 specimens in the vacuum environment. The average life in air was 167% of the average life in vacuum.

## Conclusions

Direct comparison of the numerical results of the two programs cannot be made because of the differences in the analysis of the data and the test programs. However, the conclusions based on these results can be compared, and the conclusions of this testing program confirm and enhance the conclusions of the previous research. Based on evaluation of the results of fretting fatigue and fatigue with normal load tests in this program, it is concluded that:

- Degradation of fatigue life in air due to fretting is caused primarily by wear mechanisms for 8090-T7 aluminum.
- Fretting fatigue life in air was greater than fretting fatigue life in a vacuum for the 8090-T7 aluminum under the conditions of this experimentation. No conclusion was reached as to why this occurred. A possible reason for this result could be crack closure effects facilitated by the air environment. Another reason for this result could be a possible protection of the surfaces caused by the air environment. This might involve a continuously regenerated oxide layer, protecting the metal from the damage caused by cyclic rupture of metal-to-metal welds.
- Environmental mechanisms significantly decrease fatigue life due to fretting in air for 7075-T651 aluminum.
- Fretting fatigue behavior is material dependent.

## Recommendations

The following recommendations from the previous research[4] also are applicable to this program:
- "The base line for fretting fatigue investigations to study the effects of wear and corrosion should be tests conducted in a vacuum or, possibly, an inert environment."
- "General statements of the relative effects of wear and corrosion on the fretting fatigue process are meaningless unless tied to a specific material."

These recommendations apply directly to a larger purpose of this research program, that of standardization of fretting fatigue test methods. Historically, fretting fatigue tests have been designed without the benefit of standards. The tests performed have yielded significant improvements in our understanding of the fretting fatigue process, especially when designed to simulate conditions of contact and environment found in fielded systems. These simulations will always have importance in the design of actual structures, but their usefulness in understanding the basic mechanisms of fretting fatigue often has been limited by ambiguities in the data caused by the lack of standardization. Fretting fatigue testing is extremely complicated because of the large number of process variables and because of the interactions between these variables and the self-induced changes in the tribological system. A purpose of standardization is to improve the repeatability of test data, and this may be accomplished with proper and comprehensive understanding of the sources of uncertainty, especially the interactions that continually change the process.

Investigations of the fundamental mechanisms of fatigue and fretting may contribute to standardization of fretting fatigue testing. The separate fatigue and fretting investigations should attempt to find the effects of environment on the processes, and should use vacuum (or possibly an inert environment) as the base environment.

Specific to the results of this research program, a further recommendation is experimentation to confirm the conclusion that the fretting fatigue life in air is greater than fretting fatigue life in a vacuum for the 8090 material, and determination of possible mechanisms. This will probably require further testing and analysis, including fractographic and metallographic analysis, and may involve investigation of oxide formation and properties.

## Acknowledgment

Alcoa Co. provided financial support for this project, and also the aluminum-lithium material. Their support in this research effort is greatly appreciated.

The supplies, equipment, and facilities were provided by the Quality and Integrity Design Engineering Center of the University of Utah and FASIDE INTERNATIONAL, INC.

## References

[1]    Kantimathi, A., and Alic, J. A., "The Effects of Periodic High Loads on Fretting Fatigue," *Journal of Engineering Materials and Technology*, Vol. 103, July 1981, pp. 223-228.

[2]    Reeves, R. K., and Hoeppner, D. W., "Microstructural and Environmental Effects on Fretting Fatigue," *Wear*, Vol. 47, 1978, pp. 221-229.

[3]    Poon, C., and Hoeppner, D. W., "The Effect of Environment on the Mechanism of Fretting Fatigue," *Wear*, Vol. 52, 1979, pp. 175-191.

[4]    Elliott, C. B. III, "Fretting of 8090 and 7075 Aluminum Alloys Under Interrupted Fatigue Loading Conditions in Air and Vacuum Environments," Dissertation, University of Utah, 1993.

[5]    Elliott, C. B. III, and Hoeppner, D. W., "The Importance of Wear and Corrosion on the Fretting Fatigue Behavior of Two Aluminum Alloys," submitted to *Wear*.

[6]    Elliott, C. B. III and Hoeppner, D. W., "A Fretting Fatigue System Useable in a Scanning Electron Microscope," *Fretting Fatigue*, ESIS 18, R. B. Waterhouse and T. C. Lindley, Eds., Mechanical Engineering Publications, London, 1994, pp. 211-217.

[7]    Georgeson, A. M., "Fretting Fatigue of 8090 and 7075 Aluminum Alloys in Vacuum and Air Environments," Thesis, University of Utah, 1996.

# Fretting Fatigue Crack Nucleation

Jarmila Woodtli,[1a] Oliver von Trzebiatowski,[1b] and Manfred Roth[1c]

**Influence of Ambient Air on Nucleation in Fretting Fatigue**

**REFERENCE:** Woodtli, J., von Trzebiatowski, O., and Roth, M., **"Influence of Ambient Air on Nucleation in Fretting Fatigue,"** *Fretting Fatigue: Current Technology and Practices, ASTM STP 1367*, D. W. Hoeppner, V. Chandrasekaran, and C. B. Elliott, Eds., American Society for Testing and Materials, West Conshohocken, PA, 2000.

**ABSTRACT:** Fretting damage often participates in failures caused by multiple site damage (MSD) as a nucleation process. Fretting itself is an extremely complex degradation process consisting of a combination of mechanical and chemical attacks. In order to avoid damage it is essential to understand which dominating parameter influences the crack nucleation.

The topography of the fracture surface, as well as changes in microstructure provide information about the dominant damage mechanism. An exact differentiation of the predominant damage mechanism can be established by microscopic and metallographic investigations.

Two service failures in axles for ski lifts, which suffered from fretting fatigue, are discussed. Though the macroscopic findings are very similar, the microscopic investigation indicates a completely different crack nucleation mechanism.

**KEYWORDS:** fretting fatigue, nucleation, hydrogen-induced stress corrosion cracking, crevice corrosion, atmospheric corrosion, axles for ski lifts

**Introduction**

Fretting fatigue belongs to a group of failure types caused by multiple site damage. The influence factors can be roughly divided into mechanical and chemical ones. While the mechanical aspects are often the subject of investigation with respect to improving the life cycle, the chemical parameters of service loading are usually discussed in less detail. Examples of malfunctioning of wire ropes [1], ship propeller systems [2] or surgery implants [3] have been observed in very reactive environments as for example in a hostile aqueous environment, sea water or an environment of body fluids. The environment in which fretting takes place can have a positive effect on the nature and degree of the fretting life. Sometimes the aqueous solution may act as a lubricant, thus

---

[1a] Group Leader Failure Analysis/Materials Science, EMPA Federal Laboratories for Materials Testing and Research, Surface- and Joining Technology.

[1b] Group Leader Failure Analysis/Materials Science, EMPA Federal Laboratories for Materials Testing and Research, Surface- and Joining Technology.

[1c] Section Leader Surface- and Joining Technology, EMPA Federal Laboratories for Materials Testing and Research, Surface- and Joining Technology.

reducing friction or separating the surfaces and hence reducing the wear rate [4]. Consequently, the importance of environmental effects can be understood only if the mechanism of the nucleation mode is known on a microscopic level [5, 6].

The two history cases presented show that the ambient atmospheric environment can also strongly influence the crack nucleation in fretting fatigue. A careful microscopic investigation can reveal the mechanism of nucleation.

Both cases described illustrate failures in the axle of bull wheels as used for bedding of the return idle sheave for the rope in ski lifts or gondola cables. Both axles broke by fretting fatigue after only 1 year of service which corresponds to 2,000 service hours. The investigations revealed that corrosion took place within the crack nucleation period and initiated the fatigue crack growth. Although in both cases the corrosion attack was caused by the same atmospheric environment, different mechanisms could be observed within the nucleation.

**Case history 1: Nucleation due to HISCC**

More than 100 axles of the same construction have previously been installed in ski lifts without suffering a fracture. The only change in the construction under discussion was an axial retention provided by 2 discs which were positioned over the belt and fixed by 3 screws on the bull wheel. These 3 screws (M 16) were in 120° circumferential positions relative to one another. The axle material was 34 CrNiMo6, tempered to hardness 280-315 HV10 and toughness (ISO-V-specimen) of 140 $J/cm^2$. The ultimate strength is 900-1000 $N/mm^2$.

The fracture area of the axle developed at the edge between the press fit and the free surface as indicated in Fig. 1. The press fit was the contact area between the axle and the shrink-fitted bull wheel for deflection of the rope. The fracture path was not influenced by the proximate notch effect of the keyway nor through the change of the stiffness at the belt. The residual overload fracture area was very small, less than 5% of the entire cross section thus indicating very low nominal stresses.

Visual inspection revealed that the axles failed by fatigue that was nucleated at 3 singular sites which had exactly the same positions of 120° to one another as the 3 screws (see Fig. 2). Typical for the crack nucleation areas were multiple secondary cracks and a black-brownish color of the fracture and free surface. The surface in the vicinity of each origin showed corrosion pitting. The investigation by scanning electron microscopy showed an intercrystalline fracture topography close to each fracture origin in a 15 to 20 mm deep penny-shaped zone (see Fig. 3). Then, at greater depths the fracture displayed a transgranular path with typical fatigue striations.

The metallographic sectioning was made through the fracture origin. Numerous secondary cracks could be found. The origin of these cracks was often associated with a corrosion pit and the crack path was branched. This branching appeared mostly at the sulfidic inclusions (see Fig. 4). Analysis of corrosion products did not reveal any aggressive elements, thus indicating that the corrosion was caused solely by the environmental water or humidity.

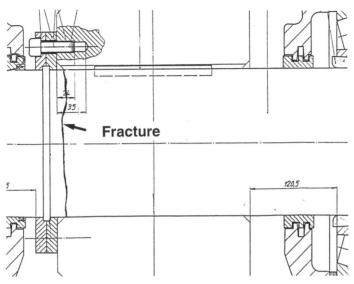

Figure 1: Fracture path of the fatigue fracture at the edge between press fit and free surface

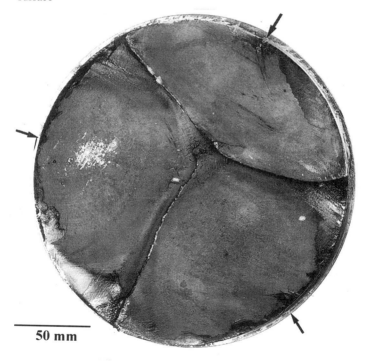

Figure 2: 120°-configuration of 3 crack origins of large fatigue fracture

Figure 3:  Intercrystalline fracture topography within nucleation area in 34CrNiMo6 due to HISCC

Figure 4:  Secondary cracks branched at sulfidic inclusions

The metallographic and fractographic results indicated that the 3 fatigue fracture zones were nucleated by hydrogen-induced stress corrosion cracking (HISCC). This damage was a result of the simultaneous action of absorbed atomic hydrogen and mechanical stress. In this case the hydrogen was a product of crevice corrosion which occurred in the gap between the axle surface and the contacting disc.

## Conclusions

The fractographic and metallographic investigations clearly showed that the fatigue cracks developed within the areas which suffered from hydrogen-induced stress corrosion cracking. This corrosion took place within the three gaps which were created between the disc and the axle surface due to the presence of the 3 screws.

The crevice corrosion was based on a localized acidic environment which differed from that of the entire surroundings. When iron corrodes, the iron dissolution reaction is:
$$Fe \rightarrow Fe^{2+} + 2e^-$$

However, there must be a cathodic reaction to consume the electrons produced. In the absence of oxygen, the cathodic reaction results in hydrogen evolution:
$$2H^+ + 2 e^- \rightarrow 2 H_{ad}$$
$$2 H_{ad} \rightarrow H_2$$

If the hydrogen can be recombined to $H_2$, it will not harm the steel. Under certain conditions the hydrogen can be absorbed. This mechanism dominates in presence of elements such as S, As or Sb, which inhibit the hydrogen recombination, thereby permitting a large fraction of the absorbed hydrogen atoms to enter the material [8]. The absorbed hydrogen interacts with the microstructure, diffuses along grain boundaries, especially into regions with a high strain field and weakens the bond of the grains.

High strength steels with an ultimate stress of 1,000 $N/mm^2$ and greater are generally susceptible to HISCC [7, 9]. Evidence of this damage is usually circumstantial. The crack origin is situated at corrosion pits, the crack path runs intercrystalline. The cracks are mostly branched at sulfide inclusions.

Though in this case the crack nucleation was caused by corrosion, the problem was corrected by modifying the design so as to avoid crevices in the region of contact stress concentration.

## Case history 2: Nucleation due to fretting under atmospheric corrosion

For this gondola the driving unit and hoisting cable tensioning were situated at the lower station. The hoisting cable was guided over the driving sheave and two adjustable deflection discs. The tension required for the hoisting cable amounted to approx. 300 kN per cable strand and was provided by a tensioning arrangement. The upper bull wheel was shrunk with a press fit onto the axle and the lower bull wheel embedded in the bearing. After one year of service which corresponds to approx. 2,000 service hours, the axle broke just at the edge of the press fit of the upper bull wheel. The lower wheel fell away and the sudden relaxation of the cable tension set the hoisting cable into abnormal vibrations.

The axle with a diameter of 280 mm was made of a low alloy steel 42CrMo4, tempered to a hardness of approx. 250 HV10, $R_m$ = 800 - 840MPa and toughness of 36J. The expected fatigue limit as found in the literature is 480MPa. However, fatigue tests which were conducted after the accident showed a lower value of 368 MPa for the fatigue limit (probability level of 95%).

The fracture zone spread mostly beneath the press fit. The origin was situated at the edge between the press fit and the free surface (see Fig. 9). The fatigue fracture showed a convex/concave shape with respect to the cross-section and the residual overload fracture occupied approx. 60% of the total cross-section. The fracture edge within the prime nucleation area was smooth. In addition, numerous secondary crack origins were observed all over the circumference . These were steplike and situated along the fracture edge. The investigation by SEM indicated an intergranular fracture path to a depth of 20 mm (see Fig. 5). Afterwards typical fatigue striations could be found all over the fatigue fracture. The proximal surface near the fracture origin showed numerous secondary cracks which were often accompanied by corrosion pits (see Fig. 6). The metallographical sections revealed that these cracks ran 0.4 mm deep. The path of these cracks was first transgranular and slant oriented with respect to the surface. When the crack reached a depth of 0.2 mm, the fracture mode changed and the crack path became intergranular.

A similar crack modus could be found in one axle, which was prematurely removed from operation. This sample had the same axle construction thus allowing the degree of damage to be studied. The axle was examined by eddy current methods; cracks could be observed at the same location as where the crack nucleation took place in the broken axle. As Fig. 7 shows, these cracks started within corrosion pits and were slant oriented and transgranular. They were filled with corrosion products, branched and 0.8 mm deep. Their morphology indicated a distinct influence of corrosion.

**Fatigue tests**

Fatigue tests were conducted to estimate the critical crack size necessary to initiate stable fatigue crack growth and the number of load cycles until residual fracture. Measurement of fatigue crack growth rates was performed in accordance with the specification ASTM E 647 (Standard Test Method for Measurement of Fatigue Crack Growth Rates) using MT specimens. The results clearly showed that an initial failure size of 0.4 mm loaded by ± 118 N/mm$^2$ would initiate a stable fatigue growth which would reach the critical overload rupture after approx. 2.5 million loading cycles.

The fractographic investigation of one specimen indicated fatigue striations (see Fig. 8). There was no region with intercrystalline fracture morphology.

**Conclusions**

The fractographic investigation showed a typical fretting mechanism at the nucleation site of the axle. Based on the metallographic results there was a significant influence of atmospheric corrosion on the mechanism of nucleation. The crack nuclei estimated metallographically amounted to 0.4 mm-0.8 mm. These were in accordance with the fatigue crack growth rate measurements and deep enough to enable the unstable crack

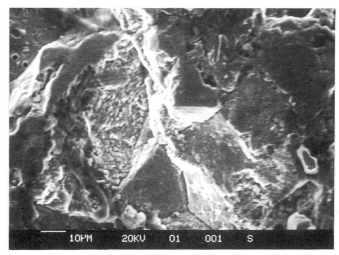

Figure 5:  Intercrystalline fracture topography within nucleation area in 34CrNiMo6 due to HISCC

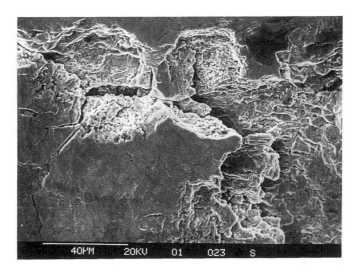

Figure 6:  Proximal surface to fracture origin with corrosion pits and secondary cracks

Figure 7:   Numerous slant oriented fretting cracks nucleated at corrosion pits and filled
with corrosion products

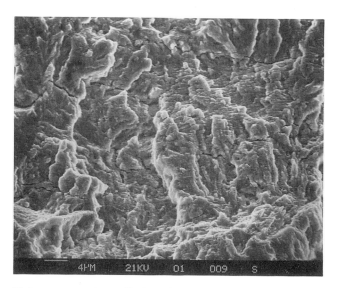

Figure 8:   Fatigue test specimen displays fatigue striation over the whole range of
crack growth rates

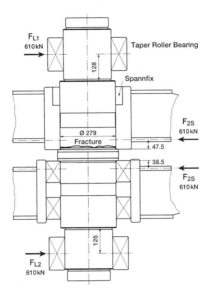

Figure 9:   Fracture path of fatigue fracture at the edge of press fit area.

growth. The purely intercrystalline crack path until the crack depth of 20 mm also indicates an influence of atmospheric corrosion in the early stage of fatigue life.

Considering the results of this investigation, corrective measures must be taken at the design stage since the influence of the atmospheric environment is inherent to the system.

## Summary

Two service failures in axles for ski lifts, which suffered from fretting fatigue, are discussed. Though the macroscopical findings are very similar, the microscopical investigation indicates completely different crack nucleation mechanisms. While in one case the hydrogen-induced stress corrosion cracking due to the crevic corrosion took place, the second case showed a typical fretting mechanism favored by atmospheric corrosion attack. In both cases the problem can be corrected by appropriate design.

## References

[1] Takeuchi, M. and Waterhouse, R. B., "The initiation and propagation of fatigue cracks under the influence of fretting in 0.64C roping steel wires in air and sea water," *Environment Assisted Fatigue, EGF 7* (Edited by P. Scott), Mechanical Eng Publi, London, 1990, pp 367-379.

[2] Heck J. W. and Baker E. J., "Marine propeller shaft casualties," *Transactions of the Society of Naval Architects and Marine Engineers*, Vol. 71, 1963, pp. 327-346.

[3] Smethurst E. and Waterhouse R. B., "Causes of Failure in Total HIP Prostheses," *J. Mater. Sci. 12*, 1977, pp. 1781-1792.

[4] Czichos, H., *"Tribology - A System Approach to the Science and Technology of Friction, Lubrication and Wear,"* Elsevier, 1978, Amsterdam.

[5] Hoeppner, D. W., "Environmental effects in fretting fatigue," *Fretting Fatigue*, edited by R.B. Waterhouse, Applied Sci Publishers, UK, 1981, pp. 143-158.

[6] Hoeppner D. W. and Gates F. L., "Material/structure degradation due to fretting and fretting initiated fatigue," *Canadian Aeronautics Space J. 27*, 1981, pp. 213-221.

[7] Stellweg, B., and Kaesche, H., "Kinetik der wasserstoffinduzierten Spannungsrisskorrosion," *Werkstoffe und Korrosion 33*, 1982, pp. 274-280.

[8] Pöpperling, R., Schenk, W. and Venkateswarlu, J., "Hydrogen induced stress corrosion cracking of steels subjected to dynamic loading involving plastic deformation in promoter free electrolyte solutions," *Werkstoffe und Korrosion 36*, 1985, pp. 389-400.

[9] Thompson, W. A., "Hydrogen-assisted fracture at notches," *Materials Science and Technology*, Vol 1, 1985, pp. 711-718.

M. P. Szolwinski,[1] G. Harish,[2] P. A. McVeigh,[2] and T. N. Farris[3]

**Experimental Study of Fretting Crack Nucleation in Aerospace Alloys with Emphasis on Life Prediction**

**REFERENCE:** Szolwinski, M. P., Harish, G., McVeigh, P. A., and Farris, T. N., **"Experimental Study of Fretting Crack Nucleation in Aerospace Alloys with Emphasis on Life Prediction,"** *Fretting Fatigue: Current Technology and Practices, ASTM STP 1367,* D. W. Hoeppner, V. Chandrasekaran, and C. B. Elliott, Eds., American Society for Testing and Materials, West Conshohocken, PA, 2000.

**ABSTRACT:** A statistically-designed experimental program engendered to validate an analytical approach for the prediction of fretting crack nucleation in 2024-T351 aluminum alloy has been completed. The test results indicate that the near-surface cyclic contact stress and strain field can be juxtaposed with a multiaxial fatigue life parameter relying on uniaxial strain-life constants to predict crack nucleation for a wide range of load intensities and conditions representative of those experienced in riveted joints. With this approach validated, efforts have been initiated to predict fretting-induced fatigue failures in riveted single lap joint structures. Research was targeted at characterizing the conditions at and around the rivet/hole interface, including finite element modeling of both the mechanics of load transfer in riveted joints and the residual stress field introduced during the rivet installation process. Model results and an ancillary set of fatigue tests of single lap joint test articles have identified a strong correlation among riveting process parameters, the mechanics of load transfer, and the subsequent tribological and fatigue degradation of the joints. Final comments are offered regarding the ability of this integrated approach to predict the fatigue performance of riveted lap joint structures.

**KEYWORDS:** fretting fatigue, crack nucleation, aluminum alloy, riveted lap joints, load transfer, aircraft structures, fatigue performance, life prediction

**Predicting Fretting Crack Nucleation**

Fretting, a synergistic damage mechanism experienced at clamped, contacting surfaces in a wide variety of mechanical systems subjected to oscillatory loading, plays a particularly critical role in the nucleation of fatigue cracks in aging aircraft systems.

[1]Assistant Professor, Department of Mechanical Engineering, Aeronautical Engineering, and Mechanics, Jonsson Engineering Center, Rensselaer Polytechnic Institute, 110 8th Street, Troy, NY 12180-3590.
[2]Research Assistants and [3]Professor and Head, respectively, School of Aeronautics & Astronautics, Purdue University, 1282 Grissom Hall, West Lafayette, IN 47907-1282.

From riveted aluminum lap joints in the fuselage to titanium dovetail/disc assemblies in rotating jet engines, the localized near-surface stresses, strains and surface microslip associated with fretting contact lead to a detrimental tripartite combination of corrosion, wear and fatigue phenomena. An understanding of the mechanics of fretting crack nucleation is crucial to those interested in not only airframe structural design but also technologies for life prediction and maintenance of aging aircraft systems, both civilian and military [1].

Traditional design-based approaches to fatigue rely on a juxtaposition of estimated or measured applied structural loads with either stress concentration factors or factors of safety to account for local increases in stress conditions due to geometric features such as cutouts, notches, and holes. Early attempts to address the decrease in fatigue performance of riveted aircraft structures [2, 3] due to fretting suggested the use of wholly empirical "knockdown" factors estimated from a limited set of laboratory tests with applied in-flight loads.

The goal of this research was to formulate a mechanics-based understanding of the localized conditions responsible for the nucleation of fretting fatigue damage at and around the rivet/hole interface in riveted lap joint structures. Steered by reports of fretting damage and fatigue cracks from inspections of both aircraft structures [4] and laboratory test articles [5, 6], initial efforts were directed at modeling the contact between rivet shank and hole in hopes of identifying a driving influence for fretting crack nucleation. A finite element model [7, 8] of a riveted lap joint incorporating the effects of load transfer and rivet/hole interaction in the absence of interference revealed that the tractions around the hole periphery are reminiscent of the classical Mindlin/Cattaneo distribution resulting from the partial slip of a cylindrical indenter on an elastic halfspace.

*Fretting Fatigue Experiments*

These results motivated the development of a well-characterized test system capable of controlled generation and automated monitoring of fretting fatigue conditions. The single-actuator design, pictured in Figure 1, applies a tangential load through the deformation of a fretting fixture attached to two cylindrical pads pressed into contact with the flat faces of a standard "dogbone" fatigue specimen. This common type of fretting fatigue test fixture relies on an in-phase bulk load applied to the lower portion of the specimen to induce the tangential load on the pads. Complete presentation of this arrangement, including the integrated array of sensors and data acquisition system used to characterize the normal, tangential, and bulk loading, is available elsewhere [9].

This setup was exercised within the context of a statistically-designed test matrix to probe the effects of factors such as normal load, contact width, stick zone width, and bulk stress magnitude on fretting fatigue failures in an aluminum alloy common to aircraft structures, 2024-T351. Failure was observed in every test at the trailing edge of contact (the edge of contact nearest the actuator end of the specimen), consistent with results reported in the literature [10] for a similar experimental setup. Figure 2 presents the observed failure lifetimes for the series of fretting tests versus applied bulk stress along with traditional unnotched and notched ($K_t$ = 2.0) S-N data for a 2024 series aluminum alloy [11]. (Note that an exhaustive summary of the test conditions, fatigue lives, and fractographic observations can also be found elsewhere [9].)

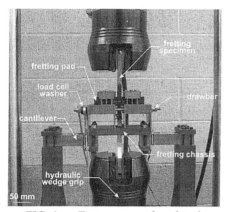

FIG. 1 — *Test setup employed in the investigation of fretting fatigue crack nucleation.*

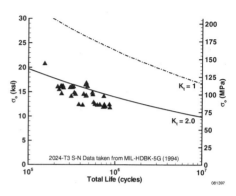

FIG. 2 — *Fretting fatigue test data presented versus applied bulk stress, with traditional S-N data for $K_t = 1$ and $K_t = 2$.*

*Fretting Fatigue Life Predictions*

This severe reduction of fatigue performance illustrates emphatically the need to incorporate the localized contact stresses into any design-based metrics for fretting fatigue failures. Furthermore, scrutinizing the distribution of the component of contact stress tangential to the surface (Figure 3a) along with a metallograph of a fretting fatigue crack (Figure 3b) provides qualitative insight into the mechanics of fretting fatigue crack nucleation. The tensile peak at the edge of contact, $x/a = +1$, coincides with both the location and orientation of the nucleated fatigue damage.

Note that the distribution of tangential stress from a finite element model of the experimental contact including the specimen bulk stress [12] agrees well at the trailing edge with the distribution from an approximate closed-form, analytical solution. In spite of the shift in distribution of stick and slip introduced by the bulk stress not included in the classical Mindlin/Cattaneo contact formulation, this agreement allows for the peak in edge-of-contact stress to be approximated as a superposition of contact stress from the Mindlin/Cattaneo problem and the bulk stress. As noted on Figure 3a, this contact stress component can be determined from the applied normal and tangential loads, maximum Hertzian pressure, and slip zone friction coefficient.

It is also noted that at the edge of contact, $x = +a$, $z = 0$, both the pressure and shear tractions are identically zero, leading to a "quasi-uniaxial" state of stress at this point, an observation also made by Fouvry, et al. [13]. Szolwinski and Farris [14] proposed the use of this edge-of-contact tangential stress in conjunction with a multiaxial fatigue life parameter to link quantitatively the near-surface cyclic stress and strain fields to the nucleation of fretting fatigue damage. The relationship between reversals to nucleation ($2N_f$) and the product of strain amplitude and maximum stress during a complete loading cycle normal to the principal plane (the plane of the nucleated crack) is:

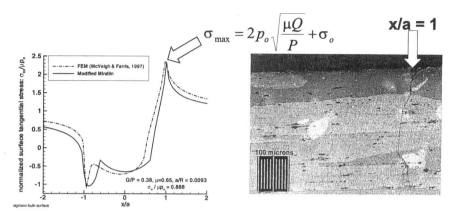

FIG. 3 — *The distribution of* $\sigma_{xx}$, *the component of stress tangential to the contact interface at the surface, z = 0, from both finite element results that incorporate the specimen bulk stress explicitly and a closed-form approximate method of superposition. The peak in tangential stress at the edge of contact coincides with the location and orientation of observed fretting fatigue damage, as indicated on the photomicrograph of a sectioned fretting fatigue specimen.*

$$\Gamma = \sigma_{max} (\Delta\varepsilon/2) = [(\sigma_f)^2/E](2N_f)^{2b} + \sigma_f\varepsilon_f(2N_f)^{b+c} \qquad (2)$$

The right-hand side of Eq. 2 is an expression of the uniaxial strain-life curve of a given material generated from strain-controlled uniaxial fatigue tests on smooth-bar specimens. With the inclusion of the $\sigma_{max}$ term, this expression is identical to the familiar Smith-Watston-Topper equation used to account for mean stress effects on crack nucleation during uniaxial loading [15] that has been shown to hold for multiaxial states of stress [16]. The leading coefficients—$\sigma_f$ (fatigue strength coefficient) and $\varepsilon_f$ (fatigue ductility coefficient); and exponents b (fatigue strength exponent or Basquin's exponent) and c (fatigue ductility exponent)—are fatigue properties determined from linear fits of applied strain amplitude ($\Delta\varepsilon/2$) and observed load reversals ($2N_f$) to failure plotted on log-log axes. For more in-depth treatment of determination of these fatigue constants and subsequent use of the strain-life approach, see references [17] or [18].

Values of the fatigue constants for the current work ($\sigma_f$ = 714 MPa, b = -0.078, $\varepsilon_f$ = 0.166, c = -0.538) have been taken from a series of strain-life tests performed on 2024-T351 specimens (E = 74.1 GPa) [19]. Coupling a two-dimensional analysis of the near-surface stress field relying on Westergaard functions [14] with these fatigue constants identifies the location and orientation of the critical plane of crack nucleation as the trailing edge of contact perpendicular to the contact surface, respectively, agreeing with the observed cracking behavior shown in Figure 3. It is noted that while these edge-of-contact cracks may have nucleated initially obliquely to the surface, they turn rapidly under the influence of the region of high tensile stress, which is a result of both the cyclic contact and bulk stresses.

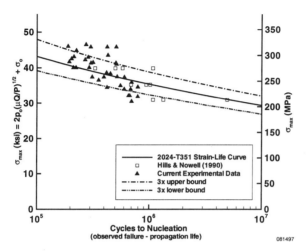

FIG. 4 — A comparison between predicted cycles to crack nucleation under
conditions of fretting fatigue as given by the multiaxial fatigue life parameter and
results from fretting fatigue tests.

Figure 4 presents the strong correlation between the results from the set of statistically
designed fretting fatigue experiments and the tensile peak in tangential stress at the
trailing edge of contact. This peak in tangential stress was calculated from the three-
component force histories recorded during the duration of each experiment and a steady-
state value of friction coefficient, $\mu = 0.65$, as determined in a separate series of
experiments [20]. The values of experimental nucleation lives presented in Figure 4 are
defined as the observed failure life minus an estimated propagation life. The solid line
represents nucleation life predictions as determined from the life criterion that accounts
for combined influence of the multiaxial state of cyclic stress induced by the oscillatory
tangential loading and cyclic bulk loading of the specimen. Data from a similar series of
fretting fatigue experiments for which the requisite load and contact data were reported is
also included on the plot [10].

The estimated propagation life is defined as the propagation of a semi-elliptical
"thumbnail" crack nucleated with minor axis length of 1 mm and major axis width
spanning the entire 12.7-mm width of the fretting specimen, a definition supported by
fractographic observations reported in [9]. Furthermore, only the influence of the cyclic
bulk stress was incorporated into this estimate, as the contact stress field decayed to
negligible levels for depths greater than 1-mm beneath the surface for the contact widths
(which ranged from approximately 2 to 4 mm) used in the series of experiments.

## Predicting Fretting-Induced Failures in Riveted Lap Joints

*Finite Element Modeling of a Riveted Single Lap Joint*

With a viable method for predicting crack nucleation under the influence of the cyclic
stresses, strains and surface microslip induced by fretting contact, attention must turn to

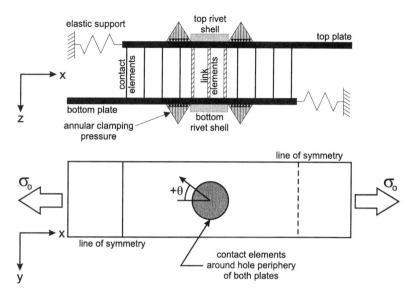

FIG. 5 — *A schematic of the three-dimensional shell model of a riveted single lap joint, including load transfer and through-thickness restraint by the rivet.*

characterizing the nature of fretting contact at and around the rivet/hole interface in riveted joints. Toward this end, a three-dimensional finite element model of a riveted joint was created using shell elements (SHELL43) as implemented in the commercial package ANSYS 5.3[†] to capture out-of-plane effects during load transfer [21]. This formulation, presented schematically in Figure 5, is comprised of two 1.8-mm 2024-T351 plates (E = 73.1 MPa, $\nu$ = 0.33) and a single 5.1-mm diameter rivet, also modeled with two shell elements for consistency with appropriate constraint equations (LINK4) ensuring displacement as a single entity. Three-dimensional point-to-point contact elements (CONTAC52) along the rivet/hole interface and at the faying surface or interface between the plates served to resolve the contact interactions. A bilinear kinematic hardening model was used to model any inelastic material response, with an initial yield stress of 331 MPa and a tangent modulus, $E_t$, of 1551 MPa.

The model dimensions, boundary conditions, and applied loads were selected to represent the area around the rivet/skin interface in the first row of a three-row joint with a large number of rivets in each row spaced 25.4 mm apart. Adjusting the stiffness of the elastic support at the end of the joint allows for the parametric study of the effects of arbitrary load transfer ratios (LTR) on the contact. Finally, the effect of restraint provided by the rivet heads was incorporated through application of an annular pressure on the skin around the rivet.

The surface tractions and hoop stress distribution along the rivet/hole interface at the shell midplane for a case of 38% load transfer through the first row are presented in

[†]    ANSYS 5.3 is through an academic license agreement with Swanson Analysis Systems, Inc., Houston, PA.

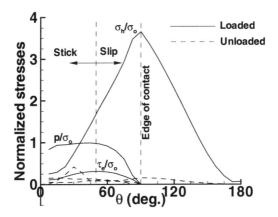

FIG. 6 — *Contact traction/stress results from the shell model for* $\sigma_o$ = 83 MPa, $\mu$ = *0.4, maximum clamping pressure of 83 MPa, and LTR of 38%.*

Figure 6. For the applied bulk load and clamping restraint considered, these results along the rivet/hole interface represent the most severe stress conditions within the model. Examining the distribution of both the pressure and shear tractions clearly reveals the partial slip nature of the contact along the hole periphery, where $\tau_e(\theta) = \mu p(\theta)$ within the regions of slip at the edges of contact. Note that in the absence of initial interference between rivet and hole, the rivet loses contact upon deformation of the hole by the applied bulk stress. Furthermore, similar to the distribution in Figure 3, the hoop stress, $\sigma_h$, has a peak at the edge of contact, ($\theta = 0°$) and decays rapidly away from the edge of contact. This hoop stress is influenced by three distinct effects: the geometric stress concentration due to the panel hole, and the components associated with the normal and shear contact tractions.

In light of the experimental results generated in the fretting fatigue experiments with the simplified cylinder/flat configuration and similar near-surface conditions at the rivet/hole interface, the edge of contact between rivet and hole must be identified qualitatively as a critical location for crack nucleation. Additional model results also point to a region of relative motion between the two panels located in a region around $\theta$ = 135° at a radial distance of 1.6 $R_{rivet}$ = 4 mm from the center of the hole. The combination of relative motion and localized shear tractions associated with frictional load transfer at the plate interface [1] drives the formation of faying surface fretting damage.

*Riveting and Fretting Fatigue Damage*

With the propensity for fretting fatigue damage to nucleate at and around the rivet/hole interface, the need to understand more accurately the initial interference induced by rivet

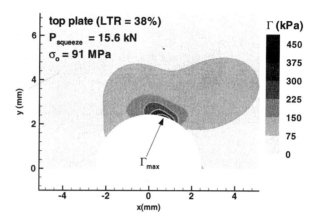

FIG. 12 — *Contours of* Γ *from the finite element model of a single riveted lap joint.*

this mapped residual stress field with the rivet/hole interference and boundary effects arising from the finite dimensions of model. Adjusting the stiffness of boundary elastic supports (similar to those incorporated into the shell model) enabled modeling of the appropriate load transfer ratio in the observed failure row. Finally, application of time-dependent traction boundary conditions to the plates corresponding to the cyclic bulk load allowed for simulation of the localized conditions at and around the rivet/hole interface.

The results from these analyses were then used in conjunction with the multiaxial fatigue life parameter presented earlier to identify the critical location and plane for crack nucleation. Figure 12 presents contours of Γ along the top plate corresponding to the test parameters for LJ003 and a coefficient of friction of $\mu = 0.50$ on all surfaces. As shown, the maximum contour levels are located near the rivet/hole periphery, off the axis perpendicular to the loading direction and behind the location of maximum contact stress. Review of the contact pressure distribution along the periphery reveals that this location corresponds roughly to the edge of contact between the rivet and hole.

Similar analyses were performed for each of the test cases in Table 1 and an auxiliary series of single lap joint tests reported by Harish and Farris [8]. The solid line in Figure 13 presents predictions for the number of cycles to nucleate a 0.5 mm crack using strain-life constants for 2024-T3 sheet reported in the literature [23] ($\sigma_f = 3.15$ GPa, $\varepsilon_f = 0.0688$, b = -0.2467, and c = -0.6339) and the multiaxial fatigue life parameter Γ. Data from the complete set of single lap joint experiments are also presented for comparison.

The abscissa or nucleation life for each of these data points was determined by subtracting an estimated propagation life from the observed cycles to failure of each joint. These estimated propagation periods, which ranged from 15-20% of the observed cycles to failure, were determined by assuming the presence of a 0.5 mm crack in one of the three rivets and tracking subsequent propagation of this crack using a numerical routine developed at Purdue University [24]. Based partially on NASA-FLAGRO solutions, this routine accounts for effects such as crack linkup and the effect of multiple-site damage installation and the resulting residual stress field around the periphery of the hole is an obvious one. The nature of this residual stress field can have a profound impact on the nucleation and subsequent propagation of fatigue cracks emanating from the edge of the hole. Perhaps less obvious is the influence of the clamp-up constraint provided by the installed rivet heads on the potential for fretting damage at the faying surface due to the influence this constraint has on frictional load transfer at the plate/plate interface.

The characterization of the residual stresses induced by rivet installation via the finite element method presented in this volume [1] and elsewhere [1, 20] motivated a series of fatigue tests with riveted single lap joints designed to forge links between rivet installation, nucleation of fatigue damage, and the subsequent fatigue performance of these joints. While several other workers cited earlier [4-6] have conducted similar investigations with both laboratory and service lap joints, none of these programs were designed to focus specifically on linking carefully-controlled conditions at and around the rivet/hole interface to subsequent fretting damage. Only Müller [22] offers results from such an approach, with some anecdotal and microscopic observations presented regarding the nucleation of fatigue damage at the faying surface under the influence of fretting.

*Experimental Details*—The single lap joint specimens used in the series of tests were composed of two 2024-T3 plates, 2.3 mm thick with no cladding. A set of fixturing plates with grip tabs were bolted to the ends of the joint allowing the specimens to be mounted securely in the load train of a 22 kip servo-hydraulic load frame, as shown in Figure 7. While the axes of the upper and lower grips were aligned with the assistance of a strain-gaged alignment specimen and alignment fixture attached to the upper half of the load train, no additional fixturing to minimize out-of-plane bending was incorporated into the setup.

The specimens were joined with MS2047AD6-6 rivets at four squeeze force levels, as summarized in Table 1 and pictured in Figure 8. These levels were based on results from the aforementioned finite element modeling of the rivet installation process to provide a wide range of consistent interference and driven-head clamping constraint, with hopes of generating varied modes of fretting fatigue failure in the joints. The tests were conducted in a laboratory atmosphere at constant-amplitude stress levels specified also in Table 1 at 10 Hz and an R-ratio of 0.03. The stress levels were calculated from the bulk applied load and the cross-sectional area of the assembled joint in the riveted test area.

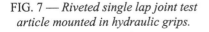

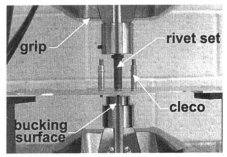

FIG. 7 — *Riveted single lap joint test article mounted in hydraulic grips.*

FIG. 8 — *Setup used for force-controlled riveting.*

TABLE 1 — Test Parameters and Results from Single Lap Joint Fatigue Tests

| Specimen | $P_{squeeze}$ lb$_f$ (kN) | $\sigma_o$ ksi (MPa) | Cycles to Failure |
|---|---|---|---|
| LJ001 | 5000 (22.2) | 13 (89.6) | 233,000 |
| LJ002 | 4250 (18.9) | 13 (89.6) | 231,450 |
| LJ003 | 3500 (15.6) | 13 (89.6) | 164,105 |
| LJ004 | 2500 (11.1) | 16 (110.3) | 41,452 |
| LJ007 | 3500 (15.6) | 10 (68.9) | 566,573 |
| LJ008 | 5000 (22.2) | 16 (110.3) | 141,169 |
| LJ009 | 5000 (22.2) | 10 (68.9) | 872,109 |
| LJ010 | 2500 (11.1) | 14 (96.5) | 69,845 |
| LJ011 | 2500 (11.1) | 10 (68.9) | 232,198 |
| LJ012 | 4250 (18.9) | 10 (68.9) | 590,980 |
| LJ013 | 3500 (15.6) | 16 (110.3) | 61,868 |
| LJ019 | 4250 (18.9) | 16 (110.3) | 72,854 |

*Analysis of Fatigue Failures* — Upon failure, each specimen was inspected carefully in an attempt to characterize the state of damage at and around the rivets in the failure row. This process relied primarily on visual inspection and photographic cataloging of any faying surface damage, the fracture surface of the plates and interior of the rivet holes in the failure row. Select specimens were also sectioned for analysis by scanning electron microscopy (SEM). An abridged summary of these observations follows.

Failure in each specimen, without exception, occurred in either the upper or lower row of rivets, consistent with the higher percentage of load transferred through these rows. The fracture surfaces of these critical rows usually exhibited clear evidence of corner cracks that originated near the rivet holes. Visible multi-site damage (cracks emanating from near multiple rivet holes in a given row) was also often noted in the opposite outer, intact row. No dominant failure mode (lower/upper row, manufactured/driven head side) was present in the series of experiments, providing some assurance in the alignment of the load train and consistency in out-of-plane bending effects.

Correlation was noted, however, between fatigue lifetimes and the controlled experimental parameters, as presented graphically elsewhere in this volume [1]. As supported by Table 1, two trends emerge from the data: (1) as expected, the fatigue life of the joints decreased as the applied load increased, and (2) fatigue life increased with increases in maximum applied squeeze force.

Concomitant with this correlation between squeeze force and life was a characteristic state of damage at and around the rivet holes. Figure 9 displays photographs of the faying surfaces of joints riveted with two different squeeze forces, 11.1 kN and 22.2 kN, respectively, and subjected to an identical bulk stress of 110.3 MPa. The first specimen (LJ015) was riveted with a lower squeeze force and failed after 38,054 cycles, a lifetime over 3.7 times shorter than joint LJ008, manufactured the higher squeeze force.

While part of this difference in life must be attributed to the compressive residual stresses induced by the increased hole expansion associated with the higher squeeze force (see [1]), a second potential source is revealed by comparing the nature of the faying

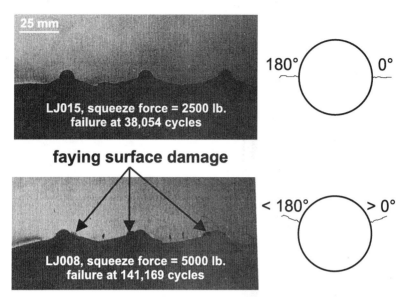

FIG. 9 — *A comparison of faying surface damage and fatigue crack paths for two lap joint specimens subjected to identical bulk loading, but manufactured with two different squeeze forces.*

surface damage in each of the photographs.  As highlighted in the photographs, the joint manufactured with a higher squeeze force shows distinct signs of fretting wear debris on the faying surface near the rivet hole, while the faying surface of the other joint exhibits no signs of such damage.  This faying surface damage was observed primarily in joints manufactured with higher squeeze forces (18.9 kN and 22.2 kN).  As conjectured earlier, this characteristic damage provides direct evidence of the impact that the riveting process parameters have on the load transfer mechanism in the joint:  larger driven heads provide increased clamping constraint at the faying surface leading to an increased amount of frictional load transfer (and fretting damage) at this interface.

A second feature of interest in each of the faying surface photos is the position of the cracks with respect to the circumference of the rivet holes.  For the two lower squeeze forces, the cracks were found to intersect the hole periphery at approximately the 0° and 180° locations (where the direction of loading was along the 90° and 270° directions).  Evidence of fretting wear through the hole thickness was also noted at these locations.  However, crack paths associated with the two higher squeeze forces were observed to intersect the holes above the axis defined by the 0° and 180° positions, away from the applied load as illustrated in Figure 9.

Steered by this macroscopic characterization of fretting damage in the riveted lap joint specimens, focus was shifted to investigating the nature of fatigue damage at and around the rivet/hole interface.  To this end, a scanning electron microscope was used to examine

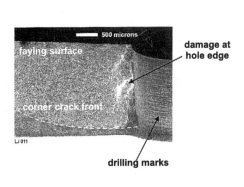

FIG. 10 —*SEM image from specimen LJ011 showing evidence of the nucleation of fatigue damage at the rivet/hole interface.*

FIG. 11 — *SEM image of a region near a hole from specimen LJ007 showing evidence of fretting damage at both the hole edge and faying surface.*

the fracture surfaces of failed joints, manufactured again at 11.1 kN and 22.2 kN squeeze force levels.

Figure 10 presents an image near the edge of a rivet hole from joint LJ011 manufactured with an 11.1 kN squeeze force. Little evidence of faying surface damage was observed upon visual examination. Likewise, little or no damage at the faying surface is noted upon review of the SEM image. Instead, most of the damage is concentrated through the thickness of the plate along the hole edge. In contrast to this image, Figure 11 offers evidence of nucleation of fatigue damage at both the faying surface and hole edges of specimen LJ001 (manufactured with a squeeze force of 22.2 kN).

*Prediction of Lap Joint Failures* — The discovery of multiple nucleation sites with increasing squeeze force further supports the notion of a link between rivet process parameters, a dual load transfer mechanism, and fretting fatigue of riveted joints. To quantify this relationship further and introduce an approach for predicting the failure of riveted aircraft structures, results from a two-dimensional submodel of the rivet/skin assembly including the residual stress field from the rivet installation process were analyzed in the context of the multiaxial fatigue life parameter introduced earlier. This model, while similar in formulation to that of Harish & Farris [21], was implemented with the commercial finite element package ABAQUS/Standard 5.6[†], relying on its contact pair approach to resolve the interactions at the rivet/hole and plate/plate interfaces.

In beginning the analysis of each experiment in Table 1, nodal results from the axisymmetric model used to analyze the rivet installation process for the appropriate squeeze force level were mapped to the nodes of the two-dimensional joint model. Before loading, an equilibrium step was added to allow the contact algorithm to rectify

---

[†]    ABAQUS/Standard 5.6 is available through an academic license agreement with Hibbitt, Karlsson & Sorensen, Inc., Pawtucket, RI.

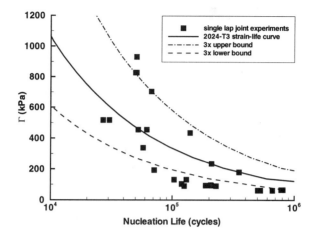

FIG. 13 — *A comparison between predicted cycles to crack nucleation in the top row of a single riveted lap joint and results from two series of lap joint fatigue tests. The lives are plotted against the multiaxial fatigue life parameter, Γ, as determined from knowledge of both the residual stress field induced during rivet installation and the contact conditions at and around the rivet/hole interface.*

present in the joint. While no exhaustive teardown analyses of the experimental articles in this series of tests were performed, the estimated propagation lives did correlate roughly with the number of cycles between the detection of one or more cracks at and around the rivet holes during periodic visual inspections and the final failure of the joints.

The agreement between the predictions and experimental data fall nearly within the same 3x scatter bands that demarcate the data from the reduced-configuration fretting fatigue tests presented previously in Figure 4. This remarkable correlation lends strong support to a design-based metric for fretting fatigue failures in riveted aircraft structures that links manufacturing process parameters (rivet installation conditions), the mechanics and intimately-related tribology of load transfer in riveted lap joints, and basic material fatigue response as captured by readily-available experimental data.

## Conclusions

The expressed goal of this research effort was to link an accurate characterization of the localized conditions at and around the rivet/hole interface in riveted aircraft structures to the fatigue performance of these structures. This goal has been realized by melding insight gleaned from a mechanics-based model of fretting contact, results from a series of fretting fatigue tests, information from finite element modeling of the conditions at and around the rivet and hole in single lap joint structures, and observations from a series of lap joint fatigue tests.

In particular, a qualitative link between rivet process parameters and the nature of fretting damage nucleated within riveted joints has been established: the combination of through-thickness clamping constraint and residual stress field associated with installed rivets impacts the fatigue performance of joints dramatically. Juxtaposing an assessment of these residual stresses with finite element models of single lap joint riveted structures has provided insight into the state of cyclic stress and strain responsible for the nucleation of fretting fatigue damage at and around the rivet/hole interface. Furthermore, this quantitative assessment of the localized conditions has been used successfully with a fatigue life parameter to predict the fatigue performance of riveted single lap joint structures manufactured under controlled conditions.

## Acknowledgment

This work was supported in part under contracts F49620-93-1-0377 and F49620-98-1-0293 from the United States Air Force Office of Scientific Research (AFOSR), issued to investigate fundamental concerns facing the fleet of aging aircraft, both civilian and military.

## References

[1]   Farris, T. N., Szolwinski, M. P., and Harish, G., "Fretting in Aerospace Structures and Materials," *Fretting Fatigue: Current Technology and Practices, ASTM STP 1367*, D. W. Hoeppner, V. Chandrasekaran, and C. B. Elliott, Eds., American Society for Testing and Materials, West Conshohocken, PA, 1999.

[2]   Harris, W. J., *Metallic Fatigue*, Pergamon Press, New York, NY, 1961.

[3]   Heywood, R. B., *Designing Against the Fatigue of Metals*, Jarrold and Sons, Norwich, Great Britain, 1962.

[4]   Piascik, R. S. and Willard, S. A., "The Characteristics of Fatigue Damage in the Fuselage Riveted Lap Splice Joint," NASA/TP-97-206257, NASA Langley Research Center, Hampton, VA, 1997.

[5]   Fawaz, S. A., "Fatigue Crack Growth in Riveted Joints," Ph.D. Thesis, Delft University, The Netherlands, 1997.

[6]   Hoeppner, D. W., Elliot, C. B., III, and Moesser, M. W., "The Role of Fretting Fatigue on Aircraft Rivet Hole Cracking," DOT/FAA/AR-96/10, Federal Aviation Administration, Salt Lake City, UT, 1996.

[7]   Harish, G., Szolwinski, M. P., and Farris, T. N., "Finite Element Modeling of Rivet Installation and Riveted Joints for Prediction of Fretting Crack Nucleation," in *Joint DoD/FAA/NASA Conference on Aging Aircraft*, Ogden, Utah, USA, 1997, pp. 647--663.

[8]   Harish, G. and Farris, T. N., "Effect of Fretting Contact Stresses on Crack Nucleation in Riveted Lapjoints," in *A Collection of Technical Papers, Proceedings of 39th AIAA/ASME/ASCE Structures, Structural Dynamics and Materials Conference*, Long Beach, CA, 1998, pp. 383-391.

[9]   Szolwinski, M. P. and Farris, T. N., "Observation, Analysis and Prediction of Fretting Fatigue in 2024-T351 Aluminum Alloy," *Wear*, Vol. 221, No. 1, 1998, pp. 24-36.

[10] Nowell, D. and Hills, D. A., "Crack Initiation Criteria in Fretting Fatigue," *Wear*, Vol. 136, 1990, pp. 329-343.

[11] MIL-HDBK-5G, *Metallic Material and Elements for Aerospace Vehicle Structures*, Vol. 1, Defense Printing Service Detachment Office, Philadelphia, PA, 1994.

[12] McVeigh, P. A. and Farris, T. N., "Finite Element Analysis of Fretting Stresses," *Journal of Tribology*, Vol. 119, No. 4, 1997, pp. 797-801.

[13] Fouvry, S., Kapsa, P., Vincent, L., and Dang Van, K., "Theoretical Analysis of Fatigue Cracking Under Dry Friction for Fretting Loading Conditions," *Wear*, Vol. 195, 1996, pp. 21-34.

[14] Szolwinski, M. P. and Farris, T. N., "Mechanics of Fretting Fatigue Crack Formation," *Wear*, Vol. 198, 1996, pp. 93-107.

[15] Smith, K. N., Watson, P., and Topper, T. H., "A Stress-Strain Function for the Fatigue of Metals," *Journal of Materials*, Vol. 5, No. 4, 1970, pp. 767-778.

[16] Socie, D., "Critical Plane Approaches for Multiaxial Fatigue Damage Assessment," *Advances in Multiaxial Fatigue, ASTM STP 1191*, D. L. McDowell and R. Ellis, Eds., American Society for Testing and Materials, Philadelphia, PA, 1993, pp. 7-36.

[17] Bannantine, J. A., Comer, J. J., and Handrock, J. L., *Fundamentals of Metal Fatigue Analysis*, Prentice-Hall, Englewood Cliffs, NJ, 1990.

[18] Hertzberg, R. W., *Deformation and Fracture Mechanics of Engineering Materials*, Wiley and Sons, New York, NY, 1976.

[19] Blatt, P. A., "Evaluation of Fatigue Crack Initiation Behavior of An Experimental Ternary Aluminum-Lithium Alloy," Master's Thesis, Purdue University, West Lafayette, IN, 1990.

[20] Szolwinski, M. P., "The Mechanics and Tribology of Fretting Fatigue With Application to Riveted Lap Joints," Ph.D., Purdue University, West Lafayette, IN, 1998.

[21] Harish, G. and Farris, T. N., "Modeling of Skin/Rivet Contact: Application to Fretting Fatigue," in *A Collection of Technical Papers, Proceedings of 38th AIAA/ASME/ASCE Structures, Structural Dynamics and Materials Conference, Volume 4*, Kissimmee, FL, 1997, pp. 2761-2771.

[22] Müller, R. P. G., "An Experimental and Analytical Investigation on the Fatigue Behavior of Fuselage Riveted Lap Joints," Ph.D., Delft University of Technology, The Netherlands, 1995.

[23] Boller, C. and Seeger, T., *Materials Data for Cyclic Loading, Part D: Aluminum and Titanium Alloys*, Materials Science Monographs, C. Laird, Ed., Vol. 42D, Elsevier, Amsterdam, 1987.

[24] Wang, H. L., Buhler, K., and Grandt, A. F., Jr., "Evaluation of Multiple Site Damage in Lap Joint Specimens," in *1995 USAF Structural Integrity Program Conference*, San Antonio, TX, 1995, pp. 21-38.

K. Kondoh[1] and Y. Mutoh[2]

Crack Behavior in the Early Stage of Fretting Fatigue Fracture

---

**REFERENCE:** Kondoh, K. and Mutoh, Y., **"Crack Behavior in the Early Stage of Fretting Fatigue Fracture,"** *Fretting Fatigue: Current Technology and Practices, ASTM STP 1367,* D. W. Hoeppner, V. Chandrasekaran, and C. B. Elliott, Eds., American Society for Testing and Materials, West Conshohocken, PA, 2000.

**ABSTRACT :** The initial stage of fretting fatigue crack growth is significantly influenced by tangential force induced by fretting action along the contact surface, where both mix-mode crack growth and short crack problems are involved. A finite element procedure for the determination of the stress field near the contact surface with friction and of stress intensity factors for fretting fatigue cracks has been introduced. Fretting fatigue tests at the stress amplitude lower than the fretting fatigue limit and mix-mode crack growth tests have been carried out. The influence of Mode II component of stress intensity factor, which accelerates the fatigue crack growth rate, is more significant for shorter and inclined cracks. The fretting fatigue cracks observed at the fretting fatigue limit conventionally defined at $10^7$ cycles were not arrested but continued propagating. Non-propagating cracks were observed at stress amplitudes much lower than the conventional fretting fatigue limit.
**KEY WORDS** : fretting fatigue, mix mode, small crack, stress intensity factor, fatigue limit

## Introduction

It is known that a fatigue crack initiates near the edge of contact in the very early stages of life in fretting fatigue compared to the case of plain fatigue (without fretting, notch, etc.) [1-4]. The initiated fretting fatigue crack is significantly accelerated under high stress concentration superimposing applied stress and tangential force (frictional force) on the contact surface [3, 4]. At the same time the initial fretting fatigue crack is under the complicated mix-mode condition combining Mode I and Mode II due to the tangential force. Various researchers have observed a fretting fatigue crack at the fatigue limit defined $10^7$ cycles, which has been often called the non-propagating (or arrested)

---

[1]Graduate School, Nagaoka University of Technology, Nagaoka-shi 940-2188 Japan.

[2]Department of Mechanical Engineering, Nagaoka University of Technology, Nagaoka-shi, 940-2188 Japan.

crack. However, the cracks observed at the fatigue limit are not always confirmed to arrest under further fatigue cycles. Although there were several reports on fracture mechanical approach for predicting fatigue life and fatigue strength, the Mode I crack was often assumed in these reports [1-6]. Recently some reports on Mode I and Mode II stress intensity factors for fretting fatigue cracks have been published [7]. In order to understand the accelerated propagation and arresting behavior of fretting fatigue cracks in the early stage of life, further efforts should be devoted to studies on stress field and stress intensity factors for a fretting fatigue crack, as well as, precise observations of fretting fatigue crack initiation and growth processes.

In the present study, Mode I and Mode II stress intensity factors for fretting fatigue cracks were analyzed using the finite element technique. Fretting fatigue tests, up to $10^8$ cycles, were carried out to observe crack growth direction and arresting behavior of fretting crack. Combining these results, propagating and arresting behaviors of fretting fatigue cracks are discussed in detail.

## Finite Element Procedure

The two-dimensional plane strain finite element analysis was carried out using the commercial code MARC. The finite element model (FEM) with contact between specimen and contact pad is shown in Fig. 1, where contact-friction elements were introduced along the contact surface. Distributions of contact pressure and tangential stress along the contact surface as well as the stress-strain distribution near the contact region were evaluated. It was confirmed that the total tangential force along the contact surface estimated by the FEM analysis almost coincided with the experimental value measured during the fretting fatigue test indicated later.

The finite element model with a fretting fatigue crack at the contact edge shown in Fig. 2, where the contact pressure and the tangential force estimated above were distributed along the contact surface, was used to estimate stress intensity factors under fretting fatigue conditions. Four kinds of crack angle, 45°, 60°, 75° and 90°, were prepared. Mode I and Mode II stress intensity factors were evaluated according to the following equations based on the crack surface displacements [8],

$$u = \frac{K_I}{2G}\sqrt{\frac{r}{2\pi}}\left\{\sin\frac{\theta}{2}\left(\kappa+1-2\cos^2\frac{\theta}{2}\right)\right\} \qquad (1)$$

$$v = \frac{K_{II}}{2G}\sqrt{\frac{r}{2\pi}}\left\{\sin\frac{\theta}{2}\left(\kappa+1+2\cos^2\frac{\theta}{2}\right)\right\} \qquad (2)$$

where r is the distance from the crack tip, G is given as E/2(1-v), E Young's modulus, v Poisson's ratio, κ=3-4v for plane strain.

In the present model, two-dimensional plane strain eight-node elements were used and a singular element using quarter-node technique was introduced at the crack tip. The material properties (shown in Table 2) and the fretting conditions (average contact pressure: 60 MPa, coefficient of friction: 0.8) in the FEM analysis coincided with those in experiments indicated in a later section. .

**Fretting Fatigue Experiments**

The materials used for the specimen and the contact pad were structural steels, JIS SM430A, the chemical composition and mechanical properties of which are shown in Tables 1 and 2, respectively. The experimental set-up is shown in Fig. 3. The specimen had a 40 mm long gauge length with two parallel flat surfaces for pressing a pair of contact pads. The contact pad had a flat, 2 mm long and 4 mm wide, contact surface. The contact pads were mounted in the contact pad holder, on which strain gauges were attached to measure the tangential force. Before testing the specimens and pads were polished in the longitudinal direction using successively finer emery paper down to a grit size of 1500 and degreased with acetone. The contact pressure of 60 MPa was applied to the contact pads using a calibrated proving ring. The frictional force was also measured using strain gauges attached to the contact pad holder. The tests were performed in air at frequencies ranging from 20 to 32 Hz and stress ratio R of -1 using a conventional servo-hydraulic testing machine. The fretting fatigue crack path on the longitudinal cross-section of the specimens tested up to $10^7$ and $10^8$ cycles was observed using a scanning electron microscope (SEM).

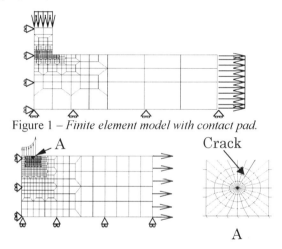

Figure 1 – *Finite element model with contact pad.*

Crack

A

Figure 2 – *Finite element model with a fretting fatigue crack.*

Table 1 – *Chemical composition of the material used. (wt%)*

|             | C    | Si   | Mn   | P    | S    |
|-------------|------|------|------|------|------|
| Specimen    | 0.19 | 0.20 | 1.04 | 0.19 | 0.02 |
| Contact pad | 0.16 | 0.16 | 0.02 | 0.64 | 0.01 |

Table 2 – *Mechanical properties of the material used.*

|             | Yield strength $\sigma_y$ (MPa) | Tensile strength $\sigma_B$ (MPa) | Elongation $\Phi$ (%) |
|-------------|------------------------------|--------------------------------|----------------------|
| Specimen    | 467                          | 559                            | 20.4                 |
| Contact pad | 270                          | 480                            | 21.1                 |

## Mix-mode Fatigue Crack Growth Experiments

A mix-mode fatigue crack growth test, modelled after Richard and Benitz [9], was carried out to discuss fretting fatigue crack growth behavior in the early stage of life. The material used was the same as the fretting fatigue specimen. The specimen had a flat, 40 mm long and 25 mm wide, gauge part with a center notch 2 mm long. The experimental set-up is shown in Fig. 4. The test was performed in air at a frequency of 20 Hz and a stress ratio R of -1 using the same testing machine as the fretting fatigue test. A crack length was measured using a traveling microscope (x100). Stress intensity factors were evaluated using the following equations [10]

$$K_I = F_I \frac{P}{2Wt} \sqrt{\pi a} \qquad K_{II} = F_{II} \frac{P}{2Wt} \sqrt{\pi a} \qquad (3)$$

where $F_I$ and $F_{II}$ are geometrical functions, values of which were estimated from the figures in Ref. [10], 2W the specimen width (25 mm), t the specimen thickness (2 mm) and P the applied load

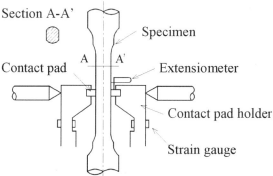

Figure 3 – *Fretting fatigue test set-up.*

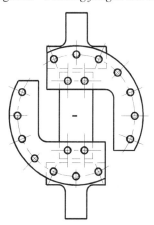

Figure 4 – *Mix-mode fatigue crack growth test set-up.*

## Results and Discussion

*Orientation of Fretting Fatigue Crack Path*

From the results of the FEM analysis, the directions and magnitudes of the principal stresses in the fretting fatigue specimen with applied stress of 175 MPa are shown in Fig. 5. The maximum principal stress is found at the contact edge, where an inclined angle of the plane normal to the principal stress is about 60°.

An S-N curve for fretting fatigue is shown in Fig. 6. The test conditions and measured values in the fretting fatigue experiments are listed in Table 3. Fretting fatigue cracks observed in the specimens tested at 175 MPa, which is lower than the conventional fretting fatigue limit (195 MPa) defined $10^7$ cycles, up to $10^7$ and $10^8$ cycles are shown in Fig. 7. From these observations, it was found that fretting fatigue cracks initiated near the contact edge and the initial inclined angle of the cracks was about 60°. From these observations and the FEM analysis, it is suggested that a fretting fatigue crack initiates at the maximum tangential stress (frictional force) point and propagates along the plane normal to the principal stress, as shown in Fig. 5. The lengths of fretting fatigue cracks were 190 μm at $10^7$ cycles and 290 μm at $10^8$ cycles. Since a fretting fatigue crack propagated at the stress amplitude of 175 MPa, which was lower than the conventional fretting fatigue limit (195 MPa), the so-called non-propagating crack found at the conventional fretting fatigue limit would not be a non-propagating crack but a propagating crack.

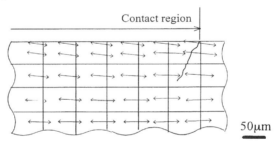

Figure 5 – *Distribution of principal stress under fretting contact region, fretting crack trace observed.*

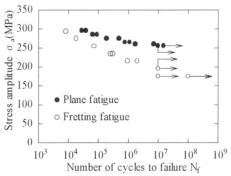

Figure 6 – *S-N curve for fretting fatigue.*

Table 3 – *Fretting fatigue test conditions and the results.*

| Stress amplitude (MPa) | Contact pressure (MPa) | Relative slip amplitude (μm) | Tangential force coefficient | Number of cycles to failure |
|---|---|---|---|---|
| 294 | 60 | 32.0 | 0.783 | $8.40*10^3$ |
| 274 | 60 | 31.3 | 0.861 | $1.84*10^4$ |
| 254 | 60 | 29.8 | 0.872 | $7.30*10^4$ |
| 235 | 60 | 28.3 | 0.753 | $2.60*10^5$ |
| 235 | 60 | 22.8 | — | $3.08*10^5$ |
| 215 | 60 | 20.6 | 0.749 | $9.56*10^5$ |
| 215 | 60 | 14.8 | 0.454 | $2.00*10^6$ |
| 196 | 60 | 9.2 | 0.368 | $>10^7$ |
| 196 | 60 | 8.6 | 0.424 | $>10^7$ |
| 175 | 60 | 12.5 | 0.415 | $>10^7$ |
| 175 | 60 | 11.5 | 0.365 | $>10^8$ |

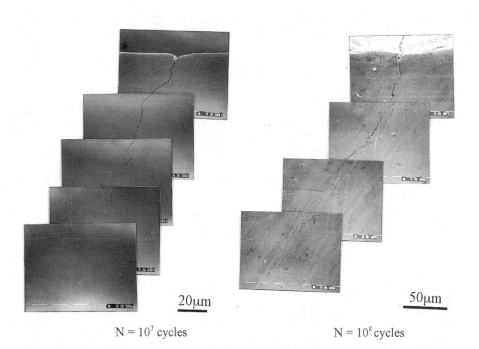

20μm                                50μm

$N = 10^7$ cycles                   $N = 10^8$ cycles

Figure 7 – *Fretting fatigue cracks observed in the specimens tested at 175 MPa, which is lower than the fretting fatigue limit, up to $10^7$ and $10^8$ cycles.*

*Stress Intensity Factors Under Fretting Condition*

From the results of FEM analysis, distributions of contact pressure and tangential stress are shown in Fig. 8. From the figure, a high concentration of contact pressure and tangential stress at the edge of contact area was observed. The relationships between crack length and stress intensity factors $K_I$ and $K_{II}$ for different crack angles are shown in Fig. 9. These are reported up to the maximum load. Based on these, it is obvious that the ratio of the $K_{II}$ component to the $K_I$ component is higher for smaller cracks and decreases with increasing crack length. However, in the case of a high inclined angle, for example 45°, the ratio of the $K_{II}$ component is high even for longer cracks. The ratio of $K_{II}$ to $K_I$ increases with increasing inclined angle of the crack. Therefore, the shorter the crack length and the larger the inclined angle of the crack, the higher the $K_{II}$ component and then the higher the influence of Mode II on fretting fatigue crack growth.

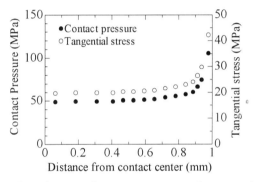

Figure 8 – *Distributions of contact pressure and tangential stress.*

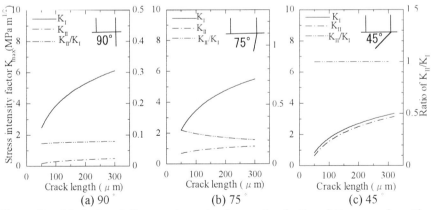

Figure 9 – *Mode I and II stress intensity factors for fretting fatigue cracks with (a) Inclined angle 90°, (b) 75° and (c) 45°.*

*Mix-mode Fatigue Crack Growth*

The fatigue crack growth curves for the pure Mode I and the mix-mode (the inclined angle: 60°, ratio of $K_{II}/K_I$: 0.81) are shown in Fig. 10. From the figure, it is obvious that crack growth rate is accelerated and the threshold stress intensity factor is decreased due to the Mode II component of stress intensity factor $K_{II}$. A similar relationship can be obtained reducing the data obtained by Quian-Fatemi [11]. Therefore, the Mode II component will accelerate the fretting fatigue crack growth rate in the early stage of life and also reduce the fretting fatigue limit.

*Fretting Fatigue Limit*

As mentioned in the previous section, a crack observed at the conventional fretting fatigue limit (195 MPa) defined at $10^7$ cycles was not a non-propagating crack but a propagating crack. Therefore, the present fretting fatigue limit of 195 MPa is the apparent fatigue limit under the conventional definition of fatigue limit. In this section, the true fretting fatigue limit with a non-propagating crack has been estimated based on the FEM analysis and the mix-mode crack growth test result.

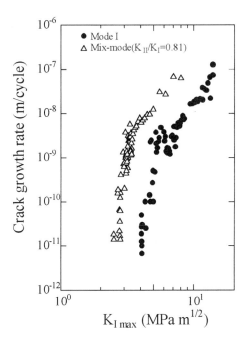

Figure 10 – *Mix-mode fatigue crack growth curve.*

From the results of the FEM analysis, relationships between stress intensity factors and crack length at the stress amplitude of 175 MPa are shown in Fig. 11, where the crack inclined angle of 60° was selected according to experimental observation. In the figure, the stress intensity factor based on the maximum tensile stress at the crack tip, $K_\sigma(\theta)$, defined by Erdogan-Sih [12] is also indicated, assuming that $K_I$ and $K_{II}$ are in phase.

$$K_\sigma(\theta) = \sigma_\theta \cdot \sqrt{2\pi r} = \cos(\theta/2)\left[ K_I \cos^2(\theta/2) - (3/2)K_{II} \sin\theta \right] \qquad (4)$$

$$\tan(\theta/2) = (1 \pm \sqrt{1 + 8\gamma^2})/4\gamma \qquad \gamma = K_{II}/K_I \qquad (5)$$

Furthermore, the threshold stress intensity factors $K_{I\,th\,(mode\,I)}$ under Mode I crack growth and $K_{I\,th\,(mix-mode)}$ under mix-mode crack growth, which are obtained from Fig. 10, are shown in the figure. From the figure, $K_\sigma(\theta)$ is smaller than $K_{I\,th\,(mode\,I)}$ at the observed crack length of 290 μm. $K_{I(mix-mode)}$ is also lower than $K_{I\,th\,(mix-mode)}$. These indicate that a fretting fatigue crack can not propagate, which does not agree with the experimental observations mentioned above.

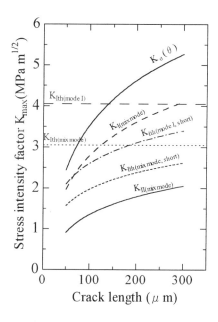

Figure 11 – *Relationships between stress intensity factors and crack length at stress amplitude of 175 MPa.*

For improving the above discussion, these threshold values have been reduced based on the short crack modification after El.Haddad et al [13].

$$\Delta K_{th(a)} = \Delta K_{th(a=0)}\sqrt{\frac{a}{a+a_0}}$$

$$a_0 = \left(\frac{\Delta K_{th(a=\infty)}}{\Delta \sigma_{w0}}\right)^2 \frac{1}{\pi}$$

(6)

The reduced threshold stress intensity factors $K_{I\,th\,(mode\,I\,.\,short\,crack)}$ and $K_{I\,th\,(mix-mode,\,short\,crack)}$ are indicated in Fig. 11. As can be seen from the figure, comparing these values to $K_\sigma(\theta)$ and $K_{I\,(mix-mode)}$, respectively, both $K_\sigma(\theta)$ and $K_{I\,(mix-mode)}$ are larger than those threshold values, which indicates that the fretting fatigue crack will not be arrested but will propagate. Therefore, it is speculated that a stress level at which a fretting fatigue crack becomes a non-propagating crack will be significantly low compared to the conventional fretting fatigue limit. In order to estimate the non-propagating condition of the fretting fatigue crack, $K_\sigma(\theta)$ for various stress levels was calculated. The results are shown in Fig. 12. It is suggested from the figure that the fretting fatigue crack will be arrested at a stress amplitude of 140 MPa, at which the length of non-propagating crack will be about 50 μm.

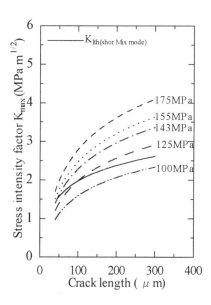

Figure 12 – Non-propagating condition for fretting fatigue crack.

**Conclusions**

Combining the results of FEM analysis and the corresponding fretting fatigue experiments, the following conclusions are summarized.
(1) A fretting fatigue crack will initiate at the point of maximum principal stress and propagate along the plane perpendicular to the maximum principal stress.
(2) A fretting fatigue crack is influenced by the mode II component of stress intensity factor, which accelerates the crack growth rate. The ratio of $K_{II}$ component to $K_I$ component increases for shorter and inclined cracks.
(3) It is confirmed from both experimental observations and FEM analysis that the fretting fatigue crack observed at the conventional fretting fatigue limit defined at $10^7$ cycles is not a non-propagating crack but a propagating crack.
(4) From the FEM analysis, the stress level for arresting the propagation of a fretting fatigue crack is estimated to be 140 MPa, which is significantly low compared to the conventional fretting fatigue limit (195 MPa).

**References**
[1] Edwards P. R., (1981) "The Application of Fracture Mechanics to Predicting Fretting Fatigue.," Fretting Fatigue, Eds. R. B. Waterhouse, Applied Science Publ.,
[2] K. Tanaka, Y, Mutoh and S. Sakota, (1985) Effect of Contact Materials on Fretting Fatigue in a Spring Steel., Trans. JSME, Ser. A, 51(464), 1200-1207.
[3] M. Takeuchi, R. B. Waterhouse, Y. Mutoh and T Satoh, (1991) The Behavior of Fatigue Crack Growth in the Fretting-Corrosion-Fatigue of High Tensile Roping Steel in Air and Seawater., Fatigue Fract. Engng. Mater. Struct., 14(1), 69-77.
[4] Y. Mutoh, (1998) Mechanisms of Fretting Fatigue, JSME Int. Journ., 38(4), 405-415
[5] T. Hattori, M. Nakamura, H. Sakata and T. Watanabe, (1988) Fretting Fatigue Analysis Using Fracture Mechanics., JSME Int. J. Series 1, 31(1), 100-107.
[6] K. J. Nix and T. C. Lindley (1985) The Application of Fracture Mechanics to Fretting Fatigue., Fatigue Fract. Engng. Mater. Struct, 8, 143,,
[7] M. A. Sheikh, U. S. Fernando M. W. Brown and K. J. M. Miller, (1994) Elastic Stress Intensity Factors for Fretting Cracks using the Finite Element Method., Fretting Fatigue, ESIS 18 Edited by R. B. Waterhouse and T. C.Linley, Mechanical Engineering Publications, London, 83-101
[8] T. L. Anderson, (1995) Fracture Mechanics., CRC Press.
[9] H. A. Richard and K. Benitz (1983) A LOADING DEVICE FOR THE CREATION OF MIXED MODE IN FRACTURE MECHANICS., Int Journ. of Fracture, 22, R55
[10]Eds. Y. Murakami, (1987) Stress Intensity Factors Handbook. Vol. 2 Pergamon Press.
[11]J. Qian and A Fatemi, (1996) Fatigue crack growth under mixed-mode I and II loading., Fatigue Fract. Engng. Mater. Struct., 19(10), 1277-1284.
[12]F. Erdogan and G. C. Sih, (1963) Trans ASME., D, 85, 519.
[13]M. H. El Haddad, K. N. Smith and T. H. Topper, (1979) Fatigue crack propagation of short cracks., Trans. ASME J. Eng. Mater. Technol. 101, 42.

# Material and Microstructural Effects

Toyoichi Satoh[1]

Influence of Microstructure on Fretting Fatigue Behavior of a Near-alpha Titanium
Alloy

**REFERENCE:** Toyoichi, S., **"Influence of Microstructure on Fretting Fatigue Behavior of a Near-alpha Titanium Alloy,"** *Fretting Fatigue: Current Technology and Practices, ASTM STP 1367,* D. W. Hoeppner, V. Chandrasekaran, and C. B. Elliott, Eds., American Society for Testing and Materials, West Conshohocken, PA, 2000.

**ABSTRACT:** To investigate the effect of microstructure on the fretting fatigue behavior of a near-α titanium alloy, fretting fatigue tests were carried out using DAT54, which is used in compressor blades and disks in aircraft gas turbine engines. Two kinds of microstructure in DAT54 were prepared using different solution heat treatment temperatures: one is the equiaxed α + α lath microstructure and the other is the transformed β structure. The plain and fretting fatigue strengths for the equiaxed α + α lath microstructure are higher than for the transformed β structure. Fretting reduced fatigue strengths by a factor of three for both materials. Sensitivity to microstructure in fretting fatigue is relatively low compared with plain fatigue. Shot peening improved fretting fatigue life, because of lower tangential force between the specimen and the contact pad and because of residual stress in compression induced by shot peening treatment.

**KEYWORDS:** fretting fatigue, near-alpha titanium alloy, microstructure, shot peening, tangential force coefficient

## Introduction

Fretting is a small amplitude oscillatory movement which may occur between contacting surfaces subjected to vibration or cyclic stress. In the aircraft gas turbine engine, surface damage caused by fretting action may occur in contact regions such as the fixing of the blade to the disk[1] and the flange of the disk[2]. Fretting damage leads to the early formation of microcracks, which spread and finally lead to the fatigue failure of the component. Therefore, fretting fatigue has become a serious problem in aircraft gas turbine engines.

Weight reduction of components is very important in obtaining higher aircraft gas turbine engine performance. Thus, titanium alloys have been used in the fan area and are in

---

[1] Research scientist, Jet Engine Division, 3rd Research Center, Technical Research & Development Institute, Japan Defense Agency, Tachikawa, Tokyo 190-8533, Japan.

competition with Ni-base superalloys for compressor blade and disk applications in aircraft gas turbine engines. In titanium alloys, it is well known that the mechanical properties and fatigue properties such as high cycle fatigue and fatigue crack growth are significantly affected by microstructure[3-5].

In the present study, the effect of microstructure on fretting fatigue properties in a near- α titanium alloy was investigated. Fretting fatigue tests using shot peened specimen were also carried out to investigate the effect of shot peening treatment on fretting fatigue life.

## Experimental Procedure

### Material

Near-α titanium alloy DAT54 used for compressor blades and disks for aircraft gas turbine engines was selected in this study. The chemical composition of the material used is shown in Table 1. The β-transus of this alloy is 1045℃. This material was subjected to solution treatment at two temperatures as listed in Table 2: one is 15℃ below β-transus (DAT54-A) and the other is 5℃ above β-transus (DAT54-B), and subsequently aged at 635℃ for 2 hours. The microstructure of the material DAT54-A subjected to solution heat treatment below β-transus consisted of the primary equiaxed α at former β grain boundaries and α lath transformed from β in the grain. On the other hand, in the microstructure of DAT54-B that was solution heat treated at a temperature above the β-transus, only α lath was observed in the former coarse β grain as shown Fig.1. The volume fraction of equiaxed α of DAT54-A is about 8%, and the mean β grain diameter for DAT54-B is about 0.8mm. The microstructures of these materials are shown in Fig.1.

Table 1 - *Chemical compositions (wt %)*

| Material | Al | Sn | Zr | Nb | Mo | Si | O | Fe | N | C | H | Ti |
|----------|------|------|------|------|------|------|------|------|-------|------|--------|------|
| DAT54 | 5.65 | 4.06 | 3.37 | 0.68 | 2.90 | 0.33 | 0.12 | 0.09 | 0.007 | 0.06 | 0.0076 | Bal. |

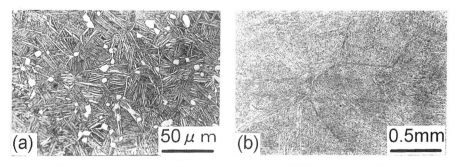

Fig.1 *Microstructures of the material used (a) DAT54-A, (b) DAT54-B*

The basic mechanical properties are shown in Table 3. Tensile strength, ultimate strength and hardness of these materials are almost identical, regardless of the microstructure, but ductility, such as elongation and reduction in area, for the material with transformed β structure was relatively low compared to the material with equiaxed α + α lath microstructure. To investigate the effectiveness of shot peening against fretting, fretting fatigue tests using shot-peened specimens (Almen 6A) were also carried out.

Table 2 - *Heat treatment condition and microstructures*

| Material | Solution heat treatment | Aging | Microstructure |
|---|---|---|---|
| DAT54-A | 1030℃, 2h, AC[1] | 635℃, 2h, AC[1] | Equiaxed α+ α lath |
| DAT54-B | 1050℃, 2h, AC[1] | 635℃, 2h, AC[1] | Transformed β |

[1]AC: Air Cooling

Table 3 - *Mechanical properties*

| Material | Yield strength $\sigma_{ys}$ (MPa) | Ultimate strength $\sigma_{UTS}$ (MPa) | Elongation $\psi$ (%) | Reduction in area $\phi$ (%) | Young's modulus E (GPa) | Hardness $H_V$ |
|---|---|---|---|---|---|---|
| DAT54-A | 1132 | 1272 | 16.0 | 26.2 | 110 | 396 |
| DAT54-B | 1102 | 1251 | 4.7 | 4.5 | 105 | 392 |

*Basic Fatigue Tests*

Plain fatigue and crack growth tests as well as fretting fatigue tests were carried out to investigate the basic fatigue properties of the materials used. The plain fatigue specimen is a cylindrical shape with 8mm-diameter gage section as shown in Fig.2(a), and the crack growth test specimen is a 0.5in-thickness compact tension specimen. Plain fatigue and crack growth tests were carried out under a load-controlled condition with stress ratios of $R=-1$ and $R=0.1$, respectively. In the fatigue crack growth test, crack length was measured using crack gage attached on the specimen surface. A servo hydraulic fatigue testing machine with a capacity of 98kN was used for all fatigue tests.

*Fretting Fatigue Test*

Fretting fatigue tests were carried out with a flat-to-flat contact configuration using a bridge type contact pad and a specimen with a flat gage part. The shape and dimensions of the fretting fatigue specimen and bridge type contact pad are shown in Fig.2(b). The contact pads were machined from the same material as the specimen. The edge geometry of the contact pad was not controlled. The gage part of the specimen and contact pad were polished using successively finer grades of emery paper. Final polishing was conducted in the longitudinal direction using 2000 grade emery paper, except for the shot peened specimen.

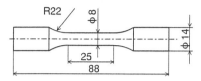

(a) Plain fatigue specimen

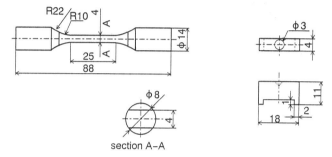

section A–A

(b) Fretting fatigue specimen and contact pad

Fig.2 *Shape and dimensions of the specimen (unit: mm)(a) plain fatigue specimen, (b) fretting fatigue specimen and contact pad.*

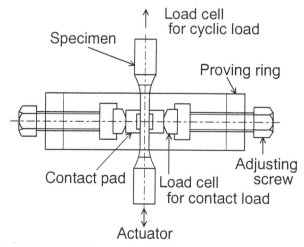

Fig.3 *Schematic illustration of the fretting fatigue test setup.*

The schematic illustration of the fretting fatigue test setup was shown in Fig.3. Fretting fatigue specimen was settled in a hydraulic grip of a fatigue testing machine. A pair of contact pads was pressed to the flat surface of the specimen and a mean contact pressure was applied using a proving ring. A mean contact pressure was measured using load cell located between contact pad and adjusting screw and that was controlled to be a constant value of 294MPa during fretting fatigue tests using a adjusting screw. Fatigue tests were carried out under a load-controlled condition with a stress ratio of $R=-1$ and frequencies from 2 to 20Hz. The tangential force between the specimen and the contact pad was measured using a strain gage attached on the inside of the bridge type contact pad. The fracture surface, fretted surface and fretting fatigue cracks on the longitudinal cross-section of the specimen tested were observed using a scanning electron microscope.

## Results and discussion

*Basic Fatigue Properties*

The S-N curves for plain fatigue are shown in Fig.4. Fatigue lives for DAT54-B are shorter than those of DAT54-A. The plain fatigue strengths $\sigma_{wp}$, are defined as the run-out stress amplitudes at $2\times10^6$ cycles, for DAT54-A and DAT54-B are 500MPa and 400MPa, respectively. The shorter plain fatigue lives and lower fatigue strength of DAT54-B compared with DAT54-A are due to the low ductility of DAT54-B. Scanning electron micrographs of the plain fatigue fracture surface for these materials are shown in Fig.5. In

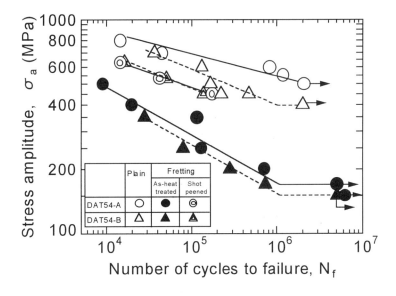

Fig.4 *S-N curves.*

the case of DAT54-B, the fracture surface morphology was strongly affected by microstructure, and the roughness of the fracture surface is large because of the large β grain size.

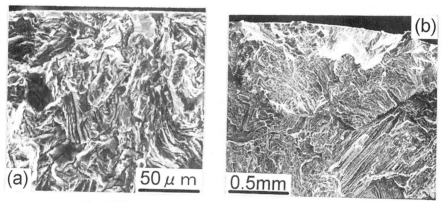

Fig.5 *SEM micrographs of plain fatigue fracture surface.*
*(a)DAT54-A, $\sigma_a$=700MPa (b)DAT54-B, $\sigma_a$=500MPa.*

The relationships between the stress intensity factor range $\Delta K$ and the crack growth rate $da/dN$ of these materials are shown in Fig.6, and these were expressed in the following Paris equation. The values of $C_0$ and $m$ for these materials used are listed in Table 4.

Table 4 - *Material constants in Paris equation*

| Material | Material constant | |
|---|---|---|
| | $C_0$ | m |
| DAT54-A | $1.34 \times 10^{-12}$ | 4.28 |
| DAT54-B | $5.58 \times 10^{-13}$ | 4.10 |

$$\frac{da}{dN} = C_0 (\Delta K)^m \qquad (1)$$

where

$da/dN$ = crack propagation rate in m/cycle
$\Delta K$ = stress intensity factor range in MPa$\sqrt{}$ m, and
$C_0$, $m$ = crack growth constants.

In DAT54, the plain fatigue strength for equiaxed $\alpha + \alpha$ lath microstructure is superior to that for transformed $\beta$ structure. On the other hand, fatigue crack propagation resistance of the former is relatively low compared with that the latter.

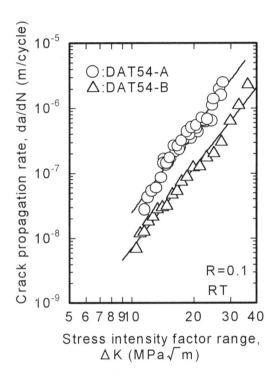

Fig.6 *Relationship between stress intensity factor range ΔK and crack growth rate da/dN*

*Fretting Fatigue Properties*

The S-N curves for fretting fatigue are also shown in Fig.4. The fretting fatigue strength $\sigma_{wf}$ is defined as the run-out stress amplitude at $5 \times 10^6$ cycles, and these values for both materials are listed in Table 5. Fretting fatigue strength and plain fatigue strength for DAT54-B is lower than for DAT54-A. It is clear from Table 5 that the effect of fretting on fatigue life and strength was significant. Fretting reduced the fatigue strength to almost one third compared with relevant plain fatigue strength, regardless of microstructure. A similar reduction of fatigue strength due to fretting has also been found in IMI829, one of the near-$\alpha$ titanium alloys, by Hamdy and Waterhouse[6].

Table 5 - *Plain fatigue and fretting fatigue strength*

| Material | Plain fatigue strength $\sigma_{wp}$, MPa | Fretting fatigue strength $\sigma_{wf}$, MPa | Reduction of fatigue strength[1] |
|---|---|---|---|
| DAT54-A | 500 | 170 | 0.66 |
| DAT54-B | 400 | 150 | 0.62 |

[1] Reduction of fatigue strength=$(\sigma_{wp}-\sigma_{wf})/\sigma_{wp}$

The relationships between the relative slip amplitude and the tangential force coefficient are shown in Fig.7. The relative slip amplitude was calculated using the stress-strain relation on the assumption of rigid pads, i.e.,

$$S_a = \frac{\sigma_a}{E}\left(\frac{l}{2}\right)$$
(2)

where

$S_a$ = slip amplitude in m
$\sigma_a$ = stress amplitude in Pa
$E$ = Young's modulus in Pa, and
$l$ = span of contact pad in m; $l$ =18 × $10^{-3}$m for as heat treated specimen and 12 × $10^{-3}$m for shot peened specimen.

The tangential force coefficient $\mu$ is defined as the ratio of the tangential force amplitude $F_a$ and the mean contact load $P$, $\mu=F_a/P$. The tangential force coefficients for both materials increased with an increase in relative slip amplitude and were almost identical regardless of the microstructure.

The difference in fatigue life and strength for fretting fatigue between DAT54-A and DAT54-B is relatively small compared with those for plain fatigue. A suggested reason for this effect is as follows. Almost the whole life of fretting fatigue is spent in propagating the fretting fatigue crack[7]. The fatigue crack growth resistance of DAT54-B is superior compared with that of DAT54-A as mentioned above. Therefore, the advantage of fatigue life and strength shown in plain fatigue for DAT54-A disappears.

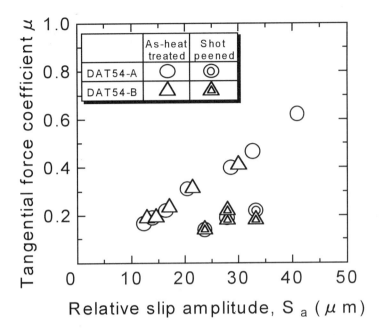

| | As-heat treated | Shot peened |
|---|---|---|
| DAT54-A | ○ | ◎ |
| DAT54-B | △ | △ |

Fig.7 *Relationship between relative slip amplitude $S_a$ and tangential force coefficient*

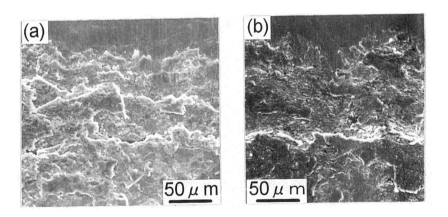

Fig.8 *SEM micrographs of the fretted surface, sliding direction is vertical.*
*(a)DAT54-A. $\sigma_a$=350MPa, (b)DAT54-B. $\sigma_a$=350MPa.*

*Observations of Fretted Surface and Fretting Fatigue Crack*

The regions significantly damaged by fretting action were found on the edge of the contact areas. Detailed observations by means of a scanning electron microscope is shown in Fig.8. The morphology of the damaged surfaces for both materials is almost the same, where accumulations of debris, wavy marks, and many microcracks in the direction transverse to the slip are observed. Fracture surfaces for fretting fatigue are shown in Fig.9. A featureless flat zone in the crack initiation area, which would be the region significantly influenced by fretting, was observed. As a crack grew inside the specimen, the features of the fracture surface changed to those observed in plain fatigue.

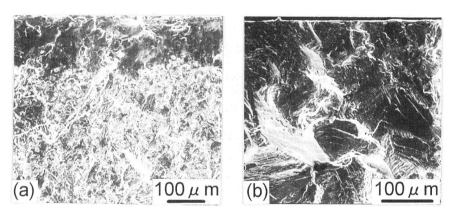

Fig.9 *SEM micrographs of fracture surface.  (a) DAT54-A, (b) DAT54-B.*

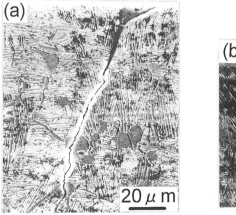

Fig.10 *SEM micrographs of fretting fatigue crack on the longitudinal cross-section of the specimen tested at $\sigma_a$=350MPa. (a) DAT54-A, (b) DAT54-B.*

The micrographs of fretting fatigue crack on the longitudinal cross-section of a fretted region of the specimen are shown in Fig.10. Many microcracks exist at the contact edge region as observed in Fig.8. As can be seen from the figure, the crack induced by fretting wear propagates at an oblique angle under the influence of the tangential force[8]. In the equiaxed $\alpha + \alpha$ lath microstructure, fretting fatigue cracks propagated transversely in the $\alpha$ lath and equiaxed $\alpha$ grain. On the other hand, in the transformed $\beta$ structure, the fatigue crack path was strongly affected by the microstructure, and crack bifurcation was also observed.

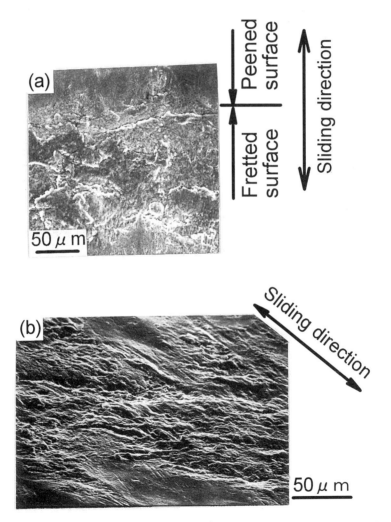

Fig.11 *Fretted surface of shot peened specimen for DAT54-B, $\sigma_a$=530MPa interrupted at $2 \times 10^4$ cycles: (a) fretting damage at contact the edge region, (b) bird's eye observation.*

*Effect of Shot Peening on Fretting Fatigue Life*

Fretting fatigue life is significantly improved by shot peening and is almost identical for both materials as shown in Fig.4. The factors, which are induced by shot peening and which influence the fretting fatigue life are work hardening in the surface layer, roughness of the surface, and residual stress.

It is generally known that the fatigue crack growth rate is significantly affected by residual stress. The residual stresses at the specimen surface measured by the X-ray diffraction method for both materials were about 500MPa in compression, regardless of the microstructure. The tangential force coefficients for shot peened specimens were lower than those for the heat treated specimens. According to the scanning electron microscope observations of the fretted surface for shot peened specimens as shown in Fig.11, many dents induced by shot peening treatment remained at the surface. It is suggested that the convex parts of the roughened surface of shot peened specimen were worn and many wear particles were produced during the fretting fatigue test. Then, the lower tangential force coefficient of shot peened specimens compared with those of heat treated unpeeded specimens was induced mainly by the higher surface roughness and tribological effect of produced wear particles. The hardness distributions in the depth direction of the specimen cross section are shown in Fig.12. The measurement of hardness was conducted using micro Vickers hardness test machine under load 1.96N and holding time 20s. Although plastic deformation of microstructure in the subsurface region of the specimen observed, significant work hardening in the surface layer was not found for the materials used. Therefore, the main factor that improves the fretting fatigue life for the materials used is the lower tangential force coefficient compared with that of unpeened specimens and the residual stress in compression induced by shot peening treatment.

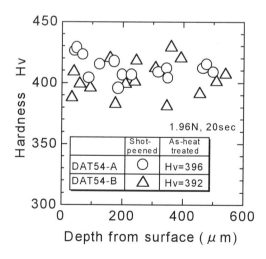

Fig.12 *Hardness distributions in the depth direction.*

## Conclusions

Fretting fatigue tests were carried out using the near-$\alpha$ titanium alloy DAT54, which has two kinds of microstructures. One is the equiaxed $\alpha + \alpha$ lath microstructure and the other is the transformed $\beta$ structure. The main results obtained are as follows:
(1) For both plain fatigue and fretting fatigue, fatigue strengths were significantly affected by microstrucure. The equiaxed $\alpha + \alpha$ lath microstructure exhibits higher strength than the transformed $\beta$ structure.
(2) Fretting reduced fatigue strength by a factor of three, regardless of microstructure. Sensitivity to microstructure in fretting fatigue is relatively low compared to plane fatigue.
(3) Shot peening improved the fretting fatigue life, because of lower tangential force between the specimen and the contact pad and because of residual stress in compression.

## References

[1] Broszeit, E., Kloos, K.H., and Schweighöfer, B., "The Fretting Fatigue Behaviour of The Titanium Alloy Ti-6Al-4V," *Titanium Science Technology*, 1985, pp. 2171-2178.

[2] Lindblom, Y. and Burman,G., "Fatigue Failure under Fretting Conditions," *High Temperature Alloys Gas Turbine*, 1982, pp.673-684.

[3] Ruppen, A.J., Hoffman, C.L., Radhakrishman, V.M., and McEvily, A.J., "The Effect of Environment and Temperature on The Fatigue Behavior of Titanium Alloys," *Fatigue, Environment and Temperature Effect*, 27th Sagamore Army Materials Research Conf.; Bolton Landing, NY, Edited by Burke, J.J., and Weiss, V., Plenum Press, New York, 1983, pp.265-300.

[4] Fujishiro, S., Fores, F.H., Matsumoto, T. and Eylon, D., "Effect of Processing on the Mechanical Properties of IMI-829 Titanium Alloys," *Titanium, Science and Technology*, Edited by Lütjering, G., Zwicker, U., and Bunk, W., DGM, 1985, pp.593-600.

[5] Okazaki, M., And Hizume, T., "Effect of Microstructure on Small Fatigue Crack Growth and The Crack Opening-Closing Behavior in Ti-6Al-4V Alloy ;Study on Materials Heat-Treated in Beta Field," *Journal of the Society of Materials Science, Japan*, Vol.43, 1994, pp.1238-1244.

[6] Hamdy, M.M., and Waterhouse, R.B., "The Fretting Fatigue Behavior of the Titanium Alloy IMI829 at Temperature up to 600°C," *Fatigue of Engineering Materials and Structures*, Vol.5, No.4, 1982, pp.267-274.

[7] Satoh, T., and Mutoh, Y., "Effect of Contact Pressure on Fretting Fatigue Crack Growth Behavior at Elevated Temperature," *Fretting Fatigue*, ESIS 18, Edited by Waterhouse, R .B., and Lindley, T.C., Mechanical Engineering Publications, London, 1994, pp.405-416.

[8] Tanaka, K., Mutoh, Y., Sakoda, S., and Leadbeater, G., "Fretting Fatigue in 0.55C Spring Steel and 0.45C Carbon Steel," *Fatigue & Fracture of Engineering Materials & Structures*, Vol.8, No.2, 1985, pp.129-142.

Alisha L. Hutson[1] and Ted Nicholas[2]

Fretting Fatigue Behavior of Ti-6Al-4V against Ti-6Al-4V under Flat-on-Flat Contact with Blending Radii

**REFERENCE:** Hutson, A. L., and Nicholas, T., **"Fretting Fatigue Behavior of Ti-6Al-4V Against Ti-6Al-4V Under Flat-on-Flat Contact with Blending Radii,"** *Fretting Fatigue: Current Technology and Practices, ASTM STP 1367,* D. W. Hoeppner, V. Chandrasekaran, and C. B. Elliott, Eds., American Society for Testing and Materials, West Conshohocken, PA, 2000.

**ABSTRACT:** A study was conducted to evaluate fretting fatigue damage of Ti-6Al-4V under flat-on-flat contact at room temperature. Results were obtained to establish the fatigue limit of the material. Axial stresses necessary to fail specimens at $10^7$ cycles for different contact radii, applied normal stresses, and stress ratios were evaluated to determine the baseline fretting fatigue behavior. Then, the effect of fretting fatigue on specimen life was quantified by conducting interrupted fretting tests for various load ratios and normal stresses followed by residual strength uniaxial fatigue tests. Fractography was used to characterize the nature of fretting damage. Results indicate that no degradation in fatigue limit is observed when the material is subjected to up to 10 percent of fretting fatigue life.

**KEYWORDS:** fretting fatigue, high cycle fatigue, Ti-6Al-4V, flat-on-flat contact, fatigue damage

## Introduction

Fretting fatigue is the damage caused by localized relative motion between contact surfaces on adjacent components, of which at least one component is under vibratory load, and may produce premature crack initiation and failure. Many studies [1-12] and literature surveys [13-16] of this phenomenon have focussed on fretting fatigue damage nucleation and crack propagation, in efforts to eliminate unanticipated failures of

---

[1] Assistant Research Engineer, Structural Integrity Division, University of Dayton Research Institute, 300 College Park, Dayton, OH 45469-0128.
[2] Senior Scientist, Metals, Ceramics & NDE Division, Materials and Manufacturing Directorate, Air Force Research Laboratory (AFRL/MLLN), Wright-Patterson AFB, OH 45433-7817.

turbine engine disks and blades subjected to fretting fatigue in the dovetail attachment region.  The manner in which the results of these studies may be applied to the understanding of the dovetail fretting problem is a subject for debate because the geometries and stresses in laboratory specimens are not usually directly comparable to those in actual components.  Turbine engine blade roots experience fretting in a flat-on-flat contact region with rather large radii at the edges of contact.  Some limited work has been performed to evaluate the dovetail geometry [1], but specimens are expensive, the stress distribution is unknown, and the results are difficult to model.  Conversely, most research has been conducted on punch-on-flat [2-8], or Hertzian [9-12], contact geometries because of the availability of closed form analytical solutions for the resulting stress distributions.  In these cases, a fretting pad is usually used to apply normal and shear loads against a thin test specimen, which most often is under axial fatigue loading.  While analytical solutions applied to ideal geometries like a half space allow the researcher to correlate test results and ultimately predict laboratory behavior, only limited progress has been made in the development of accurate life prediction models for complex components as a result of these efforts.  Development of an accurate life prediction model that bridges the gap between laboratory and service conditions requires experimentation and modeling of a more representative geometry.

The focus of this study was twofold: to evaluate the effect of prior fretting damage on fatigue strength using a test apparatus with a contact geometry similar to the dovetail, and to analyze fretting damage failure mechanisms using optical and scanning electron microscopy (SEM) techniques.

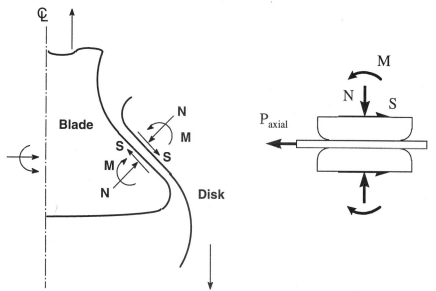

Figure 1a - *Representation of blade root loading condition.*

Figure 1b - *Representation of test configuration designed to simulate blade root loading condition.*

**Experimental Approach**

The ideal test system for a study such as this would exactly duplicate the dovetail blade root loading condition in Figure 1a. However, the results would be difficult to analyze and machine tooling costs would be exorbitant. The problem was simplified by assuming that the bending moment found in the dovetail geometry was not a significant contributor, and did not affect the magnitude of the relative motion responsible for fretting fatigue damage. This assumption resulted in the elimination of the complex dovetail geometry, thus simplifying test hardware and specimen fabrication. The test apparatus for this study, however, simulated the essential features of the blade root geometry by employing flat fretting pads, with a radius at the edge of contact, against a flat specimen. Imposed normal and shear loads on the contact surfaces were reproduced in the test geometry, as shown in Figure 1b. The bending moment present in the test geometry did not represent the moment imposed in the dovetail geometry and, in fact, has not yet been calculated. This test geometry differed from the conventional fatigue test with a fretting pad in that the stress in the specimen was zero on one end of the pad. Thus, the shear force into the pad was known by measuring only the applied stress in the specimen. As in a conventional test, only the axial and shear stresses were oscillatory: the clamping stress was static, which would not be true in a blade. Symmetry in the apparatus (see Figure 2) provided a specimen which failed on one end, leaving the other end with a fretting scar and damage obtained under nominally identical conditions corresponding to 100 percent of full life of that particular specimen.

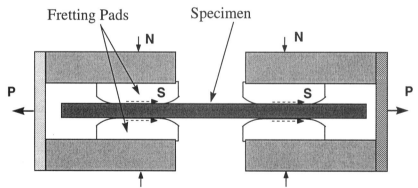

Figure 2 - *Test load train schematic. The apparatus produces two nominally identical regions of fretting for each test specimen.*

The test system used for this study was modified from a uniaxial high cycle fatigue system, which experienced a number of fretting fatigue induced failures in the specimens within the grip region. The new system was an electro-dynamic shaker system that could test at frequencies up to 500 Hz, and had a mean load capacity of 20 kN and a dynamic load capacity of 12 kN peak-to-peak at frequencies up to 300 Hz. Mean and dynamic loads were applied independently via a pneumatic cylinder and the electro-dynamic shaker, respectively, to maximize dynamic capability. System modifications were limited to the test gripping apparatus. Each grip was fitted with removable fretting pads, as shown in Figure 3, to facilitate control of surface conditions. Standard bolts, used to apply the clamping force were replaced by bolts instrumented with strain gages in the shank to allow accurate measurement of static clamping, or normal loads. Symmetric clamping of the specimen was achieved through the use of a specific torque pattern. Fretting damage occurred at the edge of contact, shown schematically in the magnified view in Figure 3.

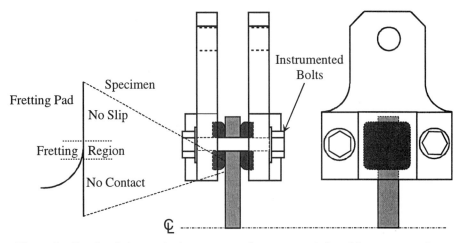

Figure 3 - *Fretting fatigue gripping system and stress state induced in contact region.*

All axial (fretting fatigue) specimens and a portion of the fretting pads were machined from vacuum annealed and hot rolled Ti-6Al-4V bar stock. The material was vacuum treated at 705°C for two hours, static argon cooled to below 149°C, vacuum annealed at 549°C for two hours, and static argon cooled to less than 149°C. This treatment produced a microstructure of 90% wide, plate-like $\alpha$ with 10% intergranular $\beta$ that is approximately equiaxed in the transverse section, with a slight elongation in the longitudinal section along the longitudinal axis of the bar. The balance of the fretting pads were machined from forged AMS4928 Ti-6Al-4V with TE01 heat treatment. All specimens and fretting pads were low stress ground to an RMS 8 surface finish.

A preliminary study of fretting fatigue parameters was conducted earlier using this apparatus [17]. An overview of the test method and results from that investigation are summarized here to provide the baseline conditions for the present investigation. A range of shear stress levels and distributions were investigated by varying contact length, contact radius, and normal contact load. The actual stress distributions have not yet been evaluated, but the contact radius is expected to have an effect on the distribution of both shear stress and normal stress. Only two values of each parameter were selected for the preliminary experiments to determine if changing the parameters would produce trends in the test results. Nominal pad lengths were 25.4 mm and 12.7 mm; contact radii were 3.2 mm and 0.4 mm; and static normal loads were 35 kN and 21 kN. Each of the resulting eight test conditions was conducted at two load ratios, R = 0.1 and R = 0.5, as shown in Table 1. The "initial axial stresses" in parentheses refer to arbitrarily selected values of axial stresses, which were used during the first block of a step loading procedure described below. Average specimen thickness was 2 mm; average width was 10 mm. Two specimen lengths were used: the bulk of the data was obtained using specimens with an average length of 100 mm; a few tests were conducted using 150 mm specimens to validate the tests described below. All tests were conducted in lab air at room temperature. Two different frequencies were required for each of the two specimen lengths to avoid harmonic resonance and bending modes in the specimen. The 100 mm long specimens were tested at 300 Hz while the 150 mm long specimens were tested at 400 Hz. No difference in test results was detectable between data for the two test frequencies. The contact areas listed in Table 1 are the nominal contact areas after allowing for the nominal contact radii and nominal specimen width of 10 mm. The results of the preliminary investigation are presented in Figure 4, which show that the maximum stress increases with increasing stress ratio or decreasing contact radius, but does not change appreciably with normal stress. The maximum axial stress is also referred to as "Goodman" stress, which is defined as the maximum (interpolated) stress corresponding to a fatigue limit of $10^7$ cycles as determined from the step loading procedure. The $10^7$ cycle fatigue limit was selected as a value commonly used in design, and as a reasonable number of cycles to obtain in laboratory tests.

All tests that were intended to run to failure were conducted using the step loading approach developed by Maxwell and Nicholas [18]. This technique employed constant amplitude loading blocks (see Figure 5) where the initial axial stress was set below the anticipated stress for a given fatigue limit (defined here as $10^7$ cycles). Each loading block

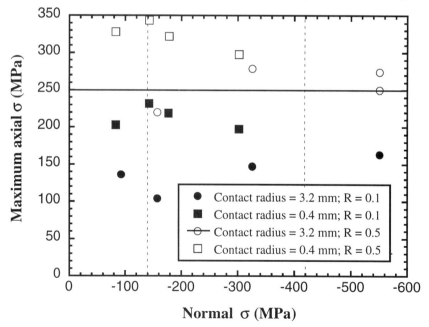

Figure 4 – *Fretting fatigue test results for a life of $10^7$ cycles.*

Table 1 - *Test matrix indicating normal loads, calculated contact areas, and load ratios. Arbitrarily selected initial axial stresses are shown in parentheses [17].*

| Normal Load | Contact Area | | | |
|---|---|---|---|---|
| | 65 mm$^2$ | 120 mm$^2$ | 190 mm$^2$ | 245 mm$^2$ |
| 35 kN | ($\sigma_{axial}$ = 150 MPa) R = 0.1 | ($\sigma_{axial}$ = 150 MPa) R = 0.1 | ($\sigma_{axial}$ = 150 MPa) R = 0.1 | ($\sigma_{axial}$ = 150 MPa) R = 0.1 |
| | ($\sigma_{axial}$ = 270 MPa) R = 0.5 | ($\sigma_{axial}$ = 220 MPa) R = 0.5 | ($\sigma_{axial}$ = 220 MPa) R = 0.5 | ($\sigma_{axial}$ = 220 MPa) R = 0.5 |
| 21 kN | ($\sigma_{axial}$ = 135 MPa) R = 0.1 | ($\sigma_{axial}$ = 100 MPa) R = 0.1 | ($\sigma_{axial}$ = 80 MPa) R = 0.1 | ($\sigma_{axial}$ = 170 MPa) R = 0.1 |
| | ($\sigma_{axial}$ = 220 MPa) R = 0.5 | ($\sigma_{axial}$ = 220 MPa) R = 0.5 | ($\sigma_{axial}$ = 220 MPa) R = 0.5 | ($\sigma_{axial}$ = 220 MPa) R = 0.5 |

was applied for the selected number of cycles ($10^7$ cycles in this case), or until the specimen failed. If the specimen did not fail in the first load block, the axial stress was raised by a percentage of the initial axial stress, and run for another $10^7$ cycles. This process was repeated for each subsequent block until the specimen failed. The fatigue limit stress, or Goodman stress was interpolated from the failure stress, the number of cycles at the failure stress, and the stress level of the previous step. The technique was validated in three separate studies where data from step loading tests were shown to coincide with those obtained using extrapolation and interpolation of conventional stress-life tests to the fatigue limit of the step tests [18-20].

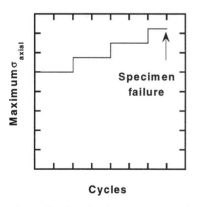

Figure 5 - *Step loading pattern used for Goodman stress determination.*

The present study built on the previous work to determine if fretting fatigue damage produced at a fraction of total fretting fatigue life for a given geometry and load condition had an effect on specimen fatigue strength. A normal load of 26 kN was selected to produce two different values of average normal stresses, when the two pad lengths were incorporated. These two normal stresses represented reasonable upper and lower limits of normal stresses calculated for an actual dovetail geometry [21]. The 3.2 mm contact radius was selected because it represented the dovetail geometry more closely than the smaller radius. Using the selected normal load and contact radius, eighteen 150 mm length specimens were tested to accumulate fretting fatigue cycles according to Table 2 conditions, under an axial stress of 250 MPa at R = 0.5 and 400 Hz. The axial stress level was selected based on the results of the preliminary work (Figure 4), given the contact radius and desired stress ratio of R = 0.5. Since no trend was observed in the preliminary Goodman stresses as a function of normal stress, the same

Table 2 - *Phase 2 test matrix indicating normal stresses, load ratio, and the number of cycles of fretting.*

| Cycles of Fretting | 140 MPa Normal stress | 420 MPa Normal stress |
|---|---|---|
| 10,000 | 3 test @ R=0.5 | 3 test @ R=0.5 |
| 100,000 | 3 test @ R=0.5 | 3 test @ R=0.5 |
| 1,000,000 | 3 test @ R=0.5 | 3 test @ R=0.5 |
| total | 9 tests | 9 tests |

axial stress was used for both values of normal stress, shown by the solid line in Figure 4. After the sub-critical fretting fatigue damage was applied, one end of the specimen was removed, and the fretted region was machined into a dogbone sample, as shown in Figure 6, where the fretted region was now within the test section of the dogbone geometry. The dogbone samples were tested in uniaxial fatigue at stress ratios of R = 0.1 and R = 0.5 using the step loading technique previously described. Because of the apparently low level of fretting damage, as observed visually, initial stresses were selected assuming no appreciable damage was present.

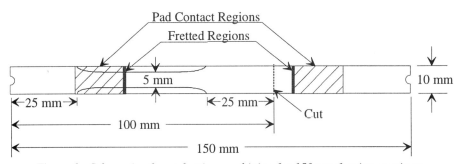

Figure 6 - *Schematic of post-fretting machining for 150 mm fretting specimen.*

### Results / Discussion

The results from a limited number of tests on specimens tested to 10 percent of the total fretting life appear to preclude the necessity of testing specimens with lower levels of damage. All eighteen specimens, which covered life ranges from 0.1 to 10 percent, were fretted and the scars were inspected optically. Representative fretting scars are shown in Figure 7 for both 140 MPa (Figure 7a) and 420MPa (Figure 7b) normal stresses. The fretting scars, indicated by the white arrows, appear approximately horizontal in the photos, and the loading axis is vertical. Three specimens subjected to one million fretting fatigue cycles (10% of life) at 140 MPa normal stress were tested to failure. The specimen tested at R = 0.5 failed at the fretting scar, but with no debit in fatigue strength compared to the baseline fatigue limit. Figure 8a shows this datum compared with data on

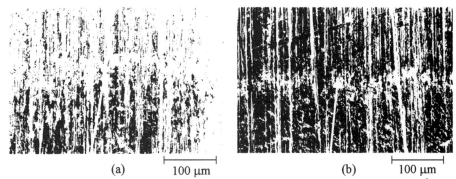

(a)    100 µm                    (b)    100 µm

Figure 7 – *Representative fretting scars from interrupted fretting tests: (a) tested for $10^6$ cycles at a 140 MPa normal stress, (b) tested for $10^6$ cycles at 420 MPa normal stress*

unfretted specimens [18], as presented on a constant life plot of mean stress vs. alternating stress. The datum falls within the scatter band of the baseline. The two remaining specimens at this damage level were tested at R = 0.1. Both specimens failed at lower stresses than the baseline (Figure 8b), but damage nucleated well away from the fretted region.

SEM inspection of the fracture surfaces from these two specimens revealed crack nucleation from embrittled features not found in the bulk of the material (Figure 9). Machining anomalies were determined to be the source of these regions. From these limited results, the authors find it reasonable to conclude that fretting damage at or below 10% of total life was not detrimental to the fatigue strength of the material. In support of this, Adibnazari and Hoeppner did not report a decrease in fatigue resistance until 20-40% of life for Ti-6Al-4V under an average normal stress of 750 MPa [9]. In spite of these results, which imply no detrimental damage, fretting scars at the edge of contact have been clearly identified.

For all the fretting fatigue tests in the preliminary study [18] failures occurred

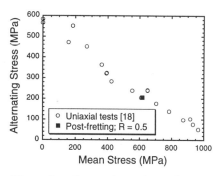

Figure 8a - *Comparison of post-fretting fatigue results indicating no change for fretting damage accumulated to ten percent of fretting life.*

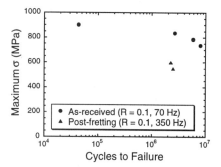

Figure 8b - *S-N curve presenting post-fretting specimens that failed prematurely compared to baseline data.*

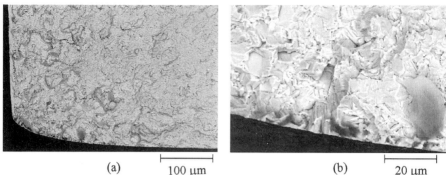

(a)    100 μm              (b)    20 μm

Figure 9 - *Crack nucleation site for a post-fretting fatigue test, which failed away from the fretted region: (a) region at 300X showing typical material structure in the upper right-hand corner, and the damaged region in the lower left-hand corner, (b) crack nucleation feature at 1500X.*

consistently at the edge of contact. In that study, the unfailed ends of the specimens, corresponding to 100 percent of life, exhibited fretting scars with varying levels of damage from specimen to specimen under both optical and SEM inspection. Figure 10 shows two representative micrographs of fretting damage observed on the unfailed ends of a typical test specimen. Fretted regions appear to the left of the dashed lines; machined surfaces are to the right. The diagonal texture throughout Figures 10 and 11 was a product of the low stress grind process, and was presumed to be non-detrimental since it was not perpendicular to the loading axis (approximately horizontal in Figure 10). Although no cracks were apparent in Figure 10, some specimens did show evidence of cracking on the unfailed end of the specimen (Figure 11). Regardless of whether or not cracking was present, fretting damage ranged from nearly invisible deformation of the machined surface near the center of the specimen (Figure 10b), to small pits produced by local adhesion and delamination near the edge of the specimen (Figure 10a). The range of damage shown in

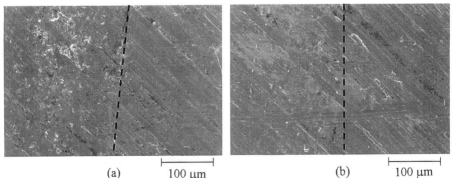

(a)    100 μm              (b)    100 μm

Figure 10 - *Two regions of a representative fretting scar from the unfailed end of a typical test specimen. a) Damage near the edge of the specimen. b) Damage near the center of the specimen. The loading axis is approximately horizontal.*

Figure 10 occurred on the same fretting scar, implying a change in stress state from the center to the edge of the specimen. This stress gradient may prove problematic if additional testing and characterization indicate primary crack nucleation at or near the edge of contact.

The crack shown in Figure 11 continued to the edge of the specimen, but did not appear to nucleate there. Rather, nucleation sites appear to have occurred 150 - 300 μm from the specimen edge. Attempts have been made to quantify the states of stress along the fretting zone using finite element analysis (FEA) to better understand this crack nucleation phenomenon, but the program selected did not have the appropriate elements for the solution of the contact problem.

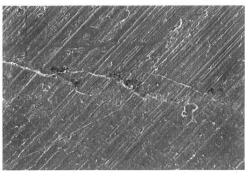

100 μm

*Figure 11 - Fretting scar with crack from the unfailed end of a failed specimen.*

SEM characterization of the fracture surfaces from the preliminary testing also revealed several unique features. Overall, a representative fracture surface was divided into four sections: oxide populated crack nucleation sites, oxide-free crack nucleation sites, a crack propagation region, and a tensile fracture region. The photos in Figure 12 exhibited debris populated crack nucleation sites (see arrow in Figure 12a). These oblique views of two different specimens displayed fretting scars on the bottom halves of the photos, and the fracture surfaces on the top halves. Spectral analysis of the white particles on the fracture surface indicated a relatively large amount of oxygen indicative of wear debris observed in many of fretting. Whether the oxide particles developed during crack nucleation or crack propagation was unclear; however, crack nucleation sites showing no debris (Figure 13) were also present. The nucleation sites with the debris were presumed to have nucleated

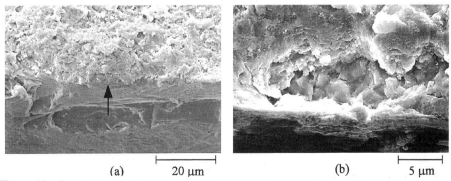

(a)    20 μm          (b)    5 μm

*Figure 12 - Fracture surface photos showing presence of (a) fretting debris particles and (b) material smearing.*

earlier that those without debris, since those with debris tended to have material smeared across them (Figure 12b), indicating a localized influence of Mode 2 as well as Mode 1 crack growth. The influence of Mode 2 crack growth will tend to diminish as the crack propagates, so cracks nucleating later will have less smeared material.

Figure 13 shows a series of four fractographs, which overlap to form a montage that illustrates the presence of five crack nucleation sites without debris (indicated by arrows) within less than 500 μm. Multiple nucleation sites were features found in low cycle fatigue (LCF) failures, but the limited results from tests reported here seemed to refute the idea that these are LCF failures. Crack nucleation under LCF is usually evident within the first ten percent of life. The fretting fatigue specimens tested here show no evidence of crack nucleation up to ten percent of life. Multiple crack nucleations might occur after some period of growth for a single, primary crack. The combination of a reduced net section and imposed bi-axial stresses may produce small secondary cracks later in life. All four photos in Figure 13 were taken near the center of the specimen. The overall crack front appeared to be propagating to the left, further evidence that these cracks were not the first to nucleate. The lack of debris on these sites indicated that the major crack driving force is coming from the Mode 1 component.

The crack propagation region of the fracture surface was generally free of debris, which left two additional features unobscured. Secondary cracking perpendicular to the dominant crack plane was evident to varying degrees throughout the crack propagation zone, as in Figure 14, where the specimen edge is shown in the lower part of the photo.

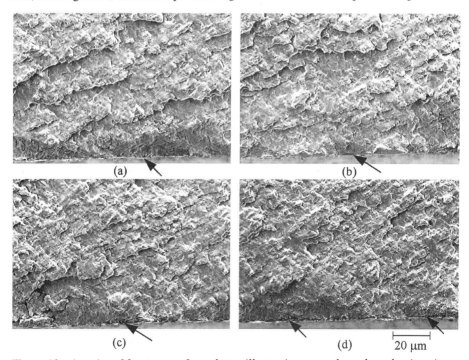

Figure 13 - *A series of fracture surface photos illustrating several crack nucleation sites.*

These cracks provided evidence of a multi-axial stress state through the thickness of the specimen. Fatigue striations with spacing on the order of $10^{-7}$ m/cycle were also visible, partially due to enhancement by secondary cracking. An example is shown in Figure 15, which was taken ~1.2 mm away from the region shown in Figure 14, where the specimen edge would appear ~20 μm below the photo as shown. Although the overall crack front was not clear, correlation of growth rates derived from these striations with uniaxial crack growth data might provide information on crack front locations and crack propagation rates under complex loading.

20 μm

Figure 14 - *Fracture surface showing secondary cracking perpendicular to the crack plane as evidence of multi-axial stresses.*

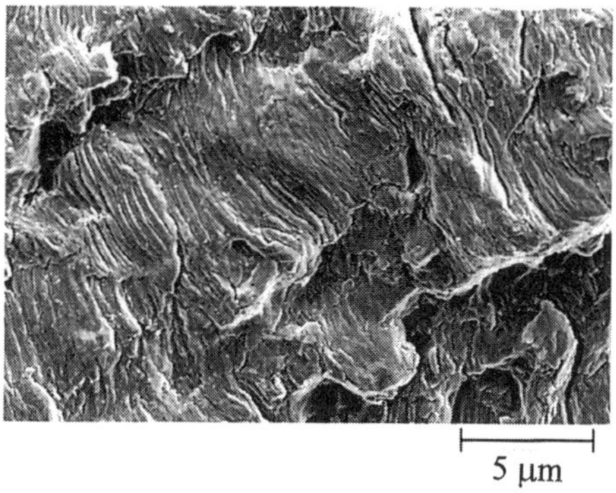

5 μm

Figure 15 - *Fracture surface showing prominent fatigue striations.*

## Conclusions

Limited tests results indicated that no measurable reduction in axial fatigue strength occurred if fretting fatigue conditions were removed at 10 percent of fretting fatigue life under a normal stress of 140 MPa. Further testing is required to statistically verify this conclusion.

Levels of fretting damage for interrupted fretting fatigue tests at life fractions below 10 percent were nearly indistinguishable from the machined surface.

Inspection of fracture surfaces revealed multiple crack nucleation sites, some fretting debris, prominent fatigue striations, and evidence of a tri-axial stress state in the form of secondary cracking.

*Acknowledgments*

This research was conducted at the Materials & Manufacturing Directorate, Air Force Research Laboratory (AFRL/MLLN), Wright-Patterson Air Force Base, OH, USA. A. Hutson was supported under on-site contract numbers F33615-94-C-5200 and F33615-98-C-5214. The authors gratefully acknowledge many helpful discussions with Dr. D. Eylon and Dr. N. Ashbaugh, at the University of Dayton, and Mr. Rick Goodman for his assistance in the experimental portion of the program.

**References**

1. Ruiz, C., Boddington, P. H. B., and Chen, K. C., "An Investigation of Fatigue and Fretting in a Dovetail Joint," *Experimental Mechanics*, Vol. 24, 1984, pp. 208-217.

2. Bryggman, U. and Soderberg, S., "Contact Conditions and Surface Degradation Mechanisms in Low Amplitude Fretting", *Wear*, Vol. 125, 1988, pp. 39-52.

3. Del Puglia, A., Pratesi, F., and Zonfrillo, G., "Experimental Procedure and Parameters Involved in Fretting Fatigue Tests," *Fretting Fatigue, ESIS 18*, R.B. Waterhouse and T.C. Lindley, Eds., Mechanical Engineering Publications, London, 1994, pp. 219-238.

4. Elkholy, A. H., "Fretting Fatigue in Elastic Contacts Due to Tangential Micromotion," *Tribology International*, Vol. 29, No. 4, 1996, pp. 265-275.

5. Fellows, L. J., Nowell, D., and Hills, D. A., "On the Initiation of Fretting Fatigue Cracks," *Wear*, Vol. 205, 1996, pp. 120-129

6. Fouvry, S., Kapsa, P., and Vincent, L., "Quantification of Fretting Damage," *Wear*, Vol. 200, 1996, pp. 186-205.

7. Vingsbo, O. and Schon, J., "Gross Slip Criteria in Fretting," *Wear*, Vol. 162-164, 1993, pp. 347-356.

8. Zhou, Z. R. and Vincent, L., "Mixed Fretting Regime," *Wear*, Vol. 181-183, 1995, pp. 551-536.

9. Adibnazari, S. and Hoeppner, D.W., "The Role of Normal Pressure in Modeling Fretting Fatigue," *Fretting Fatigue, ESIS 18*, R.B. Waterhouse and T.C. Lindley, Eds., Mechanical Engineering Publications, London, 1994, pp. 125-133.

10. Dobromirski, J. and Smith, I. O., "Metallographic Aspects of Surface Damage, Surface Temperature and Crack Initiation in Fretting Fatigue," *Wear*, Vol. 117, 1987, pp. 347-357.

11. Fayeulle, S., Blanchard, P., and Vincent, L., "Fretting Behavior in Titanium Alloys," *Tribology Transactions*, Vol. 36, No. 2, pp. 267-275, 1993.

12. Lindley, T. C. and Nix, K. J., "Fretting Fatigue in the Power Generation Industry: Experiments, Analysis, and Integrity Assessment," *Standardization of Fretting Fatigue Test Methods and Equipment, ASTM STP 1159*, M. Helmi Attia, and R. B. Waterhouse, Eds., American Society for Testing and Materials, Philadelphia, 1992, pp. 153-169.

13. Waterhouse, R.B., "Effect of Material and Surface Conditions on Fretting Fatigue," *Fretting Fatigue, ESIS 18*, R.B. Waterhouse and T.C. Lindley, Eds., Mechanical Engineering Publications, London, 1994, pp. 339-349.

14. Dobromirski, J.M., "Variables of Fretting Processes: Are There 50 of Them?," *Fretting Fatigue, ESIS 18*, R.B. Waterhouse and T.C. Lindley, Eds., Mechanical Engineering Publications, London, 1994, pp. 60-66, 1994.

15. Attia, M. H., "Fretting Fatigue Testing: Current Practices and Future Prospects for Standardization," *Standardization of Fretting Fatigue Test Methods and Equipment, ASTM STP 1159*, M. Helmi Attia and R. B. Waterhouse, Eds., American Society for Testing and Materials, Philadelphia, 1992, pp. 263-275.

16. Hoeppner, D.W., "Mechanisms of Fretting Fatigue," *Fretting Fatigue, ESIS 18*, R.B. Waterhouse and T.C. Lindley, Eds., Mechanical Engineering Publications, London, 1994, pp. 3-19.

17. Hutson, A. and Nicholas, T., "Fretting Fatigue of Ti-6Al-4V Under Flat-on-Flat Contact", *International Journal of Fatigue*, Special Issue on High Cycle Fatigue, 1999 (in press).

18. Maxwell, D.C. and Nicholas, T., "A Rapid Method for Generation of a Haigh Diagram for High Cycle Fatigue," *Fatigue and Fracture Mechanics: 29th Volume, ASTM STP 1321*, T.L. Panontin and S.D. Sheppard, Eds., American Society for Testing and Materials, 1999, pp. 626-641.

19. Bellows, R. S., Muju, S. and Nicholas, T., "Validation of the Step Test Method for Generating Goodman for Ti-6Al-4V," *International Journal of Fatigue*, Special Issue on High Cycle Fatigue, 1999 (in press).

20. Lanning, D., Haritos, G.K. and Nicholas, T., "Notch Size Effects in HCF Behavior of Ti-6Al-4V," *International Journal of Fatigue*, Special Issue on High Cycle Fatigue, 1999 (in press).

21. Van Stone, R., "Fretting and High Cycle Fatigue in Titanium," presented at the 3rd National Turbine Engine High Cycle Fatigue Conference, San Antonio, TX, 1998.

T. Nisida,[1] Mutoh Y.,[2] K. Yoshii,[3] and O. Ebihara[3]

## Fretting Fatigue Strengths of Forged and Cast Al-Si Aluminum Alloys

**REFERENCE:** Nishida, T., Mutoh, Y., Yoshii, K., and Ebihara, O., **"Fretting Fatigue Strengths of Forged and Cast Al-Si Aluminum Alloys,"** *Fretting Fatigue: Current Technology and Practices, ASTM STP 1367,* D. W. Hoeppner, V. Chandrasekaran, and C. B. Elliott, Eds., American Society for Testing and Materials, West Conshohocken, PA, 2000.

**ABSTRACT:** Plain fatigue and fretting fatigue tests were carried out using cast Al-Si aluminum alloy JIS AC4CH-T6 and forged aluminum alloy JIS 6061-T6. Plain fatigue strength of the cast aluminum alloy was lower than that of the forged alloy. The lower fatigue strength was attributed to casting defects such as pores in the cast aluminum alloy, which enhanced fatigue crack formation. On the other hand, fretting fatigue strength of the cast aluminum alloy almost coincided with that of the forged alloy. No difference of tangential force coefficient and then tangential force in fretting fatigue was observed between the two materials. From fatigue crack growth tests and fretting fatigue tests interrupted at various fatigue cycles, fatigue crack growth curves and fretting crack formation life for both materials were also found to coincide with each other. These phenomena and properties resulted in the agreement of fretting fatigue strengths between the cast and forged aluminum alloys.

**KEYWORDS:** Cast aluminum alloy, forged aluminum alloy, plain fatigue, fretting fatigue, casting defect, life prediction, tangential force

[1]Associate Professor, Department of Mechanical Engineering, Numazu College of Technology, Numazu 410-8501, Japan.
[2] Professor, Department of Mechanical Engineering, Nagaoka University of Technology, Nagaoka 940-2188, Japan.
[3]Engineer, Technical Research Department, Topy Industries, Limited , 1, Akemi-cho, Toyohashi 441-8510, Japan.

## Introduction

Forged aluminum alloy, which has superior mechanical properties and reliability compared to cast aluminum alloy, has been widely used for structural materials. However, recently casting processes have been refine and reliable alloys have been developed with low cost compared to forged alloys. Casting processes can make it easy to fabricate components with complicated shapes. Therefore, cast aluminum alloys have been increasingly utilized for automobile parts, aerocraft parts, air and oil compressors and other components.

Fretting is the small amplitude oscillatory movement which may occur between contacting surfaces that are usually nominally at rest [1]. Fretting occurs between two assembled components and hence fretting fatigue often results. Fretting significantly reduces fatigue strength and therefore, unexpected failure of the components is induced by fretting under stresses much lower than the design stress, resulting in shorter lives. Some research works on fretting fatigue of forged aluminum alloys have been reported [2-4]. However, those for cast aluminum alloys have rarely been done.

In the present study, plain fatigue and fretting fatigue tests were carried out to understand the basic characteristics of fretting fatigue for both materials. Fracture mechanics analysis combined with fatigue crack growth test results was also carried out to discuss the fretting fatigue properties for both materials in detail.

## Experimental Procedure

The materials used for specimens were cast Al-Si aluminum alloy (JIS AC4CH-T6) and forged alloy (JIS 6061-T6), chemical compositions and mechanical properties of which are shown in Table 1 and 2, respectively. The material used for the contact pad was spheroidal graphite cast iron (JIS FCD500).

The shapes and dimensions of the plain fatigue specimen, fretting fatigue specimen, bridge-type contact pad and fatigue crack propagation test specimen are shown in Fig.1. Fretting was induced by clamping a pair of bridge-type pads onto both sides of the specimen ( Fig.1(b)), using a proving ring. The clamping pressure was adjusted before and during the tests to give a constant value of 50MPa. The tangential force between the specimen and the pad during the tests was measured by strain gauges attached underneath the central part of the pad. The plain fatigue, fretting fatigue and fatigue crack growth tests were carried out using a servohydraulic fatigue test machine with a capacity of 98kN under a stress ratio R of -1 at frequencies ranging from 10 to 20Hz. The two-stage tests [5] were also carried out at a stress level of 100MPa for both materials. The threshold stress intensity factor range $\Delta K_{th}$ was evaluated on the basis of the load decreasing method [6]: During the fatigue crack growth test, $3 \sim 5\%$ reduction of load was repeated after every crack extension of $0.02 \sim 0.05$mm. Fracture surfaces of the specimens tested were observed using a scanning electron microscope. Surface roughness of the fretted region was measured using a surface roughness tester.

Table 1-*Chemical composition* ( *wt%* )

|  | Si | Fe | Cu | Mn | Mg | Cr | Zn | Ti |
|---|---|---|---|---|---|---|---|---|
| AC4CH-T6 | 6.60 | 0.12 | 0.03 | 0.02 | 0.36 | 0.03 | 0.03 | 0.13 |
| 6061-T6 | 0.57 | 0.35 | 0.29 | 0.04 | 0.87 | 0.18 | 0.03 | 0.02 |

|  | C | Mn | Si | P | S |
|---|---|---|---|---|---|
| FCD500 | 3.62 | 2.23 | 0.30 | 0.031 | 0.009 |

Table 2-*Mechanical properties*

|  | Hardness Hv(GPa) | Tensile strength $\sigma_B$(MPa) | Proof stress $\sigma_{0.2}$ (MPa) | Elongation $\phi$ (%) | Young's modulus E(GPa) |
|---|---|---|---|---|---|
| AC4CH-T6 | 1.0 | 218 | 149 | 3.0 | 78 |
| 6061-T6 | 1.2 | 354 | 336 | 11.5 | 76 |
| FCD500 | 1.7 | ---- | ---- | ---- | ---- |

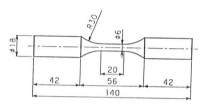

*(a)    Plain fatigue specimen*

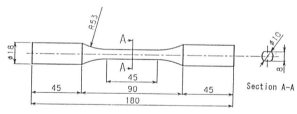

*(b)    Fretting fatigue specimen*

*(c)    Contact pad*

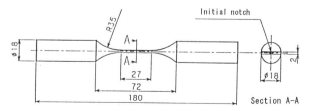

*(d)    Fatigue crack propagation specimen*

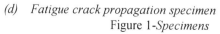

Figure 1-*Specimens*

**Results and Discussion**

*Plain Fatigue Strength*

The S-N curves for plain fatigue of cast and forged aluminum alloys are shown in Fig.2. As can be seen from the figure, the fatigue strength for the cast aluminum alloy was lower than that of the forged alloy. The plain fatigue limits of the cast and forged aluminum alloys were 80 MPa and 140MPa, respectively.

From fractographic observations for the cast aluminum alloy, the fatigue crack originated from a casting defect near the specimen surface, as shown in Fig.3. Assuming the casting defect was equivalent to that of surface crack of semi-circular shape, the depth of a semi-circular crack $a$ and aspect ratio $\lambda$ were obtained, as shown Table 3. The lower fatigue strength was attributed to defects of $50 \sim 150$ μm deep from surface, which enhanced fatigue crack formation. Therefore, to improve the plain fatigue strength of cast aluminum alloy, it is important to decrease the size of casting defects or to avoid them altogether.

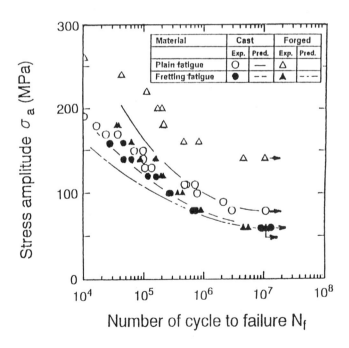

Figure 2-*S-N curve*

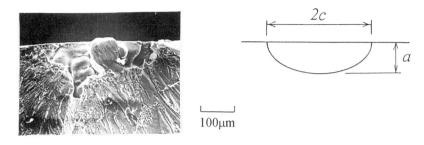

Figure 3-*Crack origin for cast aluminum alloy under plain fatigue*
*($\sigma_a$=100MPa, R=-1 )*

Table 3- *Size and aspect ratio of defects found in the crack initiation region*
*for cast aluminum alloy under the plain fatigue*

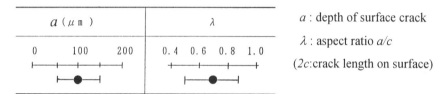

$a$ : depth of surface crack

$\lambda$ : aspect ratio $a/c$

($2c$:crack length on surface)

*Fretting Fatigue Strength*

The S-N curves for fretting fatigue of the cast and forged aluminum alloys are shown in Fig.2. The fretting fatigue limits for both materials were about 60MPa. Although tensile strength and plain fatigue strength of the cast aluminum alloy were lower than those of the forged alloy, fretting fatigue strength of the cast aluminum alloy almost coincided with that of the forged alloy.

The S-N curves, where the stress amplitude was normalized by the tensile strength $\sigma_B$, are shown in Fig.4. From the figure, the fatigue limit for plain fatigue $\sigma_{wp}/\sigma_B$ for the cast and forged aluminum alloys were 0.37 and 0.39, respectively, which coincided with each other. On the other hand, the fatigue limit for fretting fatigue $\sigma_{wf}/\sigma_B$ for the cast and forged aluminum alloys were 0.28 and 0.17, respectively. Furthermore, the ratio of the fretting fatigue limit $\sigma_{wf}$ to plain fatigue limit $\sigma_w$, $\sigma_{wf}/\sigma_w$, for the forged alloy was 0.43, which was almost equal to the values (0.4~0.5) of another report [2,3]. However, the value of $\sigma_{wf}/\sigma_w$ for the cast aluminum alloy was 0.67. This was significantly high compared to that for forged alloy. The result indicates that the cast aluminum alloy has excellent fretting fatigue resistance.

From the detailed SEM observations, a fretting fatigue crack originated near the edge of the contact region. In the cast aluminum alloy, the casting defect seemed to have no significant influence on the formation of fretting fatigue cracks.

To examine the crack formation life, a longitudinal cross section of the fretting fatigue specimen tested up to 1% fatigue life at $\sigma_a$=100MPa was observed. It was found from observations that fretting fatigue cracking would occur even at 1% of the fretting fatigue life, as shown in Fig.5.

*Tangential Force Coefficient*

It is known that tangential force between the specimen and the contact pad has a significant influence on fretting fatigue strength [7,8]. The relationship between tangential force coefficient and relative slip amplitude is shown in Fig.6. The tangential force coefficient is given as $\mu = F_a/P$, where $F_a$ is the amplitude of tangential force and $P$ is the contact pressure. The tangential force coefficient increased with increasing relative slip amplitude for both materials and attained a constant value. No significant difference of tangential force was observed between the two materials.

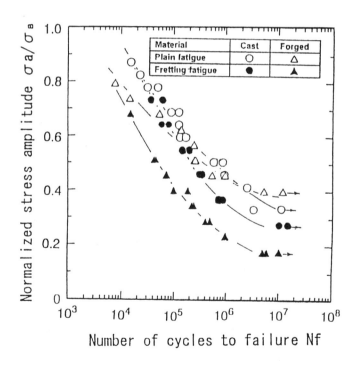

Figure 4-*S-N curves' normalized by* $\sigma_B$

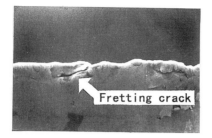

5μm

Figure 5-*SEM observation of a fretting crack on the longitudinal cross section of the specimen tested up to 1% of fretting fatigue life under $\sigma_a$=100MPa*

*Surface Roughness*

The maximum roughness of the fretted surface was also measured to examine the fretting surface damage. The relationship between maximum surface roughness and relative slip amplitude is shown in Fig.7. The maximum surface roughness increased with increasing relative slip amplitude for both materials and was saturated over 15 μm of relative slip amplitude.

*Two-stage Tests*

Two-stage tests were carried out, where a fretting device is applied for a certain number of cycles during the fatigue test and then the pads are removed, with the test continuing under the same applied alternating stress until failure occurs or a run-out is achieved. The results of the two-stage tests for the cast and forged aluminum alloys are shown in Fig.8. The horizontal axis indicates the number of fretting cycles $N_{fx}$ and the vertical axis is the total number of cycles to failure $N_T$. The line representing $N_{fx}/N_T$ =100% shows the failures $N_f$ by continuous fretting fatigue throughout the specimen life. For cast aluminum alloys, the failure occurred outside of the fretting zone in the range of $N_{fx}/N_T \leqq 0.1$, and the crack originated at casting defects similar to the case of plain fatigue. Therefore, fatigue crack growth from a casting defect dominated fracture behavior. Influence of fretting also didn't occur in the range of $N_{fx}/N_T < 0.006$ for the forged aluminum alloy, $N_T$ decreased with the increase in $N_{fx}$ in the range of $0.01 \leqq N_{fx} < 0.1$ for both materials, and $N_T$ became almost equal to $N_f$ in the range of $N_{fx}/N_f > 0.1$. Therefore, it can be considered that the crack propagation region under the influence of fretting is about 10% of fretting fatigue life for both materials in this study.

*Fatigue Crack Propagation Characteristics*

The relationship between fatigue crack growth rate d$a$/d$N$ and stress intensity factor range $\Delta K$ is shown in Fig.9. From the figure, the fatigue crack propagation

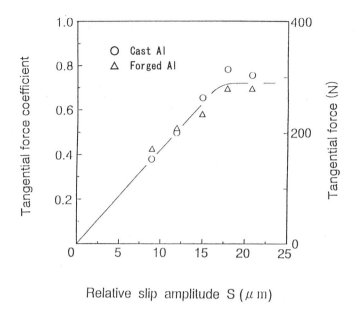

Figure 6- *Relationship between tangential force coefficient and relative slip amplitude*

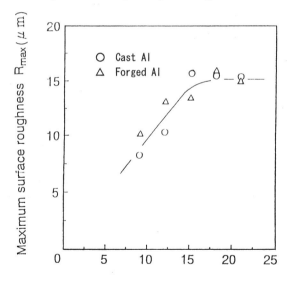

Relative slip amplitude S ($\mu$ m)

Figure 7-*Relationship between maximum surface roughness and relative slip amplitude*

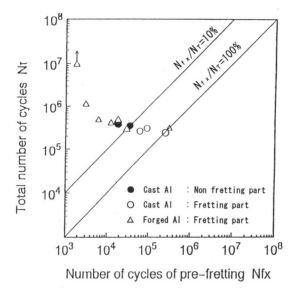

Figure 8- *Two-stage test results*

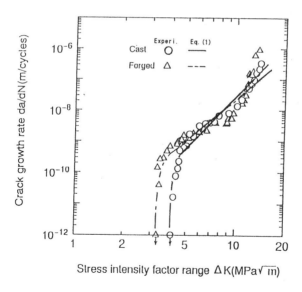

Figure 9- *Relationship between crack growth rate da/dN and stress intensity factor range* $\Delta K$

curves for both materials were almost the same except near the threshold region. Jono [9] and Shiozawa [10] have reported that stage $II_b$ in fatigue crack propagation for both materials is divided into three regions. The same tendency was also observed here. In the present study, to predict fretting fatigue lives, the crack propagation curves were expressed in the following equation [11], which was modified using the threshold stress intensity factor range $\Delta K_{th}$

$$da/dN = C_0(K^m - \Delta K_{th}^{\ m}) \qquad (1)$$

where $C_0$ and m are material parameters. The approximated curves are indicated in Fig.9. The best fitting values are $C_0$=8.702E-13, m=4.4 , $\Delta K_{th}$ =4.1MPa for the cast aluminum alloy and $C_0$=9.56E-14, m=5.4 , $\Delta K_{th}$ =3.3MPa for the forged aluminum alloy.

*Prediction of Plain Fatigue Life for Cast Aluminum Alloy*

A crack in plain fatigue for the cast aluminum alloy originated at a casting defect near the specimen surface, as shown in Fig.3. Assuming a semi-elliptical surface crack in semi-infinite body under uniform tension [12,13], prediction of plain fatigue life was carried out using the following equation,

$$K = M_f(\Phi, a/c) \, \sigma(\pi a)^{1/2} \{ \sin^2 \Phi + (a/c)2\cos^2 \Phi \}^{1/4} / E(k) \qquad (2)$$
$$E(k) = \{ 1 + 1.464(a/c)^{1.65} \}^{1/2}$$
$$M_f = 1.13 - 0.1 \, (a/c) \qquad 0.02 < a/c < 1$$
$$M_f = (1 + 0.003(c/a))(c/a)^{1/2} \qquad a/c > 1.$$

Substituting Eq. (2) into Eq.(1), the fatigue life was calculated by integrating Eq.(1). The result is shown in Fig.2. The predicted life is in good agreement with the experimental in the low stress region.

*Prediction of Fretting Fatigue Life*

It is well known that fretting fatigue cracks nucleate in the early stage of fatigue life. Therefore, fretting fatigue life is almost spent in propagating the crack [14-16]. The prediction of fretting fatigue life for sintered high speed steel (JIS SKH10) was made on the basis of the fracture mechanics approach in the previous paper [8]. The predicted lives were in good agreement with the experimental results. In the predictions, it was assumed that some part of tangential force, $\alpha F_a l$ ( $l$ : the length of contact region), was concentrated at the edge of the contact region and the remaining tangential force [17] , $(1-\alpha)F_a l$, was uniformly distributed along the contact region. The stress intensity factors of through-thickness cracks with the concentrated tangential force $(1-\alpha)F_a l$ along the contact surface in a semi-infinite body are given by the components due to the alternating applied stress $\sigma_a$ and the alternating tangential force $F_a$ [18] ,

$$\Delta K/2 = 1.12 \, \sigma_a(\pi a)^{1/2} + K_{F1} + K_{F2} \qquad (3)$$

where

$$K_{F1} = 1.29 \, \alpha F_a l (1/\pi a)^{1/2},$$
$$K_{F2} = 1.29(1-\alpha)F_a(\pi a)^{1/2}[(3/2 \, \pi)ln\{l + (l^2 + a^2)^{1/2}\}/a - (1/2 \, \pi)l /(l^2 + a^2)^{1/2}].$$

Here $K_{F1}$ is the value due to the component of tangential force concentration at the end of the contact pad and $K_{F2}$ is an approximate solution due to the uniform distribution of tangential force $(1-\alpha)F_a l$. The value of $\alpha$, at which $\Delta K$ at the local minimum point of $\Delta K(a, \alpha)$ curves coincide with $\Delta K_{th}$ was determined. The $\alpha$ value obtained for the cast aluminum alloy was 0.35, and that of the forged alloy was 0.28.

Substituting Eq. (3) into Eq.(1), the fatigue life was calculated by integrating Eq.(1). In the prediction, initial crack length was assumed to be the maximum surface roughness of the edge of contact region, and final crack length was assumed to be half of the specimen thickness. As shown in Fig.2, the predicted lives were in good agreement with the experimental results.

## Conclusions

Fretting fatigue and fatigue crack growth tests were carried out using cast and forged aluminum alloys. The main results obtained are summarized as follows:
(1) Although tensile strength and plain fatigue strength of the cast aluminum alloy were lower than those of the forged alloy, fretting fatigue strength of the cast aluminum alloy almost coincided with that of the forged alloy.
(2) In the cast aluminum alloy, a crack in plain fatigue originated at a casting defect. On the other hand, a fretting fatigue crack nucleated near the edge of contact region similar to the case of the forged alloy.

Although static strengths and plain fatigue strength for the cast aluminum alloy were lower than those of forged alloy, fretting fatigue strength of the cast alloy almost coincided with that of the forged alloy. From the experimental results mentioned above, the fretting fatigue crack nucleates in the very early stage of life and fretting fatigue life is dominated by crack propagation life. Tangential force and fatigue crack growth curves for both materials almost coincide with each other. Therefore, fretting fatigue lives of both materials also coincide with each other.

## References

[1]    Waterhouse, R.B., "Fretting Fatigue," International Materials Reviews, Vol.37, 1992, pp.77-79.

[2]    Conelius, H., Z. Metallk, Vol. 36, 1944, p. 101.

[3]    Fenner, A. J., and Field, J. E., Revue Met., Vol. 55, 1958, p. 475.

[4]    Alic, J.A., Hawley, A.L., and Urey, J.M., "Formation of Fretting Fatigue Cracks in 7075-T7351Aluminum Alloy," Wear, Vol. 56, 1979, pp. 351-361.

[5]    Wharton, M.H., Taylor, D.E., and Waterhouse, R.B., "Metallurgical Factors in the Fretting-Fatigue Behavior," Wear, Vol. 23, 1973, pp. 251-260.

[6]   ASTM Standard E647, "Test method for Constant-Load-Amplitude Fatigue Crack Growth Rates above $10^{-8}$ m/cycle," p. 662.

[7]   Endo, K., Goto, H., and Fukunaga, T., "Behaviors of Frictional Force in Fretting Fatigue," Bull. of JSME, Vol. 17(108), 1974, pp. 647.

[8]   Nishida, T., Mutoh, Y., Tanaka, K., and Nakamura, H., "Fretting Fatigue in Sintered High Speed Steel SKH10," Japan Soc. Mech. Engrs., Vol. 55(513), 1989 pp. 1073-1080.

[9]   Jono, M., Song, J., Mikami, S., and Ohgaki, M., "Fatigue Crack Growth and Crack Closure Behavior of Structural Materials," Soc. Mater. Sci., Japan, Vol. 33(367), 1984, pp. 468-474.

[10]   Shiozawa, K., Mizutani, J., Nishino, S., Ebata, S., Yokoi, N., and Haruyama, Y., "Corrosion Fatigue Crack Propagation in Squeeze-Cast Al-Si aluminum Alloy," Japan Soc. Mech. Engrs., 1991, pp. 1279-1286.

[11]   Paris, P.C., and Erdogan, F., Trans. ASME, Jour. of Basic Engng., Vol. 85, 1963,   p. 528.

[12]   Raju, I.S., and Newman,Jr. J.C., NASA Technical Paper, 1979, p. 1578.

[13]   Raju, I.S., and Newman,Jr., "Stress-intensity factors for a wide range of semi-elliptical surface cracks in finite-thickness plates," Engng. Fract. Mech., Vol. 11(4), 1979), pp. 817-829.

[14]   Endo, K. "Practical Observations of Initiation and Propagation of Fretting Fatigue Crack," Fretting Fatigue, Edit by R.B.Waterhouse, Applied Science Publ., pp. 127-141.

[15]   Mutoh, Y., "Mechanisms of Fretting Fatigue," JSME Int. Journ., Vol. 38(4), 1995, pp. 405-415.

[16]   Satoh, T., and Mutoh, Y., "Effect of contact pressure on fretting fatigue crack growth behavior at elevated temperature," Fretting Fatigue, Edit by Waterhouse, R.B., and Lindley, C., Mechanical Engineering Publications, pp. 405-416.

[17]   Edwards, P.R., "The Application of Fracture Mechanics to Predicting Fretting Fatigue," Fretting Fatigue, Edit by R.B.Waterhouse, Applied Science Publ., 1981, pp. 67-97.

[18]   Tanaka, K., and Mutoh, Y., and Sakoda, S., "Effect of Contact Materials on Fretting Fatigue in a Spring Steel," Japan Soc. Mech. Engrs, Vol. 51(464), 1986, pp. 1200-1207.

# Fretting Damage Analysis

V. Chandrasekaran[1], Young In Yoon[2] and D.W. Hoeppner[3]

Analysis of Fretting Damage Using Confocal Microscope

REFERENCE: Chandrasekaran, V., Yoon, Y. I., and Hoeppner, D. W., "Analysis of Fretting Damage Using Confocal Microscope," *Fretting Fatigue: Current Technology and Practices, ASTM STP 1367,* D. W. Hoeppner, V. Chandrasekaran, and C. B. Elliott, Eds., American Society for Testing and Materials, West Conshohocken, PA, 2000.

ABSTRACT: Fretting fatigue experiments were conducted on 7075-T6 and 2024-T3 Aluminum alloy specimens. The primary objective of this study was to quantitatively characterize fretting damage that resulted on the fatigue specimens. Fretting fatigue experiments were performed in laboratory air at various maximum fatigue stress levels at a constant normal pressure. The hypothesis of this study was that the intensity and the nature of fretting damage would vary depending upon the applied maximum fatigue stress and the three dimensional nature of the damage that would result from fretting could be quantified. Fretting fatigue experiments were interrupted at a predetermined number of cycles to analyze the damage on the fatigue specimens. Confocal microscopy was used to analyze and quantify fretting damage. Digitized images of fretting damage were obtained from the confocal microscope, using a pixel counting software package which also allowed length measurement of fretting induced cracks on the faying surface of the fatigue specimen. In addition, fretting damage was quantified in terms of material removal by characterizing the depth as well as the geometry of fretting-generated pits on the faying surface of the specimen. Pit size in terms of pit depth ($P_d$), pit area ($P_A$), and pit dimension perpendicular ($P_{Dy}$) as well as parallel ($P_{Dx}$) to the applied load also were quantified. From the confocal microscopy analysis of fretting damage, it was observed that fretting-generated *multiple cracks* on the faying surface could be responsible for the fracture of 7075-T6 aluminum alloy specimens whereas the fracture of 2024-T3 aluminum alloy specimen could be attributed to fretting-generated *multiple pits* on the faying surface. From the results, it is proposed that fretting nucleates damage of different nature depending on the material microstructure as well as its composition and the methods to alleviate fretting should consider issues pertaining to a specific material.

[1] Research Assistant Professor, Mechanical Engineering Department, University of Utah, Salt Lake City, Utah 84112.
[2] Doctoral Student, Mechanical Engineering Department, University of Utah, Salt Lake City, Utah 84112.
[3] Professor and Director, Quality and Integrity Design Engineering Center (QIDEC), Mechanical Engineering Department, University of Utah, Salt Lake City, Utah 84112.

**KEYWORDS:** fretting-fatigue, fretting-nucleated-cracks, fretting-nucleated-pits, fretting-damage.

-----------------------------------

## INTRODUCTION

Fretting fatigue is described *as the progressive damage to a solid surface that arises from fretting [1]*. Fretting is defined as a *wear phenomena occurring between two surfaces having oscillatory relative motion of small amplitude [1]*. Fretting may produce several forms of damage on the faying surface. The cause of fretting damage production has been investigated and explained by a number of investigators [2-8]. A commonly accepted view is that the first stage of fretting is adhesive contact of the asperities on opposing contact surfaces (this is supported by an increase in the coefficient of friction). These adhesive contacts are important as they are often thought to be the mechanism by which the majority of the cracks are nucleated. After this stage several things might occur: breakage of the asperities which causes production of fretting debris, oxidation/corrosion of "fresh surface" (which might cause pits on the contacting surfaces) and/or debris. These damages then accelerate the crack nucleation stage of fatigue. Cracks that nucleate may propagate at various rates and angles to the contacting surfaces and cause premature fatigue failure. The aforementioned stages are dependent on many different variables such as material, stress state, and environment, relative slip amplitude, and contact pressure. In general, fretting may result in the following forms of damage [7]:

- pits
- oxide and debris (third body)
- scratches
- fretting and/or wear tracks
- material transfer
- surface plasticity
- subsurface cracking and /or voids
- fretting craters
- cracks at various angles to the surface

The intensity and the nature of fretting damage would vary depending upon the applied maximum fatigue stress and the three dimensional nature of the damage that would result from fretting could be quantified. To characterize fretting induced damage, the confocal microscopy was used because of its versatility as well as its ability to produce high resolution images as briefly explained below.

## WHY CONFOCAL MICROSCOPE?

The popularity of confocal microscopy arises from its ability to produce blur-free, crisp images of thick specimens at various depths. In contrast to a conventional microscope, a confocal microscope projects only light coming from the focal plane of the lens. Light coming from out of focus areas is suppressed. Thus information can be collected from very defined optical sections perpendicular to the axis of the microscope. Confocal imaging can only be performed with point wise illumination and detection, which is the most important advantage of using confocal laser scanning microscopy [9]. The confocal microscope can optically section thick specimens in depth, generating stacks of images from successive focal planes. Subsequently the stack of images can be used to reconstruct a three-dimensional view of the specimen. The brightness of the pixel depends on the intensity of the light measured from that point in the specimen. In order to collect an image of the whole area of interest, either the specimen is moved on computer-controlled scanning stages in a raster scan, or the beam is moved with scanning mirrors to move the focused spot across the specimen in a raster scan. In either case, the image is assembled pixel by pixel in the computer memory as the scan proceeds. The resolution that is obtained using the confocal microscope can be better by a factor of up to 1.4 than the resolution obtained with the microscope operated conventionally.

The primary objective of this study was to quantitatively characterize fretting damage that resulted on the fatigue specimens. Fretting fatigue experiments were performed in laboratory air at various maximum fatigue stress levels at a constant normal pressure and the resulting fretting damage is quantitatively characterized as explained below.

## EXPERIMENTAL DETAILS

Fretting fatigue tests were performed using a closed loop electro-hydraulic servo-controlled testing system. A schematic representation of the fretting test set up and the specimen grips are shown in Fig. 1. As the fatigue specimen deforms during the application of the fatigue cycle, a relative movement occurs between the fatigue specimen and the fretting pad. This motion, acting under the various magnitudes of applied normal and fatigue loads, results in fretting.

Fretting fatigue tests were performed on flat fatigue specimens in contact with fretting pads. A supporting block was placed beneath the fatigue specimen test section to prevent bending of the specimen due to the normal load. The axial fatigue load was applied horizontally to the fatigue specimen. A normal pressure was applied vertically through the fretting pad which is in contact with the fatigue specimen. Fretting fatigue experiments were conducted on two aluminum alloy fatigue specimens viz. 7075-T6 and 2024-T3. The fretting pads also were made of these materials. The configurations of the fatigue specimen and the fretting pad are shown in Fig. 2. Fretting fatigue experiments were interrupted at a predetermined number of cycles to analyze the damage using the confocal microscope on the fatigue specimens. The maximum fatigue stress ($\sigma_{max}$) level was varied from specimen to specimen at a constant normal stress ($\sigma_n$ = 13.8 MPa or 2 ksi). The static normal load was calculated by multiplying the contact pad area with the normal stress. The resultant normal load was applied as shown in Fig. 2. All tests were conducted in laboratory air (room temperature) at R = 0.1 and frequency of 10 Hz. Table I shows the fretting fatigue test results including the fretting damage characterization made using the confocal microscope that will be discussed in detail in the next section.

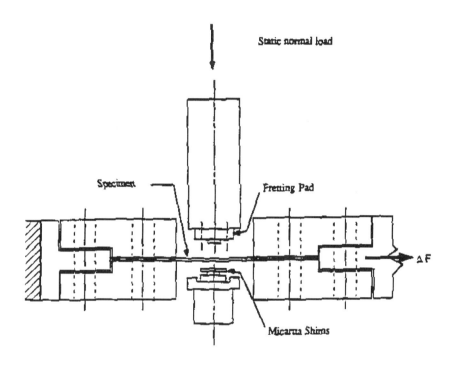

Figure 1 -- *Details of Grips and Specimen in the Fretting Fatigue Set Up.*

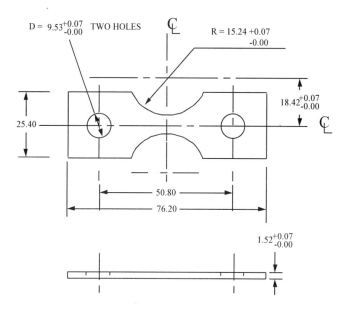

Fatigue Specimen Configuration (All dimensions in mm).

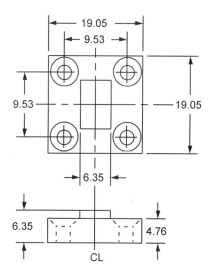

Fretting Pad Configuration (All dimensions in mm).

Figure 2 -- *Test Specimen Configuration (Not drawn to scale).*

TABLE I -- Fretting fatigue test results and the confocal microscope analysis results

| 7075-T6 Aluminum Alloy | | |
|---|---|---|
| Maximum fatigue stress | Number of fretting fatigue cycles | Observations using the confocal microscope |
| 138 MPa (20 ksi) | 51,000 cycles (test interrupted for analysis of damage) | A few black spots of debris |
| | 136,500 cycles (specimen fractured) | Observed cracks (crack length ranged from 20.64 μm - 72.05 μm) |
| 172 MPa (25 ksi) | 44,100 cycles (test interrupted for analysis of damage) | Observed debris (black color) |
| | 96,200 cycles (test interrupted for analysis of damage) | Observed material removal (depth varied from 3 - 10 μm) |
| | 128,400 cycles (test interrupted for analysis of damage) | Observed cracks (lengths varied from 20.99 - 169.06 μm) Observed two pits (pit depth of about 10 μm) |
| 241 MPa (35 ksi) | 35,500 cycles (specimen fractured) | Observed material removal (depth varied from 9 - 18 μm) Observed a couple of cracks (25.5, 35 μm) |
| 2024-T3 Aluminum Alloy | | |
| 207 MPa (30 ksi) | 81,100 cycles (fractured) | Observed multiple pits (pit depth varied from 8 - 26 μm, pit dimension perpendicular to applied load varied from 10 - 39 μm, and pit area varied from 26 - 1478 sq. μm) |

## DISCUSSION OF RESULTS

The confocal microscope analysis of the specimen faying surface revealed at least three stages in the nucleation and the development of fretting damage leading to the final fracture of the specimens. The interrupted fretting fatigue test method and the subsequent analysis using the confocal microscope have showed that the first stage is the appearance of a black color (for aluminum alloys) debris like a "smudge" on the faying surface of the specimen in the early period of fretting fatigue life. After a certain number of fretting fatigue cycles, this is followed by a removal of material as seen in the digitized images produced by the confocal microscope. Then, the third stage of the development of the damage may depend on the material. For 7075-T6, subsequent fretting fatigue cycles generated multiple cracks on the faying surface whereas in 2024-T3, it resulted in the generation of fretting nucleated multiple pits on the faying surface as illustrated in the thumbnail images as shown in Figs. 3 (a) and 3 (b).

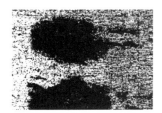  

| Stage I | Stage II | Stage III |
|---|---|---|
| Formation of debris | Removal of material | Nucleation of cracks |
| [analyzed after 44100 cycles] | [after 96200 cycles] | [after 128400 cycles] |

Figure 3(a) -- *Digitized confocal images showing the stages in the nucleation and the development of fretting damage on the faying surface of 7075-T6 Aluminum alloy specimen (X20), $\sigma_{max}$ = 172 MPa (25 ksi), $\sigma_n$ = 13.8 MPa (2 ksi).*

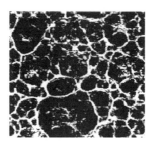

Figure 3(b) -- *Digitized image showing fretting nucleated multiple pits on the faying surface of 2024-T3 aluminum alloy specimen (X20), (analyzed after fracture, 81100 cycles).*

The most important observation from the confocal analysis of fretting damage was that fretting generated multiple cracks on the faying surfaces of 7075-T6 aluminum alloy specimen. The nucleated cracks (at the edge of the contact pad) are believed to be responsible for reduction in the residual strength of the specimens leading to the final fracture. On the other hand, there was no crack found on the faying surface of 2024-T3 specimens. However, multiple pits were observed on the faying surface (also at the edge of the contact pad where fracture occurred) that might have caused the fracture of this specimen. From this observation, it is proposed that the cause of final instability under fretting fatigue conditions is material specific.

Moreover, using the software package that is an integral part of the confocal microscope, fretting damage was quantified. Using a pixel counting software package, the length of fretting induced cracks on the faying surface of the fatigue specimen was measured. In addition, fretting damage was quantified in terms of material removal by characterizing the depth as well as the geometry of fretting-generated pits on the faying surface of the specimen. Pit size in terms of pit depth ($P_d$), pit area ($P_A$), and pit dimension perpendicular ($P_{Dy}$) as well as parallel ($P_{Dx}$) to the applied load also were quantified. Fig. 4 shows a schematic representation of the pit geometry that was characterized in this study.

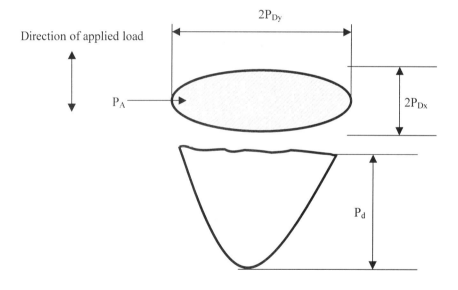

Figure 4 -- *Schematic showing pit geometry, where pit depth is $P_d$, pit dimension perpendicular to loading direction is $P_D$, and pit area is $P_A$.*

Fretting nucleated fatigue crack lengths were measured on the faying surface of 7075-T6 specimens and they were found to be in the range from 20 μm to 169 μm. It was observed that the quantified fretting nucleated cracks were the smallest at 241 MPa (35 ksi) when compared to the two lower stress levels tested in this study as illustrated in the thumbnail images shown in Fig. 5.

(a) $\sigma_{max}$ = 241 MPa (35 ksi)   (b) $\sigma_{max}$ = 172 MPa (25 ksi),   (c) $\sigma_{max}$ = 138 MPa (20 ksi),
    (crack length 25.5 μm)      (crack length = 169 μm)      (crack length = 20 - 72 μm)

Figure 5 -- *Digitized confocal images showing the size of fretting nucleated cracks at different maximum fatigue stress levels on the faying surface of 7075-T6 aluminum alloy (X20).*

The reason for longer cracks on the faying surface of 7075-T6 specimen subjected to lower maximum fatigue stress levels could be because the specimen has more fretting fatigue life (more time for crack(s) to grow) when compared to those at higher stress level. However, the material removal (in terms of depth) was greater at 241 MPa (35 ksi) maximum fatigue stress level when compared to 172 MPa (25 ksi) or 138 MPa (20 ksi). It was observed that at 35 ksi the depth of material removal was in the order 5 to 18 μm. At 172 MPa (25 ksi) it was between 3 and 10 μm. At 138 MPa (20 ksi) the material removal was found to be insignificant. Fig. 6 shows confocal images illustrating removal of material by fretting on the faying surface of 7075-T6 specimen at 241 MPa (35 ksi) maximum fatigue stress level. Fig. 6(a) shows a confocal image where the maximum material removal in terms of depth was observed. The depth of material removal at this point was quantified to be 18 μm. As well, Fig. 6(b) shows material removal on the same specimen (at different location) that was found in the range 5 - 9 μm.

Figure 6(a) -- *Depth of material removal 12 - 18 μm.*

Figure 6(b) -- Depth of material removal 5 - 9 μm.

Figure 6 -- *Digitized confocal images showing material removal on 7075-T6 faying surface (X20), $\sigma_{max}$ = 241 MPa (35 ksi), $\sigma_n$ = 13.8 MPa (2 ksi), R = 0.1, f=10 Hz, Laboratory air.*

Figs. 7 and 8 show graphs of crack size vs. maximum stress and material removal vs. maximum stress respectively. As mentioned before, one of the effects of fretting is that it may produce pits. When 2024-T3 alminum alloy specimen was tested under fretting fatigue conditions with maximum fatigue stress of 207 MPa (30 ksi) at a normal stress of 13.8 MPa (2 ksi), it fractured in 81100 cycles. Subsequently, the confocal analysis revealed multiple pits along the edge of the faying surface of the fatigue specimen where the fracture had occurred. As shown in Fig. 9, these pits varied in geometry as well as in morphology. Using the confocal microscope, the depth of these pits were quantified by scanning along the Z axis. The depth of the pits ($P_d$) varied from 8 to 26 μm. In addition, the pit dimension perpendicular to the applied load ($P_{Dy}$) was found in the range from 10 to 36.82 μm. The pit dimension parallel to the applied load ($P_{Dy}$) was found in the range from 8 to 42.07 μm. The area of the pit ($P_A$) was quantified and was in a range from 26 to 1478 sq. μm. Fig. 10 shows a confocal image revealing fretting nucleated multiple pits on the faying surface of 2024-T3 aluminum fatigue specimen.

Moreover, the quantified pit parameters also revealed some observable relationship. For example, the pit depth ($P_d$) has correlated fairly with the pit dimension perpendicular to the applied load ($P_{Dy}$) as well as with the pit area ($P_A$). The deeper the pit depth, greater the area of the pit as well as larger the pit dimension. Figs. 10 and 11 show graphs illustrating the correlation between pit depth vs. pit area and pit dimension perpendicular to the applied load respectively.

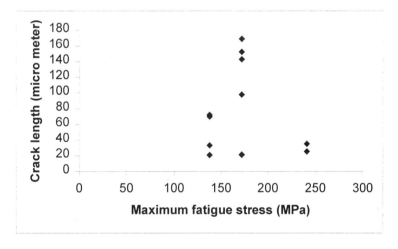

Figure 7 -- *Graph of crack length vs. maximum fatigue stress, Material: 7075-T6 aluminum alloy, $\sigma_n$ = 13.8 MPa (2 ksi), Environment: Laboratory air.*

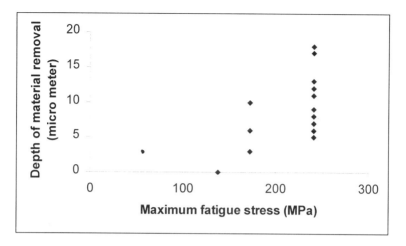

Figure 8 -- *Graph of depth of material removal vs. maximum fatigue stress, Material: 7075-T6 aluminum alloy, $\sigma_n$ = 13.8 MPa (2 ksi), Environment: Laboratory air.*

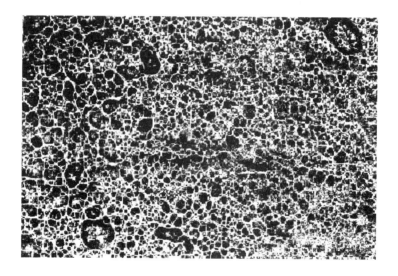

Figure 9 -- *Digitized confocal image of 2024-T3 faying surface revealing multiple pits (X20), $\sigma_{max}$ = 207 MPa (30 ksi), $\sigma_n$ = 13.8 MPa (2 ksi), R = 0.1, f=10 Hz, Laboratory air.*

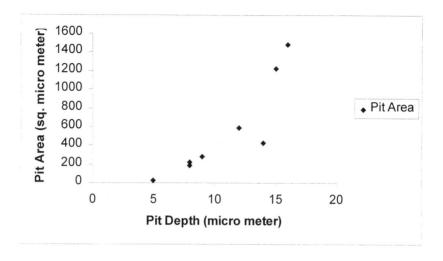

Figure 10 -- *Correlation between pit depth and pit area, Material: 2024-T3 aluminum alloy, $\sigma_n$ = 13.8 MPa (2 ksi), $\sigma_{max}$ = 207 MPa (30 ksi), Environment: Laboratory air.*

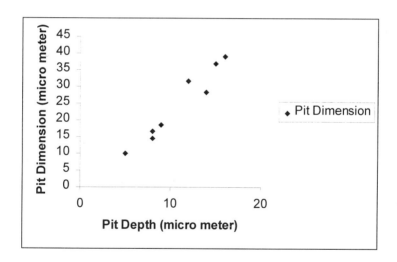

Figure 11 -- *Correlation between pit depth and pit dimension perpendicular to applied load, Material: 2024-T3 aluminum alloy, $\sigma_n$ = 13.8 MPa (2 ksi), $\sigma_{max}$ = 207 MPa (30 ksi), Environment: Laboratory air.*

It was proposed by some researchers *[10]* that the fretting damage process could be described in terms of material loss that is related to fretting wear as well as in terms of crack nucleation that is related to fretting fatigue. However, this study has revealed that fretting damage process comprises of both the material removal as well as crack nucleation on the faying surface of the specimens under fretting fatigue conditions.

To conclude the discussion, if fretting induced cracks on the surface of 7075-T6 produced greater stress concentration (loss of residual strength) that eventually would have led to fracture of the specimens, then, in 2024-T3, fretting nucleated multiple pits may be responsible for the eventual fracture of this specimen. Moreover, in general, 2024-T3 pits better than 7075-T6 because of the Cu phases. The constituent particles are CuAl compounds that are very reactive with the matrix as well. Thus, 2024-T3 is very susceptible to pitting but resistant to SCC and exfoliation. In addition, fretting motion induces faster material removal (as a result of faster oxide removal in discrete areas) resulting in the nucleation of pits on the surface. These observations suggest that even though 2000 series aluminum alloys could tolerate larger critical crack size (more tougher) but they may be susceptible to fretting induced pitting.

**CONCLUSIONS**

Based on this experimental study as well as the analysis of fretting damage using the confocal microscope, the following conclusions can be made.

- The confocal microscopy could be effectively used as a tool to quantitatively characterize fretting damage.
- For 7075-T6 aluminum alloy specimens, it was observed that the quantified fretting nucleated cracks were the smallest at 241 MPa (35 ksi) when compared to the two lower stress levels tested in this study. However, the material removal was found to be greater at higher stress levels when compared to lower stresses. From the confocal microscope analysis, it could be concluded that 7075-T6 specimens fractured because of fretting nucleated multiple cracks on the faying surface.
- For 2024-T3 specimens, the confocal microscope analysis of fretting damage suggests that fretting nucleated multiple pits are responsible for the final fracture of the specimen. Moreover, the quantified pit geometry revealed a fair correlation between the pit depth and pit dimension perpendicular to the applied load as well with the area of the pit.

Further studies are underway to quantitatively characterize the transition of pits to "small" cracks and their propagation to fracture of 2024-T3 specimens under fretting fatigue conditions.

## ACKNOWLEDGMENT

The authors gratefully acknowledge the help of Dr. Ed King at the Biology department, University of Utah in using the confocal microscope to analyze fretting damage on the specimens.

### References

1. ASM Handbook on Friction, Lubrication, and Wear Technology, Volume 18, 1992, pg. 9.
2. Hoeppner, D.W. and Goss, G.L., "Research on Mechanisms of Fretting Fatigue," Corrosion Fatigue - Chemistry, Mechanics, and Microstructure, NACE, Houston, 1971, pp. 617 - 630.
3. Waterhouse, R.B., "The Effect of Fretting Corrosion in Fatigue Crack Initiation," Corrosion Fatigue - Chemistry, Mechanics, and Microstructure, NACE, Houston, 1971, pp. 608 - 616.
4. Waterhouse, R.B. and Taylor, D.E., "The Initiation of Fatigue Cracks in a 0.7% Carbon Steel by Fretting," Wear, Vol. 17, 1971, pp. 139 - 147.
5. Endo, K. and Goto, H., "Initiation and Propagation of Fretting Fatigue Cracks," Wear, Vol. 38, 1976, pp. 311 - 324.
6. Hoeppner, D.W., Mechanisms of Fretting-Fatigue and Their Impact on Test Methods Development," Standardization of Fretting Fatigue Test Methods and Equipment, ASTM STP 1159, M.H. Attia., and R.B. Waterhouse, Eds., American Society for Testing and Materials, Philadelphia, 1992, pp. 23 - 31.
7. Hoeppner, D.W., "Mechanisms of Fretting Fatigue," Fretting Fatigue, ESIS 18, R.B. Waterhouse, and T.C. Lindley, Eds., Mechanical Engineering Publications, London, 1994, pp. 3 - 19.
8. Waterhouse, R.B., Fretting Fatigue, Applied Science Publishers, 1981.
9. Wilson, T., Confocal Microscopy, Academic Press Inc., San Diego, CA 92101, USA., 1990.
10. Vincent, L., "Materials and Fretting," Fretting Fatigue, ESIS 18, R.B. Waterhouse, and T.C. Lindley, Eds., Mechanical Engineering Publications, London, 1994, pp. 323 - 337.

Antoine Chateauminois,[1] Mohamed Kharrat,[2] and Abdelkader Krichen[2]

## Analysis of Fretting Damage in Polymers by Means of Fretting Maps

**REFERENCE:** Chateauminois, A., Kharrat, M., and Krichen, A., **"Analysis of Fretting Damage in Polymers by Means of Fretting Maps,"** *Fretting Fatigue: Current Technology and Practices, ASTM STP 1367,* D. W. Hoeppner, V. Chandrasekaran, and C. B. Elliott, Eds., American Society for Testing and Materials, West Conshohocken, PA, 2000.

**ABSTRACT:** The features of the fretting-wear behavior of glassy polymers against rigid counterfaces are reviewed through fretting maps concepts. The suitability of an elastic analysis to describe the fretting conditions in Running Condition Fretting Maps is discussed on the basis of numerical and experimental investigations of the contact zone kinematics. Material response in terms of cracking is subsequently analyzed in relation to the contact loading and to the plain fatigue properties of the polymer.

**KEYWORDS:** PMMA, epoxy, fretting-wear, cracking, particle detachment

### Nomenclature

| | |
|---|---|
| a | Radius of the contact area (sphere-flat), Semi-width of the contact (cylinder-flat) |
| G | Shear modulus |
| h | Maximum crack depth |
| $N, N_i$ | Number of cycles, Number of cycles for crack initiation |
| P | Normal load |
| $p_m$ | Mean contact pressure |
| $x_o$ | Tangential displacement of the center of the contact area |
| $Q, Q^*$ | Tangential load, maximum tangential load |
| $\delta, \delta^*, \delta_t$ | Relative displacement, displacement amplitude, critical displacement amplitude |
| | at the transition from Partial Slip (PS) to Gross Slip (GS) condition |
| $\nu$ | Poisson's ratio |
| $\mu$ | Coefficient of friction (COF) |
| $\sigma_{xx}$ | Maximum value of the tensile stress at the edge of the contact |

[1] Research Fellow, Laboratoire IFoS, UMR 5621, Ecole Centrale de Lyon, Ecully 69131, France.

[2] Lecturers, Laboratoire LPMM, Ecole Nationale d'Ingénieurs de Sfax, Sfax 3038, Tunisia.

## Introduction

Polymers as coatings or in their bulk form are widely used as palliatives against fretting damage, particularly in quasi-static assemblies subjected to vibration. Previous studies on the fretting of polymers rubbing against metal counterfaces have demonstrated that small oscillatory contact motions are able to induce wear of metal surfaces together with cracking on, or particle detachment from, the polymer surface [1-9]. The occurrence of either of these processes depends in a complex manner on the loading conditions (normal load, tangential displacement, frequency), on the contact geometry, on the material properties and on the environmental conditions. In order to rationalize the effects of these various parameters, the use of fretting maps concepts emerged as a potential route [10]. This methodology is based upon the use of two sets of maps: the Running Condition Fretting Maps (RCFMs), which define the contact conditions as a function of the loading parameters and the number of cycles, and the Material Response Fretting Maps (MRFMs), which delimit different domains related to the main initial degradation (contact cracking or particle detachment). The association of RCFMs and MRFMs has proved its efficiency in relating the fretting damage of metals to the local loading on the surfaces and to the intrinsic properties of materials [11].

The topic of this paper is to review the possibilities of applying this methodology to the fretting of polymers against rigid counterfaces. On the basis of investigations carried out in the Laboratory over the past few years, the following points will be addressed:

*(i)* the analysis of the fretting conditions by means of experimental investigations of the contact zone kinematics. These results have been used to validate the analytical and numerical contact mechanics tools which are required to establish the RCFMs and to calculate contact stresses.

*(ii)* the contact cracking processes in relation to the contact loading and the plain fatigue properties of the polymer material. Results regarding the particle detachment domain have been reported elsewhere [9], and the analysis enclosed herein is thus restricted to the cracking domain of the MRFM.

## Test Methodology

The methodology which has been developed closely associates the experiments and numerical simulations of the fretting tests. The emphasis was especially put on the in-situ visualization of the contact during the test. This provided useful information on the contact zone kinematics and the various stages of the damage development in the polymer bodies (cracking, particle detachment). Due to the limitations of the analytical contact mechanics tools for contacts involving dissimilar materials, most of the experimental results were interpreted in the light of more realistic FEM computations.

### *Materials*

The analysis was focused on the fretting behavior of glassy amorphous polymers rubbing against rigid counterfaces. Two different polymer systems have been studied:

*(i)* an epoxy thermoset which was found to be very appropriate for investigating the cracking processes in the contact. The selected epoxy was a DGEBA/IPD system which is currently used as a matrix for polymer matrix composites. Details on the processing conditions are given in Reference [8].

*(ii)* a commercial grade of cast PMMA which was mainly used for the validation of FEM simulations.

At the test temperature(22°C), both polymers were in the glassy state and exhibited similar Young's moduli (2.8 GPa and 3.5 GPa for epoxy and PMMA, respectively).

In order to allow in-situ visualization of the contact, glass counterfaces have been used. A sphere-flat and a cylinder-flat configuration have been selected for the epoxy and the PMMA conterfaces, respectively. In the cylinder-flat arrangement, the axis of the glass rod was oriented parallel to the surface of the PMMA specimen. This latter configuration was used for the comparison with F.E.M simulations because it involved two-dimensional computations which were easier to handle than the three-dimensional ones required by the sphere-flat configuration.

*Fretting Tests*

Fretting tests were carried out using a specific rig mounted on a tension-compression hydraulic machine which has been described elsewhere [8]. The polymer specimen was mounted on a rod actuator and the glass counterface was held in a fixed holder. The normal force P was set between 100N and 200N. It ensured that the polymers were strained in their elastic range throughout the fretting tests. An oscillating (triangular shaped) tangential displacement was applied to the polymer counterface. Displacement amplitudes $\delta^*$ have been selected in order to investigate contact conditions ranging from Partial Slip (PS) to Gross Slip (GS). The tangential load Q and the relative displacement $\delta$ were monitored continuously during the test in order to allow the recording of fretting cycles. Tests were performed at frequencies ranging from 0.1 to 10 Hz, which ensured that no substantial heating of the polymer substrate occurred during cycling. In-situ microscopic observations of the contact were performed during the tests through the glass counterface by means of a microscope linked to a CCD camera and a video recorder. The resolution of the optical device was on the order of a few microns, the dimensions of the contact areas being in the millimeter range.

*Numerical Simulations*

Two-dimensional simulations of the fretting test were carried out using an already developed commercial F.E.M. package (SYSTUS®). Details of the model have been reported in a previous paper [6]. Due to the modulus ratio between the glass and PMMA (around 20), the cylindrical indentor was considered to be perfectly rigid and only the PMMA specimen was meshed. Computations were performed assuming a linear elastic behavior for the polymer sample and using a Coulomb friction law at the contact interface. Computations provided the stresses and strains in the polymer substrate, together with the micro-slip distribution at the contact interface during a whole fretting cycle.

**Analysis of the Contact Conditions**

*Contact Zone Kinematics*

The first step in the analysis of the fretting conditions was to investigate the contact zone kinematics during the cyclic tangential loading by means of optical observations through the glass counterface. These investigations were carried out during the first fretting cycles, i.e. before the appearance of any damage in terms of cracking or particle detachment. Whatever the contact configuration (cylinder-flat or sphere-flat), a slight tangential shift of the contact area on both sides of its initial position was systematically observed during the fretting cycle (Figure 1). As an example, Figure 2 shows the tangential displacement $x_0$ of the center of the contact area as a function of time in the case of the glass/epoxy contact. For the lowest displacement amplitude, the alternate shift of the contact area follows the same triangular shape as the tangential displacement signal. At higher displacements, a plateau value is reached, which can be related to the achievement of stationary gross slip conditions in the contact.

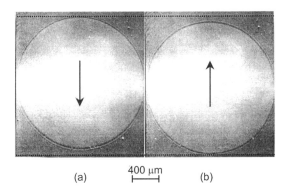

(a)    400 µm    (b)

Figure 1 - *Displacement of the contact area during a fretting cycle*
*(glass/epoxy contact)*
*(a) : Downward displacement of the epoxy substrate ($\delta = - \delta^*$)*
*(b) : Upward displacement of the epoxy substrate ($\delta = + \delta^*$)*

If it is kept in mind that the glass counterface and the microscope are fixed onto the frame of the fretting rig, these observations can be attributed to the alternate accumulation of the moving epoxy body on two opposite sides of the counterface along the sliding direction. Such a process can have two origins:
    *(i)* the first one is related to the polymer viscoelasticity which causes the material to relax more slowly than it is compressed during sliding of the rigid counterface. Such effects have been analyzed in detail [12] in the context of rolling or lubricated sliding of rigid bodies on elastomers, where they have been found to be related to the magnitude of the loss factor tan $\delta$. Although these processes are beyond simple analysis in the context of fretting, it can be emphasized that both the PMMA and the epoxy are strained in their glassy range where they exhibit a reduced viscoelastic behavior.

*(ii)* the alternate displacement of the contact area can also be attributed to the coupling effects between the tangential and normal tractions. Such processes arise from the vertical displacements of the surface of the counterbodies which are induced by the shear tractions [*13,14*]. Since the shear traction distribution is mutual, the vertical direction displacements are the same if the materials have the same elastic properties. As a result, the size and shape of the contact area are fixed only by the profiles of the two surfaces and the normal force. If the counterbodies are dissimilar, the warping of one surface no longer conforms with that of the other, and the shape and location of the contact area are thus modified by the application of the tangential force.

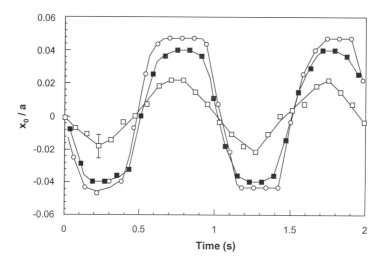

Figure 2 - *Normalized displacement $x_0$ of the center of the contact area during a cyclic tangential loading at 1 Hz*
*(Glass/epoxy contact; □ $\delta^* = \pm 15 \ \mu m$; ○ $\delta^* = \pm 35 \ \mu m$; ■ $\delta^* = \pm 60 \ \mu m$)*

This coupling effect has been theoretically studied in detail by Büfler [*15*] for a sliding cylinder and by Nowell *et al.*[*16*] in the case of fretting contacts. In the present study, linear elastic F.E.M. computations have been carried out to simulate the contact zone kinematics during a fretting cycle. In Figure 3, the computed displacement $x_0$ of the contact area has been reported as a function of the imposed displacement for the glass/PMMA contact under PS condition. The observed hysteresis can be attributed to the occurrence of micro-slip within the partial slip strips, which induced some irreversibility in the displacements of the PMMA surface during the oscillating tengential loading.

Experiments carried out under identical loading conditions indicated that the maximum measured displacement of the contact strip (about 10 percent of the semi-width a of the contact) agreed well with the computed one. This tends to demonstrate that the contact zone kinematics can be satisfactorily described using an elastic analysis of the contact, without necessarily taking into account the viscoelastic properties of the glassy polymers.

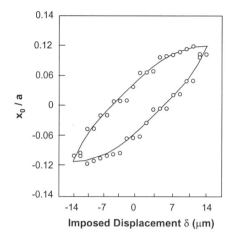

Figure 3 - *F.E.M. simulation of the displacement of the contact strip during a fretting cycle (cylinder- flat configuration, glass/PMMA contact, $\delta^* = \pm 15 \mu m$ )*

*Transition Between Partial Slip and Gross Slip*

The determination of the boundary between PS and GS is essential to predict the loading conditions associated with either cracking or debris formation [11]. The theoretical predictions for the transition from PS to GS conditions are widely based on the use of Mindlin's elastic analysis [17]. For a sphere on flat configuration, this approach predicts that the critical displacement $\delta_t$ at the transition can be related to the elastic properties of the bodies, the normal load and the coefficient of friction (COF):

$$\delta_t = K \mu \, a \, p_m \qquad (1)$$

with:

$$K = \frac{3}{16}\left( \frac{2-v_1}{G_1} - \frac{2-v_2}{G_2} \right) \qquad (2)$$

where

$v_1$ and $v_2$ = Poisson's ratios of materials 1 and 2, respectively
$G_1$ and $G_2$ = Shear elastic moduli of materials 1 and 2, respectively
$\mu$ = coefficient of friction
$p_m$ = mean contact pressure

This theory, however, assumes that there is no coupling effects between the normal and tangential tractions. The numerical and experimental results reported above clearly demonstrate that this assumption is not verified in the context of the fretting of glassy polymers against rigid counterfaces. In order to quantify the induced discrepancies with Mindlin's analysis of the transition, an experimental analysis has been carried out using the glass/epoxy contact. Following a procedure described elsewhere [8], the critical displacement at the transition between PS and GS has been detected for various normal loads by means of the in-situ detection of the displacement of a slight mark located in the middle of the contact. For partial slip conditions, the central portion of the polymer surface in the contact remained stuck to the fixed glass counterface. As a result no displacement of the mark was observed in the middle of the contact area. On the other hand, a displacement of the overall mark was detected in the contact area when gross slip conditions were reached. The results, which are summarized in Figure 4, indicated that only a minor deviation from Mindlin's analysis was induced by both viscoelastic and coupling effects.

The validity of the elastic description of the transition was subsequently investigated for the cylinder-flat configuration. The theoretical values of $\delta_t$ were obtained from more accurate F.E.M. simulations of the contact zone kinematics [6]. The results (Figure 4) indicated a very good agreement with the experimental transition for the glass/PMMA contact. One of the reasons for this good accuracy of the numerical simulations arises from the fact that they take into account the interactions between the normal and shear tractions at the contact interface.

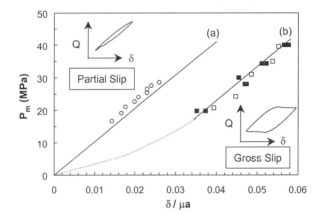

Figure 4 - *Experimental and theoretical boundary between the PS and the GS domains.*
*(a) Sphere-flat, glass/epoxy contact*
○ : *Experimental determination of the transition; Solid line: equation (1)*
*(b) Cylinder-flat, glass/PMMA contact*
■ : *Experimental determination of the transition*; □ : *FEM simulations*

*Running Condition Fretting Maps*

The information regarding the contact conditions can be summarized in Running Condition Fretting Maps (RCFM) which delimit the boundary between PS and GS conditions as a function of time and loading parameters (normal load, tangential displacement). A schematic description of such a RCFM is given in Figure 5. As it was reported above, the boundary between the PS and the GS domains can be satisfactorily described using elastic contact mechanics tools such as Mindlin's analysis or numerical simulations assuming a Coulomb friction law at the interface. By virtue of changes in the COF, this boundary may, however, evolve during the test. In many situations, it was observed that a progressive increase in the COF caused an extension of the PS domain during the early stages of the test. This change occurred before the appearance of any detectable damage and it was therefore attributed to increased adhesion at the glass/polymer interface. According to the definition which was proposed by Vincent [10], three fretting regimes may thus be defined in the RCFM: a Partial Slip Regime (PSR), a Gross Slip Regime (GSR) and a Mixed Slip Regime (MSR), the latter being characterized by a change in the contact conditions from GS to PS as the number of cycles is increased. The MSR is therefore enclosed between the transition boundaries which correspond to the initial and to the stabilized values of the COF respectively.

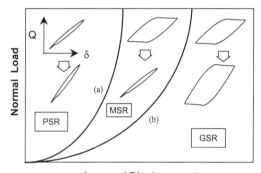

**Imposed Displacement**

Figure 5 - *Schematic representation of a Running Condition Fretting Map (RCFM)*
*PSR: Partial Slip Regime; MSR: Mixed Slip Regime; GSR: Gross Slip Regime*
*(a) initial PS/GS boundary; (b) stabilized PS/GS boundary*

**Material Response**

Under fretting loading, the material response in terms of degradation can involve wear and/or cracking. For metal to metal contacts, it is now recognized that these two kinds of damage can be encountered whether or not an external loading is superposed to the contact loading [10]. The situation is similar for polymers which can exhibit both particle detachment and cracking under fretting-wear loading [1,5,8,9]. In order to analyze the fretting degradation, it appears very important to identify the first damage which can strongly affect the velocity accommodation mechanisms and the subsequent development of the degradation. This early damage is usually described in the MRFMs as a function of the normal load, the imposed displacement and the number of cycles [10]. Three different domains are defined in the MRFMs, which are associated with absence of damage, cracking, and particle detachment, respectively. In the last part of this paper, the analysis will be focused only on the cracking domain of the MRFMs. The results reported herein were obtained using the epoxy thermoset, which proved to be a very appropriate model system to study contact fatigue. Cracks propagated in this substrate over a wide range of displacement amplitudes, before the occurrence of any significant particle detachment. As a result, it was possible to study cracking processes without the complications arising from third body accumulation in the contact.

*Description of Cracking Processes in Epoxy Substrates*

The first damage in the polymer body was associated with the early propagation of two main cracks at opposite contact edges along the loading direction (Figure 6a). Cross-sections of the contact showed that the crack depths were of the same order of magnitude as the radius of the contact area (Figure 6b). The initial propagation of these deep cracks occurred in a direction close to 90° from the surface, which indicates that the contact traction $\sigma_{xx}$ predominates during the nucleation stage.

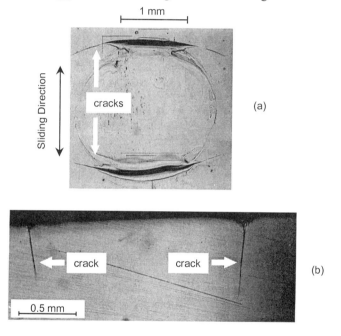

Figure 6 - *Propagation of deep cracks at the edges of the contact (epoxy)*
*(a)Top view of the contact*
*(b) Cross-section of the contact area along the sliding direction*

An investigation of the crack initiation times ($N_i$) and the crack depths (h) has been carried out in the different fretting regimes which were delimited in the RCFM. For a given normal load, three successive domains have been identified as the displacement amplitude was increased (Figure 7):

*(i)* a non degradation domain (I) where no crack nucleation was detected after $10^5$ cycles,

*(ii)* a second domain (II) where crack initiation occurred between approximately $3.0 \times 10^3$ and $5.0 \times 10^4$ cycles. In this domain, both crack initiation times and crack depths

were found to be dependent upon the displacement amplitude. Moreover, in-situ observations of the contact during the test showed that these cracks propagated in a progressive way.

*(iii)* in the last domain (III), crack propagation occurred within a few fretting cycles after an initiation time close to 400 cycles. No significant effect of the displacement amplitude upon $N_i$ and h was detected.

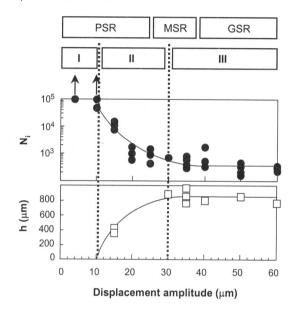

Figure 7 - *Number of cycles to crack initiation ($N_i$) and crack depth (h) as a function of the displacement amplitude (glass/epoxy contact, P=100 N, N=5.0x10⁴ cycles) PSR : Partial Slip Regime; MSR : Mixed Slip Regime; GSR : Gross Slip Regime; see comments in the text for domains I, II, III*

The propagation of these cracks at the edges of the contact strongly affected the velocity accommodation mechanisms. Owing to the fact that their depth is of the same order of magnitude as the size of the contact, a significant amount of the imposed tangential displacement was taken up through the opening and the closing of the cracks. As a result, the micro-slip amplitude at the interface was decreased, and this in turn reduced the particle detachment processes. This mechanism can explain why cracking was observed over a wide part of the gross slip regime, i.e. in a domain where particle detachment is usually reported at the main degradation [6,10,11].

In addition to the two deep cracks at the edges of the contact, short cracks were also detected within the contact area in the gross slip regime (Figure 8). These cracks extended in a direction which lay approximately at + or - 45 degree to the surface, i.e. in a direction of maximum shear, but their lengths did not exceed 50 μm. At high numbers of cycles, they were associated to a particle detachment process which indicated the occurrence of some competition between material loss and cracking.

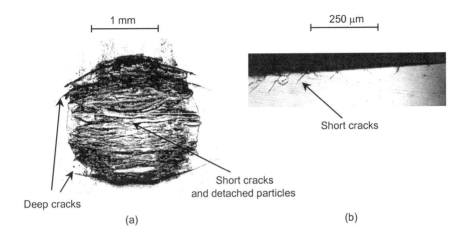

Figure 8 - *Cracking and particle detachment in the gross slip regime (epoxy)*
*(a) Top view of the contact area after 5.0x10^4 cycles*
*(b) Magnified cross-section of the epoxy substrate in the middle of the contact patch*

*Relationships Between Contact Cracking and Plain Fatigue Properties*

An analysis of the contact cracking behavior can be envisaged on the basis of a knowledge of the contact stresses and the plain fatigue properties of the polymer. The use of constant stress (or strain)-amplitude fatigue data in the context of contact fatigue is, however, not straightforward due to the continuous change in the contact loading which often precedes crack initiation. Two different processes must in particular be taken into account :

(i) The first one is related to the formation of a third body layer which can accommodate the velocity and thus reduce the stresses transmitted to the polymer substrate. As was mentioned above, third body effects can be discarded in a large part of the cracking domain of the epoxy thermoset. On the other hand, third body accumulation needs to be taken into account in the glass/PMMA system where particle accumulation at the contact interface was observed from the very first strokes. To some extent, this can explain why no deep crack propagated in the PMMA body under fretting conditions similar to those achieved in the glass/epoxy contact [9].

(ii) Changes in the contact conditions may also be induced during the fretting test by virtue of the development of adhesion in the contact area. As was mentioned above, such effects were particularly relevant to the glass/epoxy contact. In the gross slip regime, experiments showed that crack nucleation was preceded by a progressive increase in the COF from 1.0 up to values close to 1.5 [8]. These processes occurred during most of the crack initiation period, so that a time dependent loading must be taken into account in the analysis of the contact cracking. The latter can be deduced from contact mechanics tools provided that the kinetics of the changes in the coefficient of friction is known. The basis for such a methodology is detailed below for the epoxy thermoset, the discussion being restricted to the analysis of the cracks which were induced at the edges of the contact.

Plain fatigue data of polymer materials can be obtained from conventional cyclic tests using un-notched specimens, provided that the fatigue failures resulting from hysteretic heating are avoided. For the epoxy thermoset studied, the temperature rise during fatigue testing at moderate frequency (10 Hz) was, however, found to be limited by virtue of the low damping capacity and the high $T_g$ (165°C at 1 Hz) of the material. Three point bending fatigue data using a stress ratio R equal to 0.1 are reported in Figure 9. Monitoring of stiffness loss during the tests indicated that crack propagation occurred over a limited part of the fatigue life. Crack nucleation times are thus very close to the time to failure reported in Figure 9.

From this S-N curve, it is possible to differentiate between two fatigue domains:

*(i)* a high cycle fatigue domain where failures occurred between $10^2$ and $10^6$ cycles. This domain can be assimilated to a horizontal scatter band which is bounded in the low stress range by the endurance limit of the epoxy (about 40 MPa).

*(ii)* a low cycle fatigue domain characterized by the early breakage of the specimen in less than $10^2$ cycles. This domain corresponds to tensile stresses greater than 70 MPa.

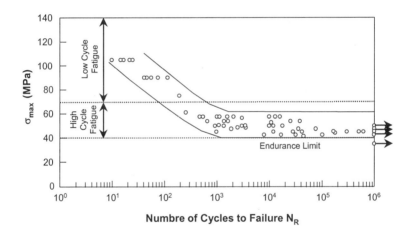

Figure 9 - *S-N curve of the bulk epoxy polymer under three point bending.*
*10 Hz, stress ratio R=-1. Arrows correspond to run-outs.*

The next step is to compare these fatigue data to the maximum values of the tensile stress $\sigma_{xx}$ which were induced at the edges of the contact during the tangential loading. As a first order approximation, the latter can be assessed in GS conditions by means of Hamilton elastic analysis [*18*] for a sliding sphere. Under PS conditions, the stress at the contact edges can be estimated by combining the elliptic Cattaneo-Mindlin shear stress distribution [*14*] with Hamilton's expressions. Using these approaches, the values of $\sigma_{xx}$ have been calculated as a function of the imposed displacement amplitude for the two limiting values of the coefficient of friction which correspond to the boundaries of the mixed regimes. It was implicitly assumed that the increase in the coefficient of friction occurred to the same extent whatever the imposed tangential displacement. This latter point was experimentally verified under gross slip conditions [*8*].

If the plain fatigue data of the epoxy are reported in this graph, it becomes possible to justify the existence of the three cracking domains which were identified experimentally (cf Figure 7):

*(i)* for displacement amplitudes less than 6 μm, the values of $\sigma_{xx}$ remain below the endurance limit of the epoxy resin during the whole fretting test. This first domain thus corresponds to the non-degradation domain (I).

*(ii)* For tangential displacements greater than 16 μm, the tensile stress at the contact edges moves from the high cycle fatigue domain to the low cycle fatigue domain of the bulk epoxy as the number of cycles is increased. As a result, short crack initiation times may be expected. This range of displacements can thus be assimilated to the domain III, where early crack nucleation and propagation was observed.

*(iii)* for intermediate displacements (between 6 μm and 16 μm), $\sigma_{xx}$ remains in the high cycle fatigue domain during the whole fretting test. This third domain will therefore be associated with long crack initiation times and progressive crack propagation, as was experimentally observed in domain (III). It is of interest to note that the range of displacements associated with this last domain is quite narrow. This means that contact conditions can move very rapidly from a domain where no cracks are initiated to a domain where deep cracks can propagate very rapidly during the early stages of the fretting test. This emphasizes the necessity to define very accurately the contact conditions in the RCFM in order to predict the contact cracking behavior.

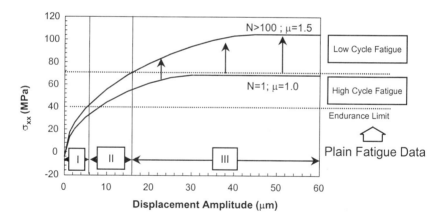

Figure 10 - *Maximum tensile stress $\sigma_{xx}$ at the edge of the contact as a function of the displacement amplitude for the initial ($\mu=1$) and the stabilized ($\mu=1.5$) values of the COF (glass/epoxy contact, N denotes the number of fretting cycles, P=100 N)*

The methodology presented above is clearly based on some crude assumptions regarding the calculation of contact stresses and the kinetics of the changes in the coefficient of friction. In a more detailed analysis, it would be necessary to compute the stresses using numerical methods which take into account the coupling effects between normal and tangential tractions. Some limitations also arise from the fact that the stress ratio at the edges of the contact is probably much less (about −1) than the value used for plain fatigue tests. Despite these restrictions, these results demonstrate that a semi-quantitative relationship can be established between different plain fatigue domains (i.e. low cycle and high cycle fatigue) and existence of distinct contact cracking regimes.

## Conclusion

The use of fretting maps concepts to describe the fretting of polymers against rigid counterfaces has been considered. Model tribological systems where the degradation was restricted to the polymer body have been investigated by means of experiments and FEM simulations of the contact loading. The analysis of the contact zone kinematics showed that elastic approaches using a Coulomb friction law can provide a good prediction of the boundaries between the various fretting regimes defined in Running Condition Fretting Maps (RCFM). In-situ observations of the contact area revealed, however, the occurrence of coupling effects between the tangential and normal tractions. Analytical contact mechanics tools generally neglect such effects, which means that more complex numerical simulations are required to accurately compute the contact stresses induced in the various fretting regimes.

Cracking processes in the Material Response Fretting Map (MRFM) have been investigated in relation to the contact loading and the material fatigue properties. For epoxy thermosets, it was observed that this domain can extend over a large part of the Gross Slip Regime, where particle detachment processes are more generally found to be the main damage. This reduction of the wear domain was related to the pronounced change in the velocity accommodation mechanisms, which resulted from the early propagation of deep cracks at the edges of the contact. From a knowledge of the time dependent contact loading, it was possible to establish relationships between the contact cracking behavior and the bulk fatigue properties of the polymer. For epoxy substrates , the narrow stress range associated with the high cycle fatigue domain was able to explain the sharp transition from a non-degradation domain to a cracking domain with an increase in the displacement amplitude.

## References

[1]    Higham, P.A., Bethune, B. and Stott, F.H., "Changes in the Surface Morphology of Polycarbonate induced by Fretting," *Journal of Material Science*, Vol. 12, 1977, 2503

[2]    Higham, P.A., Stott, F.H. and Bethune, B., "Mechanisms of Wear of the Metal Surface During Fretting Corrosion of Steel on Polymers," *Corrosion Science*, Vol.18, 1978, pp. 3-13

[3]    Gaydos, P.A., Eiss, N.S., Furey, M.J. and Mabie, H.H., "Fretting Wear of Polymeric Coatings", *Wear of Materials*, K. Ludema Ed., American Society of Mechanical Engineers, New York, 1989, pp. 529-535

[4]    Kang, C. and Eiss, N.S., "Fretting Wear of Polysiloxane-Polyimide Copolymer Coatings as a Function of Varying Humidity," *Wear*, Vol. 58, 1992, pp. 29-40

[5] Dahmani, N., Vincent, L., Vannes, B., Berthier, Y. and Godet, M., "Velocity Accommodation in Polymer Fretting," *Wear*, Vol. 158 , N°1-2, 1992, pp 15-27

[6] Krichen A., Kharrat M. and Chateauminois A., "Experimental and Numerical Investigation of the Sliding Behaviour in a Fretting Contact Between Poly(MethylMethacrylate) and a Glass Counterface," *Tribology International.*, Vol. **29**, N° 7, 1996, pp 615-624

[7] Yan, F.Y. and Xue, Q.J., "Study of Fretting Wear Behaviors of FEP," *J. of Applied Polymer Science*, Vol. 67, N° 6, 1998, pp. 1119-1125

[8] Kharrat, M., Krichen, A. and Chateauminois, A., "Analysis of the fretting conditions in a contact between an epoxy thermoset and a glass counterface," to appear in *Tribology Transactions*

[9] Krichen, A., Bradai, C., Chateauminois, A. and Kharrat, M., " Surface damage of PMMA under fretting loading," to appear in *Wear*

[10] Vincent, L., "Materials and Fretting," *Fretting Fatigue ESIS 18*, R.B. Waterhouse and T.C. Lindley Eds, Mechanical Engineering Publications, London, 1994, pp. 323-337

[11] Fouvry, S., Kapsa, Ph. and Vincent, L., "Quantification of Fretting Damage," *Wear*, Vol. 200, N°1, 1996, pp. 186-205

[12] Moore, D., *The Friction and Lubrication of Elastomers*, Academic Press, London,1972

[13] Hills, D.A., Nowell, D. and Sackfield, A., *Mechanics of Elastic Contacts*, Butterworth-Heinemann Ltd, Oxford, 1993

[14] Johnson, K.L., *Contact Mechanics*, Cambridge University Press, Cambridge, 1985

[15] Büfler, H., "Zur Theorie der Rollenden Reibung," *Ing. Arch.*, Vol. 27, 1959, 137

[16] Nowell, D., Hills, D.A. and Sackfield, A., "Contact of Dissimilar Elastic Cylinders under Normal and Tangential Loading," *Journal of Mechanics and Physics of Solids*, Vol. 36, N°1, 1998, pp. 59-75

[17] Mindlin, R.D., "Compliance of Elastic Bodies in Contact," *ASME Transactions, J. of Applied. Mechanics, Series E*, Vol. 16, 1953, pp. 327-344

[18] Hamilton, G.M., "Explicit Equations for the Stresses Beneath a Sliding Spherical Contact,"*Proceedings of the Institution of Mechanical Engineers*, Vol. C 197, 1983, pp.53-59

# Life Prediction

Richard W. Neu,[1] John A. Pape,[1] and Dana R. Swalla[1]

## Methodologies for Linking Nucleation and Propagation Approaches for Predicting Life Under Fretting Fatigue

**REFERENCE:** Neu, R. W., Pape, J. A., and Swalla, D. R., **"Methodologies for Linking Nucleation and Propagation Approaches for Predicting Life Under Fretting Fatigue,"** *Fretting Fatigue: Current Technology and Practices, ASTM STP 1367,* D. W. Hoeppner, V. Chandrasekaran, and C. B. Elliott, Eds., American Society for Testing and Materials, West Conshohocken, PA, 2000.

**ABSTRACT:** During the past 25 years, models for predicting either the nucleation or propagation of fretting fatigue cracks have been developed and verified to a limited extent. Current nucleation models can predict the location of the crack nucleus by correlating the level of fatigue damage with the local state of stress. Fracture mechanics methods are used to predict fretting fatigue crack growth by assuming an initial crack length and orientation. This paper describes multiaxial fatigue criteria based on critical plane approaches that can potentially bridge current nucleation and fracture mechanics methods. Conventional and critical plane models are compared in light of some recent experiments on PH 13-8 Mo stainless steel. A critical plane damage model can predict the orientation and growth direction of crack nuclei. The link between critical plane and fracture mechanics approaches for predicting the early stages of crack propagation is discussed.

**KEYWORDS:** fretting fatigue, life prediction, crack nucleation, critical plane approaches, fracture mechanics, PH 13-8 Mo stainless steel

### Introduction

The prediction of fretting fatigue damage for determining the life of structural components is currently divided into methods that are used to predict the nucleation of a fretting fatigue crack and methods to predict crack growth based on fracture mechanics. Nucleation involves all of the processes leading to the formation of a crack. The crack growth process describes the intensification of the crack driving force that is present under fretting fatigue due to the local cyclic shear tractions. One can argue that crack nucleation accounts for a small fraction of the total life of a component and that the total

---

[1] Assistant professor, graduate research assistant, and graduate research assistant, respectively, The George W. Woodruff School of Mechanical Engineering, Georgia Institute of Technology, Atlanta, GA 30332-0405.

life can be well predicted using crack growth modeling based on fracture mechanics. However, in a complex structure it is often not known whether fretting fatigue will be a problem, or in other words, whether a fretting fatigue crack nucleus will form or a fretting fatigue limit or damage threshold will be reached [1-3]. In design, one attempts to reduce the likelihood of the formation of a crack nucleus that can sustain crack growth. In addition, the size and orientation of the crack nuclei can have strong implications as to whether a crack will grow to catastrophic failure or simply become non-propagating.

The application of fracture mechanics for predicting the crack growth rate in the presence of combined frictional force and alternating load has been somewhat successful. However, models that describe the nucleation process and are able to bridge nucleation and fracture mechanics approaches are still needed. As will be shown later, critical plane approaches can potentially build this bridge. To begin, a description of state-of-the-art nucleation models is presented.

Modeling the nucleation process can be approached from two fronts: empirical approaches based on continuum field variables, and approaches based on the mechanics of nucleation at a smaller scale (e.g., dislocation level in the formation of extrusions) [4]. The empirical approach combines the relevant macroscopic variables such as stress, strain, and slip range, and correlates them with observed sites and duration of crack nucleation. In the micromechanics approach, a particular nucleation mechanism is hypothesized and modeled, and is then used to predict the likelihood of crack nucleation based on a characteristic description of the local conditions. Even though the latter approach is more desirable from a fundamental point of view, it is not sufficiently well-developed to be used for general design purposes. Fortunately, there remains much potential for refining the empirical approach by bringing additional physics of crack nucleation into the analysis.

The conventional approach for predicting fatigue crack nucleation involves correlating the number of cycles to the development of a crack of a certain size with a parameter dependent on the local state of cyclic stress or strain. The simplest approach, which is partially based on the stress-life approach to plain fatigue, is to correlate fretting fatigue life with the dominant stress component, the maximum tangential stress, $\sigma_T$, which is the maximum normal stress in the direction parallel to the surface [3]. The magnitude of the tangential stress is strongly influenced by the frictional force developed at the interface during loading.

In the mid 1980's Ruiz and co-workers [5, 6] proposed two fretting fatigue criteria based on an analysis of a dovetail joint between blades and discs in gas turbine engines. They postulated that the extent and intensity of the surface damage from fretting depends on the work done by the frictional force between the contacting bodies. Since their load ratio was near zero, this work is represented by the product of the shear stress, $\tau$, at the maximum load and the relative amplitude of slip between the bodies, $\delta$. This leads to a Fretting Damage Parameter: $FDP = \tau\delta$. The peak value of this parameter along the interface was found to correspond well with the most heavily damaged areas. This parameter can be thought of as a measure of the frictional energy expenditure density [7]. However, the maximum value of this parameter along the interface does not correspond to the location of the main fretting fatigue crack. Since crack nucleation in fretting fatigue has been found to depend on the maximum tangential stress component, the FDP was combined with the tangential stress component, $\sigma_T$, to obtain a combined Fretting Fatigue Damage Parameter: $FFDP = \sigma_T \tau \delta$. The FFDP has been shown to successfully predict the location of the fretting fatigue crack along the interface [5, 7, 8], and it has been shown that the nucleation time is related to the accumulated incremental strain using the FFDP [7].

The FFDP predicts the location of the primary crack nucleus along a fretting interface quite well; however, it gives no relevance to the observed orientation of the nucleated crack and therefore misses the physical observation that the crack tends to initially grow at an oblique angle. In addition, it does not take into account the local multiaxial state of stress. These arguments also apply to the maximum tangential stress parameter.

Another approach is to extend parameters that have been shown to predict lives under plain fatigue to fretting fatigue. The simplest parameter is the maximum shear stress amplitude [9, 10]. One of the planes of maximum shear stress amplitude at the point of observed crack nucleation usually correlates to the initial oblique angle of the crack nucleus. To account for the mean stress, other multiaxial fatigue parameters have been used. For example, a multiaxial fatigue criterion described by Dang Van [11] has been used to predict whether fretting fatigue crack nucleation occurs [12, 13]. This local critical slip plane failure criterion of Mohr-Coulomb-type is invoked throughout the loading path,

$$\tau_{max} + k\sigma_H = F(N_i),$$

where $\tau_{max}$ is the local maximum shear stress and $\sigma_H$ is the hydrostatic stress. If this criterion is satisfied for a given initiation life, $N_i$, at any point, then failure occurs. The cyclic range is not directly incorporated into the criterion, but it does influence the loading path. A Mohr-Coulomb-type failure criterion takes mean stress into account in an average sense. A drawback is that there are two possible planes where the shear stress is maximum, yet cracking is observed to occur preferentially on only one of those planes [14]. A criterion dependent on hydrostatic stress can not distinguish between the damage planes. However, the advantage of this method is its ease of implementation.

Based on the physical observations of crack nucleation and early growth, the hypothesis of so-called critical plane approaches is that critical damage accumulation is dependent on a combination of the range in normal or shear strain (or stress under high cycle fatigue conditions) on a specific plane along with the normal stress acting perpendicular to this plane. These models reflect the influence of the normal stress on the crack plane under mixed-mode crack growth rates during Stage I transgranular growth [14]. Discriminating experiments conducted by Socie [14] support the critical plane approach. The prediction of fretting fatigue crack nucleation using a critical plane approach has shown some promise [15, 16].

In this paper, the theory behind critical plane approaches is presented. Then using recent experimental data, along with a finite element analysis, the results of using critical plane approaches are shown and compared to other nucleation models. A discussion of the merits of critical plane approaches in comparison to these other methods follows, along with some discussion of how these approaches can potentially be used to link nucleation and propagation methods.

**Critical Plane Approaches**

The earliest formulations of critical plane approaches were aimed at predicting high cycle fatigue [17-19]. The distinguishing characteristic of these approaches was that they not only included the maximum shear stress amplitude but also the amplitude of the normal stress acting on the plane of maximum shear stress amplitude. A similar criteria can be written in terms of the shear and normal strains to give more emphasis to low cycle fatigue [20]. Over the past 20 years various damage models based on the critical plane approach have been introduced. A common thread of the more recent and most successful damage models is that they consider both cyclic stresses *and* strains.

Understanding the physical nucleation process is the key to employing critical plane approaches. The models must incorporate governing parameters consistent with the damage. The cracking behavior is dependent on material, strain amplitude, and loading mode under multiaxial fatigue [14]. In the simplest sense, materials can be divided into two classes: those in which cracks nucleate by predominantly tensile cracking (very short Stage I cracks), and those in which cracks nucleate by shear cracking (sufficiently long Stage I cracks). For each class the controlling cyclic stress/strain components have a different influence. The damage model must therefore be consistent with the observed size and orientation of the crack nuclei.

Fatemi and Socie [21] proposed a shear cracking damage model that is dependent on the maximum shear strain range ($\Delta \gamma_{max}$) acting on a plane coupled with the influence of the maximum normal stress to the plane ($\sigma_n^{max}$),

$$\frac{\Delta \gamma_{max}}{2}\left[1 + k\frac{\sigma_n^{max}}{\sigma_y}\right] = F_1(N_i),$$

where $\sigma_y$ is the yield strength and $k$ is a constant which approaches unity at long lives and is reduced for shorter lives [22]. $F_1(N_i)$ is an empirical function relating the parameter (left side of the equation) to the cycles needed to initiate a crack (i.e., nucleate and grow a crack to a certain size). The ratio $\sigma_y / k$ is often very close to the value of the fatigue strength coefficient ($\sigma_f'$). This model evolved from experimental observations demonstrating that shear strain amplitude alone will not correlate data from tension and torsion tests even if the crack grows in a predominantly shear mode. If the fracture plane is determined by shear strain alone, then two planes 90° from each other will be equally damaged. However, the crack usually prefers one of the planes. Using roughness-induced crack closure arguments, the normal stress acting perpendicular to the plane controls the preferred cracking plane.

For materials that exhibit tensile mode cracking (i.e., materials that exhibit virtually no Stage I fatigue crack growth), the Smith-Watson-Topper parameter [23], which accounts for the mean stress normal to the plane of the maximum principal strain amplitude, works very well [14]. This parameter can be written:

$$\left(\frac{\Delta\varepsilon}{2}\right)\sigma_{max} = F_2(N_i),$$

where $\Delta\varepsilon / 2$ is the maximum strain amplitude normal to the plane and $\sigma_{max}$ is the peak stress on this plane during a cycle.

Since the critical plane is not typically known *a priori*, one must compute the damage value for all possible planes. As an engineering approach, the damage parameters for both shear and tensile cracking failure modes are calculated on all planes. The lower of the two estimates indicates the mode controlling nucleation and gives the expected life.

**Experiments**

The subsequent fatigue prediction analyses and discussion will concentrate on the effect of mean stress by focusing on two tests conducted at the same fatigue stress amplitude ($\sigma_a = 217$ MPa) but different stress ratios (R = 0.1 and −1). The complete test details can be found in Ref. [24, 25]. The tests involved fretting PH 13-8 Mo stainless steel against itself. PH 13-8 Mo stainless steel (UNS S13800) is a precipitation hardenable martensitic stainless steel with a low carbon content. The material was solution treated and then aged for 4 hours at 566°C and air cooled. The grain size was 20±10 μm. The yield strength was 1286 MPa and the ultimate tensile strength was 1325 MPa. The applied stress amplitude was approximately one-half the plain fatigue limit of the material [26]. Additional microstructural and mechanical property data for this material can be found in Ref. [26].

A dogbone-shaped plate fatigue specimen, with a gage section 19.0 mm wide and 3.81 mm thick, was used to conduct the fretting fatigue experiments. The fretting took place through bridge-type pads located on opposite edges of the plate specimen. The feet of the pads were cylindrical with a 15 mm radius of curvature. The fatigue specimens and fretting pads were machined using a low stress grind process. The areas of fretting were polished to a 600 silicon carbide grit finish and cleaned in acetone. Residual polishing marks were oriented along the length of the specimen and pads.

A normal load of 686 N was applied to the pad through a proving ring. This resulted in a contact load of 343 N at each foot and created a maximum Hertzian contact

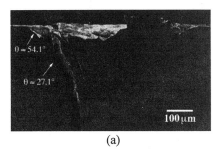
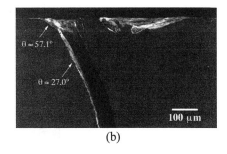

(a)                                    (b)

Figure 1 – *Microphotographs of the primary cracks.  (a) R = 0.1.  (b) R = -1.*

pressure of 449 MPa. The pad span, defined as the distance between the centers of contact, was 16.5 mm. Based on the pad span and stress amplitude, the nominal relative slip range was calculated to be 18.7 μm. It should be noted that, due to friction, the local relative slip range predicted through modeling is much smaller. Hertzian theory gives a contact half width of 0.13 mm. Measurements after the test indicated that the contact line did not go to the corners of the specimen due to some curvature of the specimen edge. Therefore, the actual initial half width was estimated to be closer to 0.18 mm, and the maximum contact pressure was closer to 640 MPa.

The fatigue specimen was cycled at 10 Hz on a uniaxial servohydraulic test system. The dimensions and loading conditions selected initially resulted in a gross slip condition which transitioned to a partial slip condition in less than a thousand cycles (based on the shape of the frictional force – displacement hysteresis) [24]. The half width measurement of the fretting scar was 0.50 mm, which includes the effects of relative slip and some wear.

Tests were stopped after the primary crack grew about one-third the way across the specimen (about 6 mm) and catastrophic failure was imminent. Two tests were run at R = 0.1. They were stopped at 184,000 and 123,000 cycles, respectively. The test run at R = -1 was stopped at 368,000 cycles. If failure is defined as the cycle at which the frictional force range diverges from its steady state condition, indicating an increase in compliance due to a crack of significant size, then failure at R = 0.1 occurred at 171,000 and 110,000 cycles, respectively, and failure at R = -1 occurred at 320,000 cycles.

The specimens were sectioned through the primary crack along their length (Figure 1). Three main regimes of crack growth were identified at both stress ratios. The first regime was within about 50 μm of the surface. The crack angle was near 55° as measured from a line perpendicular to the surface. Our angle convention is shown in Figure 2. The next regime was between about 50 μm and 200 μm from the surface. The crack angle was near 27°. The third regime was between 200 μm and 800 μm. During this regime, the crack angle gradually reduced to 0°, which is the direction normal to the fatigue loading. The crack grew nearly perpendicular to the direction of the fatigue loading further than 800 μm below the surface. The crack surfaces of the R = -1 test appear to be smoother than those of the R = 0.1 test. A crack less than 10 μm in length can also be seen on the leading edge of the contact in the R = 0.1 case. These cross sections are quite representative of the fretting fatigue crack anywhere along the surface [24]. There does not appear to be a significant microstructural influence on crack growth direction in this relatively fine-structured high strength steel.

**Finite Element Analysis**

A finite element analysis was used to estimate the local state of stress in the fatigue specimen near the region of contact. The specimen and pad were modeled as an elastic-plastic material with bilinear kinematic hardening in plane strain using ABAQUS CPE8 elements for the main body and ISL22 slide line elements at the pad/specimen interface (see Figure 2). The pad was fixed in the x-direction along the vertical plane of

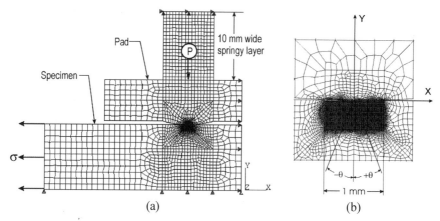

Figure 2 – (a) Finite element mesh. (b) Mesh in contact region.

symmetry. The specimen was fixed in the x-direction and the y-direction along the vertical and horizontal planes of symmetry, respectively. The normal contact force was applied at a single node on the top surface of the pad. A 10 mm wide springy layer of elements was added over the top of the pad to eliminate rigid body motion in the early stages of the analysis and to model the compliance of the proving ring assembly in the experimental configuration. The fatigue load was modeled as a pressure distribution along one edge of the specimen. Plane strain conditions were considered more relevant because the specimen width is much greater than the size of a nucleated crack. Since critical plane approaches being evaluated were strain-based, an elastic-plastic analysis was performed. However, the location of crack nucleation was found to be similarly predicted using either an elastic or elastic-plastic analysis.

CPE8 elements are 8-noded quadrilateral parametric elements. A very fine mapped mesh density (element width = 0.01 mm) was used across a width of 1 mm. This width encompassed the entire range of contact observed throughout the cycle. Earlier analysis suggested that changes in mesh density across the contact resulted in oscillation of stress and pressure at the surface. Mesh densities finer than 0.01 mm also resulted in stress and pressure oscillation.

ISL22 slide line elements are used to represent the contact and sliding between two surfaces. Ideally, the Lagrange multiplier friction formulation should be used because it forces the relative motion in the absence of slip to be exactly zero. However, this method made convergence difficult or impossible. Therefore, the elastic Coulomb friction model was used. In this case, the contact stiffness is chosen such that the relative motion from the position of zero shear stress is bounded by a value $\gamma_{crit}$, which is the allowable elastic slip. This value, by default, is assumed to be 0.5% of the average length of all contact elements and can be changed by the user. For this model, the value of $\gamma_{crit}$ was set to $5 \times 10^{-5}$.

Both the R = 0.1 and R = -1 experiments were simulated. An average coefficient of friction of 0.75 was used based on the steady-state response during one of the R = 0.1

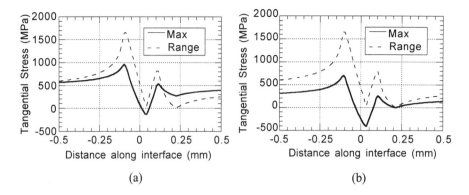

Figure 3 – *Maximum and range in tangential stress along interface.*
*(a) R = 0.1.  (b) R = -1.*

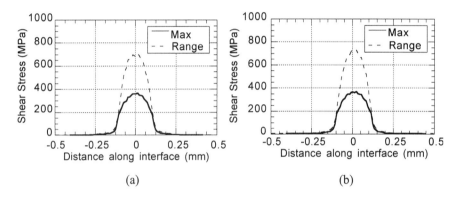

Figure 4 – *Maximum and range in shear stress component along interface.*
*(a) R = 0.1.  (b) R = -1.*

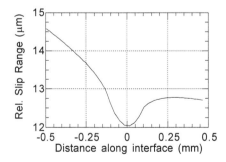

Figure 5 – *Local relative slip range between points along interface.*

experiments [24]. Two full cycles were modeled by incrementing load steps during the analysis. Shakedown occurred by the second cycle, and the response during this representative cycle was used in the fatigue analysis.

The maximum tangential stress ($\sigma_T = \sigma_x$) and tangential stress range at points along the interface are shown in Figure 3. The maximum shear stress ($\tau_{xy}$) and shear stress range along the interface are shown in Figure 4. The stresses plotted in Figures 3 and 4 represent the values after the cyclic response stabilized. The peak values in both maximum tangential stress and tangential stress range correspond very well to the observed crack location which was estimated to be near –0.10 ± 0.03 mm by assuming that the center of the wear scar observed in the SEM was the center of contact at the beginning of the test [24]. Predicted values anywhere in that range are essentially the same location. The location of the maximum shear stress and maximum shear stress range components along the interface does not correspond to the location of crack nucleation (Figure 4). In fact, the mean stress had little influence on these values since they were mostly influenced by the local friction.

The local relative slip along the interface is shown in Figure 5. Since the fatigue stress amplitude was the same for both test cases, the plots for R = 0.1 and R = -1 are identical. The model predicts that the relative slip range in the contact region is around 12 μm. Therefore, the model indicates that the stabilized response at the maximum and minimum load remains gross slip. However, when observing the shear stress distribution throughout the cycle, conditions of partial slip exist during the initial loading and unloading stages of each cycle.

**Fatigue Analyses**

Fatigue analyses were performed at different points along the contact interface using both traditional and critical plane theories. First, consider correlating fretting fatigue damage with either the local maximum tangential stress or tangential stress range (see Figure 3). Both parameters predict the location of the primary crack. Since the stress amplitude (or range) is known to correlate the lives of plain fatigue data better, it may be more appropriate to correlate life with the tangential stress range. However, the tangential stress range does not capture the effect of the mean stress on fatigue life. The tangential stress range is nearly independent of mean stress. The maximum tangential stress parameter varies with mean stress but indicates nothing about the amplitude of the loading. So it may only be useful to use the maximum tangential stress parameter for stress ratios near zero [5].

The FDP is plotted in Figure 6. It is defined as the local relative slip range multiplied by the maximum interfacial shear stress during one representative cycle. The FDP indicates that the fretting damage is independent of the mean stress. Cross sections of the contact patch, as shown in Figure 1, and the fretting scars [24] support this distribution and its invariance with mean stress. The non-zero value of the parameter outside of the contact area is due to numerical error since the tangential shear stress must be zero outside the contact region.

The FFDP is shown in Figure 7. It was determined by multiplying the FDP with the maximum tangential stress occurring during the cycle. This criterion correctly predicts the location of the crack. Also, the magnitude of the parameter for R = 0.1 is greater than that for R = -1, which is consistent with the trend in fatigue lives. Both cases also show a secondary peak on the leading edge of the contact where small cracks were also observed. Both the maximum tangential stress and the FFDP peak at the same location.

Moving on to multiaxial parameters based on the local stress-strain response, the maximum shear strain amplitude at points along the interface is given in Figure 8. The shear-cracking Fatemi-Socie (F-S) parameter is plotted in Figure 9. The normalizing factor, $\sigma_f'$, which is the fatigue strength coefficient, is not yet known for our material. Therefore, we used a value of $\sigma_f' = 1879$ MPa, which is for 4340 steel with a similar hardness. The tensile-cracking SWT parameter is plotted in Figure 10. All three parameters peak at the experimentally-determined location of the crack (i.e., -0.10 ±

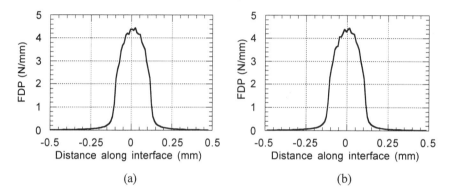

(a)                                                    (b)

Figure 6 – *Fretting damage parameter (τδ) along interface.*  *(a) R = 0.1.  (b) R = -1.*

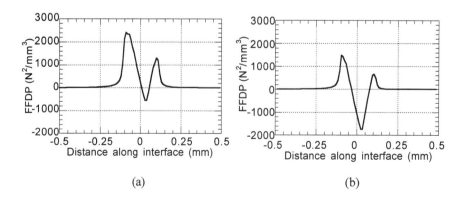

(a)                                                    (b)

Figure 7 – *Fretting fatigue damage parameter ($\sigma_T \tau \delta$) along interface.*
*(a) R = 0.1.  (b) R = -1.*

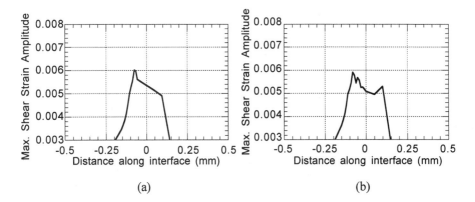

Figure 8 – *Maximum shear strain amplitude along interface. (a) R = 0.1. (b) R = -1.*

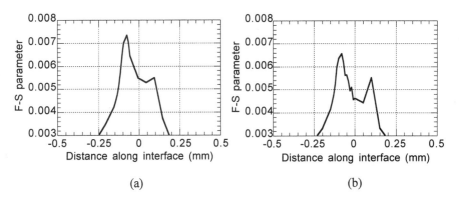

Figure 9 – *Fatemi-Socie parameter along interface. (a) R = 0.1. (b) R = -1.*

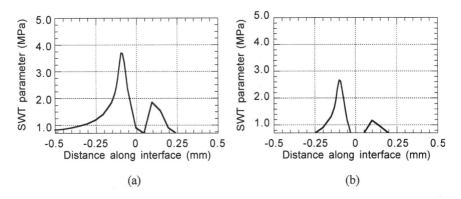

Figure 10 – *SWT parameter along interface. (a) R = 0.1. (b) R = -1.*

0.03 mm). In addition, the F-S and SWT parameters exhibit the desired trend with mean stress, whereas the maximum shear strain amplitude is nearly independent of mean stress.

The predicted planes of maximum damage at points along the interface are shown in Figure 11. Because of the symmetry of the current fretting fatigue problem, only damage on the planes perpendicular to the wide surface of the specimen at angles between −90° and 90°, as shown in Figure 2, need to be computed. For the maximum shear strain amplitude, two planes have equal damage. Since the F-S parameter is strongly associated with the shear strain amplitude, there are two orientations where this parameter is maximized. Although, in most cases the parameter is higher on one of the planes (*F-S* orientation in Figure 11). The other plane is labeled *F-S c* to denote the conjugate damage plane. Both the F-S parameter and the maximum shear strain amplitude predict the initial crack angle at the site of crack nucleation, though the orientation of the observed crack is consistent with the conjugate plane of maximum F-S damage instead of the plane of maximum F-S damage. However, the value of the parameter on the conjugate plane is almost the same as that on the maximum plane. When the coefficient of friction used in the finite element analysis is increased, which may be more representative of local friction conditions, the plane of maximum F-S damage becomes consistent with the observed orientation of the crack [27].

The distribution of damage on planes at the node of maximum damage as predicted by the F-S parameter for R = 0.1 is shown in Figure 12(a). The distribution of the maximum shear strain amplitude and the SWT parameter are also shown. In addition, the distribution of damage for R = -1.0 is shown in Figure 12(b). The peak values of these damage parameters occur over a range of ±5°. Therefore, variations in crack angle on this order are likely. Based on the F-S parameter, the normal stress acting perpendicular to the crack plane has the effect of shifting the critical plane orientation from that predicted by the planes of maximum shear strain amplitude by about 10° toward the direction of the plane with the larger normal stress.

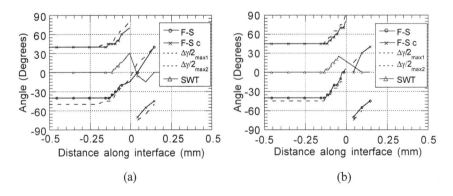

(a)                                              (b)

Figure 11 – *Predicted planes of maximum damage. (a) R = 0.1. (b) R = -1.*

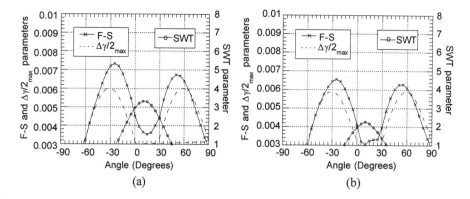

Figure 12 – *Values of the $\Delta\gamma/2_{max}$, F-S, and SWT parameters as function of angle (-90° to 90°) for (a) R = 0.1 at point of maximum damage (-0.075 mm) and (b) R = -1.0 at point of maximum damage (-0.08 mm).*

## Discussion

Among the nucleation parameters examined, the maximum tangential stress, the maximum shear strain amplitude ($\Delta\gamma/2_{max}$), and the FFDP, F-S, and SWT parameters all predict the general location of the primary crack within ±0.03 mm. All but $\Delta\gamma/2_{max}$ clearly predict that greater damage occurs in the R = 0.1 case. In addition, $\Delta\gamma/2_{max}$ and the F-S parameter predict the orientation of the crack nucleus observed in the first regime of crack growth in PH 13-8 Mo stainless steel. Therefore, the F-S parameter is the most promising of these criteria because it is the only parameter that predicts both the location and orientation of the crack nucleus and captures the correct trend in life with mean stress.

Critical plane approaches account for the directionality of mean stress effects, which overcomes a weakness of equivalent stress and strain theories such as a correlation with the maximum shear strain amplitude and Mohr-Coulomb approaches. The F-S parameter is based on a representative cycle. It does not matter when the maximum normal stress occurs within the cycle, so it may not always predict the maximum plane of damage, although in that case it will most likely be the conjugate damage plane. It should be noted that the range in normal strain on the conjugate damage plane was greater at the primary crack location for R = -1.

The results of these exercises support the hypothesis that fretting fatigue crack nucleation is controlled by mechanisms similar to those observed in multiaxial low cycle fatigue (LCF), in contrast to mechanisms of crack nucleation under high cycle fatigue and corrosion fatigue. Therefore, the fatigue damage parameter must capture the state of cyclic strain, with a possible secondary influence of stress, as opposed to a parameter that is completely stress-based. Also consider the following evidence: (1) Nowell and Hills [7] report that when the peak load is kept the same and the minimum load is changed, life decreases with decreasing minimum load as well as mean stress. They attribute the lower life to the increased slip amplitude and note that a fracture mechanics correlation can not predict this trend. Alternatively, their observations can be explained by nucleation models dependent on the shear strain amplitude, since the local strain amplitude increases with decreasing minimum load. (2) Multiple cracks are often found in the highest stressed regions. Multiple crack nuclei are often observed under LCF conditions, but are rarely observed under high cycle fatigue (HCF) conditions where a single crack typically dominates. (3) The mean stress has a minimal effect on fretting fatigue crack nucleation but a strong effect on crack growth [28, 29]. A mean stress under constrained conditions in the fatigue process volume quickly relaxes under reversed plasticity, resulting in little influence on the nucleation process. However, outside the process volume the mean stress does not relax and therefore strongly influences crack growth. Similarly, imposing a constant compressive bulk stress in a specimen undergoing fretting alone, and then removing the fretting condition and performing fatigue on the specimen, led to a lower life compared to specimens that experienced a constant tensile bulk stress during fretting [30]. This is explained by the shakedown of the local compressive stress due to cyclic plastic deformation during fretting. When the remote compressive stress is removed, a local residual tensile stress remains. This result is consistent with the notion that nucleation is not strongly influenced by mean stress, since shakedown with cyclic plasticity relaxes this stress in the fatigue process volume.

Recent computational microplasticity analyses [31] provide evidence that a critical plane model, such as the F-S parameter, likely governs small crack behavior at the scale of the grain size. The distribution of the F-S parameter employed at the grain size scale in a polycrystalline material undergoing a low amplitude tensile and shear loading followed Weibull statistics consistent with typical HCF life statistics. Neither a maximum shear strain amplitude or a Mohr-Coulomb parameter led to a damage distribution consistent with fatigue life statistics.

The initial crack growth to a depth of less than 50 μm can be described with a shear-cracking critical plane model. Once the crack is well-developed, say 50 μm, the stress intensity range is well within the Paris regime [32, 26] and continued growth is appropriately predicted by fracture mechanics methods. Because PH 13-8 Mo has an extremely fine microstructure and high strength, linear elastic fracture mechanics can be successfully used to approximate the subsequent crack growth behavior for cracks greater than 50 μm [26]. The transition in angle from about 55° to about 27° can be explained by the growth being controlled by mixed-mode fracture mechanics rather than by the cyclic stress and strain field. Crack growth can not continue at 55° because the crack grows into a region of crack tip closure. The 27° crack angle observed experimentally corresponds closely to the angle of a mixed mode crack that would just avoid crack closure [33].

The most important attributes of critical plane approaches are that they provide a framework for linking crack nucleation models to traditional crack growth models based on fracture mechanics [34]. The success of critical plane approaches appears to be closely linked with the shear-dominated Stage I propagation of small cracks. Consequently, a propagation criterion such as [35]

$$\frac{da}{dN} = A_I(\Delta\delta_I)^{m_I} + A_{II}(\Delta\delta_{II})^{m_{II}}$$

governs crack growth and can be considered an extension of a critical plane approach, since it is understood that $\Delta\delta_I$ and $\Delta\delta_{II}$ refer to Mode I and Mode II crack tip opening and sliding displacement ranges, respectively, at a specific distance behind the crack tip. A generalization of such a mixed-mode propagation law that can account for heterogeneity and constraint effects, important for microstructurally and mechanically small cracks, has been proposed [36]. Such a formulation admits a set of length scales which can potentially account for the microstructural influence on crack growth, which is significant to the growth of small fatigue cracks [37]. It is also interesting to point out that both the F-S and SWT parameters involve energy-like products of stress and strain range, which are analogous, but not equivalent, to hysteresis energy and $\Delta J$-integral approaches to fracture mechanics [34].

Before critical plane approaches can be universally applied, much work is still needed. The critical plane parameters described here can be used in parametric studies and relative damage can be compared. However, life prediction requires the availability of both tension and torsion fatigue properties, along with an extensive description of experimental observations. Most axial and multiaxial fatigue constants currently available are based on the life required to nucleate a 1 mm crack in a body under a uniform state of cyclic stress. However, the low cycle fatigue process volume is more than an order of magnitude smaller, and the stress and strain gradients in this volume may be significant. Therefore, it is not appropriate to simply use conventional fatigue constants. Either new tests must be done to determine constants suitable for cracks this small, or approximations for extending conventional parameters to the prediction of smaller crack nuclei is needed. It should be kept in mind that the relationship between crack length and cycles may be highly nonlinear in the nucleation regime, and the number of cycles to nucleation is most likely not proportionally scalable to smaller crack lengths.

Predictions using multiaxial fatigue criteria in fretting fatigue problems rely heavily on the accuracy of the local cyclic stress and strain analysis. Due to the severe stress and strain gradients, obtaining an accurate analysis is quite challenging. With increasing computational power, the effects of plasticity and level of work hardening or softening, the evolution of the local friction, microstructural changes, surface roughness, the generation of heat and its influence on microstructure and friction, and the influence of local wear can begin to be incorporated. As cracks develop, the stress redistributes due to nonpropagating microcracks. An understanding of the effect of surface treatments on the nucleation life will need to consider these various factors. Advances in computational modeling will ultimately lead to better predictions.

An overwhelming number of experiments suggest that the evolution of the local friction coefficient should be modeled in more detail, since it plays an important role in the magnitude of the local stress and strain states and dictates the stick and microslip regions. Some approximations of the non-constant frictional forces generated have been made [38]; however, either a friction evolution model or constitutive model that describes the evolution of the frictional behavior at the interface that can be used in fretting fatigue stress and strain analyses is more desirable.

It is also possible that the critical plane may be influenced by microstructural texture which is inherent to the material or developed during fretting. If important, computational polycrystalline plasticity modeling may need to be employed to describe this microstructural evolution [31].

## Conclusions

Critical plane approaches for the prediction of multiaxial fatigue damage show promise as a methodology for capturing the physical characteristics of crack nucleation under fretting fatigue conditions. In particular, the shear-cracking Fatemi-Socie parameter was able to predict not only the location of the primary fretting fatigue crack, but also its orientation in tests performed on PH 13-8 Mo stainless steel. The maximum shear strain amplitude predicted the crack location and the orientation of the nucleated crack, but did not clearly separate mean stress differences in the fatigue life. The combined fretting fatigue damage parameter (FFDP) predicted the crack location and is easy to apply, but it can not predict the crack orientation. The tensile-cracking SWT critical plane parameter also predicted the crack location, but did not predict the initial crack orientation. Evidence suggests that nucleation is a multiaxial low cycle fatigue process which takes place within a small process volume, and as a consequence, a parameter that is strain-based is most appropriate for predicting fretting fatigue crack nucleation.

The critical plane parameter is suitable for predicting the nucleation of a crack in PH 13-8 Mo stainless steel to a depth on the order of 50 μm from the surface plane. The growth of a crack larger than 50 μm in this steel can be described by fracture mechanics. The subsequent growth of an oblique crack occurs under combined Mode I and II at an angle sufficient to just avoid crack closure.

Because shear-cracking models based on critical plane approaches can potentially predict the size and orientation of the crack nucleus, these models offer a link between conventional nucleation approaches and crack propagation approaches based on fracture mechanics. Recent work on computational microplasticity, aimed at describing microstructural influences on fatigue crack growth (i.e., microstructural fracture mechanics), has shown that shear-cracking critical plane models such as the Fatemi-Socie parameter correctly capture the crack nucleation and early crack growth characteristics.

*Acknowledgments*

This research is funded by the Office of Naval Research through research grant N00014-95-1-0539 entitled Integrated Diagnostics. Dr. Peter Schmidt serves as Program Manager. The assistance of Jon Wallace and Professor D.F. Socie with the fatigue analysis is appreciated.

# References

[1] Wharton, D.E., Taylor, D.E., and Waterhouse, R.B., "Metallurgical Factors in the Fretting-Fatigue Behavior of 70/30 Brass and 0.7% Carbon Steel," *Wear*, Vol. 23, 1973, pp. 251-260.

[2] Hoeppner, D.W. and Goss, G.L., "A Fretting Fatigue Damage Threshold Concept," *Wear*, Vol. 27, 1974, pp. 61-70.

[3] Endo, K. and Goto, H., "Initiation and Propagation of Fretting Fatigue Cracks," *Wear*, Vol. 38, 1976, pp. 311-320.

[4] Hills, D. A., and Nowell, D., *Mechanics of Fretting Fatigue*, Kluwer Academic Publishers, Dordrecht, The Netherlands, 1994.

[5] Ruiz, C., Boddington, P. H. B., and Chen, K. C., "An Investigation of Fatigue and Fretting in a Dovetail Joint," *Experimental Mechanics*, Vol. 24, No. 3, 1984, pp. 208-217.

[6] Ruiz, C. and Chen, K.C., "Life Assessment of Dovetail Joints Between Blades and Disks in Aero-Engines," *Proc. Int. Conf. on Fatigue of Engineering Materials and Structures*, Vol. 1, Institution of Mechanical Engineers, London, 1986, pp. 187-194.

[7] Nowell, D., and Hills, D. A., "Crack Initiation Criteria in Fretting Fatigue," *Wear*, Vol. 136, 1990, pp. 329-343.

[8] Kuno, M., Waterhouse, R. B., Nowell, D., and Hills, D. A., "Initiation and Growth of Fretting Fatigue Cracks in the Partial Slip Regime," *Fatigue and Fracture of Engineering Materials and Structures*, Vol. 12, No. 5, 1989, pp. 387-398.

[9] Sato, K., "Damage Formation During Fretting Fatigue," *Wear*, Vol. 125, 1988, pp. 163-174.

[10] Fellows, L. J., Nowell, D., and Hills, D. A., "On the Initiation of Fretting Fatigue Cracks," *Wear*, Vol. 205, 1997, pp. 120-129.

[11] Dang Van, K., "Macro-Micro Approach in High-Cycle Multiaxial Fatigue," *Advances in Multiaxial Fatigue, ASTM STP 1191*, D.L. McDowell and R. Ellis, Eds., American Society for Testing and Materials, Philadelphia, 1993, pp. 120-130.

[12] Dang Van, K. and Maitournam, M.H., "Elastic-Plastic Calculations of the Mechanical State in Reciprocating Moving Contacts: Application to Fretting Fatigue," *Fretting Fatigue*, ESIS 18, R.B. Waterhouse and T.C. Lindley, Eds., Mechanical Engineering Publications, London, 1994, pp. 161-168.

[13] Petiot, C., Vincent, L., Dang Van, K., Maouche, N., Foulquier, J., and Journet, B., "An Analysis of Fretting-Fatigue Failure Combined with Numerical Calculations to Predict Crack Nucleation," *Wear*, Vol. 181-183, 1995, pp. 101-111.

[14] Socie, D.F., "Critical Plane Approaches for Multiaxial Fatigue Damage Assessment," *Advances in Multiaxial Fatigue, ASTM STP 1191*, D.L. McDowell and R. Ellis, Eds., 1993, pp. 7-36.

[15] Szolwinski, M.P. and Farris, T.N., "Mechanics of Fretting Fatigue Crack Formation," *Wear*, Vol. 198, 1996, pp. 93-107.

[16] Farris, T.N., Grandt, A.F., Jr., Harish, G., and Wang, H.L., "Analysis of Widespread Fatigue Damage in Structural Joints," 41st International SAMPE Symposium, March 24-28, 1996, pp. 65-79.

[17] Guest, J.J., *Proc., Institute of Automobile Engineers*, Vol. 35, 1940, pp. 33-72.

[18] Stulen, F.B. and Cummings, H.N., "A Failure Criterion for Multiaxial Fatigue Stresses," *Proc. ASTM*, Vol. 54, 1954, pp. 822-835.

[19] Findley, W.N., "A Theory for the Effect of Mean Stress on Fatigue of Metals under Combined Torsion and Axial Load or Bending," *Journal of the Engineering Industry*, 1959, pp. 301-306.

[20] Brown, M.W. and Miller, K.J., "A Theory for Fatigue Failure Under Multiaxial Stress-Strain Conditions," *Proc. Institute of Mechanical Engineers*, Vol. 187, No. 65, 1973, pp. 745-755.

[21] Fatemi, A. and Socie, D., "A Critical Plane Approach to Multiaxial Fatigue Damage Including Out-of-Phase Loading," *Fatigue and Fracture of Engineering Materials and Structures*, Vol. 11, No. 3, 1988, pp. 145-165.

[22] Fatemi, A. and Kurath, P., "Multiaxial Fatigue Life Predictions under the Influence of Mean Stresses," *Journal of Engineering Materials and Technology*, Vol. 110, 1988, pp. 380-388.

[23] Smith, R.N., Watson, P., and Topper, T.H., "A Stress Strain Function for the Fatigue of Metals," *Journal of Materials JMLSA*, Vol. 5, No. 4, 1970, pp. 767-778.

[24] Pape, J.A. and Neu, R.W., "Influence of Contact Configuration in Fretting Fatigue Testing," *Wear*, Vol. 225-229, April 1999.

[25] Pape, J.A., "Design and Implementation of an Apparatus to Investigate the Fretting Fatigue of PH 13-8 Mo Stainless Steel," M.S. Thesis, Georgia Institute of Technology, 1997.

[26] Patel, A.M., Neu, R.W., and Pape, J.A., "Growth of Small Fatigue Cracks in PH 13-8 Mo Stainless Steel," *Metallurgical and Materials Transactions A*, Vol. 33A, May 1999.

[27] Swalla, D.R., "Fretting Fatigue Damage Prediction Using Multiaxial Fatigue Criteria," M.S. Thesis, Georgia Institute of Technology, 1999.

[28] Nishioka, K. and Hirakawa, K., "Fundamental Investigations of Fretting Fatigue (Part 4, The Effect of Mean Stress)," *Bulletin of JSME*, Vol. 12, No. 51, 1969, pp. 408-414.

[29] Sato, K., Fugii, H., and Kodama, S., "Effects of Stress Ratio and Fretting Fatigue Cycles on the Accumulation of Fretting Fatigue Damage to Carbon Steel S45C," *Bulletin of JSME*, Vol. 29, No. 255, 1986, pp. 2759-2764.

[30] Collins, J.A., "Fretting-Fatigue Damage-Factor Determination," *Journal of Engineering for Industry*, 1965, pp. 298-302.

[31] Bennett, V. and McDowell, D.L., "Polycrystal Distribution Effects on Microslip and Mixed Mode Behavior of Microstructurally Small Cracks," *Mixed-Mode Crack Behavior, ASTM STP 1315*, K.J. Miller and D.L. McDowell, Eds., American Society for Testing and Materials, West Conshohocken, PA, 1999.

[32] Gardner, B. and Qu, J., personal communication, 1998.

[33] Nowell, D., Hills, D.A., and O'Connor, J.J., "An Analysis of Fretting Fatigue," Proc. Conf. on Tribology, Friction, Lubrication and Wear – 50 Years On, London, I. Mech. E., 1987, pp. 965-973.

[34] McDowell, D.L., "Basic Issues in the Mechanics of High Cycle Metal Fatigue," *International Journal of Fracture*, Vol. 80, 1996, pp. 103-145.

[35] Hoshide, T. and Socie, D., "Mechanics of Mixed Mode Small Fatigue Crack Growth," *Engineering Fracture Mechanics*, Vol. 26, No. 6, 1987, pp. 842-850.

[36] McDowell, D.L. and Berard, J.-Y., "A ΔJ-based Approach to Biaxial Fatigue," *Fatigue and Fracture of Engineering Materials and Structures*, Vol. 15, No. 8, 1992, pp. 719-741.

[37] McDowell, D.L. and Bennett, V.P., "Microcrack Growth Law for Multiaxial Fatigue," *Fatigue and Fracture of Engineering Materials and Structures*, Vol. 19, 1996, pp. 821-837.

[38] Rooke, D.P. and Courtney, T.J., "The Effect of Final Friction Coefficient on Fretting Fatigue Waveforms," *Fatigue and Fracture of Engineering Materials and Structures*, Vol. 12, No. 3, 1989, pp. 227-236.

# Experimental Studies

*Leroy H. Favrow,[1] David Werner,[2] David D. Pearson[3] Kenneth W. Brown,[4] Michael J. Lutian, Balkrishna S. Annigeri,[6] Donald L. Anton[7]*

**Fretting Fatigue Testing Methodology Incorporating Independent Slip and Fatigue Stress Control**

**REFERENCE:** Favrow, L. H., Werner, D., Pearson, D. D., Brown, K. W., Lutian, M. J., Annigeri, B. S., and Anton, D. L., **"Fretting Fatigue Testing Methodology Incorporating Independent Slip and Fatigue Stress Control,"** *Fretting Fatigue: Current Technology and Practices, ASTM STP 1367,* D. W. Hoeppner, V. Chandrasekaran, and C. B. Elliott, Eds., American Society for Testing and Materials, West Conshohocken, PA, 2000.

**ABSTRACT:** A fretting fatigue apparatus was designed, built and tested incorporating independent computer control of fretting fatigue slip distance and fatigue stress. This was accomplished through the utilization of two coaxial servo-hydraulic test actuators controlled in real time by computer. The central hydraulic actuator applies the fatigue load to the test specimen, while the outer concentric hydraulic actuator moves the fretting pin carrier apparatus. Independent control of slip displacement is achieved with the use of a capacitance displacement gage attached to the specimen fret pin carrier in such a manner that relative displacements of <5µm can be controlled. Capacitance gage measurements indicate the relative motion of the fatigue specimen surface caused by loading with respect to the fret pins. The fret pin carrier is subsequently moved to accommodate this motion plus its own-programmed motion. Load cells are provided both above and below the fatigue specimen allowing for measurement, by difference, of the forces applied by the fretting pins. These forces can be used to calculate the dynamic coefficient of friction during test operation. Finally, a 3-D finite element analysis model was constructed of the fatigue specimen and the fret pins to determine analytically the slip occurring at the fatigue specimen surface within the bounds of the test operation.

**KEYWORDS:** fretting fatigue, fret pin, fret pin carrier, slip displacement, proportional, derivative, offset, integral, error, coefficient of friction (COF), system files.

[1] Senior Research Engineer, United Technologies Research Center, East Hartford, Connecticut USA.
[2] President, Epsilon Technologies, Jackson, Wyoming USA.
[3] Senior Materials Scientist, United Technologies Research Center, East Hartford, Connecticut USA.
[4] Senior Analytical Engineer, Computer Aided Structural Analysis, Tolland, Connecticut USA.
[5] Senior Materials Engineer, Sikorsky Aircraft Corporation, Stratford, Connecticut, USA.
[6] Principal Analytical Engineer, United Technologies Research Center, East Hartford, Connecticut USA.
[7] Principal Scientist, United Technologies Research Center, East Hartford, Connecticut USA.

The ability to control the relative motion of fretting fatigue test hardware such as pins, bridges, rods and balls versus the test specimen has been of concern for many years, Reference [1,2,3], in the process of attempting to determine the "appropriate" fretting fatigue debit to be applied in various life prediction models employed in commercial as well as military applications. Fretting fatigue damage life debits have been reported ranging from approximately a 20% to 60% depending test type, test conditions, wear couples as well as environmental factors, References [4,5,6]. Application of fatigue life debits in design allowable limits is dependent upon the particular user design limit systems, which take into account the actual service conditions experienced by the components at issue.

This paper reports on a fretting fatigue system developed in order to perform closed loop fret pin amplitude control fretting fatigue tests at United Technologies Research Center of East Hartford, Connecticut. The fretting fatigue system resident at United Technologies Research Center has the ability to control axial specimen loading, relative fret pin motion of 15μm peak-to-peak to 100μm peak-to-peak as well as fret pin force of 2615 Newtons of applied load maximum, Figure [1]. Test conditions typically monitored are dual load cells (above and below specimen in load train), fret pin vibratory amplitude, fret pin applied force feedback signals as well as cycle count. Under and over peak limit detection is employed on the fret amplitude and load feedback signals as early indicators of sample crack formation.

Figure 1: UTRC Fretting Fatigue System

The objective of obtaining a testing system such as the system described in this paper was

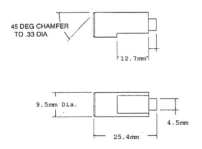

Figure 2: UTRC Cylindrical Fret Pin

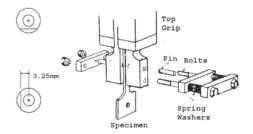

Figure 3: Example of Fixed Fret Pin and Spring Force Loading Device

to enable the investigator to control as many test variables as possible. The fretting fatigue system resident at UTRC is able to control axial load, fret pin amplitude (fret pin carrier actuator concentric with the axial actuator), fret pin applied load, cycle frequency and loading waveform. Most fretting systems used in the past and for the most part currently, do not actively control the fret pin relative motion, but rely upon test coupon compliance. Movement of the pin contact point relative to the test coupon active gage section can adjust the amount of relative motion using this method. Fret pin configurations can take the form of a pin (flat, hemispherical, circular, rectangular or cylindrical in nature) or bridge type device is used, Figures [2], with Figure [3] an example of both a fixed fret pin assembly and a spring washer load application for fret stress. The ability to fully control fret pin relative motion is a key control parameter. Literature has shown Reference [7], that fretting damage causing debit to fatigue life is not solely related to relative motion of the contact surfaces, but to a combination of fret pin relative motion of approximately 25 to 40μm coupled with fret stresses that do not cause "lock-up". "Lock up" is a condition, which in essence means a solid couple between the fret pin and the test coupon. The solid coupling elastic motion of the fret pin with no stick/slip, or the converse, an all slip/sliding condition may in fact be far less severe than the stick/slip condition.

**Test System Description-Mechanical**

The fretting test system discussed in this paper has two main sections, the main actuator which controls the axial load applied to the test coupon, Figure [4], and the fretting section which is comprised of two sections, the fretting actuator and the fret pin carrier

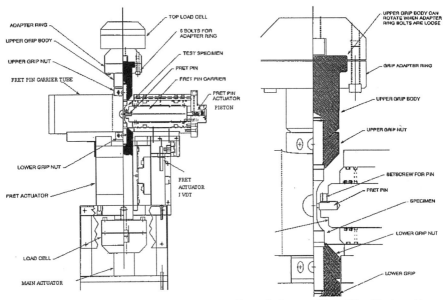

*Figure 4:Fretting  System Major Components*        *Figure 5: Cutaway View of Fret Pin Assembly*

assembly. The fret pin carrier assembly has a 50mm diameter fret pin carrier on each side, Figures [4,5,6], contained in a single horizontal tube. The fret pin carriers are supported in a common tube by means of hydrostatic bearings. These bearings float the fret pin carriers on a high pressure oil film and prevent the carriers from touching the tube walls, coupled with the nature of the design the bearings provide high radial stiffness. At the end of each fret pin carrier there is a small dual acting hydraulic cylinder which provides the fret pin contact force. These cylinders are supplied by system pilot high pressure hydraulic oil providing the lateral force against the test specimen. This force is converted to a stress depending upon the fret pin configuration. Fret pin loading is controlled through the use of a manual four-way valve controlling the activation of the dual acting actuators previously mentioned. Applied pressure is monitored with a full bridge strain gage pressure transducer. This supply pressure is converted to applied load based upon the fret pin carrier loading cylinder geometry and recorded throughout the test.

The fret pin carrier relative motion is monitored with a capacitance-sensing device, Figure [6]. The capacitive probe has a target, which is mounted to an arm equipped with zero backlash bearings and is spring loaded against the specimen while being held in place by means of a conical point. The capacitive probe itself, is mounted directly to one fret pin carrier shaft, Figure [7]. The fret pin carrier has anti-rotate devices in place as well as the fret pin having lockdown flats for setscrew locking as anti-rotation devices. The anti-rotation feature is crucial as pertains to wear scare integrity. The capacitance device allows for very accurate (Δ1.25um resolution) relative displacement measurements. The UTRC system is equipped with a device, which has a full-scale range of 100 microns or 4000 microinches (0.004"). The

*Figure 6: Capacitance Gage In Situ*

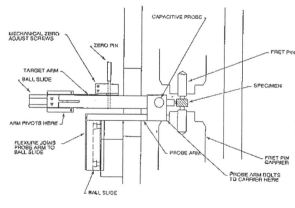

*Figure 7: Schematic Cutaway of Capacitance Gage*

capacitance device is powered by a signal conditioner, which produces an analog output which is in turn fed to a DC conditioner in the closed loop control circuit. This signal is used in part, for input to the control system associated with the fret carrier relative motion control, as well as supplying enunciation signals via under and over peak detection of possible crack formation under either of the fret pins. The detection of apparent crack formation is noted in a change in fret pin carrier wave form and amplitude.

Control of fret pin slip amplitude is accomplished using three variables, the programmed sine wave slip amplitude, the main rate and the fret error. The control of the fret pin carrier motion relative to the specimen motion is non-trivial due to the interaction of the fret pin pressure on the fret pins. The fret pins will tend to clamp the specimen, Figure [8], and impart specimen relative motion to the fret pin carrier if not counteracted by active machine controls. This can be understood more clearly, if one envisions a case where the fret pin carrier was commanded to hold zero slip. The zero slip command would require the fret pin carrier to move in perfect concert with the specimen to preclude relative slip of the fret pins, which means in some cases cause large excursions of the fret pin carrier due to specimen compliance. On the other hand, 100% slip conditions would require the fret pin carrier to be held at a zero motion position allowing the specimen to "slide/slip" by the locked in place pins. Actual and accurate control

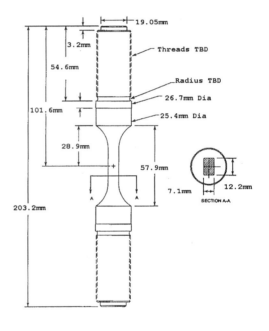

Figure 8: Typical UTRC Fretting Test Coupon Geometry

of the fret slip amplitude is accomplished by using the cross-coupling signals from the main actuator rate (the first derivative of the main command signal), as opposed to the main actuator command signal. Another important cross-coupling parameter is the fret actuator error signal which has a portion of the error signal fed to the fret servovalve circuit. The logic involved relates to the amplitude of error signal being directly proportional to the correction signal fed to the fret servovalve.

**Test System Description-Control**

The UTRC fretting fatigue system provides closed loop servohydraulic control through the use of a modified BASIC-like language specifically designed for Materials Testing applications. This system provides on-board analog PIDO control (proportional, integral, derivative an offset functions), amplitude optimization, function generation including ramp, sine, triangle, block and disk function generation, data acquisition stored in the ASCII format to include timed data and level crossing, 6 channel scope display, digital display of twin load cells (above and below specimen), Figure [4], fret transducer motion, main LVDT, fret LVDT, fret pressure, under and over peak detection indicators, limit detection enunciation and actions, which include program stop-hydraulics on or off, cycle count and various function buttons for tuning and graphical displays.

Typical control in closed loop systems relies upon the fact that there is always an error between command and the feedback of actual system transducer responses. The error signal that arises as a result of this difference is the signal, which is amplified and conditioned through the *proportional* circuitry in order to boost the error signal. The next processing is in the *integration* circuit in order to minimize servovalve imbalance. The *derivative* circuit provides damping for control of high proportional boost, with the final tuning dealing with the *offset*, which helps control mechanical offsets present in most servovalves.

The UTRC fretting fatigue system software control and hardware package consist of three main segments, transducers, such as the twin load cells, capacitance motion sensor assembly, fret LVDT and main LVDT, designated movers, which in the case of the UTRC fretting fatigue system are the main actuator and the fretting actuator, and groups which in the UTRC Fretting Fatigue System consist of a combination of the main actuator and fretting actuator (movers), working in concert with transducers, the twin load cells, fret LVDT and capacitance displacement device.

Fretting fatigue system operators typically enter maximum and minimum axial load levels, fret pin motion amplitude, test frequency, tuning parameters ( PIDO, fret error and main rate values), cycle data acquisition information to include cycle number to take data, such as cycles 1 through 10, 20, 30..., 100, 200, 300..., 1000, 2000, 3000..., 10,000, 20,000, 30,000..., 100,000, 200,000, 300,000..., 1,000,000, 2,000,000, 3,000,000... to test stop or failure. Limits such as under and over peak on the fret transducer and load cells are by necessity, set at test start up and stabilization. Data is typically stored on a mini diskette (100MB capacity or greater) in ASCII format for post processing.

The system files consist of the following:

1) The main program, comprised of two sections, the(*.EXE) and the (*.OVR file), which is used to minimize the memory requirements of the lower 640K of RAM.

2) The transducer files (*.XDR) which contain the information concerning the transducer type, units the transducer is calibrated in, the ID of the PC series card the a transducer is attached to, the full scale range values and calibration constants as well as the mouse command, toggle and slew rates.

3) The configuration file (*.CFG), which is used to tell the main program what hardware is present in the test system, how fast to update the PC cards (typically 250 to 1000 times per second), limit check intervals (typically 100 to 250 time per second), digital polling time intervals (typically 10 to 100 times per second), peak and valley check time intervals usually 250 to 1000 times per second, PC cards present and appropriate addresses, digital input and out connections on the PC cards, such as hydraulic power supplies, on/off solenoids and any other limit switches present, power supplies present, such as air or oil, transducers present and definition of the calculated channels such as fret pressure. The configuration file also identifies the movers (actuators) present in the system along with the associated PC cards, transducers associated with the particular movers and whether or not the transducer outputs are displayed, the feedback transducers which can be used for the mover and energy source associated with the mover.

4) The initialization file (*.INI), which is the boot up sequence for the system on start up contains the parameters "set" upon system "wake up", including such items as customer identification, group information, transducer filter settings, PIDO settings, tare and range data, initial mover feedback transducer, initial mover command and dither settings, the mover stop and hold conditions, initial transducer limit settings, initial polling settings, initial CRT scope settings and monitor "box" settings.

5) Numerous other files are involved in the UTRC Fretting Fatigue System, which will not be described in detail , but are listed here for reference. The Scope file, Data Text file, History file, Timed Data Acquisition file, Peak/Valley Data Acquisition file, Printer Driver file, and others.

## Finite Element Analysis

The UTRC fretting fatigue system as previously noted, can independently control fret pin motion relative to that of the test coupon. The authors and other investigators have made use of commercially

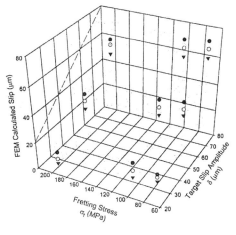

Figure 9: Command Versus FEA Calculated Slip Amplitude and Fret Stress

available FEA codes (ANSYS 5.4, using 8 node meshing and 3 dimensional brick elements as one example), References [8,9,10], in order to determine the "elastic wind up" of fret pin carrier and fret pin combinations. This type of analysis allows the operator to be supplied with appropriate corrections in input test parameters, in order to ensure that the desired fret pin relative motion is occurring with respect to the specimen surface during actual test conditions. Figure [9], is key to understanding the relationship of the relative motion of the fret pin-test coupon combination modeled for the UTRC flat circular17-4PH stainless steel fret pins versus a flat rectangular Ti-6-4 test coupons. The y-axis, FEM Slip Calculated (um) has three symbols associated with it, • which is the calculated relative "top" of the fret pin motion, o the calculated relative "middle" of the fret pin motion and ▼ the calculated relative "bottom" of the fret pin motion. The fact that there is lack of symmetry of relative pin motion around the center of the fret pin centroid is most likely due to the nature of the axial loading profile selected, which in the particular FEA cases shown, used a mean axial stress of 150 MPa and a vibratory stress of 200 MPa, with varying "command/target" slip amplitudes versus fretting stress sequences displayed. Figures [9,10], both had running COF values of 0.7 used in the FEA.

It is certainly beyond the scope of this paper to properly characterize this FEM data, but rather to note the complex relative motion state at the frct pair intcrfaccs tcstcd at UTRC and noted elsewhere in the literature, References [8,9,10]. Figure [10], represents the FEM depiction of the relative fret pin displacement versus time eleven step sequence. Step one is zero time with mean stress applied and the fret pins loaded against the specimen. Step two is the full tensile axial applied load position. Step 3 is zero and step four is full compression. This sequence is continued for the duration of

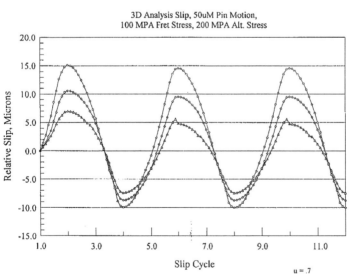

Figure 10: FEM Fret Pin Motion

the test, with the bottom of the pin indicated by the o symbol, the middle of the pin indicated by the ◊ symbol and Δ the symbol for the pin top location. Top of the pin is referenced in this paper as the top of the pin looking directly at the test machine while facing the machine as in Figure [1]. The time step sequence is derived from the FEM inputs incorporating fret pin stress, fret pin material properties such as Young's Modulus and Poisson's Ratio, fret pin and fret pin carrier geometry's and calculated local stress/strain states of the fret pin and test coupon. Also included in the FEA is the plastic strain component.

## Data Reduction and Graphical Representation

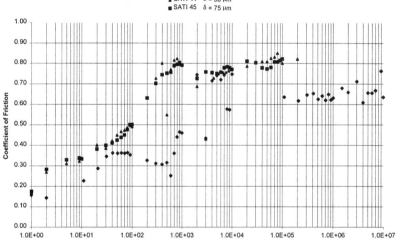

Figure 11: Average Coefficient of Friction

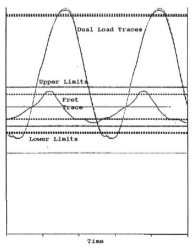

*Figure 12: Real Time Test Control Display*

Data reduction for a typical data set will include start up coefficient of friction values as well as running coefficient of friction changes throughout the test duration is indicative of a typical test from start to finish. Experience at UTRC and at other institutions, Reference [9], h as shown that in a substantial number of fretting cases, the coefficient of friction, Figure[11] will exhibit an initial breakaway value followed by a gradual increase of the COF, at which point a leveling off of the COF occurs and may meander up and down until specimen failure or cessation of testing. Figure [11], also demonstrates the gradual increase in COF for fret pin amplitudes of 25, 50 and 75μm.

Figure [12], is the operator monitor display during test in progress. The information displayed is operator chosen and in the case of the Figure [12], the operator has chosen upper and lower load cells (large sine waves) and fret carrier motion traces versus time. Of note are the parallel dotted lines seen above and below the fret transducer trace (smaller sine wave), which are indicators of under and over peak detection turned on. The under and over peak detection will hold the test system at mean load and zero fret transducer position if tripped and allow the operator the opportunity to verify visually, the

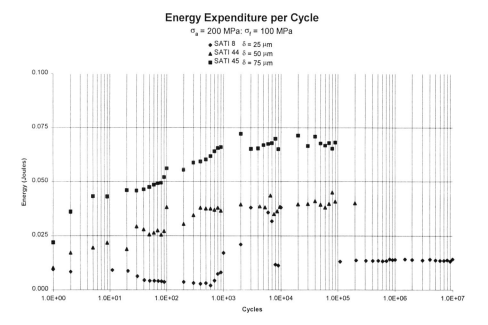

*Figure 13: Joules versus Fret Pin Amplitude*

presence of a substantial crack under either fret pin location. Data reduction is for the most part, accomplished through the use of a spreadsheet type format, with area under curves used to extract "the work-of-fretting" in the form of Joules versus Cycle Count, Figure [13], Reference [9]. The point of interest to the investigators is the leveling out of the energy expended per cycle (Joules) at approximately 2000 cycles in each of the three cases shown, varying only the fret pin relative motion amplitudes of 25, 50 and 75μm peak-to-peak, while maintaining the same mean stress level, vibratory stress level and applied fretting stress level for each test. Each fretting peak-to-peak amplitude level reached a "Joule" plateau value varying in amplitude from 0.0125 Joules at 25 μm peak-to-peak, 0.0375 Joules at 50μm peak-to-peak and 0.065 Joules at 75μm peak to peak.

The average coefficients of friction, Figure [10], Reference [9], for the test coupons discussed previously, also demonstrate an increasing COF up to the 2000 cycle point and then leveling out until failure and or runout. Both the energy per cycle and the increasing COF are most probably related to the "wearing in" of the fretting couple. The "wearing in" period produces third body material and the build up of substantial plastic strain damage. When sufficient amounts of third body material and plastic strain damage accumulates, a period of relative stability ensues which will either remain until test stoppage ($10^7$ cycles typically) or crack growth under the pin contact areas produce a crack(s) which transition from small (25 to 50μm) mixed Mode I and Mode II to Mode I type cracks, with failure imminent at this point.

## Conclusions

1. Current state-of-the-art "fast" computer systems make servo-hydraulic closed loop control of high rate multi axis systems feasible.
2. Finite Element Analysis is required in order to verify machine command and actual machine responses. Finite Element Analysis is also required to understand interface stress/strain states, as load over area and Hertzian stress states are dynamic and complex.
3. Data acquisition and manipulation are simplified using ASCII formats and allows for relatively easy reproduction of hysteretic loops and basic data reduction efforts.
4. The automated control of fret pin relative motion helps the investigator more precisely control this key variable and adds more freedom in fret couple characterization work

*Acknowledgments*

Technical tasks described in this document include tasks supported with shared funding by the U.S. rotorcraft industry and the Government under the RITA/NASA cooperative agreement number NCCW-0076, Advanced Rotorcraft Technology, August 16, 1995. The authors wish to express their gratitude for the dedication of Ronald I. Holland Jr. and Higinio Roman, United Technologies Research Center Senior technicians, who learned the use and fine tuning of the UTRC fretting fatigue test system. Ralph E. Skoe of Northome, Minnesota, is to be thanked and commended for the fine job he performed on the machine control and data acquisition software for the UTRC fretting system..

# References

[1]  Waterhouse, R. B., "The Problems of Fretting Fatigue Testing," *Standardization of Fretting Fatigue Test Methods and Equipment, ASTM STP 1159,* M. Helmi Attia, R. B. Waterhouse, Ed., American Society for Testing and Materials, West Conshohocken, PA, 1992, pp. 13-19.

[2]  Rayaprolu, D. B., Cook, R., "A Critical Review o Fretting Fatigue Investigations at the Royal Aerospace Establishment," *Standardization of Fretting fatigue Test Methods and Equipment, ASTM STP 1159,* M. Helmi Attia, R.B. Waterhouse, Ed., American Society for Testing and Materials, West Conshohocken, PA, 1992, pp. 129-152.

[3]  Hoeppner, D. W., "Mechanisms of Fretting-Fatigue and Their Impact on Test Methods Development," *Standardization of Fretting Test Methods and Equipment, ASTM STP 1159,* M. Helmi Attia, R.. B. Waterhouse, Ed., American Society for Testing and Materials, West Conshohocken, PA, 1992, pp. 23-32.

[4]  Attia, M. H., "Frretting Fatigue Testing: Current Practice and Future Prospects for Standardization," *Standardization of Fretting Test Methods and Equipment, ASTM STP 1159,* M. Helmi Attia, R. B. Waterhouse, Ed., American Society for Testing and Materials, West Conshohocken, PA, 1992, pp. 263-275.

[5]  Vincent, L., Berthier, Y., and Godet, M., " Testing Methods in Fretting fatigue: A Critical Appraisal," *Standard of Fretting Test Methods and Equipment, ASTM STP 1159,* M. Helmi Attia, R. B. Waterhouse, Ed., American Society for Test and Materials, West Conshohocken, PA, 1992, pp. 33-48.

[6]  Labedz, J., "Adaptation of Servohydraulic Testing Machine to Investigate the Life of Machine Components Operating under Fretting Conditions," *Standardization of Fretting Test Methods and Equipment, ASTM STP 1159,* M. Helmi Attia, R. B. Waterhouse, Ed., American Society for Testing and Materials, West Conshohocken, PA, 1992, pp. 190-195.

[7]  Vingsbo, O., and Soderberg, D., "On Fretting Maps," *Wear,* 1988, 126 131-147.

[8]  Hills, D. A., Nowell, D., "The Development of a Fretting Fatigue Experiment with Well-defined Characteristics," *Standardization of Fretting Test Methods and Equipment, ASTM STP 1159,* M. Helmi Attia, R. B. Waterhouse, Ed., American Society for Testing and Materials, West Conshohocken, PA, 1992, pp. 69-84.

[9]  Anton, D. L, Favrow, L. H., Logan, D., Lutian, M. J., Brown, K., Annigeri, B., "The Effects of Contact Stress and Slip Distance On Fretting Fatigue Damage in Ti-6Al-4V/17-4PH Contacts," *Fretting Fatigue: Current Technology and Practices, ASTM*

*STP 1367,* David W. Hoeppner, Ed., American Society for Testing and Materials, West Conshohocken, PA, 1999.

[10] Lindley, T. C., "Fretting Fatigue in Engineering Alloys," *International Journal of Fatigue*, Vol. 19, Supplement No. 1, 1997, pp. S39-S49-S49.

M. Ciavarella,[1] G.Demelio,[2] and D.A. Hills[3]

# An Analysis of Rotating Bending Fretting Fatigue Tests Using Bridge Specimens

**REFERENCE:** Ciavarella, M., Demelio, G., and Hills, D. A., **"An Analysis of Rotating Bending Fretting Fatigue Tests Using Bridge Specimens,"** *Fretting Fatigue: Current Technology and Practices, ASTM STP 1367,* D. W. Hoeppner, V. Chandrasekaran, and C. B. Elliott, Eds., American Society for Testing and Materials, West Conshohocken, PA, 2000.

**ABSTRACT:** The rotating-bending type of fretting fatigue apparatus is analyzed by treating the 'feet' of the bridge as flat pads with rounded corners. This permits a closed-form solution for the contact pressure—indeed, even more complex geometries could be considered, providing the corners are not sharp—and stick/slip zones to be found, allowing for: (a) shearing force, (b) tension within the specimen and (c) tilting of the specimen with respect to the bridge. Example solutions for each of these separate effects are displayed, together with certain combinations of effects.

**KEYWORDS:** rotating-bending fatigue test, fretting fatigue, contact problems, crack initiation

## Introduction

The rotating-bridge type fretting fatigue apparatus has seen many years of use [1,2]. Its principal advantages over more sophisticated forms of test apparatus are that the machine itself is relatively inexpensive to construct, and the tests themselves may be conducted at high cyclic speed, as the rotor assembly is in balance, Figure 1(a), and there is relatively little energy expended during rotation. The principal drawback of the device is that it is extremely difficult to determine with any certainty what the contact pressure distribution is if bridges with flat-ended feet are employed, as the nature of the contact problem is one where a half-plane formulation is inappropriate. Previous attempts to get around this problem have relied on the use of boundary elements or finite elements to solve the problem [3], but an alternative approach, advocated in a recent paper by the authors [4], is to use a bridge specimen where the feet have flat pads but with *well-defined* radii of curvature at each end. An even simpler solution would be to use Hertzian geometry, but the proposed geometry has the

---

[1] Senior Researcher (Primo Ricercatore), CNR-IRIS, Str. Crocifisso 2/b, 70126 Bari (ITALY)

[2] Associate Professor, Dip.di Prog.e Prod. Ind.le, Politecnico di Bari, Viale Japigia 182, 70126 Bari, ITALY.

[3] Professor, Dept. of Eng.Sci., Univ.Oxford, OX1 3PJ Oxford UK.

additional advantage that the pressure distribution achieved is very representative of that arising in both dovetail joints of aero-engine fan blades, and in the involute form spline connections found in shafts. Further, the contact problem is not significantly more complex to solve. In this paper we will assume that the basic solution for the normal contact loading for a punch pressed normally into the surface is known [5], and we will focus on the elements of the problem specific to a model of the test apparatus.

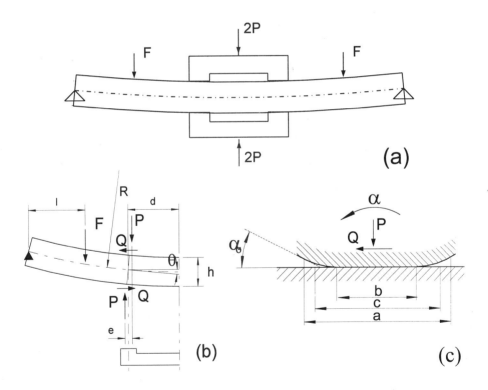

*Fig. 1 Schematics of the rotating bending apparatus (a, b), and of the contact geometry (c)*

The working conditions and the punch geometry are shown in the schematics of Figure 1(a,b,c). Assume, for simpler visualization, that the bridge itself is relatively rigid, and that the stress state in the shaft consists of bending stresses described by elementary bending theory alone. The behavior of the system will then depend on three main effects:

1. Two points, A & B on the shaft surface, beneath the centres of the feet will move together (in the compression half of the cycle) and apart (in the tension half of the cycle) as the shaft rotates. The effect will be to introduce equal and opposite shearing forces at each foot, and providing that the displacement imposed can be accommodated within the compliance of the contact (and, if appropriate the remainder of the bridge) partial slip alone will be present (i.e. full sliding is avoided, as it would not be representative of usual fretting conditions). If this were the only effect present, this contact would be a form of the well-known Cattaneo-Mindlin contact [see for example, ref. 2].

2. Locally present surface tension or compression will mean that, even in the absence of an external force, conditions of slip will arise over the contact. In practice this effect will be superimposed on the effect of the external shear force, and a complicated unsymmetrical shear traction distribution will be present.

3. The slope of the shaft at the contact will mean that, as the shaft rotates, the bridge pad will tilt with respect to the shaft. This affects the contact pressure distribution, and hence subsequently modifies the interfacial stick-slip problem.

It will be appreciated that the sequence in which the tilting, shearing, and stretching occurs is critical to the solution of the problem, as the presence of friction means that the precise history of loading critically influences the stick-slip system. Here, if we consider a point on the surface of the test specimen but beneath the pad, and lying initially on the neutral plane of the beam, the shearing force, tilt and surface direct stress all change *in proportion*, while the contact force alone remains, in principle, constant. At the moment a complete solution in which all three effects and the interaction between them are considered would seem impossibly difficult. In this paper we treat the case where the attitude of the punch is fixed- either upright or tilted, and where the shearing force and bulk tension elements of loading are increased in proportion.

**Formulation**

*Pressure distribution*

The geometry of the contact between one of the feet and the main specimen is shown in detail in Figure 1(b,c). It will be treated as a classical contact between two half-planes supposing the contact half-width, $a$, does not extend too far into the radiused portion of the foot, *i.e* $.a - b \ll r$, where $b$ is the half-width of the flat portion of the foot, and $r$ is the edge radius of the foot.

Consider a plane contact problem, under the half-plane approximation, and define the function $h(x)$ as the amount of overlap if the bodies were allowed to interpenetrate each other freely, as

$$h(x) = T_y + \alpha x - [f_1(x) - f_2(x)] \tag{1}$$

where $f_1(x)$, $f_2(x)$ describe the profile of the upper and lower (1 & 2 respectively) contacting bodies, and $T_y$, $\alpha$ are the normal and rotational components, respectively, of the rigid body motion that brings the two bodies into contact. This first integral equation relates to normal displacements

$$\frac{1}{A}h'(x) = \frac{1}{\pi}\int_S \frac{p(\xi)\,d\xi}{x - \xi}, \qquad x \in S \tag{2}$$

where $A$ is the composite compliance of the two bodies, and in plane strain

$$A = \frac{1 - \nu_1}{\mu_1} + \frac{1 - \nu_2}{\mu_2}. \tag{3}$$

Here, $\mu_i$ is the modulus of rigidity of body $i$, and $\nu_i$ its corresponding Poisson's ratio. The integral equation gives an explicit relation between the slope of the profile and the contact pressure, $p(x)$. The normal contact pressure distribution for this geometry has been solved before [5], and will not be treated here, for brevity. This form, using a conventional explicit representation of the shape of the indenter, is to be preferred for simplicity, but it does not permit a ready extension to the case when the punch is tilted.

A recent extension to the original solution, allowing for unsymmetrical indenter geometries, and therefore, implicitly, for the influence of tilt, has also recently been published [6]. Briefly, it is assumed that the contact occurs over a connected area $S$, where the gap function is given as a general spline function ($n$ parabolic functions), so that in the $i$−th interval ($i = 1, .., n$), $h_0'(x) = h'(x) - \alpha$, will be given by

$$h_0'(x) = m_i x + D_i, \qquad x_i < x < x_{i+1}. \tag{4}$$

The solution reads

$$p(\varphi)\frac{b}{P} = -\frac{1}{\pi \cos\varphi}\left[1 + \frac{b}{AP}(\alpha\pi\sin\varphi - I(\varphi))\right] \tag{5}$$

where the non-elementary integral $I(\varphi)$ depends only on the geometry and the actual position of the contact (i.e. its size *and* the offset of the contact area, $\delta$, and the form of the gap function), and is given in reference [6]. The coordinate, $\varphi$, is an auxiliary variable defined by $\sin\varphi = (x - \delta)/b$. Further, the following implicit relation, also given in reference [6] defines the offset, $\delta$:

$$\alpha\pi - b\sum_{i=1}^{n} m_i \Delta\cos\varphi_i + \sum_{i=1}^{n} D_i\Delta\varphi_i + \delta\sum_{i=1}^{n} m_i\Delta\varphi_i = 0, \tag{6}$$

where $\Delta\varphi_i = \varphi_{i+1} - \varphi_i$, and $\Delta\cos\varphi_i = \cos\varphi_{i+1} - \cos\varphi_i$. The load is determined by the following relation

$$-\frac{AP}{b} = \frac{b}{2}\sum_{i=1}^{n} m_i\left(\Delta\varphi_i - \frac{\Delta\sin 2\varphi_i}{2}\right) - \delta\sum_{i=1}^{n} m_i\Delta\cos\varphi_i - \sum_{i=1}^{n} D_i\Delta\cos\varphi_i \tag{7}$$

where $\Delta \sin 2\varphi_i = \sin 2\varphi_{i+1} - \sin 2\varphi_i$. Finally, the moment of the pressure with respect to the contact area center, $M_0$, is

$$\frac{AM_0}{b^2} = \frac{\alpha\pi}{2} + \frac{1}{2}\sum_{i=1}^{n}\left(D_i + m_i\delta\right)\ \Delta\varphi_i + \frac{\Delta \sin 2\varphi_i}{2}\right] - \frac{b}{3}\sum_{i=1}^{n}m_i\Delta \cos^3\varphi_i \qquad (8)$$

*Shearing tractions*

The second integral equation relates the relative displacement of particles parallel with the surface, $g(x)$, to the interfacial shearing traction, $q(x)$. It reads, again using displacement derivatives,

$$\frac{1}{A}g'\left(x\right) = \frac{1}{\pi}\int_S \frac{q\left(\xi\right)d\xi}{x - \xi}, \qquad x \in S. \qquad (9)$$

On applying a general increment of loading (which may include both tangential loading and surface strain elements), the function $g$ must satisfy the condition

$$g\left(x\right) = \Delta g\left(x\right) + g_0\left(x\right), \qquad x \in S_{stick}, \qquad (10)$$

where $S_{stick}$ is the domain in which stick occurs, $\Delta g\left(x\right)$ is the increment of surface relative tangential displacement, and $g_0\left(x\right)$ the value before entering the stick area. Problems in the tangentially loaded regime invariably reduce to one of finding the correct size and location of stick and slip areas, $S_{stick}$, $S_{slip}$ as well as the *direction* of slip, which in general must be postulated and checked *a posteriori*. The conditions in terms of equalities and inequalities are given by (10) and the classical Coulomb friction law, which can be summarized by writing

$$q\left(x,t\right)\dot{g}\left(x,t\right) + fp\left(x,t\right)\left|\dot{g}\left(x,t\right)\right| = 0, \qquad x \in S_{slip} \qquad (11)$$
$$q\left(x,t\right) + fp\left(x,t\right) < 0, \qquad x \in S_{stick} \qquad (12)$$

where $t$ indicates a fictitious time and the dot partial differentiation with respect to $t$. The solution of the problem is, in general, incremental, whereas the solution to be derived will, instead, impose certain restrictions on the *sequence* of loading, for the results displayed to be appropriate, and these will be stated at the corresponding points. These restrictions will obviate the need for an incremental loading regime to be followed.

**Loading Path**

In this section we will consider the general formulation for the influence of the different elements of loading, *viz.*, normal load, shearing force, and surface straining, on the interfacial shearing traction distribution, and the location of stick and slip zones.

*Normal load (1)*

On applying the normal load, $P$, alone[2], the initial stick zone coincides with the entire contact area $S$, as it is assumed that the specimen and bridge are made from

---

[2] Any pre-existing strain (constant in time) in the contacting bodies clearly does not affect the displacement condition.

elastically similar materials, and hence

$$g'(x) = g'_0(x) = 0, \quad x \in S. \tag{13}$$

Therefore, as $Q = 0$, the only solution to (8) is obviously the trivial solution $q(x) = 0$, $x \in S$. The only relevant characteristics are therefore the ones related to the pressure distribution, its size and offset.

*Tangential load (2)*

Upon applying now a monotonically increasing shearing force, $Q > 0$, with fixed normal load, advancing slip results, and using Cattaneo's superposition principle, we may write

$$q(x) = \begin{cases} fp(x) - q^*(x), & x \in S_{stick} \\ fp(x), & x \in S_{slip} \end{cases} \tag{14}$$

The integral equation for relative displacements in the tangential direction states, using (9) again, and substituting (2) for the full sliding component,

$$0 = \frac{1}{\pi} \int_S \frac{q(\xi)\,d\xi}{x - \xi} = \frac{f}{A}h'(x) - \frac{1}{\pi} \int_{S_{stick}} \frac{q^*(\xi)\,d\xi}{x - \xi}, \quad x \in S_{stick} \tag{15}$$

Then, $q^*(x)$ is the solution of the following integral equation

$$\frac{1}{\pi} \int_{S_{stick}} \frac{q^*(\xi)/f}{x - \xi}\,d\xi = \frac{1}{A}h'(x), \quad x \in S_{stick} \tag{16}$$

which has the same kind of solution as (2) formulated with a spline profile, i.e. equations (4) - (8). These requirements are completely general, and apply even when the pressure distribution, $p(x)$, is not symmetrical. It follows that the corresponding indenter profile does not have to be symmetrical, which means, in turn, that the results may be applied to a tilted punch of the profile in Figure 1(b). However, the pressure distribution must be constant in time, which means that the punch must have a constant angle of tilt, as the shearing force is applied. Notice also that an interesting consequence of this property is that wear in the sliding regions due to an oscillating tangential load will not affect the profile within the adhesive region. In cases where wear is predicted to occur at a fast enough rate, this should be taken into account, perhaps using Archard's law for the rate of wear [7], and the formulation provided permits this to be done without any further complication. In the limit of wear after a very long time the contact is purely adhesive and a singularity develops in the normal traction at the edge of this region [7].

*Surface Strains (3)*

The next type of loading we wish to consider is the effect of a monotonically increasing direct strain developed parallel with the free surface. In the rotating-bending test this would originate in the bending load, but the same principles would apply if an applied tension were imposed; the variation with depth of the strain parallel with the surface is unimportant, as only the surface values are of relevance to the solution. Here only the case where the surface strain is constant with position is considered, *i.e.* in the fretting geometry, the variation in bending moment over the extent of the pad is neglected, and hence

$$g'(x) = \Delta g'(x) = e_x. \tag{17}$$

As a first step in the solution it is assumed, tentatively, that full adhesion occurs, so that integral equation (9) reads,

$$\frac{e_x}{A} = \frac{1}{\pi} \int_S \frac{q(\xi) d\xi}{x - \xi}, \quad x \in S_{stick} = S \tag{18}$$

This has the solution

$$q(t) = -\frac{e_x}{A} \frac{t}{\sqrt{a^2 - t^2}}. \quad -a < t < a \tag{19}$$

where $t = x + \delta$, thus taking into account the possiblity of an unsymmetrical pressure distribution (having an offset $\delta$). As an infinitely large shearing traction clearly cannot be sustained at the edges of the contact slip regions *must* arise for finite friction.

Next, assume the slip direction is opposite in the two slip regions at the edges of the contact area, which may be incorporated into the solution by writing

$$q(t) = q_{slip}(t) = -fp(t) sign(t), \quad t \in S_{slip} \tag{20}$$

where the function $sign(t)$ gives the correct sign of the shearing traction in the slipping areas. Moreover, note that $q_{slip}(t) = 0$ for $t \notin S_{slip}$. Now, it cannot be assumed that the size of the slip regions are equal, so that an unknown offset $\delta^*$ of the stick area must be introduced. It follows that integral equation (9) reads

$$\frac{1}{\pi} \int_{S_{stick}} \frac{q(\xi) d\xi}{t - \xi} = F(t) \equiv \frac{e_x}{A} + \frac{f}{\pi} \int_{S_{slip}} \frac{p(t) sign(t) d\xi}{t - \xi}, \quad t \in S_{stick} \tag{21}$$

Integral equation (21) can be solved in quadrature with standard technique, together with the two appropriate consistency conditions (Q=0 and that the shear traction is bounded at the edges of the stick region), although analytical solutions will not be possible in general, or else with standard numerical techniques. However, the case of symmetrical pressure is easier and will here be considered separately.

*Symmetrical Pressure*
Here the contact area is symmetrical ($\delta = 0$), and so, if $a$ is the contact half-width, integral equation (18) is

$$\frac{e_x}{A} = \frac{1}{\pi} \int_{-a}^{a} \frac{q(\xi) d\xi}{x - \xi}, \quad -a < x < a \tag{22}$$

It follows, from the inherent antisymmetry in the shearing tractions present in the problem, that $S_{stick} = [-c, c]$ and $S_{slip} = [-a, -c] \cup [c, a]$, and hence $q_{sl}(x) = -fp(x)\,sign(x)$, for $x \in S_{slip}$. It follows that the integral equation (22) for $q$ is

$$\frac{2}{\pi} \int_0^c \frac{q(\xi)\,\xi d\xi}{x^2 - \xi^2} = F(x) \equiv \frac{2f}{\pi} \int_c^a \frac{p(\xi)\,\xi d\xi}{x^2 - \xi^2} + \frac{e_x}{A}, \qquad x \in [-c, c] \qquad (23)$$

A recipe for the inversion of this kind of integral equation is given in reference [8], and hence

$$q(x) = \frac{2x}{\pi} \sqrt{c^2 - x^2} \int_0^c \frac{F(\xi)}{\sqrt{c^2 - \xi^2}\,(x^2 - \xi^2)} d\xi, \qquad x \in [-c, c] \qquad (24)$$

where use has already been made of the two consistency conditions, one implicitly imposed by the antisymmetric assumption for $q(x)$ while the other one is

$$\int_0^c \frac{F(\xi)}{\sqrt{c^2 - \xi^2}} d\xi = 0 \qquad (25)$$

This condition is also needed to compute the size of the stick area. As an example in the use of these results they will here be applied to a Hertzian contact. In this case [8]

$$p(x) = -p_0 \sqrt{1 - \left(\frac{x}{a}\right)^2} \qquad (26)$$

so that $F(x)$ acquires a slightly simpler form. However, a closed form solution of (24) is possible in only a limited number of cases. On the other hand the consistency condition for the Hertzian case has the relatively simple form

$$f[K'(b) - E'(b)] = \frac{e_x}{Ap_0} \frac{\pi}{2} \qquad (27)$$

where $b = c/a$, $K(\cdot), E(\cdot)$ are elliptic integrals, and the prime denotes replacing the argument $(b)$ by $\sqrt{1 - b^2}$. This equation explicitly gives $b$. However, the shearing traction distribution can be obtained only in terms of special functions. A previous solution to this problem [8] was given in an entirely numerical form.

*Surface Strain and Tangential Loading* (4)

If a shearing force and a surface tension are applied simultaneously, the problem becomes more complicated; the application of a shearing force alone gives rise to slip zones of the *same* sign, while the application of surface tension alone causes slip zones of *opposite* sign to occur. The question therefore arises of which of these two effects dominates the problem. If the loading is proportional, such that

$$\frac{de_x}{e_x} = \frac{dQ}{Q}$$

it is clear that the initial condition must be stick everywhere (when $e_x = Q = 0$), and that the stick zones must recede monotonically. If attention is restricted to this class of problem, so that 'marching-in-time' solutions are avoided, a solution for any instant can be found by 'jumping' straight to that point, providing that no reversals of load occur. The following cases arise:

a) moderate surface strain

Suppose that only *moderate* surface strains are present, so that slip zones of the same sign must occur, i.e. the problem looks similar to Case (2) -shearing force alone, but where, now,

$$\Delta g'(x) = \Delta g'(x) = e_x. \qquad (28)$$

Cattaneo's superposition principle [9] may be used, and the integral equation transformed to

$$\frac{1}{A}e_x = \frac{1}{\pi}\int_S \frac{q(\xi)\,d\xi}{x-\xi} = \frac{f}{A}h'(x) - \frac{1}{\pi}\int_{S_{stick}} \frac{q^*(\xi)\,d\xi}{x-\xi}. \qquad x \in S_{stick} \qquad (29)$$

Then, $q^*(x)$ is the solution of the following integral equation

$$\frac{1}{\pi}\int_{S_{stick}} \frac{q^*(\xi)/f}{x-\xi}\,d\xi = \frac{1}{A}h^{*'}(x), \qquad x \in S_{stick} \qquad (30)$$

which can be recognized as being of the same *form* as the original equation for normal contact, equation (2). Now, the slope is modified to be of the form

$$h^{*'}(x) = h'(x) - \frac{e_x}{fA} = \left(\alpha - \frac{e_x}{fA}\right) - [f_1'(x) - f_2'(x)].$$

Hence we can define a fictitious rotation, $\alpha^* = \left(\alpha - \frac{e_x}{fA}\right)$, i.e. the *actual* rotation reduced by $e_x/(fA)$.

The partial slip problem generalizing the one associated with the names of Cattaneo and Mindlin was re-solved above. The analogy between the integral equations for normal loading and tangential loading has been proved rigorously already by Ciavarella [9], for the case of zero imposed surface strain. In that paper it is proved that given the solution for normal pressure, the 'corrective' shear within the stick zone is always obtained in a similar way. Here, an extension of that analogy is developed: an alternative way of viewing the corrective solution is that it is given by examining the normal contact loading equation for a lower normal load, but with an imposed rotation. In the case of a pre-tilted punch the angle of rotation is simply *changed* by the corresponding amount. Note, however, that again the solution will be valid only if the pressure distribution is constant in time, and hence if the pre-existing 'tilt' of the punch also remains constant in time.

The condition for this 'shifted' solution to hold is that the stick zone has to be contained entirely within the contact area. If it lies at least partly outside, case (b), below must be considered.

b) large surface strain

Here the case is similar to case (3), no particular analogy can be worked out with the normal contact problem, and a numerical solution must be used.

*General Loading Path*

In a recent paper Jager [10] has independently established similar results for more general loading cases, thus permitting in principle to solve more realistic fretting configurations with the same kind of procedure. The results carry over to the general three-dimensional contact problem strictly only for the case where Poisson's ratio is zero [11]. In all other cases, the 'Cattaneo' superposition approach may satisfy the limiting friction equations (in the case of circular of elliptical contact area), but even then, it predicts a mismatch between the direction of the traction and the direction of slip. The effects of this error have been shown to be small for the Hertzian contact case. With this caveat, the results of [9] can probably be extended to the general loading case using the arguments of Jager [10]. The resulting "Ciavarella-Jager theorem" will prove to be a very powerful tool in the understanding of fretting problems for half-spaces of similar materials, at least for cases when the direction of the tangential force does not change with time.

## Results

*Effect of Non-Symmetrical Normal Loading*

Note, first, that the normal pressure distribution, when rotation is absent, lies between the limiting solutions for a flat-ended punch and a Hertzian contact, depending on the ratio $b/a$. Example solutions are given in [5,6], and here example results will be displayed, as they are useful in interpreting the shearing traction distribution. Figure 2 displays the principal results.

The dimensionless ratios defining the geometry are taken as $b/a$, where $b$ is the contact half-width and $a$ is the semi-width of the flat portion of the contact. Note that $b$ is, of course, a dependent variable of the problem, insofar as it depends on the contact load, whilst the blend radius at the edge of punch is fixed. However, it is very difficult to produce a convenient dimensional variable except by introducing the contact half-width. The rotation of the punch, $\alpha$, is non-dimensionalized with respect to the slope of the punch at the edge of the contact, $\alpha_a$, (i.e. $x = \pm b$, Fig.1(c)). Three representative values of $b/a = 0.25, 0.5, 0.75$ are taken in these results. Clearly, the salient features of these figures are:

1. The gradual migration from Hertzian pressure ($b/a \to 0$) to singular punch pressure as $b/a \to 1$.

2. The increasing skewness of the plots with increasing angle of tilt.

These results are repeated in Figure 3 in an absolute frame of reference: in Fig. 2 the frame of reference was fixed with respect to the contact area. Clearly, the skewed pressure distribution continues to show asymmetry, and the other feature of the results displayed explicitly here is the bulk offset of the contact patch ($\delta$ in the analysis), compared with a line drawn normally through the centre of the contact pad. Note that in both Figures 2 and 3 the pressure distribution is plotted with a dashed line when contact has been lost from the curved part of the punch and extends only part way along the flat region.

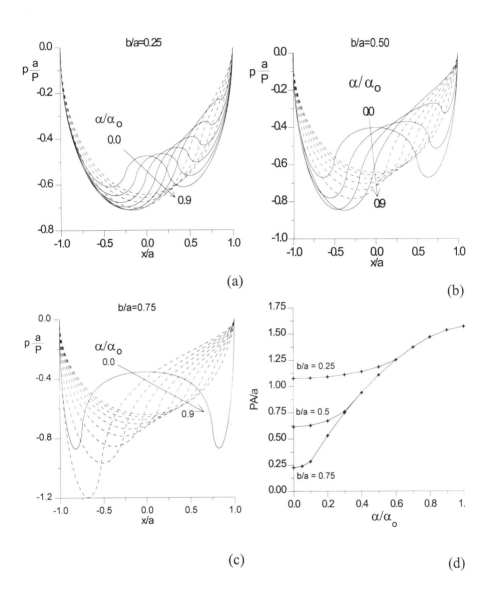

*Fig. 2 Pressure distributions for the frame of reference fixed with the contact area for b/a = 0.25, 0.5, 0.75 respectively in (a, b, c), and variation of the load with rotation (d)*

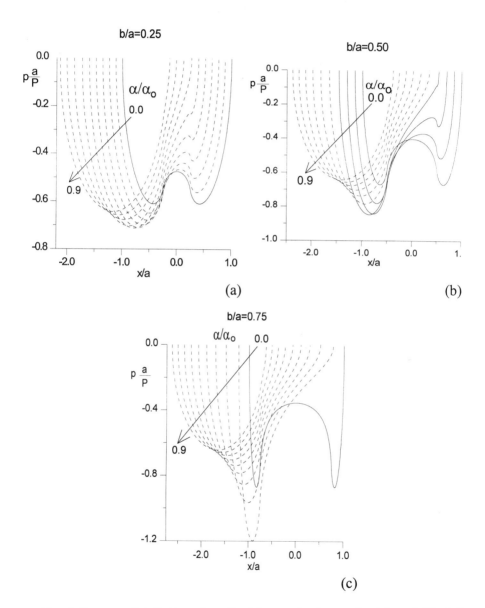

Fig. 3 Pressure distributions for a fixed frame of reference for $b/a = 0.25,\ 0.5,$

$0.75$ respectively in $(a,\ b,\ c)$

*Effect of Shearing Force*

Figure 4 shows the shearing traction distribution for a rounded flat punch: in order to reduce the number of independent variables attention is focused on a punch giving $b/a = 0.75$.

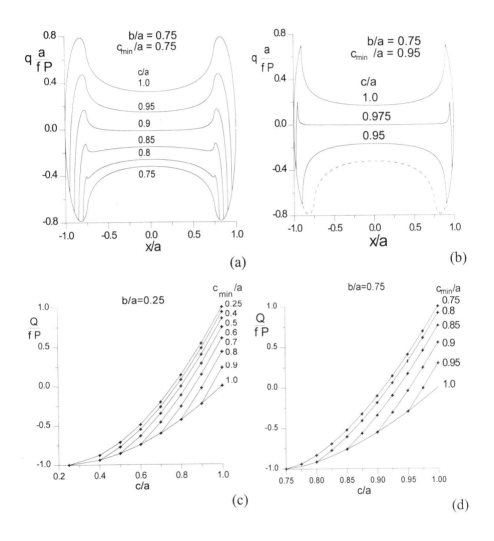

Fig. 4 *Variation of shear tractions in Cattaneo-Mindlin loading conditions, for two representative cases having $b/a = 0.75$ and $c_{min}/a = 0.75$ (i.e. up to the full sliding limit), or $c_{min}/a = 0.95$ (a, b, respectively). Variation of the transmitted tangential load with stick zone size for all possible cases with , Pressure distributions for a fixed frame of reference for $b/a = 0.25$, $0.5$, $0.75$ respectively in (a, b, c)*

Figure 4a shows the case when, with a fixed normal load, the tangential load is first increased almost to the limit of sliding ($c/a = 0.75$ corresponds to incipient sliding), and then reduced until the reverse full sliding limit is attained. It is clear that the reverse loading produces an abrupt variation in the shear traction at the edges, where the *sign* of the shear tractions changes suddenly (note that their magnitude does not change significantly, as the sliding limit is approached, as this is governed by the local value of the contact pressure). In the subsequent Figure 4b, the case illustrated corresponds to a lower value of initial applied shearing force, giving $c_{min}/a = 0.95$. These figures show that it is possible to have a significant variation of the tangential shearing force with very little variation of the slip region dimension. The dashed line in the figures shows the corresponding normal pressure distribution.

Additional information about this problem is given in Figure 4c,d, which shows the stick zone size as a function of the instantaneous normalized maximum shearing force, for the three particular geometries $b/a = 0.25, 0.75$. It is evident that the higher the ratio $b/a$, the smaller the sliding area tends to be. This result is very relevant for fretting damage assessment, as the damage is clearly localized within the slip regions. The lines at $c_{min} = const$ indicate the instantaneous size of the stick zone when an oscillating load is present.

*Effect of Surface Strain*

Figure 5 shows the most general kind of solution which can arise with an imposed surface strain, and with the simultaneous application of a shearing force. As before, the three representative geometries are taken as $b/a = 0.25, 0.5, 0.75$, and here there is no rotation.

The dimensionless ratio $\alpha^*/\alpha$ gives the effect of the surface strain, where $\alpha^* = \alpha - \frac{e_x}{fA}$ and when this is zero the conventional, symmetrical Cattaneo solution is recovered. As the bulk tension is increased, the stick zone gradually migrates across the contact patch, one slip zone extends, whilst the other becomes very small, such that the fretting damage will be extremely localised. The corresponding tangential load transmitted is plotted in Figure 5d.

Lastly, in Figure 6a,b,c,d results for the case when a fixed initial rotation of the punch is also present are shown.

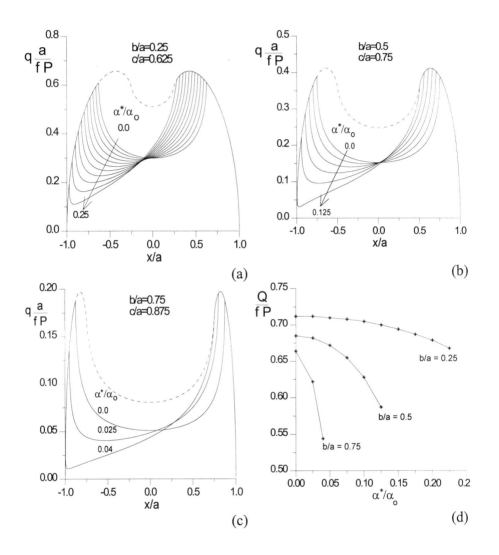

Fig. 5 *Effect of surface strain applied simultaneously to tangential load (a, b, c) and variation of the transmitted tangential load for the cases considered (d)*

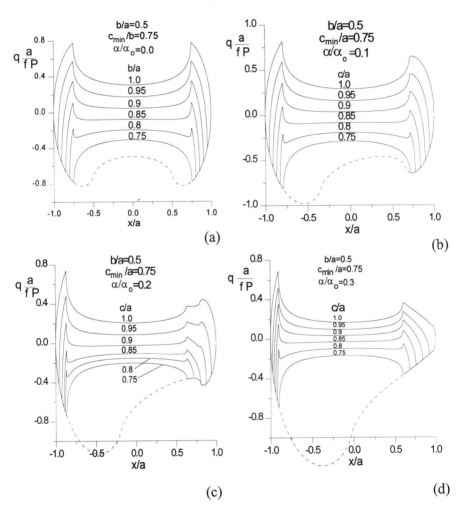

*Fig. 6 Variation of shear tractions when tangential load is applied to a normal contact which is originally rotated. Representative cases having $b/a = 0.5$. The initial rotation is $\alpha/\alpha_0 = 0.0$ (a), $\alpha/\alpha_0 = 0.1$ (b), $\alpha/\alpha_0 = 0.2$, and $\alpha/\alpha_0 = 0.3$.*

## Stress Field and Crack Initiation

The solutions presented would seem to indicate the limit of development of an analytical solution to this class of problem, and further refinement will inevitably mean a recourse to numerical methods. However, when also considering that the interior stress field is needed if the crack initiation and propagation conditions are to be quantified, together with the surface relative displacement within the slip zone, it appears clear that an analytical treatment, although complex, gives much more confidence and practicality of use. It is clear from the results in the previous section that all that is needed is the stress state corresponding to normal pressure and *full sliding* shearing tractions, as the influence of partial slip shear distributions can be found by superposition and appropriate scaling. A Muskhelishvili potential approach is appropriate, expanding the traction distribution as a series in terms of Chebyshev polynomials as proposed in [4,5]. Interpolation fitting is needed in general for the coefficients of the Chebishev expansion, but this is easily incorporated in the code with library routines. An alternative approach could be to use a set of overlapping triangles, but this will give only a piecewise linear representation of the traction distribution, so that the Chebishev expansion method should be preferred.

From the potential, the stresses as well as the displacement derivatives can be obtained from standard relations [2,4,5,6,8]. Of course, the half-plane formulation could again be called into questioned if the specimen is not very thick compared with the dimension of the contact area, but it has been shown [12] that the traction distributions predicted with this assumption are very representative of the actual values even for a ratio of the width of the specimen to the width of contact of 5. On the other hand, the interior stress field differs quite remarkably already for a ratio such as 10 and in [12] it is proposed to consider only the component of the stress parallel to the surface to be different from the half-space idealization, and indeed to take this difference as constant. Although this approach may be oversimplified, a more refined model, as proposed in [13] could be used, where the difference term only is computed as Fourier integrals. Regarding initiation parameters, many models have been proposed in the literature over the past years [see 14,15 and the references therein] which are based on computing the surface energy produced in the slip zones. These energy models are actually compatible with more fundamental studies on initiation, like the models by Mura [16], giving the critical fatigue cycle number beyond which the dislocation dipoles locally cumulated by some kind of ratchetting mechanism are annihilated to form a crack or microvoid. Although the model seems to catch the mechanics of the phenomenon, the predictions are not always of the correct order of magnitude, and anyway S-N curves depend on many physical parameters such as grain size; surface energy; yield strength; width of the persistent slip bands etc. Therefore, limiting attention to fretting fatigue experiments, the continuum mechanics models are still seen to be of greater practical use than micromechanics ones.

Finally, notice that the crack, once initiated, propagates in a rapidly decaying stress field (given by the contact), but the presence of the bending stress field plays a role. Depending on the relative intensity of the two factors, conditions of self-arrest could be found, as was done in [17] for the constant tensile stress field. For the crack propagation study, techniques of distributed dislocations seems the most appropriate, as discussed in [2,13,15,17].

## Discussion

If the half-plane idealization is to be retained- and to do otherwise would certainly involve a very detailed numerical model -the model presented is a good compromise, as it permits the salient variables to be examined clearly, and in closed form. One feature which should be noted is that as the dub-off radius is reduced ($b/a \to 1$), and hence the feet become more flat-punch like, the transition from partial slip to full sliding is very abrupt, so that extremely careful control of both the pad geometry (the quality of manufacture) and the stiffness of the whole arrangement are needed to ensure reliability. Further, the presence of tension tends to exacerbate the inherent instability of the problem, and to make the transition even more difficult to control. The first of these remarks follows directly from Ciavarella's theorem [9], which, as a corollary, states that in any partial slip contact problem a stick zone can never lie entirely *within* a region of zero punch slope.

Notice that the complexity of the loading path introduced by the rotating bending setup is not limited to the rounded flat punch; indeed, even using a Hertzian geometry, the tilting of the feet, although not producing a change of the pressure distribution, does cause a change in position, and therefore the Cattaneo-Mindlin solution is not valid.

The presence of a surface tension, giving rise to surface strains, has a complex effect. In cases where one of the slip zones is very small, the localization of shear within that band is very severe, while the other slip zone is usually long. A very large tension will give rise to reverse slip (not treated here), but it would seem probable that, in such cases, the bulk of the damage was incurred at the larger slip zone, where relative surface displacements are much larger.

## Conclusions

Some results have been shown for the solution of the complex frictional contact problems arising in the commonly used rotating-bridge type fretting fatigue apparatus, with slightly radiused feet. The main effects of the contact have been highlighted, using analytical methods only, although the coupling of the various phenomena does not permit at present a complete, general solution. Numerical solutions are still an open possibility, and indeed several methods are available, ranging from commercial FE or BE packages to specialized algorithms. However, it is not easy to obtain satisfactory accuracy for complex frictional patterns by this approach.

## Acknowledgments

Michele Ciavarella is pleased to acknowledge the support from CNR-Consiglio Nazionale delle Ricerche (Borsa per l'estero 203.07.26 del 12.9.96), for his visit to Oxford University in the period Feb.-Aug. 1998, permitting the completion of the present work.

## References

[1] Bramhall, R., Studies in Fretting Fatigue. D. Phil thesis, University of Oxford, (1973).

[2] Hills, D.A. and Nowell, D. *Mechanics of Fretting Fatigue,* Kluwer, Dordrecht, pp. 154-158.

[3] Sato, K., Determination and control of contact pressure distribution in fretting fatigue. *Standardization of Fretting Fatigue Test Methods and Equipment,* ASTM STP 1159, Ed. M.Helmi Attia and RB Waterhouse, American Society for Testing and Materials, Philadelphia, (1992), pp. 85-100.

[4] Ciavarella, M., Monno, G. and Hills, D.A. The use of almost complete contacts for fretting fatigue tests. Presented at the ASTM conference, Stanford, June1997, to be published in a ASTM publication.

[5] Ciavarella M, Hills DA, Monno G, The influence of rounded edges on indentation by a flat punch, *Proceedings of Institution of Mechanical Engineers part C - Journal of Mechanical Engineering Sci*ence, 1998, Vol.212, No.4, pp.319-328.

[6] Ciavarella M, Demelio G, On non-symmetrical plane contacts, *International Journal of Mechanical Sci*ence, in press.

[7] Ciavarella M, Hills DA, Brief Note: Some observations on the Oscillating Tangential Forces and Wear in General Plane Contacts, *Europ. J. Mech. A-Solids,* in press.

[8] Hills, D.A., Nowell, D. and Sackfield, A. *Mechanics of Elastic Contacts,* Butterworth-Heinemann, Oxford (1993).

[9] Ciavarella, M. The generalised Cattaneo partial slip plane contact problem. Part I Theory, Part II Examples. *Int. J. Solids, Struct,* Vol. 35, (1998), 2349-2378.

[10] Jager, J. A new principle in contact mechanics. *ASME J.Tribology,* 1998, 120 (4), 677-684.

[11] Ciavarella M, Tangential loading of general 3D contacts, *J. of Appl. Mech,* 1998, 65 (4), 998-1003.

[12] Fellows, LJ, Nowell, D, Hills, DA, Contact stresses in a moderately thin strip (with particular reference to fretting experiments, *Wear,* 1995, Vol.185, No.1-2, pp.235-238

[13] Kelly, PA, Hills, DA, OConnor, JJ, Stress state and modelling of mode I cracks in layered materials suffering normal or near-normal contact loading, *Proceedings of Institution of Mechanical Engineers part C - Journal of Mechanical Engineering Sci*ence, 1997, Vol.211, No.4, pp.301-311

[14] Fellows, LJ, Nowell, D, Hills, DA, On the initiation of fretting fatigue cracks, *Wear,* 1997, Vol.205, No.1-2, pp.120-129

[15] Fellows, LJ, Nowell, D, Hills, DA, Analysis of crack initiation and propagation in fretting fatigue: The effective initial flaw size methodology, *Fatigue and Fracture of Engineering Materials and Structures,* 1997, Vol.20, No.1, pp.61-70

[16] Mura, T, A Theory of Fatigue-Crack Initiation, *Materials Science and Engineering.,* A, 1994, Vol.176, No.1-2,pp.61-70

[17] Moobola, R., Hills, D.A. and Nowell, D. Designing against fretting fatigue: crack self-arrest. *Journal of Strain Analysis,* Vol. 33,1, (1998), 17-25.

G. Harish,[1] M. P. Szolwinski,[2] T. N. Farris,[3] and T. Sakagami[4]

## Evaluation of Fretting Stresses Through Full-Field Temperature Measurements

**REFERENCE:** Harish, G., Szolwinski, M. P., Farris, T. N., and Sakagami, T., "Evaluation of Fretting Stresses Through Full-Field Temperature Measurements," *Fretting Fatigue: Current Technology and Practices, ASTM STP 1367,* D. W. Hoeppner, V. Chandrasekaran, and C. B. Elliott, Eds., American Society for Testing and Materials, West Conshohocken, PA, 2000.

**ABSTRACT:** The near-surface stress field in fretting has long escaped experimental characterization due in large part to the fact that the friction coefficient in the slip zones associated with the partial slip contacts cannot be evaluated from measured forces. Attempts at circumventing this through measurements of microslip or extent of the slip zones have been inconclusive. However, newly available infrared detector technology is capable of resolving finely, both spatially and temporally, subsurface temperatures near the fretting contact. These temperature changes are induced by both frictional heating at the surface due to microslip as well as the coupled thermoelastic effect arising from the strains in the material. A finite element model has been developed for fretting that includes the heat generation due to sliding and partial slip; and the coupled thermoelastic effect. The model also incorporates heat conduction, thermal deformation, and contact. The correlation between the temperature changes measured by the infrared camera and those predicted by the finite elements is remarkable. During gross sliding, a patch of heating throughout the contact length, attributed to frictional heating, is observed. As the friction coefficient rises and the contact transitions to a partial slip regime, the temperature changes are more clearly associated with strain through the coupled thermoelastic effect. The excellent agreement of the finite element results with the experiments demonstrates the ability of the model to provide validated values for fretting-induced stresses and microslip.

**KEYWORDS:** fretting fatigue, coupled thermoelasticity, near-surface temperatures, coupled finite element analysis, thermal imaging

[1]Research Assistant; [2]Professor and Head, School of Aeronautics & Astronautics, Purdue University, 1282 Grissom Hall, West Lafayette, IN 47907-1282.
[2]Assistant Professor, Department of Mechanical Engineering, Aeronautical Engineering & Mechanics, Rensselaer Polytechnic Institute, 110 8th St., 2046 Jonsson Engineering Center, Troy, NY 12180-3590
[3]Associate Professor, Department of Mechanical Engineering & Systems, Graduate School of Engineering, Osaka University, 2-1 Yamadaoka, Suita, Osaka 565-0871, Japan.

## Introduction

Fretting is the tribological phenomenon observed in nominally-clamped components subjected to vibratory loads or oscillations. Associated with fretting contacts are regions of small-amplitude relative motion or microslip that occur at the edges of contact. The distribution and amplitude of the microslip have been difficult to study due to the change in the friction coefficient with fretting cycles. As the friction coefficient increases with the number of loading cycles, the contact transitions from global sliding, where the tangential force, $Q$, is equal to the global frictional resistance, $\mu_{avg} P$, to partial slip, where $Q < \mu_{avg} P$. Consequently, quantitative insight into the frictional characteristics at the contact interface is no longer available from knowledge of the tangential force history alone. A newly available, full-field infrared technology capable of resolving temperatures finely both spatially and temporally provides a powerful tool for characterizing the near-surface conditions associated with fretting contacts. The temperature fields observed with this setup are influenced by both the frictional heating due to the relative motion at the contact interface and the coupled thermoelastic effect arising from mechanical strains in the material.

## Experimental Setup

The experimental setup, presented in Figure 1, consists of a fretting fatigue test fixture mounted on a standard servo-hydraulic testing machine. Two cylindrical pads, fastened to the fixture, contact a standard fatigue dogbone specimen. Both the pads and specimen are machined from 2024-T351 rolled bar stock. The pads transmit a constant normal

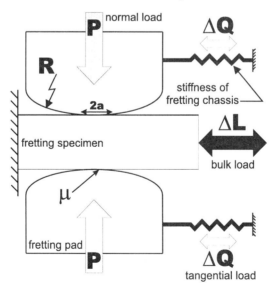

FIG. 1 — *A schematic of the fretting fatigue fixture highlighting the applied loads.*

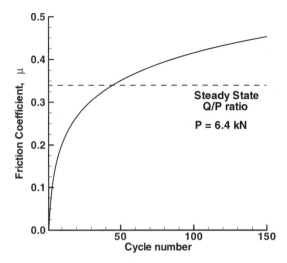

FIG. 2 — *Friction coefficient evolution curve for the 2 Hz experiment. This curve is used as input to the finite element program presented later in the document.*

force onto the specimen while a cyclic bulk stress, $\sigma_o$, is applied by the servo-hydraulic testing machine. Through the action of the fretting fixture, the applied force is partitioned into a tangential force, Q, transmitted as friction through the pads/specimen interface, and a bypassed force at the top of the specimen. The design of the rig allows accurate application and control of the normal force on the pads. A complement of analog and digital sensors allows for real-time tracking of the requisite applied loads. Complete details of this fretting fixture may be found in the literature [1].

At the onset of each experiment, the coefficient of friction is low, resulting in gross slip between the pad and specimen. This gross sliding is manifested in the nature of the continuously tracked force data. As the number of fretting cycles increases, the gross motion at the interface causes wear, which increases the coefficient of friction and produces partial slip conditions. This transition point can be ascertained from force data. The evolution of the friction coefficient beyond this point is the subject of ancillary experiments on the evolution of friction coefficient under partial slip conditions [2]. Figure 2 presents a representative result from these experiments, showing the evolution of the friction coefficient with cycle number. The dotted line shows the maximum ratio of tangential force to normal force in the partial slip regime. As will be presented later in this document, this data is used as input to the finite element model of the fretting contact.

*Temperature Measurement*

Temperature measurements are collected with a forward looking infrared (FLIR) sensor. Thermal radiation from the plane of focus is detected by a focal plane array of indium-antimonide sensors with a sensitivity of 0.025 K. To increase the emissivity of

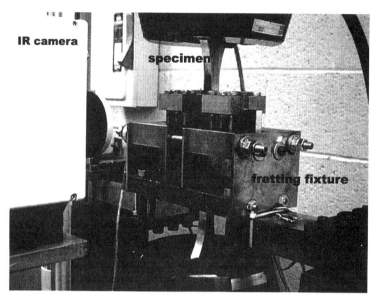

FIG. 3 — *Photograph of the experimental setup used in the current investigation, including fretting fatigue fixture and thermal imaging system.*

the source, a thin coating of commercially available black paint with a high emissivity is applied to the exposed surfaces of the specimen and the pads. The paint has a negligible effect on the sensor response at the frequencies used and does not penetrate the contact interface. The working distance between the lens and faces of the pad/specimen pair is 90 mm, leading to an image spatial resolution of about 75 μm. A photograph of the setup is shown in Figure 3.

A calibration between measured intensity and temperature is achieved by placing a thin aluminum plate outfitted with a thermocouple in front of the camera at the same focal distance used in the fretting experiments. A calibration curve between measured intensity and temperature is obtained by varying the plate temperature and matching thermal images with thermocouple readings. With calibration completed, experiments to verify the ability of the camera to detect the coupled thermoelastic effect are performed. These involved simple cyclic loading of a fatigue specimen at various frequencies and subsequent comparison to the resulting temperature changes predicted by coupled thermoelastic theory. For isotropic elastic material behavior, it can be shown [3] that elastic deformations lead to finite changes in temperature. Under adiabatic conditions, a relation between the sum of principal stresses, $\sigma_{kk}$, and temperature, T, can be written as:

$$T = T_o - \frac{\alpha T_o}{\rho C_p} \sigma_{kk} \qquad (1)$$

where $\alpha$ is the thermal expansion coefficient, $\rho$ is the mass density, $C_p$ is the specific heat at constant pressure, and $T_o$ is the nominal temperature. Note that Eq 1 is a linearized

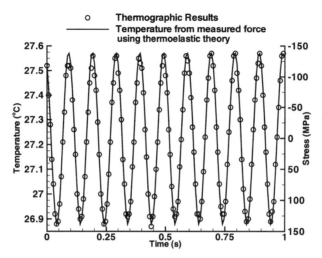

FIG. 4 — *A comparison between measured temperatures from the thermal imaging system during a fully-reversed uniaxial tension/compression test and those predicted by thermoelastic theory (Eq 1).*

equation valid for small changes in temperature from the nominal value. To evaluate the ability of the camera to detect this effect, Eq 1 is evaluated with handbook values of material properties and the value of stress in the specimen calculated simply as the load as monitored by the system load cell divided by specimen cross-sectional area. Figure 4 shows the comparison between the theoretical temperature values and those detected by the camera. The remarkable agreement establishes confidence in both the experimental setup and calibration. Additional details of the experimental technique and imaging parameters and a review of other work on the thermomechanical aspects of fretting contacts can be found elsewhere [4].

*Near Surface Temperature Measurements*

With a calibration between temperature and intensity determined, fretting experiments are conducted at frequencies between 2 Hz and 10 Hz under fully-reversed (R-ratio = $L_{min}/L_{max}$ = -1) loading conditions. Estimates of both the contact zone halfwidth, a, and stick zone halfwidth, c, are made by exercising available expressions for the mechanics of fretting contacts [5] using a value for coefficient of friction in the slip zones, $\mu_{slip}$. These relations ignore the effect of the bulk loading on the distribution of stick and slip. The evolution of $\mu_{slip}$ to a steady-state value of 0.65 has been investigated in a separate series of experiments on friction evolution [2]. In order to ensure complete contact, the pad thickness is reduced in one experiment. This modification eliminated any effects of incomplete contact at the edge where the camera is focused.

After focusing the camera is focused on an area centered on the right pad/specimen interface, ten thermographic images are captured per cycle for each experiment. Force and position data is gathered via the real-time force sensors. The image size is 128 x 128 pixels with an integration time of 1 ms. The ambient temperature is assumed to be the laboratory temperature at the beginning of the experiment. The two-dimensional array of intensity values captured is then filtered using a Gauss filter with a 3 x 3-pixel window. These filtered values are subsequently converted into temperature values using the aforementioned calibration curve.

## Results:  Temperature Measurements

The temperature histories collected to date for contacts between flat and cylindrical aluminum surfaces show clearly the transition from global sliding to partial slip under constant amplitude loading as the friction coefficient increases with number of loading cycles.  During the first few cycles, when the contact is in the gross sliding regime, the images reveal a patch of heating throughout the contact length.  These friction-induced temperatures are superposed on the thermoelastic heating and cooling caused by the change in the applied loads.  The patch of heating is caused by the frictional heating during sliding and is observed twice in a cycle.  This double-frequency effect is expected as sliding occurs twice in a cycle.  Figure 5 displays the sequence of thermoelastic images from an area around the right pad/specimen interface during the gross sliding regime.

The specimen portion of the images also clearly elucidates the thermoelastic temperature changes.  As bulk tension increases, the bulk specimen temperature drops, while as bulk compression increases, the bulk specimen temperature rises.  Note that the thermoelastic temperatures change have a frequency equal to that of the cyclic loading.  It must be recognized that the temperature changes are very small, a result due in part to the high thermal conductivity of aluminum.  A material with different thermal and mechanical properties may show a much more significant temperature change during sliding.

As the experiment progresses, the friction coefficient increases forcing the contact into a partial slip regime.  Figure 6 shows a sequence of images collected during one cycle in the partial slip regime.  The central patch of heating is absent, indicating no gross slip. The microslip regions do not contribute to observable frictional heating.  The dominant cause of the temperature change is now the thermoelastic effect.  The existence of localized tensile and compressive peaks of contact stress is captured distinctly by the images.  Also, the bulk temperature change away from the contact is lower at the top of the specimen when compared to the bottom of the specimen.  This is an indicator of the tangential load transferred into the pad, causing lower bulk stress magnitude and (hence smaller temperature changes) at the top of the specimen.

## Finite Element Model

### Coupled Thermoelasticity

The dependence of stress on temperature is widely known and is manifested as the last term in the constitutive relation for an isotropic elastic body, which can be written as

$$\sigma_{ij} = \lambda\varepsilon_{kk}\delta_{ij} + 2\mu\varepsilon_{ij} - (3\lambda + 2\mu)\alpha\Delta T \tag{2}$$

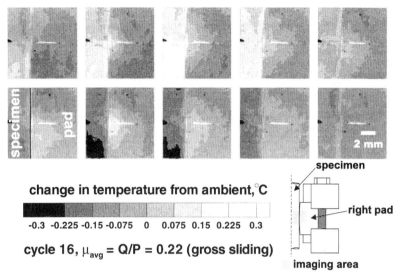

FIG. 5 — *Sequence of thermal images during gross sliding.*
*Contact parameters:  2 Hz, $\sigma_o$ = 100 MPa (R-ratio = -1), Q/P = 0.22, P = 6.4 kN,*
*and R = 178 mm.  Center image on first row is at maximum bulk compression and*
*center image on second row is at maximum bulk tension.*

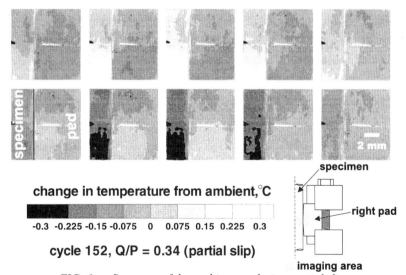

FIG. 6 — *Sequence of thermal images during partial slip.*
*Contact parameters:  2 Hz, $\sigma_o$ = 100 MPa (R-ratio = -1), Q/P = 0.34, P = 6.4 kN,*
*and R = 178 mm.*

where $\sigma_{ij}$, $\varepsilon_{ij}$, and $\Delta T$ are the stress components, strain components, and change in temperature from ambient, respectively. The terms $\lambda$ and $\mu$ are the isothermal Lamé constants. The converse effect of a change in stress causing temperature change is manifested in an additional term in the heat conduction equation. This augmented expression is now known as the coupled thermoelastic equation, written as

$$k\nabla^2 T = \rho C_v \frac{dT}{dt} + T(A\frac{d\varepsilon_{kk}}{dt} + B \cdot \frac{d\varepsilon_i}{dt})$$ (3)

$$A = 3K\alpha - \frac{\partial\lambda}{\partial T}\frac{d\varepsilon_{kk}}{dt}, \quad B = 2\frac{\partial\mu}{\partial T}\frac{d\varepsilon_i}{dt}$$

where k, $\rho$, $C_v$, and K are the thermal conductivity, mass density, specific heat at constant deformation, and bulk modulus, respectively. Note the use of principal components of strain. Eq 1 may be obtained from Eq 3 by using the constitutive relation (Eq 2), linearizing the resulting equation and finally assuming adiabatic conditions. Most thermographic applications use an adiabatic approximation (requiring high frequency), thereby eliminating the left-hand side of Eq 3. However, the complexities of frictional heating and conduction effects preclude this assumption in the current analysis. This necessitates the simultaneous solution of the constitutive relation and coupled thermoelastic equation given by Eqs 2 and 3 and motivates the development of a finite element model that includes both frictional heat generation and coupled thermoelastic effect.

*Finite Element Model*

The finite element model consists of a symmetric plain strain model with half of the specimen and one pad as shown in Figure 7. The use of spring elements and an elastic support layer at the top of the specimen enable the accurate modeling of the testing machine compliance as well as the fretting fixture stiffness. This feature allows the

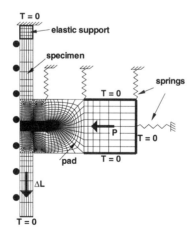

FIG. 7 — *Finite element mesh used for solution.*

TABLE 1 — *Material properties of 2024-T351 Al used in experiments and model*

| | |
|---|---|
| Young's modulus, E | 74.1 GPa |
| Poisson's ratio, $\nu$ | 0.33 |
| Yield strength, $\sigma_y$ | 310 MPa |
| Coefficient of thermal expansion, $\alpha$ | 22.5 x $10^{-6}$ / $^\circ$C |
| Specific heat of constant pressure, $C_p$ | 900 J/kg $^\circ$C |
| Mass density, $\rho$ | 2800 kg/m$^3$ |

correct load transfer into the rig to be obtained in the model without extensive modeling of the fretting fixture. The commercial finite element package ABAQUS/Standard 5.6[†] is used for the solution. The model uses four-noded coupled temperature-displacement quadrilateral elements (CPE4T) for the specimen and pad. The spring elements are two-noded elements (SPRING1) and the contact is resolved with small sliding contact formulation (*CONTACT PAIR). The material properties used are that of 2024-T351 aluminum (Table 1). The coefficient of friction during each cycle is determined from Figure 2.

Since the finite element code does not include the coupled thermoelastic effect directly, user subroutines are used to incorporate the volumetric heat flux generated due to the strain rate in any increment. These flux values are applied directly to the element integration points in the subsequent increment. This approach is an explicit formulation that places a restriction on the largest usable time increment.

As conditions of plane stress prevail on the surface where the measurements are acquired, a conversion is necessary to determine the correct volumetric flux from the results of the plane strain model. A relationship between the strains under plane stress and plane strain conditions is employed, after noting that the in-plane stresses are identical for each condition. A rigorous verification of this type of model can be found elsewhere [6].

*Results*

The results for gross sliding are shown in Figs. 8 and 9. Temperatures are presented as changes from ambient. A mechanically etched horizontal mark along the centerline of the specimen and pad is used for tracking the relative displacement and does not represent temperature change in each image. The comparisons of finite element results with experiments are shown at maximum tensile and compressive bulk loading. The comparison is good for both the frequencies. In addition, both the frictional heating and thermoelastic effect are captured adequately. For the 2 Hz experiment, in which the pad thickness is reduced, no steady state heating is observed due to the low frequency of the applied loads and high thermal conductivity of aluminum. However, results of the 5 Hz experimental results reflect a small steady state temperature rise during the time the

---

[†]   ABAQUS/Standard 5.6 is available at Purdue University through an academic license agreement with Hibbett, Karlsson & Sorensen, Inc., Pawtucket, RI.

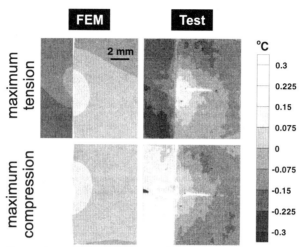

FIG. 8 — *Comparison of finite element and experimental results for gross sliding case. Contact parameters: 2 Hz, $\sigma_o$ = 100 MPa (R-ratio = -1), Q/P = 0.34, P = 6.4 kN, and R = 178 mm. Cycle number 16.*

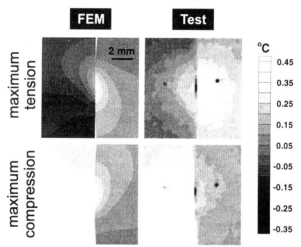

FIG. 9 — *Comparison of finite element and experimental results for gross sliding case. Contact parameters: 5 Hz, $\sigma_o$ = 100 MPa (R-ratio = -1), Q/P = 0.30, P = 6.1 kN, and R = 178 mm. Cycle number 10.*

contact is in a gross slip regime; this gross temperature rise persists until conditions of partial slip are established.

Figure 10 shows comparisons after the onset of partial slip. Once again, good agreement exists between experimental and model results. As the frictional heating is negligible during this contact regime, the entire spectrum is attributed to the temperature changes arising from the thermoelastic effect. Note that conduction must still be considered.

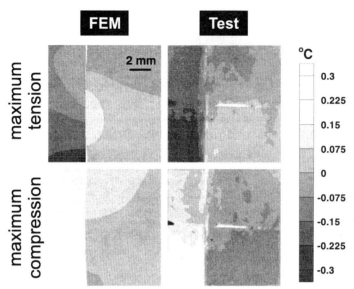

FIG. 10 — *Comparison of finite element results and experimental measurements for partial slip case. Contact parameters: 2 Hz, $\sigma_o$ = 100 MPa (R = -1), Q/P = 0.34, P = 6.4 kN, and R = 178 mm. Cycle number 152.*

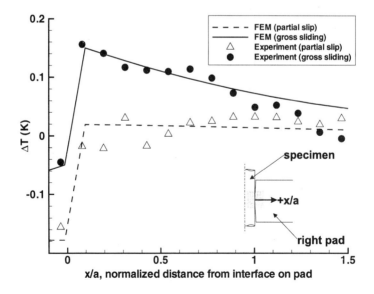

FIG. 11 — *A comparison of measured and modeled temperature amplitude on the pad surface along a line through the center of the contact. Contact parameters: 2 Hz, $\sigma_o$ = 100 MPa (R-ratio = -1), Q/P = 0.34, P = 6.4 kN, and R = 178 mm.*

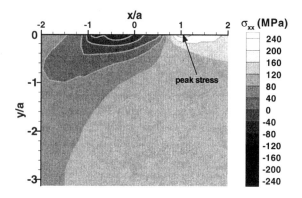

FIG. 12 — *Contours of the component of stress parallel to the contact surface, $\sigma_{xx}$, at maximum tensile bulk load from the finite element model showing a sharp peak at trailing edge of contact. Contact parameters: 2 Hz, $\sigma_o$ = 100 MPa (R = -1), Q/P = 0.34, P = 6.4 kN, and R = 178 mm.*

Figure 11 shows the temperature profile into the pad along a line at the center of the contact. This graph demonstrates the ability of the finite element analysis to discern quantitative details of the temperature distribution accurately.

At the trailing edge of contact, application of the maximum tensile bulk stress results in a high tensile peak on the specimen and a high compressive peak on the pad. At the maximum compressive bulk stress, these peaks are reversed. Each of these effects is captured accurately by the finite element results. Furthermore, a global comparison of these results clearly highlight the differences in temperature profile during gross sliding and partial slip conditions.

The agreement between experimental and calculated temperatures serves to validate the finite element model that includes both the compliance of the experimental rig and results from the ancillary experiments on evolution of friction coefficient. Consequently, the finite element results also contain validated values of the fretting-induced stresses and microslip. Figure 12 shows a contour plot of $\sigma_{xx}$, with the peak at the edges of contact and the sharp decay into the bulk of the specimen. A detailed finite element analysis of this contact configuration has been performed by McVeigh & Farris [6]. This contact stress distribution has been used subsequently to predict successfully fretting crack nucleation [1].

## Conclusions

A new approach of using infrared measurements of near-surface temperatures associated with fretting contacts has been introduced. The thermal images provide an understanding of the evolution of friction coefficient at the contact interface and the nature of the stress distribution. The transition from gross sliding to partial slip regime is captured. The temperature changes are small, due in part to the load levels and the thermal properties of aluminum.  A finite element analysis using coupled temperature-displacement elements, incorporating the coupled thermoelastic effect, frictional heating and generalized contact has been developed.  The results show excellent agreement between the temperatures measured in the experiment and the numerical predictions. These results demonstrate convincingly the ability to exploit cutting edge computational techniques and state of the art infrared technology to develop a robust model to assess fretting contact stresses accurately.  The knowledge of contact stresses enables the prediction fretting crack nucleation, resulting in a better understanding of fretting fatigue, which in turn can lead to better design principles.

## Acknowledgments

This research was supported in part by AFOSR through contract #F49620-93-1-0377. T. Sakagami would also like to acknowledge the Japanese Ministry of Education for support during his stay at Purdue University.

## References

[1]    Szolwinski, M. P. and Farris, T. N., "Observation, Analysis and Prediction of Fretting Fatigue in 2024-T351 Aluminum Alloy," *Wear*, Vol. 221, No. 1, 1998, pp. 24-26

[2]    Szolwinski, M. P., Harish, G., Farris, T. N., and Sakagami, T., "An Experimental Study of Fretting Fatigue Crack Nucleation in Airframe Alloys: A Life Prediction and Maintenance Perspective," in *1st Joint DoD/FAA/NASA Conference on Aging Aircraft*, Ogden, Utah, USA, 1997, in press.

[3]    Boley, B. A. and Weiner, J. H., *Theory of Thermal Stresses*, Wiley and Sons, New York, 1960

[4]    Szolwinski, M. P., Harish, G., Farris, T. N., and Sakagami, T., "In-Situ Measurement of Near-Surface Fretting Contact Temperatures in an Aluminum Alloy," *Journal of Tribology*, 1998, in press.

[5]    Johnson, K. L., *Contact Mechanics*, Cambridge University Press, Cambridge, 1985

[6]    McVeigh, P. A. and Farris, T. N., "Finite Element Analysis of Fretting Stresses," *Journal of Tribology*, Vol. 119, No. 4, 1997, pp. 797-801

Marie-Christine Dubourg [1] and Valérie Lamacq[1]

Stage II Crack Propagation Direction Determination Under Fretting Fatigue Loading: A New Approach in Accordance with Experimental Observations

REFERENCE: Dubourg, M. C., Lamacq, V., "Stage II Crack Propagation Direction Under Fretting Fatigue Loading: A New Approach in Accordance with Experimental Observations", *Fretting Fatigue: Current Technology and Practices, ASTM STP 1367,* D. W. Hoeppner, V. Chandrasekaran, and C. B. Elliott, Eds., American Society for Testing and Materials, West Conshohocken, PA, 2000.

ABSTRACT: Cracking is a dangerous degradation mode under fretting loading and the understanding of crack initiation and propagation is thus a necessity. A double experimental and theoretical approach has been undertaken to deal with crack initiation during Stage I [1,2], Stage I/Stage II transition and Stage II propagation. The work presented here is related to the latter point and aims at determining the direction and the propagation mode of cracks. A new approach is proposed here to account for non-proportional mixed mode conditions such as those encountered under fretting conditions at crack tips. Propagation directions during Stage II are derived from $\Delta\sigma_{\theta\theta}{}^*{}_{max}$, the maximum effective amplitude of the tangential stress perpendicular to the crack trajectory. They correlate well with experimental data. The stress field analysis shows that the trajectory of cracks borders the tensile-compressive and the tensile zones existing around the crack tip over a loading cycle.

KEYWORDS: fretting, crack growth direction, non-proportional loading, contact, friction

Introduction

The two main degradation responses under fretting that are very often in the literature related to fretting wear and fretting fatigue are particle detachment and cracking whereas both can coexist in the same contact [3]. The first main degradation has been related to a fretting regime during fretting tests through the fretting map concept proposed by

---

[1] Research Scientist and Doctor, respectively, Laboratoire de Mécanique des Contacts, UMR-CNRS 5514, INSA, 20, Av A. Einstein, 69621 Villeurbanne Cédex, France.

Vincent et al. [*4*]. Hence, the Running Conditions Fretting maps (RCFM) and the Material Response Fretting Maps (MRFM) (Figure 1) describe, respectively, the local fretting regime and the corresponding contact kinematics conditions (sticking, partial slip, gross slip) and the main fretting damage (non-degradation, cracking, and particle detachment) for normal load-displacement pairings. Cracking induced by fretting is a dangerous degradation mechanism and has been related, thanks to these maps, to partial slip conditions that occur under both the Mixed Fretting Regime and the Partial Slip Regime.

Experimental and theoretical work has been undertaken to improve our knowledge of crack initiation and propagation under fretting conditions. First, tests were conducted on 7075 aluminum alloys under Partial Slip and Mixed Fretting Regimes to obtain experimental data on crack behavior. Second, issues such as the location of crack initiation, initial crack growth direction during a macroscopic "Stage I", the conditions governing crack branching (Stage I/Stage II transition), crack growth direction during propagation (macroscopic "Stage II") were theoretically addressed. It is shown that a new approach based on the continuum stress field analysis and on the Linear Elastic Fracture Mechanics is able to predict correctly the crack behavior and allows us to identify the respective influences of the contact and the external loading on crack behavior.

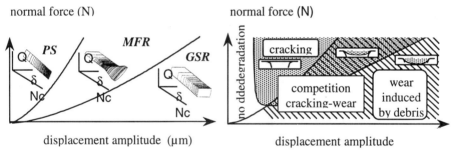

Figure 1 - *Running Condition Fretting Map (RCFM) and Material Response Fretting Map (MRFM) (PS: Partial Slip characterized by quasi-closed Q-δ loops), MFR: Mixed Fretting Regime characterized by quasi-closed, parallelepipedic and elliptic Q-δ loops, GSR: Gross Slip Regime characterized by parallelepipedic Q-δ loops) [5].*

**Experiments and Observations**

A new device called a Pre-stressed Fretting Wear (FWPSS) [6] rig has been developed to study crack initiation and propagation under fretting conditions. The displacement amplitude being a governing parameter in crack initiation is here controlled. Whereas it is not possible during fretting fatigue testing as the displacement varies according to the elongation of the test specimen which itself varies with the applied oscillatory fatigue stress in the specimen. It is very difficult in that case to correlate the crack geometry with the local contact conditions. The FWPSS test has the great advantage that all mechanical parameters can be controlled during loading. The external load is fixed during testing and the relative displacement between contacting pads is imposed as in the case of the fretting wear (FW) (Figure 2). Further the RCFMs being almost the same under FW and FWPSS [7], it is possible to study separately the effects of fretting and external loading on cracking behavior.

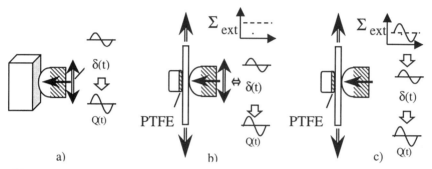

Figure 2 - *a) Fretting Wear (FW), b) Fretting Wear on pre-stressed specimen (FWPSS) ($\Sigma ext$=constant) and c) Fretting Fatigue (FF) (varying $\Sigma ext$) tests.*

The mechanical properties of 7075 aluminum alloy used are summarized here: Young modulus E=73000 MPa, $R_{0,2}$ = 460 MPa, $R_m$ = 540 MPa, $\sigma_D$ = 230 MPa, Hardness: 160 VH, A%=11.

Contact conditions corresponding to a Mixed Fretting Regime have been imposed as cracks initiate earlier than under the Partial Slip Regime. But those conditions are close enough to those encountered under Partial Slip Regime so that negligible wear surface through particle detachment occurs. A spherical aluminum alloy pad of radius 0.3m is pressed against the planar tested sample of the same alloy by a constant normal force, $F_n$=1000N. The spherical body is submitted to a reciprocating microdisplacement, $\delta$ =+/-25µm at a frequency of 5Hz. This induces an oscillating tangential force, $Q(t)$=+/-500N, in the contact area. The constant external $\Sigma_{ext}$ is equal to 230MPa. The ratio between the stick contact zone size, c and the contact zone diameter, a is equal to c/a=0.9 (Figure 3(a)). The cracking behavior in both the spherical and the planar pads experiencing respectively FW and FW combined with an external constant load are analyzed in order to identify the contact and external loading influences.

**Fretting Crack Morphology**

Multiple cracks initiate in the microsliding zone of the contact area (Figure 3), as revealed by metallographic examination. Crack sites, initiation angles, branching depths, Stage II propagation angles are identical in both FW and FWPSS specimens. The only difference is that smaller and less cracks are observed in the FWPS specimen. The crack network is symmetrical with respect to the contact center. Two main cracks develop from this network while most of the other cracks self-arrest. At the contacting surface, cracks propagate along a semi-elliptical trajectory while in the sample depth they progress along two distinct macroscopic directions during respectively macroscopic "Stage I" and "Stage II". The transition between these two periods is revealed by a crack branching along the Stage II propagation direction. Two types of crack are observed during Stage I:

♦ **Type I crack** grows initially at a shallow angle to the specimen surface with a propagation direction ranging from **15° to 35°** to the surface and occurs mainly in the middle of the contact microsliding zone,

♦ **Type II crack** grows along a direction approximately perpendicular to the surface with a direction ranging from **75° to 90°** to the surface and appears near the edge of the contact area in the microsliding zone, inside and/or outside the contact patch.

Furthermore, it was observed that the nearer to the contact center cracks initiate, the shallower to the surface the initial crack growth direction, for both crack types.

Then during Stage II, both types of cracks in both specimens propagate along a direction of **approximately 65°** to the surface. The branching depth corresponds roughly to a maximum crack length of 160μm.

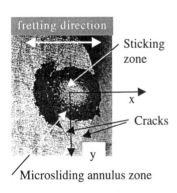

Figure 3(a) - *Elliptical cracks at the contacting surface*

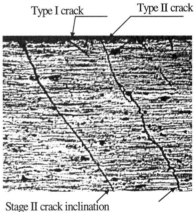

Figure 3(b) - *Type I and II cracks in the meridian plane y=0*

## Theoretical Approach

As a starting point, the cracking behavior is analyzed in the meridian plane y=0 and an equivalent two-dimensional cylinder on plane problem is defined. As crack initiation is critically linked to the contact patch dimension [8], the line contact parameters (normal and tangential loads per unit length, cylinder radius) are defined in order that 3D and 2D contact area size, 2a, and maximum hertzian pressure, Po, are identical (Figure

4). The contact problem is solved as a unilateral contact problem with friction. Coulomb's law is used. An incremental description of the tangential loading is used to account for hysteresis phenomena. Contact area size, sequence of stick and microsliding zones within the contact area, normal and tangential traction distributions are determined.

Crack propagation during Stage I and crack branching mechanisms are investigated on the basis of the continuum stresses induced by the loading. As the stresses in the surface layer change steeply, average stresses are used [9], calculated along planes of length 20μm [2]. These results are briefly recalled before focusing our attention on Stage II propagation.

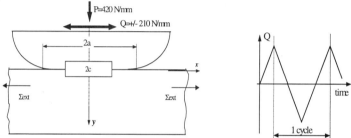

Figure 4 – Cylinder/plane contact model (contact area 2a=3,57 mm, central stick zone of varying diameter, maximum value 2c=2,54 mm) and tangential variations over a loading cycle.

*Stage I [1,2]*

The crack initiation mechanisms have been identified theoretically by employing a simple dislocation dipole model and taking into consideration the reverse sliding along two dislocation layers [9,10] and the influence of the distance between these layers [11]. It is shown that cracks may initiate either by an extrusion-intrusion mechanism or a fatigue tensile process. Two parameters, based on the amplitude of the average shear stress $\Delta\tau_m$ and the average tensile stress perpendicular to crack path, $\sigma_m$, have been proposed to predict initiation domains and a single growth direction at each initiation site [1] whereas most of the existing criteria predict initial crack growth direction in the shear mode [9] along two theoretical initial directions.

Type I crack is a shear mode fatigue crack. Its growth occurs macroscopically in the direction $\alpha$ (Figure 5) along which the value of the shear driving force, i.e. the amplitude of the average shear stress, $\Delta\tau_m$ is maximum $\Delta\tau_m = \Delta\tau_{m,max}$ and such as the average value $\sigma_a$ of the average tensile stress perpendicular to that direction, $\sigma_m$ ($\sigma_{a=}0.5*(\sigma_{m,max} + \sigma_{m,min})$) is minimum (an absolute value of 40 MPa is here considered for this minimum). Type I crack initiation domain in the contact area satisfies these two conditions. Type I initiation risk is assumed to be the highest where $\sigma_a$ is nil.

Type II crack is a tensile mode fatigue crack. Its nucleation is due to the presence of an initial flaw in the material, either preexisting or formed as a result of dislocation movements. The propagation driving force is assumed to be the maximum amplitude of crack opening. The crack extension angle $\alpha$ (Figure 5) is therefore defined by the direction along which the *effective* amplitude of the average stress normal to crack trajectory $\Delta\sigma_m*$ is maximum, $\Delta\sigma_m*=\Delta\sigma_m*,_{max}$. Further the amplitude of the average shear stress tends to a minimum value along that direction. Type II crack location is therefore near the edge of the contact area, inside or outside of it, where a high level of tensile stress occurs.

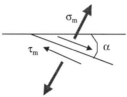

Figure 5 - *Average tensile stress $\sigma_m$ and shear stress $\tau_m$ with respect to $\alpha$ direction*

The predicted crack locations and initial growth directions derived from these two criteria are summarized in Figure 6 and are in very good agreement with experimental results.

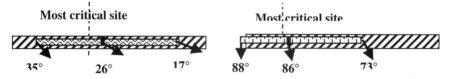

Figure 6 - *Theoretical results. Type I crack initiation zone* ▧▧▧▧ *and type II crack initiation zone* ▦▦▦▦ *in the microsliding zone* ▨▨▨▨. *Extension angle $\alpha$.*

*Stage I/ Stage II Transition*

The conditions leading to transition from Stage I to Stage II are identified [*13*]. Stage I/Stage II transition corresponds to crack branching along a direction of 65° for both crack types under FW and FWPSS. Further the branching depths are of the same order of magnitude for both specimens. These observations highlight the contact influence.

For the planar specimen submitted to FW combined with an external constant load, the crack growth mode for both crack types is progressively turned to a mixed I+II mode within the specimen depth. Thus, the Stage I propagation driving force decreases ($\Delta\tau_m$ and $\Delta\sigma_m*$ for type I and II respectively), the crack locking effect increases as $\sigma_m$ perpendicular to crack path becomes gradually more and more compressive and hinders crack slips and opening. Crack self-arrest is therefore unavoidable unless branching occurs. Concomitantly, $\Delta\sigma_m*$ perpendicular to a direction $\beta$ of 65° increases and along that direction $\Delta\tau_m$ is further minimum. Hence, branching and initial propagation during

Stage II occurs along a direction β along which the amplitude of the crack opening is maximum (and not the opening). Therefore, type I crack branch from a shear mode growth to a tensile mode growth, whereas type II crack branch from a tensile mode growth to a tensile mode growth in order to enhance crack propagation.

For the spherical specimen submitted to FW only, type I and II cracks branch also in the 65° direction. The decreasing of the initial driving force ($\Delta\tau_m$ or $\Delta\sigma_m^*$) along the direction α under the contact patch is also combined with crack locking effect raising due to the increase of the compressive value of $\sigma_m$. The 65° direction is such as along it $\Delta\tau_m$ is minimum and perpendicular to it $\Delta\sigma_m^*$ is small, much more small than under FWPSS as the external load is here nil. Therefore, branching occurs in the direction β of 65° that preserves crack opening in order to protect its propagation, whatever the growth mode is.

The contact influence on cracking behavior during Stage I and Stage I/Stage II transition is predominant: crack locations and profiles, branching depths are identical in both specimens. Further the fatigue part of the loading is linked to the contact tangential load.

*Stage II*

The fretting conditions induce non-proportional mixed mode I+II loading conditions at crack tips. Non-proportional loading is characterized by a varying $K_I/K_{II}$ ratio during a loading cycle while a proportional one is characterized by a constant ratio. As a general rule a crack branches under mixed mode. Classical criteria such as those of Erdogan and Sih (MTS), Tirosh or the S-theory of Sih predict that a crack bifurcates in a direction $\theta_0$ where a specific parameter reaches an extreme, respectively the maximum of the tangential stress $\sigma_{\theta\theta}$, the maximum of ($\sigma_{\theta\theta}^2-\sigma_{rr}^2$)) and the minimum of the strain energy density. The bifurcation angle, $\theta_0$, is in good agreement with experimental results. These criteria are no longer valid under non-proportional loading. New criteria or extensions of these criteria are therefore required. Hourlier et al. [12] proposed 3 extensions of the MTS criterion, assuming that a crack propagates in a direction corresponding to $k_1^*{}_{max}$, $\Delta k_1^*{}_{max}$ or to $(da/dN)_{max}$ (Figure 7). Bower [14] proposed also two extensions of the

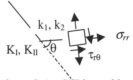

Figure 7 - SIF $K_I$ and $K_{II}$ at initial crack tip, SIF $k_1$ and $k_2$ at branched crack tip, tangential stress $\sigma_{\theta\theta}$ and shear stress $\tau_{r\theta}$ ahead of crack tip.

MTS criterion. He stated that a crack extends in a direction along which $K_\sigma$ ($K_\sigma=\sigma_{\theta\theta}\sqrt{2\pi r}$) or its amplitude $\Delta K_\sigma$ is maximum over the load cycle, $K_\sigma=K_{\sigma max}$ or $\Delta K_\sigma=\Delta K_{\sigma max}$. The criterion proposed here is also an extension of the MTS criterion. It is assumed that the crack extension angle is linked to the maximum effective amplitude of the tangential stress at crack tip over a load cycle $\Delta\sigma_{\theta\theta}^{*}{}_{max}$ [13].

The objective here is to test the applicability of these different approaches. The Stage II propagation is analyzed for the two main cracks situated symmetrically with respect to the contact center (Figure 8). For the cracks situated in the spherical pad, the attention is focused on the conditions leading to self-arrest as they stop very rapidly in the specimen depth while for the cracks situated in the planar specimen the growth angles derived from the different criteria are compared to experimental observations. During Stage II propagation, the cracks extend in the direction of their initial growth at branching leading to a crack extension angle almost equal to zero.

The fretting regime considered inducing a predominant cracking degradation response combined with negligible wear surface through particle detachment allows us to assume that the contacting surfaces are still smooth. Dubourg's fatigue crack model, [15,16] based on the continuous distribution of dislocations coupled with unilateral contact analysis with friction is used. The two main kinked cracks, located in the microsliding zone, where the type I crack initiation risk is important (x=+/-1.65 mm), are macroscopically modeled by using two straight segments [17] of inclination $\alpha=29°$ and $\beta=67°$ (Figure 8). Seventy-seven discretization points are distributed along each crack, and the loading cycle is described into 81 steps. Stress intensity factors (SIFs) $K_I$, $K_{II}$ at crack tips and stress fields along crack interfaces are calculated in both specimens. Due to symmetry, the results are presented for the left crack only.

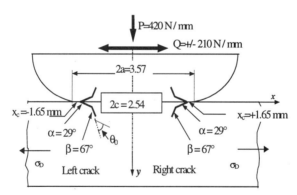

Figure 8 - *Loading conditions and main crack locations and profiles.*

*Cracks Induced by Fretting Wear*

The contact loading induces shear stresses and compressive and tensile subsurface stresses depending on the direction of the surface traction Q(t) during the loading cycle. The two 500 μm long cracks experience mixed mode with a dominant mode II ($\Delta K_I/\Delta K_{II}=0.84$) (Figure 9). It may be concluded that crack growth mode is a shear mode, although branching was governed by $\Delta\sigma_m*$. It is further shown that maximum $K_{II}$ values are obtained when the crack is totally open as slips are not restrained and then hindered by the compressive stresses acting on crack faces. Crack self-arrest is therefore linked to crack closure over the whole loading cycle.

$K_I$ variations have been computed for crack lengths ranging from 300 to 800 µm (Figure 10) under these loading conditions. 800 µm is the crack length corresponding to complete closure over a loading cycle. This "critical length" correlates very well with the experimental observations. This length is further linked to the contact patch dimensions and is smaller than a half of the half contact size as shown in Figure 11 where the critical length variations are computed versus the half size of the contact area, the maximum hertzian pressure being constant.

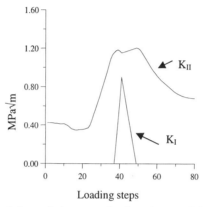

Figure 8 - *SIFs at left crack tip over a load cycle. Crack length of 500 µm.*

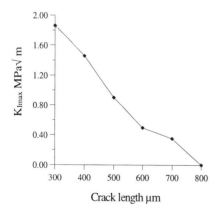

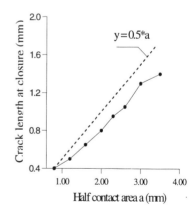

Figure 10 - $K_{Imax}$ *variations over a loading cycle versus crack length. 2a=3.57mm, Po= 150 MPa.*

Figure 11 - *Crack length at closure versus varying half-contact area a. Po= 150 MPa.*

*Cracks Induced by Fretting Wear Combined with a Constant External Load*

Crack behavior in the FWPSS is different. The contact loading and the external load act differently but complementary. The contact loading through the tangential loading variation produces the fatigue part of the loading while the constant external load induces average opening and sliding along crack faces that strengthen and extend the influence of the contact. As seen above, the contact influence is limited to the first crack segment corresponding to Stage I propagation. But fatigue crack growth in mode I after branching is only feasible through slips (contact role) acting along the first crack segment that are passed on the open crack tip (external load influence) leading to a varying crack opening.

The crack tips experienced non-proportional mixed mode conditions with a dominant mode I (Figure 12). The ratio $\Delta K_I/\Delta K_{II}$ is equal to 6.35 for 1,2 mm long crack.

The crack extension angles are determined according to the classical MTS criterion and the different extensions of this MTS criterion cited above. The results are checked at the left crack tip with regards to experimental results.

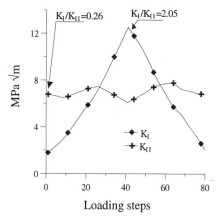

Figure 12 - SIF *variations at left crack tip over a load cycle. 1,2 mm crack length.*

*MTS criterion* – This criterion states that a crack extends in the direction normal to the direction of the maximum tangential stress at crack tip $\sigma_{\theta\theta max}$. The results presented in Figure 13 demonstrate its inapplicability under non-proportional conditions. As $K_I/K_{II}$ varies over the cycle, varying $\theta_0$ values are obtained, about 40° for the right crack and about –40° for the left one. Such inclinations do not correlate with the experimental results.

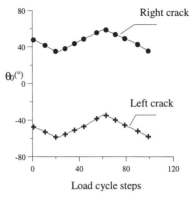

Figure 13 - *Crack extension angle $\theta_0$ for both cracks according to the MTS criterion.*

$k_1^*{}_{max}, \Delta k_1^*{}_{max}$ *criteria* – 3 extensions of the MTS criterion are proposed by Hourlier and Pineau [12]. They assume that a crack propagates either in the direction along which $k_1^*$ ($k_1^*$ and $k_2^*$ values are linked to $K_I$ and $K_{II}$ calculated at crack tip before branching [18]) or $\Delta k_1^*$ or the crack growth rate (da/dN) is maximum. The latter criterion is not tested here by lack of experimental data. $k_1^*{}_{max}$, $k_1^*{}_{min}$ and $\Delta k_1^*{}_{max}$ variations at left crack tip versus the bifurcation angle $\theta$ defined with respect to the crack direction $\beta$ are presented in Figure 14 over a load cycle.

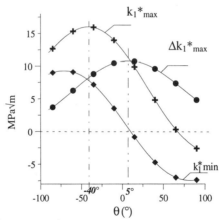

Figure 14 - *$k_1^*{}_{max}$, $k_1^*{}_{min}$ and $\Delta k_1^*{}_{max}$ variations versus $\theta$ at crack tip over a load cycle.*

The angles derived from $k_1^*{}_{max}$ and $\Delta k_1^*{}_{max}$ criteria are very different and equal to -40° and +5° respectively. The former approach cannot account for the experimental observations while the latter is in good agreement with the experiments. A crack extends in the direction along which its opening amplitude is maximum and not its opening.

$K_{\sigma max}$, $\Delta K_{\sigma max}$ *criteria* – These approaches [14] derive also from the MTS criterion and state that a crack extends in the direction along which $K_\sigma$ respectively $\Delta K_\sigma$ is maximum over a load cycle. $\sigma_{\theta\theta max}$ and $\Delta\sigma_{\theta\theta max}$ variations corresponding, respectively, to $K_\sigma$ and $\Delta K_\sigma$ variations versus $\theta$ are determined under the FWPSS conditions over a load cycle and are presented in Figure 15.

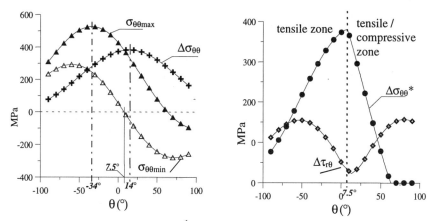

Figure 15 - $\sigma_{\theta\theta min}$, $\sigma_{\theta\theta max}$, $\Delta\sigma_{\theta\theta}$, $\Delta\sigma_{\theta\theta}{}^*$ and $\Delta\tau_{r\theta}$ variations at crack tip over a load cycle versus the crack extension angle $\theta$.

$\Delta\sigma_{\theta\theta}{}^*$ *criterion* – This criterion rests on the stress field analysis as performed during the Stage I and Stage I/Stage II transition. The maximum and the minimum tangential stress, the tangential stress amplitude, the effective tangential stress amplitude and the shear stress amplitude respectively $\sigma_{\theta\theta max}$, $\sigma_{\theta\theta min}$, $\Delta\sigma_{\theta\theta}$, $\Delta\sigma_{\theta\theta}{}^*$ and $\Delta\tau_{r\theta}$ are presented in Figure 15 versus $\theta$ over a load cycle for the left crack.

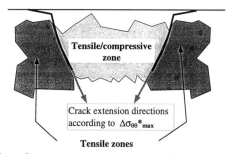

Figure 16 – *Tensile and compressive zones surrounding the cracks in the FWPSS.*

It appears that $\Delta\sigma_{\theta\theta}{}^*$ gives the best prediction $\theta=+7.5°$. Further, the crack path follows a boundary between zones that experience alternatively tensile-compressive and compressive stresses over a load cycle (Figure 16). This extension angle optimizes the crack opening amplitude. This criterion is therefore equivalent to those proposed at Stage I/Stage II transition and based on $\Delta\sigma_m{}^*$, the only difference between these two parameters is that the crack response is accounted for in the stress field calculation during Stage II. The mechanisms governing crack growth are therefore identical.

*Discussion* –The results are summarized in Table 1. The inapplicability of the MTS criterion under non-proportional conditions is clearly established. The crack extension angles derived from the $k_1^*{}_{max}$, $K_{\sigma max}$ approaches can also not account for the experimental results. $\Delta K_{\sigma max}$ gives reasonable results but the crack extension angle $\theta_0$ is such as the crack tip enters a zone that experiences alternatively compressive and tensile $\sigma_{\theta\theta}$ stresses (Figure 16). Further these compressive stresses will close the crack faces, lead to a decreasing of the opening amplitude over a load cycle and cause a drop in the crack growth speed. The $\Delta k_1^*{}_{max}$ and $\Delta\sigma_{\theta\theta}^*{}_{max}$ criteria give coherent results as they are both based on the concept of the maximum amplitude of crack opening.

Table 1 – *Crack extension angles $\theta_0$ derived from the different criteria*

| Criterion | $\theta_0$ for left crack | $\theta_0$ for right crack |
|:---:|:---:|:---:|
| $K_1^*{}_{max}$ | -40° | +40° |
| $\Delta k_1^*{}_{max}$ | +5° | -5° |
| $\sigma_{\theta\theta max} \Leftrightarrow K_{\sigma max}$ | -34° | +34° |
| $\Delta\sigma_{\theta\theta max} \Leftrightarrow \Delta K_{\sigma max}$ | +14° | -14° |
| $\Delta\sigma_{\theta\theta}^*{}_{max}$ | +7.5° | -7.5° |

**Conclusion**

The objectives of this work were to improve our understanding at a macroscopic scale of crack initiation and propagation mechanisms under fretting conditions. A combined experimental and theoretical approach has been developed. Thanks to the fretting map concepts, experiments have been performed under running conditions corresponding, to Mixed Fretting Regime and Partial Slip Regime that led to a predominant cracking degradation response.

Reproducible data on crack location, crack inclination within specimen depth, crack branching depth have been obtained in Fretting Wear specimens and Fretting Wear Pre-Stressed Specimens. Further the crack network is identical in both specimens with respect to crack sites and initial growth directions. Macroscopic Stage I and Stage II periods have been defined, corresponding to two different macroscopic crack growth directions. During Stage I, two different crack types have been observed. Type I crack is a shear mode fatigue crack while type II crack is a tensile mode fatigue crack. Crack branching occurs at the Stage I/Stage II transition and then both crack types, in both specimens, propagate in a direction of 65° with respect to the specimen surface, but the cracks induced by Fretting Wear (the spherical specimen) self-arrest rapidly.

Theoretically, crack initiation and propagation during Stage I, Stage I/Stage II transition and Stage II have been investigated. A new approach has been proposed following [9,11] to predict the crack location and the crack growth initial angle during Stage I and contrary to general initiation criteria. This model has enabled us to predict a single initiation plane for shear mode fatigue crack [1]. The conditions governing the crack branching have also been identified. It has been shown that the contact loading has a predominant influence during this stage.

A new criterion has been proposed to predict the crack extension angle during Stage II. As cracks experience non-proportional mixed mode fatigue under fretting loading, the classical criteria proposed for the crack path determination are no longer valid. The MTS criterion and different extensions derived from this criterion proposed by Houlier and Pineau ($k_1^{*}{}_{max}$, $\Delta k_1^{*}{}_{max}$) [12], Bower ($K_{\sigma max}$, $\Delta K_{\sigma max}$) [14] and the authors ($\Delta \sigma_{\theta\theta}^{*}{}_{max}$) have been tested and checked with regards to the experimental results. It is shown that $\Delta \sigma_{\theta\theta}^{*}{}_{max}$ and $\Delta k_1^{*}{}_{max}$ can account for the experimental results. They are further coherent as they both state that a crack propagates in a direction along which its opening amplitude is maximum. It has been further demonstrated that this mode I propagation is feasible through the contact loading and the constant external loading combination. The external loading induces a constant opening at crack tip. Without it, cracks self-arrest (FW specimen) and the corresponding crack length is linked to the contact patch dimension. In the FWPSS, varying opening at crack tip is indeed due to the slips acting along the initial crack direction that are passed on the open crack tip.

## References

[1]    Reybet Degat, P., Lamacq, V., Dubourg, M. C., Zhou, Z. R., Vincent, L., "Experimental and Theoretical Approach to Fretting Crack Nucleation on Pre-stressed Aluminum Alloy", *ECF 11, Mechanisms and Mechanics of Damage and Failure*, 1996, pp 1443-1445.

[2]    Lamacq, V., Dubourg, M. C., Vincent, L., "A Theoretical Model for the Prediction of Fretting Fatigue Crack Initial Growth Angles and Sites", *Tribology International*, Vol. 30, n° 6, 1997, pp. 391-400.

[3]    Vincent, L., Berthier, Y., Dubourg, M. C. and Godet, M., "Mechanics and Materials in Fretting", *Wear*, Vol. 153, 1992, pp. 135-148.

[4]    Vincent L., Berthier Y. and Godet M., "Testing Methods in Fretting Fatigue: A Critical Appraisal", *ASTM-STP* 1159, 1992, pp. 33-48

[5]    Fouvry, S., Kapsa, P., Vincent, L., "Quantification of Fretting Damage", *Wear* 200, 1996, pp. 186-205.

[6]    Reybet Degat, P., Zhou, Z. R., Vincent, L., "Fretting Behavior on Pre-stressed Aluminum Alloy Specimen", *Tribology International*, Vol. 30, n°3, 1996, pp 711-720.

[7]    Zhou, Z., R., Vincent, L., "Mixed Fretting Regime", *Wear* 181-183, 1995, pp. 531-536.

[8]    Nowells, D., Hills, D.A., "Crack Initiation Criteria in Fretting Fatigue", *Wear*, Vol. 135, 1990, pp. 329-343.

[9]    Yamashita, N. and Mura, T., "Contact Fatigue Crack Initiation under Repeated Oblique Force", *Wear*, 91, 1983, pp. 235-250.

[10]   Tanaka, N., Mura, T., "A Dislocation Model For Fatigue Crack Initiation", *Journal of Applied Mechanics*, 47, 1981, pp. 111-113.

[11]   Mura, T., Nakasone, Y., "A Theory of Fatigue Crack Initiation in Solids", *Journal of Applied Mechanics*, Vol. 57, 1990, pp 1-6.

[12]   Hourlier, F., Pineau, A., "Fatigue Crack Path Behavior under Complex Mode Loading", in *Advances in Fracture Research, Proceedings 5th International Conference on Fracture*, Pergamon, Oxford, Vol. 4, 1981, pp. 1841-1849.

[13]   Lamacq, V., " Amorçage et Propagation de Fissures de Fatigue sous Conditions de Fretting", Thèse: Doctorat, INSA, 1997, 251 p.

[14]   Bower, A., F., "The influence of Crack Face Friction and Trapped Fluid on Surface Initiated Rolling Contact Fatigue Cracks", *Journal of Tribology*, Transaction of the ASME, 110, 1988, pp. 704-711.

[15]   Dubourg, M. C., Villechaise, B., "Analysis of Multiple Cracks - Part I: Theory", *ASME, Journal of Tribology*, Vol. 114, 1992, pp 455-461.

[16]   Dubourg, M. C., Godet, M., Villechaise, B., "Analysis of Multiple Cracks - Part II: Results", *ASME, Journal of Tribology*, Vol. 114, 1992, pp. 462-468.

[17]   Dubourg, M.C., Villechaise, B., "Stress Intensity Factors in a Bent Crack: a Model", *European Journal of Mechanics, A/Solids*, 11, n°2, 1992, pp. 169-179.

[18]   Amestoy, M., Bui H. D., Dang Van K., "Déviation Infinitésimale d'une Fissure dans une Direction Arbitraire", *Compte Rendu Académie des Sciences* Paris, t. 289, Série B, 1979, pp. 99-102

Michael P. Blinn[1] and Jane M. Lipkin[2]

Development of a High-Temperature-Steam Fretting Wear Test Apparatus

REFERENCE: Blinn, M. P. and Lipkin, J. M., "Development of a High-Temperature-Steam Fretting Wear Test Apparatus," *Fretting Fatigue: Current Technology and Practices, ASTM STP 1367*, D. W. Hoeppner, V. Chandrasekaran, and C. B. Elliott, Eds., American Society for Testing and Materials, West Conshohocken, PA, 2000.

ABSTRACT: Modern power generation technology requires the development and use of advanced materials to withstand aggressive environments. Candidate materials must be durable and provide resistance to fretting wear damage; this includes both coatings and the base materials used in the machinery. A prototypical, multi-specimen test apparatus was developed for use in this fretting wear test (FWT) program. The FWT apparatus was designed to generate and contain steam at temperatures up to 595 °C (1100 °F). Contact stresses over 3.45 MPa (500 psi), displacements of 0.1 mm (4 mil), and a frequency of 160 Hz are possible using this test machine.

KEYWORDS: fretting wear, wear testing, high temperature, steam, power generation

Introduction

Power generation requires machinery that can withstand the effects of an aggressive operating environment. In particular, the effects of fretting wear damage on machine parts can be a concern in design, especially if the fretting wear leads to down-time of mechanical equipment. Obtaining realistic information on the fretting wear characteristics of candidate materials used in machinery is vital for selecting the best materials for the final design. Overall, research and development of fretting wear resistant materials for power generating equipment can reduce operating and maintenance time, leading to more cost efficient systems.

Numerous machines are available for all types of wear testing, as described in some of the more recent tribology texts and standards [1,2]. However, many of these machines are limited by various test parameters, such as temperature, relative displacement, sliding velocity, load, and environment. It is often necessary to design, develop, prove, and subsequently use prototypical test machines to achieve the objectives of a research project. Developing such a test apparatus allows the designer to customize

[1] Senior engineer, Materials Characterization Laboratory, 704 Corporations Park, Scotia, New York 12302.
[2] Material scientist, General Electric Power Systems, Bldg. 40 - Room 136M, 1 River Road, Schenectady, New York 12345.

the test program to suit the specific needs of the machinery under consideration. Further, a custom designed test apparatus allows for evaluating various combinations of test parameters. A study of various combinations of multiple test parameters might lead to the discovery of an interesting synergy among these test parameters that might have been unnoticed in earlier, single parameter test programs.

Of particular interest to this test program, in terms of design information, was the generation of fretting wear data for candidate materials and coatings. This design information would ultimately determine the materials and coatings most suitable for components used in power generating equipment. To achieve this objective, a high-temperature-steam FWT apparatus was required to test various coated and uncoated wear couples under aggressive test conditions. This prototypical FWT system was designed, fabricated, and proven to give meaningful results by the authors and their co-workers.

## Test Methods and Parameters of Interest

To correctly model the fretting wear that might occur in the power generating equipment under consideration, wear samples had to be exposed to a high-temperature-steam environment, up to 595 °C (1100 °F). Further, contact stresses of 3.45 MPa (500 psi) and double amplitude displacements of up to 0.1 mm (4 mil) also had to be mechanically imposed on the wear samples. Finally, the test would be conducted at a frequency of greater than 100 Hz for a duration of well over one million cycles. In general, the FWT apparatus had to be designed and developed to apply the same aggressive environment that the power generating equipment would be exposed to in service (Table 1). It was important to maintain control of the test parameters under consideration, especially due to a potential synergy among these parameters.

Table 1 - *Test Parameters, Design Limits, and Instrumentation*

| Test Parameter | Description (design limit) | Transducer |
|---|---|---|
| Normal Force | Force imposed on the samples by each pneumatic cylinder (140 N) | Load cell and pressure gauge |
| Displacement | Relative displacement of the samples; double amplitude distance of sliding motion (0.1 mm) | Verified optically using a measuring microscope (1.25 mm range with 0.025 mm divisions) |
| Frequency | Test frequency (160 Hz) | Verified with a frequency standard |
| Temperature | Nominal temperature inside the test chamber (maximum 595 °C) | Thermocouple (Type K) and digital readout |
| Cycles | Number of fretting wear cycles; a measure of endurance (> 10 million) | Time based |
| Steam Flow | Amount of liquid into the system to be used as steam (300 ml/hr) | Flow meter (visual sightglass) |

## Test Equipment - Development and Description

Based on a review of wear test systems available from various suppliers, it became evident that a suitable FWT apparatus was not readily obtainable for the aggressive conditions required. However, a sliding wear test (SWT) system developed by the authors in a previous test program proved to be an excellent candidate for modification as a FWT apparatus [3]. Design issues that would have to be addressed in modifying the SWT system into a FWT apparatus were as follows: controlling relatively small test (specimen) displacements, providing high test frequencies, and measuring and verifying the modified test system to suit the requirements of the FWT program.

In general, the high-temperature-steam SWT apparatus, available for modification as a FWT machine, is a stainless steel test chamber. All openings to the test chamber are sealed mechanically, or by the use of gaskets. Side doors, with high temperature glass inspection ports, are provided for access inside the test chamber. Heat is supplied by cartridge and strip heater units, mounted internal and external to the test chamber. A specially designed system generates steam external to the test chamber. Steam enters the chamber through an inlet tube and exits to a condensing tube. The condensing tube can be cooled by the use of an external fan or by the use of a heat exchanger. The heat exchanger can be a simple water filled tube that the condensing tube must pass through before it ends in the outflow.

Normal force loading of the test specimens is applied by pneumatic cylinders, mounted on top of the steam chamber (Figure 1). The normal force load is transferred to the top table, which is supported laterally by the use of flex-plates. The flex-plates and top table bellows allow limited travel in the vertical direction (important for specimen

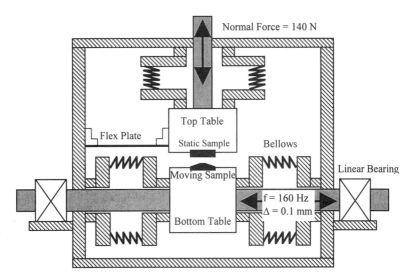

Figure 1 – *Loading System of the FWT Apparatus*

installation). Finally, the friction force load is supplied by a 5.34 kN (1 200 lb) shaker, which excites the bottom table at frequencies of up to 160 Hz. Re-circulating ball type linear bearings support the bottom table, allowing for "controlled" lateral motion.

To provide for the high temperatures required the FWT apparatus was designed as a large steam chamber (Figure 2). Removable covers on both sides of the machine were provided for the installation of two sets of fretting wear specimens. The internal temperature of the steam chamber was measured and servo-controlled by thermocouples, connected to a series of temperature controllers. The top tables, bottom table, steam chamber, and steam generator all had their own heaters and temperature controllers for optimum control of the system. Resistive cartridge heaters were used inside the top and bottom tables; strip heaters were mounted on the floor and walls of the steam chamber.

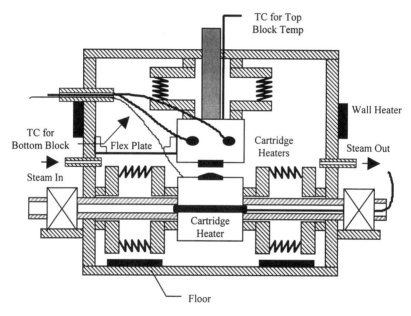

Figure 2 - *Heating and Steam System of the FWT Apparatus*

In addition to high temperature capabilities, introduction of steam into the test chamber was an integral part of the design. The use of welded metal bellows to prevent steam leakage out of the test chamber is a unique feature. These bellows were fabricated from stainless steel to provide corrosion resistance at relatively high temperatures. High temperature graphite gaskets were used on the two side doors and on the bellow flanges to help seal the test system.

A steam generation system was designed and built for the FWT program, with the capabilities of heating deionized water at a fast rate to produce super-heated steam (Figure 3). A thermocouple was used to servo-control the steam generator (a small pipe, surrounded by a tubular heater unit) with an electrical temperature controller. The steam generator would continuously fill with deionized water, which was quickly heated to boiling, thus producing a steady flow of superheated steam. During initial heat-up of the system, nitrogen is carried through the steam system lines as a "purge" gas, to prevent oxidation of the test samples.

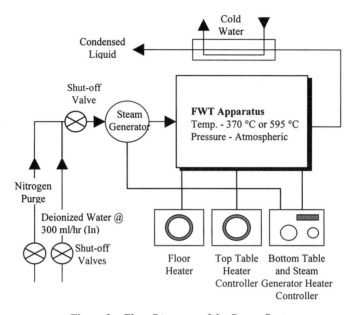

Figure 3 - *Flow Diagram of the Steam System*

While a basic overview has been presented, the FWT apparatus involves numerous other distinctive features. These will be presented from the "outside-in," starting with the fully assembled FWT apparatus and continuing with detailed "feature" photos of the test rig. The normal force pneumatic devices (located on top), the steam chamber, and the friction force actuator (shaker) are shown in the first photo (Figure 4). Stainless steel covers with observation ports are shown mounted to the FWT apparatus.

The second photo highlights some internal features unique to the FWT apparatus (Figure 5). In this photo, the side covers of the steam chamber are removed, exposing the top and bottom tables. The two top tables are the end of the normal force load train, where the static specimens are mounted. Similarly, the bottom table is the end of the friction force load train, where the two moving specimens are mounted. A total of six bellows are mounted to the top and bottom tables by the use of standard size stainless steel pipe flanges. The bellows allow for motion of the tables and provide sealing of the steam chamber.

In addition to the top and bottom tables, flex-plates are provided to restrain the top table from lateral motion. These flex-plates keep the top tables in line with the bottom table during the course of a test run. While providing lateral constraint, the flex-plates allow for unrestrained vertical motion of the normal force load train during testing. To prevent excessive vibration of the flex-plates during testing, weights were added to the center of the plates to dampen the system. The flex-plate system is adjustable, allowing for various weights to be used. This feature is important for future fretting wear test programs, where the damping mass is determined on-the-fly.

Figure 4 - *FWT Apparatus with the Side Covers in Place*

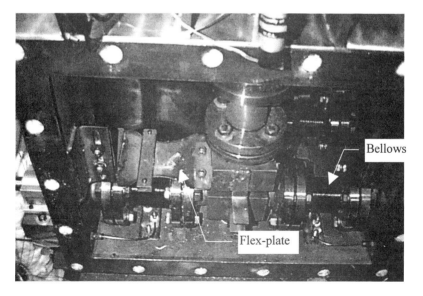

Figure 5 - *FWT Apparatus with the Covers Removed*

Movement of the bottom table is provided by an electro-mechanical shaker, capable of forces up to 5.34 kN (1 200 lb) and frequencies greater than 1 kHz, depending on the resonant frequency of the driven system (Figure 6). A special yoke was required for the friction force actuator, which is attached to the two bottom table shafts by a clevis and yoke. To ensure travel of the bottom table is in a forward-reverse, high frequency type of motion, the dedicated electronics of the electro-mechanical shaker were calibrated and set to control the 0.1 mm (4 mil) double amplitude of displacement (fretting wear travel).

Verification of the double amplitude displacement during any test run is by an optical microscope, with a range of 1.27 mm (50 mil). A reference "arm" was welded in place on the top table; this arm could be readily viewed through the inspection port by the microscope. The motion of the reference arm during testing shows up under the microscope as a solid line, which can be gauged against the 0.025 mm (1 mil) tick marks viewed in the microscope. Any required adjustments to the double amplitude displacement or frequency during a test run could be achieved on-the-fly, by adjusting the dedicated electronics of the shaker.

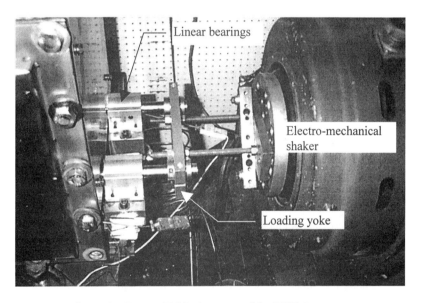

Figure 6 - *Bottom Table Actuator of the FWT Apparatus*

To conclude this discussion, it is appropriate to discuss the specimen size and geometry used for the FWT program, and how the samples are mounted into the test apparatus. The static samples (Figure 7) are mounted into the top tables by clips that hold them in place. Similarly, the moving samples (Figure 8) are mounted into the bottom table by the use of slotted holders. The actual wear surface of the static samples is much larger than the wear surface of the moving samples; this allows for any misalignment of the wear couples during heat-up of the test system. In summary, the wear surface area used for computing the "nominal" contact stress is the area of the moving "shoe" specimen.

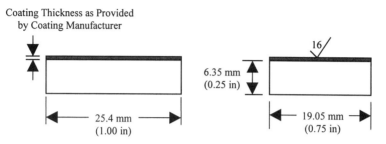

Figure 7 - *Static Sample, Showing Basic Dimensions*

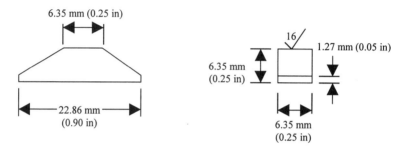

Figure 8 - *Moving Sample, Showing Basic Dimensions*

## Sample Data and Comparative Results

Thirteen wear couples were tested in this program, with eleven of the wear couples using the FWT apparatus. In general, most wear couples consisted of coated static specimens and uncoated moving specimens. The most meaningful data, from the perspective of the machine designer, are the test frequency, duration, specimen weight loss, and wear rate. The data would be evaluated to determine which wear couples are the best candidates for fretting wear resistance in the power generating equipment being designed.

During the initial stages of the FWT program, some interesting results were observed, regarding testing at various temperatures. A baseline test condition was first established, testing uncoated Inconel wear couples at room temperature in a laboratory air environment. Following baseline tests, the temperature was increased to 370 °C (700 °F), while maintaining the laboratory air environment. In these test runs, fretting wear damage at the higher temperatures unexpectedly decreased significantly. These initial results caused a re-evaluation of the test methods used in the FWT program, due in part to the issue of whether the decrease in fretting wear damage was due to a suspect test apparatus, or it was indeed a material phenomenon.

When designing and proving-out a proto-typical test apparatus, it is often difficult to establish whether the data obtained is suspect or accurate. To calibrate specific transducers on the test apparatus separately, and verify their accuracy, is usually an acceptable start. However, there is always concern that the test machine may not produce results in exact accordance with the "sum of the parts" (i.e. the individual transducer readings). The biggest problem with the "shake-down" of a proto-typical test apparatus is commonly the lack of a similar test machine against which to compare the preliminary results. Further, if two test machines are compared, if any parameter is altered, such as specimen geometry, frequency, temperature, displacement, or loading, there is always a concern that the two comparative tests are independent of one-another.

For the case of the FWT apparatus, the authors fortunately had a test machine available to prove-out the system by comparing like results. The specimen geometry, means of actuating, heating, and loading were all very similar for both test machines. The comparative test machine, known as the multi-sample wear test (MSWT) machine, was used to test additional wear couples at room temperature and 370 °C (700 °F) to determine if the FWT apparatus was indeed producing meaningful results. In general, the comparative results showed weight losses of both static and moving specimens and wear rates that were within 40% of each other (Table 2). Further, the phenomenon that the bare Inconel specimens exhibited a decrease in fretting wear at 370 °C (700 °F) was also proven using the MSWT apparatus.

Table 2 - *Comparative Test Results for the FWT Apparatus at Room Temperature*

| Specimen Type and Test Apparatus | Freq. (Hz) | Duration (hr) | Cycles ($10^6$) | Wear (mm) | Wt. Loss (mg) | Wear Rate ($\mu m/10^6$ cycle) |
|---|---|---|---|---|---|---|
| Static Sample | | | | | | |
| FWT | 160 | 55 | 31.68 | 0.108 | 42.5 | 3.41 |
| MSWT | 180 | 24 | 15.55 | 0.084 | 32.5 | 5.40 |
| % Difference | | | | | | 37% |
| Moving Sample | | | | | | |
| FWT | 160 | 55 | 31.68 | 0.116 | 40.7 | 3.66 |
| MSWT | 180 | 24 | 15.55 | 0.089 | 33.3 | 5.72 |
| % Difference | | | | | | 36% |

Visual inspection of the specimens, following testing, was also important in evaluating the fretting wear damage that occurred under the various test conditions. Static specimens that were tested at room temperature showed a degree of fretting wear damage on their faces, in the form of loss of material in an area that mimicked the shape of the moving samples (Figure 9). Fretting wear damage was readily apparent on the room temperature wear couples; at the higher test temperatures of 370 °C and 595 °C (700 °F and 1100 °F), fretting wear markings were almost non-existent.

Static sample

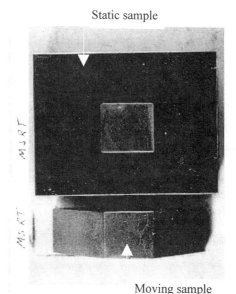

Moving sample

Figure 9 - *Tested Specimen Pair, Showing Fretting Wear*

## Conclusions

A test machine was designed and developed to produce meaningful results for screening candidate materials in a high-temperature-steam environment, subject to the effects of fretting wear. The use of unique features in the FWT apparatus, such as a steam generating system, flex-plates, special heaters, and bellows, allowed for testing wear couples in an aggressive environment. The FWT apparatus developed in this program was shown to produce test results that compared well against a similar test machine (the MSWT apparatus). Finally, there was a definite temperature effect exhibited by uncoated Inconel specimens, tested in laboratory air. Fretting wear decreased considerably at the elevated temperatures of 370 °C and 595 °C (700 °F and 1100 °F), as opposed to room temperature. However, as the focus of this test program provided screening information, additional research and testing could be conducted to evaluate what appears to be a material and temperature based phenomenon. It should be noted that this apparent temperature dependence of fretting wear, when testing Inconel specimens, may not be exhibited when tested other wear couples at elevated temperatures.

## Acknowledgments

The authors would like to thank Christine Furstoss, Kevin Jerwann, Ming Li, and Peter Schilke of the General Electric Company for their support in this fretting wear research program. The technical advice and input of Ray Englehart, John Griffin, Roger Haskell, Stuart Heitkamp, and Rod Holmstrom of MCL has also been greatly appreciated.

## References

[*1*]  Budinski, K.G., "Laboratory Test Methods for Solid Friction," in *Friction, Lubrication, and Fretting Wear Technology*, ASM Handbook Vol. 18, ASM, Metals Park, Ohio, USA, 1992, pp. 45-58.

[*2*]  *Friction and Fretting Wear Testing Source Book: Selected References from ASTM Standards and ASM Handbooks*, Published jointly by ASTM, West Conshohocken, Pennsylvania, USA and ASM, Metals Park, Ohio, USA, 1997.

[*3*]  Blinn, M.P. and Lipkin, J.M., "Development Of A High-Temperature-Steam Sliding Wear Test Apparatus," *Tribology 98, Sixth International Tribology Conference of the South African Institute of Tribology*, March, 1998.

# Surface Treatments

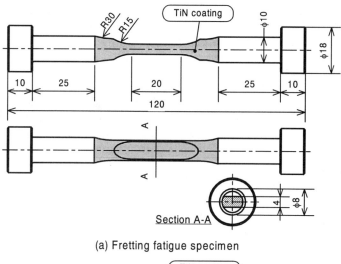

(a) Fretting fatigue specimen

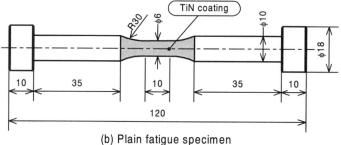

(b) Plain fatigue specimen

Fig.1-*Shapes and dimensions of the fatigue specimen.*

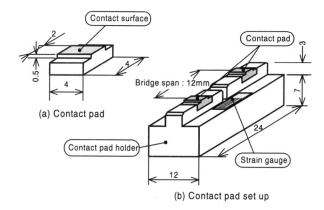

Fig.2-*Shapes and dimensions of the contact pad and contact pad holder.*

discharge in nitrogen into which titanium was evaporated at a constant substrate temperature of 623K. The thickness of the TiN film was about 3 μm. Mechanical properties of the TiN coated specimen are also shown in Table 2.

*Contact pad*

The shapes and dimensions of the contact pad and contact pad holder are shown in Fig. 2. The contact pads were machined from the same material as the specimen and were used without TiN coating. Contact surface of the pad was polished by using alumina slurry with a grain size of 0.3 μm. For the fretting fatigue tests, contact pads were set in the pad holder as shown in Fig.2 (b).

*Fretting Fatigue Test and Plain Fatigue Test*

Fretting fatigue tests and plain fatigue tests were carried out using an electric hydraulic fatigue testing machine. The fretting fatigue test apparatus is schematically illustrated in Fig.3. The contact pads setting into the contact pad holder (see Fig.2) were pressed onto the

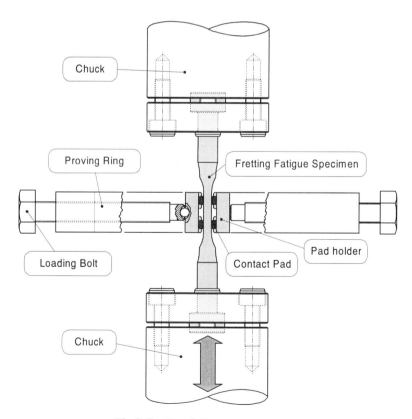

Fig.3-*Fretting fatigue test apparatus*

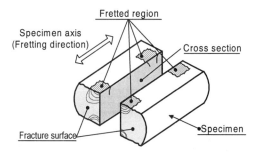

Fig.4-*Cutting procedure of the specimen for cross sectional observation.*

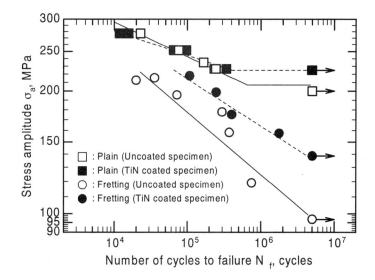

Fig.5 - *S-N curves for fretting fatigue of TiN coated steel*

gauge section of the fretting fatigue specimen by using the proving ring. Frictional force between the specimen and the contact pad during the fretting fatigue test was measured using a strain gauge attached to the contact pad holder as shown in Fig.2.

Plain and fretting fatigue tests were carried out under a load-controlled sinusoidal wave form condition with a stress ratio $R(\sigma_{min}/\sigma_{max})$=-1, and frequency f=20Hz in air at room temperature. The contact pressure was controlled at constant value of 100 MPa.

Fretted surfaces were observed in detail on a specimen tested up to a specified number of cycles, using a scanning electron microscope (SEM). The cross section of the specimen around the fretted region was also observed using SEM to discuss initiation and growth of the fretting cracks. Cutting procedure of the specimen for cross sectional observation is schematically illustrated in Fig.4.

**Results and Discussion**

*Plain and Fretting Fatigue Strength*

The relationship between applied stress amplitude and number of cycles to failure (S-N curve) is shown in Fig.5. In the figure, data points with an arrow show that no failure occurred up to the testing cycles of $5\times10^6$ cycles.

It can be seen from this figure that TiN coating film does not affect the plain fatigue strength in a finite-life region. But the fatigue limit of the TiN coated specimen which is defined at $5\times10^6$ cycles increases about 20 MPa as compared with that of the uncoated specimen. The main reason for the improvement in fatigue strength by TiN coating was that fatigue crack initiation is delayed by the hard coating film on the specimen surface, which can act as a barrier to the egress of dislocations [1-3].

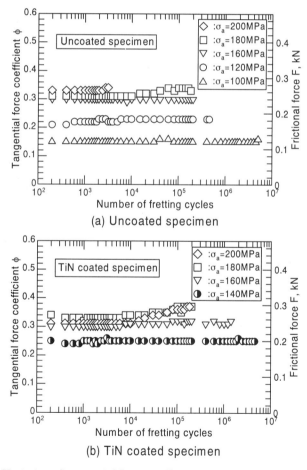

(a) Uncoated specimen

(b) TiN coated specimen

Fig.6-*Variation of tangential force coefficient during fretting fatigue test.*

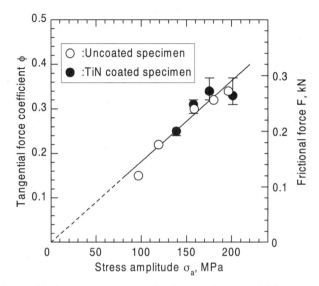

Fig.7- *Relationship between stress amplitude* $\sigma_a$ *and tangential force coefficient* $\phi$.

Fretting fatigue strengths of the uncoated specimen (○) and TiN coated steel (●) were significantly reduced when compared with the plain fatigue strength. However, the fretting fatigue strength was clearly improved by TiN coating. The fretting fatigue strength of the uncoated specimen at $5 \times 10^6$ cycles was around $\sigma_a = 100$ MPa and that of the TiN coated specimen was around 140 MPa. About 40% increase in fretting fatigue strength was observed in the TiN coated specimen, as compared with the uncoated specimen.

*Tangential Force Coefficient (Frictional Force)*

Variation of tangential force coefficient $\phi$ (=F/P) during fretting fatigue tests is shown in Fig.6, where F is the frictional force between specimen surface and contact pad and P is the contact load. Under the condition of stress amplitude $\sigma_a$ below 180 MPa, $\phi$ was almost constant during fretting fatigue tests. On the other hand, $\phi$ was slightly increased with the number of fretting cycles in both uncoated and TiN coated specimens for the region in the high stress amplitude.

The relationship between the stress amplitude $\sigma_a$ and the mean values of the tangential force coefficient $\phi$ during the fretting fatigue test is shown in Fig.7. The tangential force coefficients of both uncoated and TiN coated specimens show a proportional relationship with applied stress amplitude. And there was no difference in the values of $\phi$ between uncoated and TiN coated specimen.

*Observation of Fretted Surfaces and Fretting Fatigue Cracks*

Damage on the fretted specimen surface during the fretting fatigue process was observed by SEM. The results obtained under a condition of $\sigma_a = 200$ MPa are shown in Fig.8. Figure

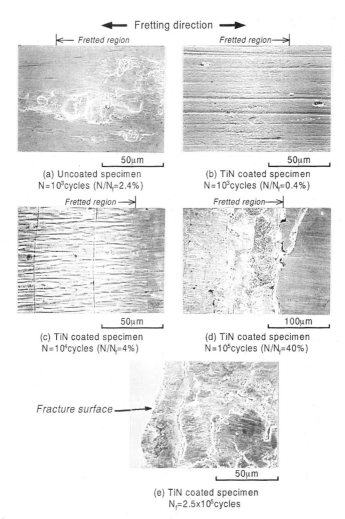

Fig.8-*SEM micrographs of fretted surface for the specimen tested at* $\sigma_a$ =200 *MPa.*

8(a) shows a fretted surface of the uncoated specimen interrupted at N=$10^3$ cycles which is 2.4% of the fretting fatigue life $N_f$. The microcracks induced by fretting action were observed on the specimen surface. Figure 8(b) shows a fretted surface of the TiN coated specimen interrupted at $10^3$ cycles (N/$N_f$=0.4%), where the observation was conducted from a 45-degree oblique direction in order to clearly observe the marks. Though slight fretting wear damage was found on the TiN film, there were no microcracks as found in Fig.8(a). Figure 8(c) shows a fretted surface of the TiN coated specimen interrupted at $10^4$ cycles (N/$N_f$=4%). It is clearly found that the fretting wear damage is severe as compared with that at $10^3$ cycles, and there are some flaws in the TiN film induced by fretting action. Figure 8(d) shows a fretted surface of the TiN coated specimen interrupted at $10^5$ cycles (N/$N_f$=40%). At the edge

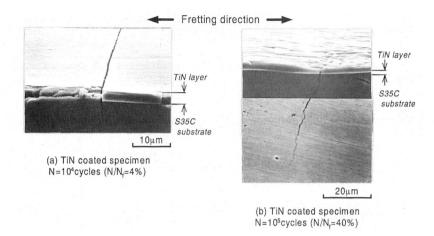

Fig.9-*Observation of fretting cracks for TiN coated specimens tested at* $\sigma_a=200$ *MPa.*

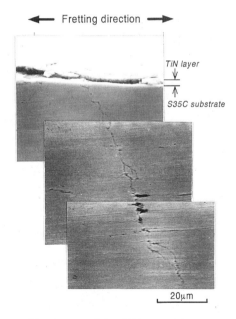

Fig.10 - *Observation of fretting crack for TiN coated specimen tested at* $\sigma_a=140$ *MPa.*

of fretted region, more severe wear and spalling of the TiN film occurred, and some fretting cracks were observed. Figure 8(e) shows a typical example of specimen surface after a fretting fatigue fracture of the TiN coated specimen. The fracture occurred from one of the surface cracks as found in Fig.8(d).

SEM micrographs of a cross section (see Fig.4) of the TiN coated specimen around the fretted region tested under a condition of $\sigma_a$=200 MPa are shown in Fig.9. Figure 9(a) is an observation of the specimen interrupted at $10^4$ cycles (N/N$_f$=4%). A small crack in the substrate was seen just below flaw in the TiN coated film. Figure 9(b) shows a specimen interrupted at $10^5$ cycles (N/N$_f$=40%). The fretting crack length of about 35 μm was observed at the edge of the fretted region where the fretting wear was severe.

Figure 10 shows a cross sectional observation of the TiN coated specimen run-out under a condition of $\sigma_a$=140 MPa. It is found that the fretted region was severely worn and fretting crack length of about 90 μm was observed clearly.

*Fretting Fatigue Fracture Process in TiN Coated Steel*

From the experimental results and observations of the fretted region in the specimen, the fretting fatigue fracture process of uncoated steel will be discussed. The fretting cracks initiate at a very early stage of fatigue life and propagate in consequence of cyclic stress and frictional force. Therefore, almost the whole of fretting fatigue life is spent in propagating the fretting cracks as is observed with many metallic materials. It is well known that the tangential force (frictional force) is the most significant mechanical factor influencing the fretting fatigue properties, and the fretting fatigue life decreases with increasing the frictional force. In other words, when the frictional force is the same level the fretting fatigue lives will be similar.

In this study, there was no difference in the tangential force coefficient (frictional force) between the uncoated specimen and the TiN coated specimen as shown in Fig.7. However, the fretting fatigue lives of the TiN coated specimen were increased 4-10 times compared to those of uncoated specimen. It is suggested from this fact that fretting fatigue properties of the TiN coated steel are not explained by only the magnitude of frictional force. The fretting fatigue cracks in the uncoated specimen were initiated directly on the specimen surface by an accumulation of fretting damage. On the contrary, TiN coating film protects the substrate from fretting damage and improves crack initiation life on the substrate. The crack initiation life in the TiN coated specimen was not equivalent to that in the uncoated specimens, even when the frictional force is equal with both materials.

From the observations of the fretting fatigue process in the TiN coated specimen as shown in Fig.8 and 9, it was observed that the fatigue crack in the substrate occurred from the flaws in the TiN film which were induced by fretting action. The flaws in the TiN film occurred during fatigue due to an accumulation of the deformation of substrate metal. The fretting damage and fatigue of the coating film act as notches with crack initiating on the substrate. Initiation of fretting fatigue cracks may depend on the fracture of TiN film, and the mechanism of fretting fatigue crack initiation in TiN coated steel was different from that in uncoated steel. Under a testing condition that the fracture in TiN film did not occur during fretting fatigue, it is considered that fretting fatigue crack in substrate is induced by the mechanism on uncoated metal mentioned above after wearing out the TiN coating film, and the superior improvement of fatigue life time is expected.

The fretting fatigue fracture process in TiN coated steel is summarized as follows. A fretting crack initiates at a certain stage of fatigue life from the flaws or worn region in TiN film damaged by fretting action, and propagates until final unstable fracture occurs. The improvement of fretting fatigue strength by TiN coating resulted mainly from the retardation of fretting fatigue crack initiation due to the existence of hard TiN film on the contact surface of the specimen.

## Conclusions

To investigate the effect of ceramic coating on fretting fatigue behavior, fretting fatigue tests were carried out using 0.37wt.% carbon steel coated with TiN by the PVD method. The main results obtained are summarized as follows:

1. The fretting fatigue strength of 0.37wt.% carbon steel was significantly reduced as compared with the plain fatigue strength. The fretting fatigue strength was improved about 40% by TiN coating onto the specimen.
2. The fretting fatigue fracture process in TiN coated carbon steel is as follows. Fretting action induces the flows in TiN film which are the origins of fatigue cracks in substrate metal and the cracks propagate until final unstable fracture occurs.
3. The main reason for improving the fretting fatigue strength by TiN coating is the retardation of fretting fatigue crack initiation due to the existence of hard TiN film on the contact surface of the specimen.

## *Acknowledgments*

The authors greatly thank Dr. K. Kanda, Fujikoshi Co.Ltd., for kindly supplying coating treatment for the specimen. This study was supported by a Grant-in-Aid for Encouragement of Young Scientists (No.09750106) from the Ministry of Education, Science and Culture, Japan.

## References

[1]    Shiozawa, K. and Ohshima, S., "Effect of TiN Coating on Fatigue Strength of Carbon Steel," *Journal of The Society of Materials Science Japan*, Vol.39, No.442, 1990, pp.927-932.

[2]    Shiozawa, K., Nishino, S. and Handa, K., "The Influence of Applied Stress Ratio on the Fatigue Strength of TiN-Coated Carbon Steel," *JSME International Journal, Series A*, Vol.35, No.3, 1992, pp.347-353.

[3]    Shiozawa, K., Tomosaka, T., Han, L. and Motobayashi, K., "Effect of Flaws in Coating Film on Fatigue Strength of Steel Coated with Titanium Nitride," *JSME International Journal Series A*, Vol.39, No.1, 1996, pp.142-150.

[4]    Edited by Waterhouse, R.B., "Fretting Fatigue," *Applied Science Publishers Ltd. London*, 1981.

[5]    Attia, M.H. and Waterhouse, R.B., editors, *Standardization of Fretting Fatigue Test Methods and Equipment, ASTM STP 1159*, 1992.

[6]    Nakazawa, K., Takei, A., Kasahara, K., Ishida, A. and Sumita, M., "Effect of Ni-TiC Composite Film Coating on Fretting Fatigue of High Strength Steel," *J. Japan Inst. Metals*, Vol. 59, No.11, 1995, pp.1118-1123.

[7]    Sato, K. and Kodama, S., "Effect of TiN Coating by PVD and CVD Processes on Fretting Fatigue Characteristics in Steel, " *Fretting Fatigue, ESIS 18* (Edited by R.B.Waterhouse and T.C.Lindley) 1994, Mechanical Publications, London, pp.513-526.

[8]    Batchelor, A.W., Stachowiak, G.W., Stachowiak, G.B., Leech, P.W. and Reihold, O., "Control of Fretting Friction and Wear of Roping Wire by Laser Surface Alloying and Physical Vapour Deposition Coatings," *Wear 152*, 1992, pp.127-150.

[9]    Blanpain, B., Mohrbacher, H., Liu, E., Celis, J.P. and Roos, J.R., "Hard Coatings Under Vibrational Contact Conditions," *Surface and Coating Technology*, Vol.74, No.75, 1995, pp.953-958.

Masanobu Kubota,[1] Kentaro Tsutsui,[2] Taizo Makino,[3] and Kenji Hirakawa[4]

**The Effect of the Contact Conditions and Surface Treatments on the Fretting Fatigue Strength of Medium Carbon Steel**

**REFERENCE:** Kubota, M., Tsutsui, K., Makino, T., and Hirakawa, K., **"The Effect of the Contact Conditions and Surface Treatments on the Fretting Fatigue Strength of Medium Carbon Steel,"** *Fretting Fatigue: Current Technology and Practices, ASTM STP 1367,* D. W. Hoeppner, V. Chandrasekaran, and C. B. Elliott, Eds., American Society for Testing and Materials, West Conshohocken, PA, 2000.

**ABSTRACT:** Fretting fatigue tests characterized with employing a contact bridge pad clamped to a fatigue specimen were conducted to find the effect of shapes of contact pads. The shapes of the contact edge of the pad used in the test were sharp and round. The relative slip amplitude during fatigue testing was measured and the effect of the location of the displacement sensor was discussed with the result of the stress analysis around the contact area. The effect of molybdenum coating on the fretting fatigue strength was also investigated.

It was found that by using the round edge pad, reproducible fretting scars could be obtained and also the round edge pad prevented the fatigue failure at the contact edge. A remarkable difference of the measured slip amplitude was found by the location of the displacement sensor. The molybdenum coating increases the fretting fatigue limit about 1.5 times that of the fretting fatigue strength of uncoated specimens. Fretting fatigue cracks always initiate at the inner site of the contact area. The results were explained from the results of contact stress analysis where the maximum shear stress occurred at the inner surface of the contact area in both types of the pad.

**KEYWORDS:** fretting fatigue, bridge pad, contact edge shape, stress analysis, sprayed molybdenum coating

## Introduction

The surface damage produced in service by fretting of mechanical components and structures takes various appearances depending on the configuration and load conditions. It has been considered very difficult to design these components based on the fretting

[1] Technical Assistant, Mechanical Science and Engineering Department, Kyushu University, 6-10-1 Hakozaki, Higasi-ku, Fukuoka, 812-8581 Japan.
[2] Graduate School of Kyushu University.
[3] Corporate Research and Department Laboratories, Sumitomo Metal Industries, Ltd., 1-8 Fuso-cho, Amagasaki, Hyogo, 660-0891 Japan.
[4] Professor, Department of Mechanical Science and Engineering, Kyushu University.

fatigue data obtained by standard fretting fatigue testing of which some simple and small size specimens were employed. Therefore, to get a precise fretting fatigue strength, large scale fatigue testing has been conducted by using actual components or near-full-scale models [1-4]. Large scale fatigue testing has a merit to simulate the fretting fatigue conditions such as slip amplitude, contact stress distribution and counterpart materials. Since large scale fatigue testing is time consuming and expensive, it has been the objective of many researchers to simulate the fretting fatigue conditions of the actual components by a simple and small size fatigue testing [5-8]. Several investigators have conducted fretting fatigue testing using a bridge type pad which is particularly convenient for studying the effect of slip amplitude and contact stress. With these small size fretting fatigue tests, one can easily control and measure the fretting conditions such as relative slip, contact pressure and fatigue loading; however, it is not sure that if the combination of these factors simulates the conditions of the large scale actual component. For example, many researchers reported that fretting fatigue cracks initiate at the contact edge of the pad but in-service fretting cracks usually initiate at the inner side of contact. This is because the bridge type pad testing did not simulate the contact pressure distribution. To overcome these problems, it is necessary to analyze the stress conditions of the contact pad type fatigue testing and join the results of small-modeled testing to that of a full-scale test.

The authors have made a fretting fatigue test using a press-fitted axle assembly 40mm in diameter [9]. In this large-scale fatigue testing, fatigue cracks always initiated at the inner surface of the fretted area. The bridge pad type fretting fatigue test was also made with the pad of sharp edge. In the test, fatigue cracks frequently initiated at the edge of the contact area. Furthermore, it was found that it is difficult to control precisely the contact condition in this type of testing. The shape of the fretting scar observed after the fretting fatigue test was not always the same as that expected from the geometry of the specimen and it varied by each experiment.

From the point stated above, a fretting fatigue test was performed by using a bridge type pad with different contact edge shapes, and the two-dimensional stress analysis was carried out to clarify the fretting fatigue conditions of this type of testing.

Press-fitted axle such as wheel-sets for railway is one of the typical components which suffers from fretting damage. Many countermeasures have been made to prevent the fatigue failure of axle assemblies such as material heat-treatment, shape of contact part, and inspection of fatigue cracks. One of the candidates to improve the fretting fatigue strength of axle assemblies may be to separate the contact using a coating on the surface of the axle. Taylor and Waterhouse have investigated the effect of the sprayed molybdenum coating against fretting fatigue [10]. The result showed that the coating proved to be quite successful. To confirm the effect of sprayed molybdenum coating, bridge type fretting fatigue was conducted with specimens having flats machined parallel where coated with two kinds of thicknesses.

**Procedures**

Fretting fatigue tests were carried out using a servo-hydraulic fatigue machine (load capacity 98kN). Two types of fretting bridge pads used in the experiment are shown in Fig. 1. A conventional pad designated square type in this study has sharp corners at the

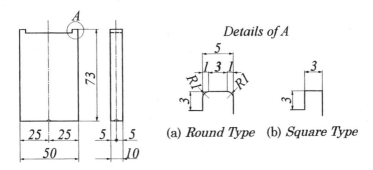

Details of A

(a) *Round Type*    (b) *Square Type*

Figure 1 — *Shapes of Fretting Bridge Pads*

foot of the pad. The round type pad which has a round contact edge was also used. The corner rounding of contact edge was precisely milled by an end mill tool. The tolerance of length of feet was within 2%. Since the contact length is large enough as compared with corner radii, the change of contact area due to fretting wear or the change of nominal contact pressure is very small. The fatigue specimen is shown in Fig. 2. The specimen consisted of a 15mm diameter bar with two parallel flats. The flat portions on the specimen and the contact surface of the fretting bridge were ground.

The material of the fretting bridge pad was high carbon steel whose detailed chemical composition is given in Table 1. Mechanical properties are shown in Table 2. Both tables also show the properties of the material of the fatigue specimen. The fatigue specimen is medium carbon steel. This combination of pad and specimen corresponds to the Japanese high-speed railway axle assembly.

All the tests were conducted with axial loading and a stress ratio ($R$) of –1 using a sinusoidal wave at a frequency of 20Hz in air at room temperature. Contact pressure was kept constant at 70MPa.

The sprayed molybdenum specimen was made by the flame spraying method. The coating was ground to the dimensions required to 0.25mm and 0.5mm in the thickness. The fretting bridge pad for the coated specimen is the round type.

In order to estimate shear stress and contact stress along the contact surface between the fretting pad and the fatigue specimen, two-dimensional stress analysis was

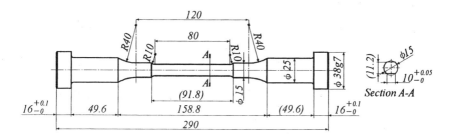

Figure 2 — *Dimensions of Fatigue Specimen*

Table 1 — *Chemical Compositions of the Materials*

| | C | Si | Mn | P | mass% S |
|---|---|---|---|---|---|
| Specimen (JIS S38C*) | 0.38 | 0.28 | 0.76 | 0.009 | 0.008 |
| Bridge (JIS SSW-1*) | 0.63 | 0.18 | 0.74 | 0.017 | 0.005 |

*JIS S38C and SSW1 are corresponding to AISI 1038 and AISI 1065, respectively.

Table 2 — *Mechanical Properties of the Materials*

| | Y.P.(MPa) | T.S.(MPa) | El.(%) | R.A.(%) |
|---|---|---|---|---|
| Specimen (JIS S38C) | 433 | 679 | 25.8 | 70.4 |
| Bridge (JIS SSW-1) | 361 | 779 | 19.5 | 34.7 |

done by the finite element method. For symmetry reason only one-quarter of the fretting assembly is studied. The boundary conditions in the finite element model are shown in Fig. 3. For the contact load, a concentrated force was loaded on to the center of the fretting pad normal to the contact surface. The value of the load corresponded to the load of 70MPa in nominal contact pressure. Mechanical loading is applied as an axial fatigue load over the left side of the specimen. The materials of the fretting bridge pad and the specimen were assumed to be elastic. The elastic modulus was $2.1 \times 10^5$MPa and Poisson's ratio was 0.3.

**Experimental Results**

*S-N Curves*

Figure 4 shows the results of the fretting fatigue tests and the unfretted fatigue test. Each data point represents one test. The plain fatigue limit was 280MPa. The fatigue limit under fretting was 90MPa, which was lower than one-third of the plain fatigue. There is little difference in the *S-N* curves between the round and the square type pad conditions.

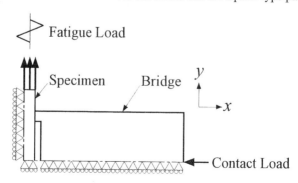

Figure 3 — *Boundary Conditions in Finite Element Model*

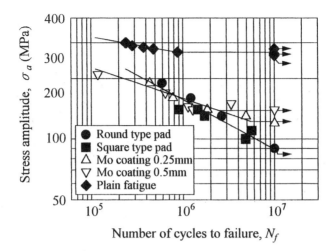

Figure 4 — *Results of Fretting Fatigue Tests and Plain Fatigue Test*

The fretting fatigue results of the sprayed molybdenum coatings are shown in the same figure. The fatigue limits of 0.25mm and 0.5mm coating thicknesses were 120MPa and 140MPa, respectively. The thicker coating showed a higher fatigue limit but the endurance cycles of both specimens were nearly the same. The sprayed molybdenum coating increases the fretting fatigue strength up to 140MPa, and over 50% improvement could be obtained. The endurance cycles of both coated specimens are the same as that of the uncoated specimen. This shows that the porous microstructure of the coating has no effect on the endurance cycles under fretting condition. It is considered that the coating has little effect to prevent crack growth, but can prevent crack initiation.

*Fretting Scar*

Figure 5 shows the photographs of the fretting damage with a low magnification microscope. Figure 5(a) is the fretting scar using the square type pad on the uncoated steel after $1.45 \times 10^6$ cycles. The fretting scar was localized in the intended contact surface. The fatigue cracks which grow to failure were at the boundary between the fretted and unfretted area. When the fretting scars cover the whole contact area, the cracks developed at the inner surface of the fretted area. Although the damaged area is different from each experiment, the *S-N* data of the square type bridge did not scatter remarkably. It is considered that the development of high stress at the boundary between the damaged and undamaged area, namely the stress concentration at the edge of the contact area, dominates the initiations of the fretting fatigue cracks.

Figure 5(b) shows a typical fretting scar after $1.21 \times 10^6$ cycles of fretting using the round type pad. The boundary of the fretted area is corresponding to the edge of the pad. When the round type pad was used, the fatigue crack did not initiate at the boundary but initiated at the inner surface of the fretted area. Suitable contact was obtained and good

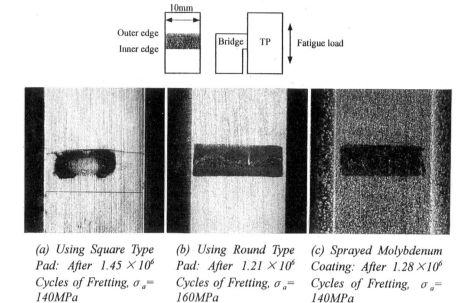

*(a) Using Square Type Pad: After 1.45 × 10⁶ Cycles of Fretting, σ ₐ= 140MPa*

*(b) Using Round Type Pad: After 1.21 × 10⁶ Cycles of Fretting, σ ₐ= 160MPa*

*(c) Sprayed Molybdenum Coating: After 1.28 × 10⁶ Cycles of Fretting, σ ₐ= 140MPa*

Figure 5 — *Specimen Surface after Fretting*

reproducibility also by the test using the round type pad.

Figure 5(c) shows a fretting scar appeared on the sprayed molybdenum coating after approximately $1.28 \times 10^6$ cycles of fretting. The scar is quite clear, but it can be seen that the coating itself has suffered a little damage. The discoloration area was made by fretting wear of the coating surface and the adherence of wear particles from the bridge.

**Discussions**

*Fretted Surface Profiles*

Figure 6 shows the surface roughness of the fretted area. Figure 6(a) is a profile using the square type pad after approximately $1.45 \times 10^6$ cycles when the fretting scar is localized in the contact area. It is observed that fretting wear is less in the center of the damaged area than on both sides of the contact area. The inner edge of the contact was raised from the profile before testing due to plastic flow [11].

The surface profile when using the round type pad is shown in Fig. 6(b) after approximately $1.21 \times 10^6$ cycles of fretting. In this case, the whole geometrical contact zone is fretted. The profile is different from the one obtained by the sharp corner pad.

The contact surface of the sprayed molybdenum coating after approximately $1.28 \times 10^6$ cycles is shown in Fig. 6(c) when the round type pad was used. This profile is very rough due to the porosity of the coating. The surface roughness of the coated specimen did not change by fretting. The crevice became shallow due to the accumulation of the fretting debris. Since the hardness of the coating is 4 times harder than that of the pad material, the pad surface suffered from fretting wear more than the coated surface.

*Amount of Relative Slip*

Relative slip between the pad and the specimen was measured by an eddy current type displacement sensor. The sensor was fixed to the pad at 10mm up from the contact surface. The axial distance from this sensor point to the specimen surface at the end of

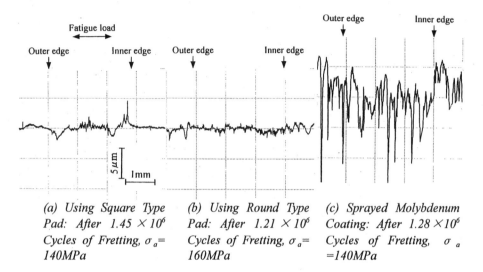

*(a) Using Square Type Pad: After 1.45 × 10⁶ Cycles of Fretting, σₐ= 140MPa*

*(b) Using Round Type Pad: After 1.21 × 10⁶ Cycles of Fretting, σₐ= 160MPa*

*(c) Sprayed Molybdenum Coating: After 1.28 ×10⁶ Cycles of Fretting, σₐ =140MPa*

Figure 6—*Surface Profiles of the Specimens after Fretting*

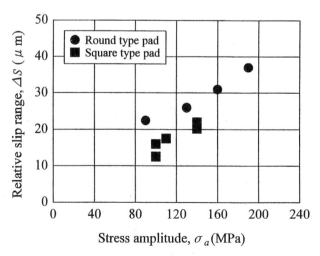

Figure 7—*Relationship between Relative Slip Range and Stress Amplitude*

contact was measured. The results are shown in Fig. 7. The range of the relative slip was approximately 10μm up to 40μm when the nominal stress of the specimen was between 90MPa and 190MPa.

However, as shown in Fig. 8, the amount of the relative slip calculated by the finite element method is much lower than that obtained by the experiment.

Assuming the friction coefficient being zero, the relative slip should be almost equal to the elongation of the specimen. The calculated elongation is approximately 12μm under loading of 100MPa, and 11μm was measured by the sensor. When the friction coefficient is taken into account, the relative slip must be smaller than 11-12μm

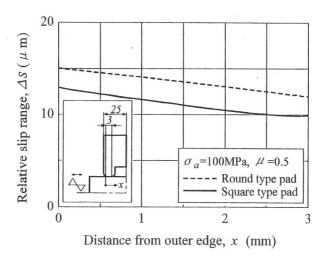

Figure 8 — *Relative Slip Range Estimated from Finite Element Analysis*

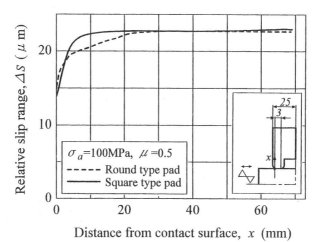

Figure 9 — *Nominal Increase of Relative Slip Range Related to Sensor Fixed Position*

for applied stress being 100MPa. Since the measured 10-40μm is too large, some modification from the stress analysis is needed.

Figure 9 shows the relative slip calculated as a function of the distance from the contact surface to the location of the displacement sensor. As shown in the figure, the relative slip calculated increased as the distance increased. This means that the bridge foot is deformed by frictional force. When the measured relative slip is modified by the results of Fig. 9, the relative slip concurs with that calculated with 2% tolerance.

*Distributions of Contact Stress*

The distribution of the contact stress along the contact surface is shown in Fig.10, where the round type and square type pads were used. As shown in the figure, the contact

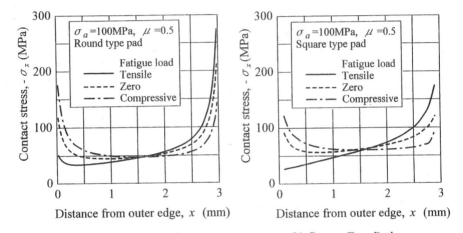

(a) *Round Type Pad*                    (b) *Square Type Pad*

Figure 10 — *Distributions of Contact Stress along Contact Surface of Specimen*

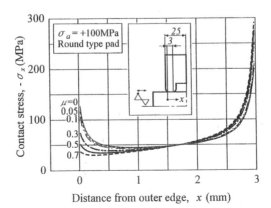

Figure 11 — *Effect of Friction Coefficient on Contact Stress*

stress takes its peak value at the edge of the contact surface when the round type pad was used. When tensile axial stress is applied to the specimen, the peak contact stress at the inner edge of the bridge foot increases and it decreases at the outer edge of the foot. On the contrary, the peak contact stress decreases at the inner edge of the bridge foot and it increases at the outer edge when compressive axial stress is applied to the specimen. When the square type pad was used under a tensile axial stress, the contact stress gradually decreases as the distance from the outer edge of the bridge decreases.

Figure 11 shows the distribution of the contact stress along the contact surface where the friction coefficients are taken as the parameter. It is noted that the distribution does not change by applying axial stress when the friction coefficient is zero. The peak contact stress at the outer contact edge decreases with the increase of the friction

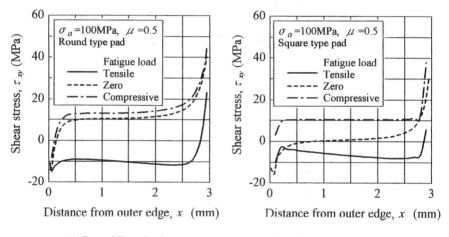

(a) *Round Type Pad*          (b) *Square Type Pad*

Figure 12 — *Distributions of Shear Stress along Contact Surface of Specimen*

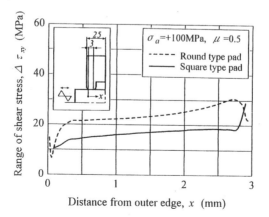

Figure 13 — *Relationship between Range of Shear Stress and Distance from Outer Edge*

coefficient, however, the reverse effect can be observed at the inner edge of the bridge.

*Distribution of Shear Stress*

Shear stress distributions of the specimen along the contact surface are shown in Fig.12. The shear stress rapidly increases at the end of the contact for both types of contact pads. The ranges of shear stresses during push-pull loading are illustrated in Fig. 13. As shown in the figure, the shear stress ranges are higher in the middle of the contact zone than those at the edges of contact. It is noted that fatigue cracks could initiate at the

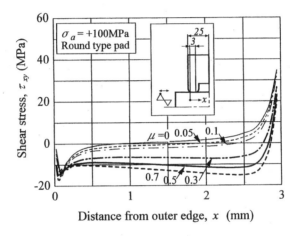

Figure 14 — *Effect of Coefficient of Friction on Shear Stress*

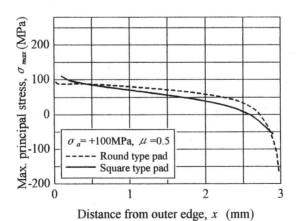

Figure 15 — *Distribution of Maximum Principal Stress along Contact Surface of Specimen*

inner side of contact assuming the fatigue crack initiation is first related to the shear stress. The experimental results show that fatigue cracks initiate at the inner side of the contact zone about 1000μm-1500μm from the outer contact edge. Figure 14 shows the effect of friction coefficient on the shear stress distributions. As the friction coefficient increases the absolute value of the shear stress increases.

The maximum principal stress distribution along the contact surface of the specimen is shown in Fig. 15. The maximum principal stress rapidly decreases at the inner edge of contact pad.

*Fatigue Crack Initiation Behavior of the Coating*

Figure 16 shows the fatigue crack obtained by fretting fatigue testing of molybdenum coated specimens. The fatigue crack initiated at the surface of the coating and propagated obliquely into the specimen until it reached the boundary of coating and base metal. The crack then propagated into the base metal perpendicular to the axial stress. The delamination or other cracks of the coating could not be observed.

The elastic modulus of the molybdenum coating was estimated from the tensile test of the coated specimen. It was known that the value could be about 1/200 of the elastic modulus of steel. The results can be explained from the porous structure of the coating. Because of the low elastic modulus of the coating, the contact surface stress could be reduced. Fretting fatigue strength of the coated specimen increased over 50% of the strength of non-coated specimen. The reason for the high resistance of molybdenum coating against fretting fatigue can be attributed to the stress relief and high resistance against fretting wear.

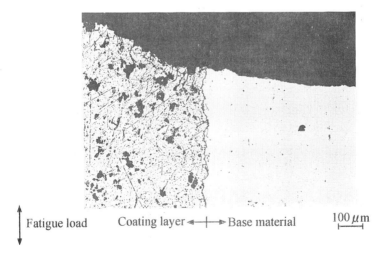

Fatigue load    Coating layer ◄──┼──► Base material    $\underset{\vdash\!\!-\!\!-\!\!\dashv}{100\,\mu m}$

Figure 16 — *Fatigue Crack of Molybdenum Coated Specimen Observed by Scanning Electron Microscope*

**Conclusions**

Fretting fatigue tests employing a contact pad clamped to a fatigue specimen was conducted to find the effect of shapes of contact pads. The shapes of the contact edge of the pad used in the test were sharp and round. The effect of molybdenum coating on the fretting fatigue strength was also investigated. Two-dimensional stress analysis of the fretting area was made to interpret the meanings of the bridge type fretting fatigue testing. The results obtained are summarized as follows.

1.  By using the round type bridge pad, reproducible fretting scar can be obtained in the experiments.

2.  Fretting fatigue cracks initiate at the inner side of the contact by the round type pad.

3.  The fretting fatigue limit of the sprayed molybdenum coating specimen was improved to 1.5 times that of the uncoated specimen. In the specimen fractured by fretting fatigue, no delamination or other cracks in the coating was observed. The reason for the higher performance of the molybdenum coating against fretting fatigue is attributed to the stress relief effect and high resistance against fretting wear.

4.  The contact stress takes its peak value at the edge of the contact surface when the round type pad was used. The shear stress rapidly increases at the end of the contact for both types of contact pads. The ranges of shear stresses during push-pull loading are higher in the middle of the contact zone than those at the edges of contact.

**References**

[1]   Hoger, O. J. "Influence of Fretting Corrosion on the Fatigue Strength of Fitted Members", *ASTM STP 144*, ASTM, Philadelphia, 1952, pp.40-51.

[2]   Hirakawa, K., Toyama, K., and Kubota, M., "The Analysis and Prevention of Failure in Railway Axles", *International Journal of Fatigue*, 20-2, 1998, pp. 135-144.

[3]   Nakamura, H., Tanaka, S., Hatsuno, K., Yaguchi, S., and Mori, B., "Investigation of Fretting Fatigue using Real Axle Assembly – Part 1", *Transaction of the Japan Society of Mechanical Engineers*, *Series A*, 34-268, pp. 2092-2102, Japanese.

[4]   Makino, T., Yamamoto, M., and Hirakawa, K., "Fracture Mechanics Approach to the Fretting Fatigue Strength of Axle Assemblies", *Proceedings of International Conference on Materials and Mechanics '97*, Materials and Mechanics Division, The Japan Society of Mechanical Engineers, 1997, pp. 33-38.

[5]   Mutoh, Y. and Tanaka, T., "Fretting Fatigue in Several Steels and a Cast Iron", *Wear*, 125, 1988, pp.175-191.

[6]   "The Survey of Present Condition of High Cycle Fatigue and Prevention of Fatigue Failure", RC-130, The Japan Society of Mechanical Engineers, 1994, pp. 118-140, Japanese.

[7]   Nix, K. J. and Lindley, T. C., "The Influence of Relative Slip Range and Contact Material on the Fretting Fatigue Properties of 3.5NiCrMoV Rotor Steel", *Wear*, 125, 1988, pp.147-163

[8]   Kamaraj, M. and Kitsunai, Y., "Fretting Fatigue Strength of SNCM439 Steel", *Proceedings of International Conference on Materials and Mechanics '97*, Materials and Mechanics Division, The Japan Society of Mechanical Engineers, 1997, pp. 39-43.

[9]   Kubota, M., Nakahashi, J., Makino, T., and Hirakawa, K., "Fretting Fatigue Properties of Press-Fitting Axle under Variable Load", *Pre Print of the 74th JSME Spring Annual Meeting, The Japan Society of Mechanical Engineers*, pp. 11-12, Japanese.

[10]  Taylor, D. E. and Waterhhouse, R. B., "Sprayed Molybdenum Coatings as a Protection against Fretting Fatigue", *Wear*, 20, 1972, pp.401-407.

[11]  Nishioka K. and Hirakawa, K., "Fundamental Investigation of Fretting Fatigue - Part 3", *Transaction of the Japan Society of Mechanical Engineers, Series A*, 34-266, 1968, pp. 1635-1643, Japanese.

Swati Chakravarty,[1] Jason P. Dyer,[1] Joseph C. Conway, Jr.,[2] Albert E. Segall,[3] and Prakash C. Patnaik[1]

## Influence of Surface Treatments on Fretting Fatigue of Ti-6242 at Elevated Temperatures

**REFERENCE:** Chakravarty, S., Dyer, J. P., Conway, Jr., J. C., Segall, A. E. and Patnaik, P. C., **"Influence of Surface Treatments on Fretting Fatigue of Ti-6242 at Elevated Temperatures,"** *Fretting Fatigue: Current Technology and Practices, ASTM STP 1367,* D. W. Hoeppner, V. Chandrasekaran, and C. B. Elliott, Eds., American Society for Testing and Materials, West Conshohocken, PA, 2000.

**ABSTRACT:** Where fretting fatigue of titanium alloy gas turbine engine component is concerned, the effect of temperature on fretting processes can be quite significant. Service exposed third stage compressor blades, operating at 454°C or 850°F, showed severe signs of fretting, wear, CuNiIn coating delamination, oxidation and pitting on the dovetail pressure surfaces. Under cyclic loading conditions fretting fatigue cracks, initiated in these damaged areas, can cause dramatic reduction in fatigue life of these components leading to potential catastrophic failure. A study has been carried out to evaluate the influence of various surface treatments, such as ion treatments, coatings and shot peening, on the fretting fatigue life of titanium alloy gas turbine engine components at elevated temperature. Results from high temperature fretting fatigue and pin-on-disk wear tests are discussed in this paper. Significant contrast between the wear and fatigue test results illustrates the potential dangers of using sliding wear tests to rank surface treatments for fretting fatigue amelioration.

**KEYWORDS:** fretting fatigue, high temperature wear, gas turbine engines, titanium alloys, surface modification techniques, ion implantation, ion plating, oxidation

### Introduction

Titanium alloy components in the hot section of the gas turbine engines suffer degradation due to fretting wear, fatigue and fretting fatigue at elevated temperatures. One of the principal characteristics of fretting is that it is caused by a combination of

[1]Project manager, Aerospace engineer and Senior manager, respectively, Advanced Material and Energy Systems, Orenda Aerospace Corporation, 1420 Blair Place, Suite 608, Gloucester, Ont. Canada K1J9L8.
[2]Professor, Applied Research Laboratory, The Pennsylvania State University, 227 Hammond Building University Park, PA 16802-1401.
[3]Associate Professor of Mechanical Engineering, Washington State University-Vancouver, 14204 NE Salmon Creek Ave, Vancouver Washington, 98686-9600.

adhesion, abrasion and oxidation of the mating surfaces. In fretting fatigue, cracks are initiated within the abraded pits on the fretted surface and then propagate to failure as a result of the cyclic loading. This premature fatigue crack initiation can have a significant effect on the life of the rotating components, since for alloys such as titanium, the largest part of the fatigue life is taken up in the crack initiation stage.

The effect of temperature on the fretting process can be quite significant in the fretting fatigue of titanium alloy gas turbine engine components. For instance, mechanisms of fretting in the fan blades operating at near ambient temperatures are different than compressor blades which operate at temperatures ranging from 400 – 600 °C. As the operating temperature increases, the abrasion by oxide debris becomes the predominant (and severe) degradation mechanism. Many premature fretting fatigue failures have been observed in the titanium high-pressure compressor components of gas turbine engines. Service exposed third-stage compressor blades, which operate at 454 °C (850 °F), have been found to exhibit severe signs of fretting, wear, delamination of CuNiIn coating, as well as oxidation and pitting on the blade dovetail pressure surfaces.

Recognizing that the fretting process is initiated by local adjesion provides a fundamental tool in minimizing fretting [1]. In the case of high temperature fretting, it has long been understood that the surface oxide films that form naturally on metals can be very effective at reducing wear between contacting metal surfaces. Success depends on the spontaneous generation of thin, tenacious surface oxide films. Whatever the increased tenacity mechanism may be, it is reasonable to suppose that it would offer increased resistance to high temperature fretting in an oxidizing environment [2]. However, it is known that at temperatures less than 550 °C, titanium does not form a stable adherent oxide [3]. It is also likely at this temperature the oxide film which forms may be detrimental to fretting where it may detach, forming additional abrasive debris. Combination of surface treatments and lubricants is thus required to provide protection to titanium in the temperature range of 400 – 550 °C under fretting fatigue conditions.

The titanium alloy under study is Ti-6242 alloy used in the compressor section of the gas turbine engine at an operating temperature of 454 °C. This paper summarizes the results of a study where efforts were made to improve the fretting fatigue resistance of titanium alloys at elevated temperatures. Emphasis in this paper is placed on revealing the effect of a combination of various surface treatments on the high-temperature fretting fatigue and wear characteristics of Ti-6242 titanium alloy and evaluating the potential for improving the fretting fatigue life of the compressor components using these surface modification techniques.

## Selection of Surface Treatments for Fretting Fatigue Life Improvement of Ti-6242 Titanium Alloy at Elevated Temperature

Fatigue failure is usually characterized by cracks initiated at the surface. Accordingly, any benefit from surface treatments may produce in improving the fatigue life of a material, must stem only from the retardation of crack initiation. At high temperature under fretting fatigue condition the key to provide protection to titanium alloys is to develop adherent oxides with good frictional properties and low wear rate. Previous investigations have shown that, where a given situation resulted in a low coefficient of friction and low wear rate, the fretting fatigue strength of the material also showed a significant increase [4].

To prevent or reduce the impact of fretting on the fatigue life, several approaches may be taken. These approaches may range from treatments which improve the fatigue life, such as shot peening or surface hardening, to treatments such as coatings which reduce the friction or surface roughness between the two contact surfaces, thereby delaying the formation of wear debris. However, since most processes have their own unique advantages several can be used in combination to maximize their affect. Special consideration has to be taken in selecting the combination of coatings and treatment as it has been observed in a previous investigation that certain processes may reduce wear or increases fatigue life individually but when used in conjunction with another treatment produce detrimental effects [5].

Table 1 – *Potential Surface Treatments for High Temperature Wear and Fretting Fatigue Life Improvement of Ti-6242 Alloy*

| Pin-On-Disk Wear Test [1] | Fretting Fatigue Test [1] |
| --- | --- |
| Ti-6242 (Ti) | Ti-6242 (Ti) [2] |
| $Cr^+$(1.8 E17) Ion Implant (Cr-1) | Pt IBAD + Orenda lubricant (PtIBAD/OL) [2,3] |
| $Cr^+$(3.0 E17) Ion Implant (Cr-2) | $CuNiIn + MoS_2$ (CuNiIn/ DAG) [2,4] |
| $Ni^+$(1.5 E17) Ion Implant (Ni-1) | NiFeCr + Solid Film Lubricant1 (NiFeCr/SFL1) [2] |
| $Ni^+$(2.5 E17) Ion Implant (Ni-2) | NiMo + Solid Film Lubricant2 (NiMo/SFL2) [2] |
| Pt IBAD ( Pt-IBAD) | Orenda Chemical Treatment + Orenda lubricant (OCT/OL) [2,3] |
| Pt IBAD + Orenda lubricant (PtIBAD/OL) | |
| $CuNiIn + MoS_2$ (CuNiIn/ DAG) | |
| NiFeCr + Solid Film Lubricant1(NiFeCr/SFL1) | |
| NiMo + Solid Film Lubricant2 (NiMo/SFL2) | |
| Orenda Chemical Treatment + Orenda lubricant (OCT/OL) | |

[1] All specimens for both the tests were shot peened prior to surface treatments.
[2] Fretting fatigue specimens were surface treated and the bridges were just shot peened.
[3] Fretting bridges were shot peened and OL treated.
[4] Fretting bridges were shot peened and DAG treated.

The matrix of surface treatments evaluated in the current investigation includes ion treatments, coatings and solid film lubricants, (Table 1). All of the treatments were provided with commercial level processes. Shot peening has been established as a standard treatment for improving the fretting fatigue life of titanium alloy fan and compressor blades and disks by inducing compressive residual stresses and increased surface hardness. In this study, therefore, all the above surface treatments have been applied to shot peened Ti-6242 test specimens.

*Ion Treatments*

*Ion Implantation* - Ion implantation may be used to modify the surface chemistry and structure and may offer the means of improving a material's resistance to fatigue crack initiation. The damage induced by the ion beam is sufficient to develop dislocation tangles and residual stresses that contribute to the surface hardening. Also, since the stress amplitudes in high-cycle fatigue are usually much lower than the yield stress, the surface residual compressive stresses created in an implanted layer may increase the fatigue life [5]. On ferrous material a smooth oxide layer known as the "glaze oxide" develops in high temperature unidirectional sliding. It has been shown in the case of nickel-base alloy, the formation of this glaze oxide, results in the reduction of the wear rate and a great improvement in the fatigue strength [6]. Previous work by Waterhouse et al [4] has shown that $Ba^+$ implantation on titanium alloy, results in the formation of $BaTiO_3$, which has the perovskite structure similar to the spinal structure of the glaze oxide. Further oxidation of Ti alloy is inhibited by $Ba^+$ ions, which diffuses towards the dislocations and blocks the inward diffusion of oxygen. Ultimately, this is beneficial in that it inhibits fretting wear. BaTiO precipitates also lead to hardening by pinning the dislocation movement in the substrate material [7]. Overall, implantation with barium reduced the coefficient of friction and increased the fretting resistance of titanium alloys in the temperature range of 400 – 500°C [4]. Similar behavior is expected out of $Cr^+$ and $Pt^+$ ion treated titanium surfaces. Beneficial changes in terms of low coefficient of friction and wear rate, has also been observed by Waterhouse [4] with $Ni^+$ or $Bi^+$ ion implants in the temperature range of 400 – 600 °C.

Based on these observations, $Cr^+$ and $Ni^+$ ions have been considered for the ion implantation on Ti-6242 alloy. On the other hand, $Ba^+$ and $Pt^+$ ions cannot be considered, as they are not commercially viable. However for the case of platinum, ion plating using the Ion Beam Assisted Deposition (IBAD) process has been considered.

*Ion Plating* - Platinum ion plating of high temperature titanium alloys was originally developed to increase their creep and oxidation resistance. Further studies of the effect of this coating on other properties showed that it could also improve the high cycle fatigue strength at room and especially at elevated temperature [8]. Work by Fujishirio [9] demonstrated that a very thin layer of ion plated noble metal protects the alloy from oxidation and in addition improves both high temperature fatigue and creep strengths. Advantages of ion plating over conventional coating techniques are two-fold. First, the implanted elements become integrated with the substrate alloy surface over which coherent plating can subsequently be built up. Secondly, ion impingement by the coating materials may produce surface hardening which will improve the fatigue strength of the substrate alloy. Ion plating of Pt was selected because it forms a sound and effective coating without risk of hydrogen contamination or the formation of brittle inter-metallic compounds. Research showed that the Pt ion plating has excellent oxidation resistance and high thermal stability [9]. It is effective at very high temperature because of its low solubility in titanium. It provides an effective barrier for oxygen diffusion and results in improved creep resistance strength as well as high cycle fatigue strength of $\alpha+\beta$ titanium alloys in the range of 450 – 550 °C temperature range. The process of Ion Beam Assisted Deposition (IBAD) has been selected for Pt ion plating in this study. With concurrent ion bombardment of both the substrate surface and the target material to be sputtered, thin films with improved properties are produced. This technique has been shown to improve step coverage, modify thin film stress, increase packing density, minimize pinholes, promote crystalline growth, reduce absorbed residual gas contaminants and improve film adhesion and hardness.

*Coatings and Solid-film Lubricants* – Plasma-sprayed coatings of metals that form adhering oxide films and show lubricating properties have given favorable results in the past [2]. However, the major drawback of plasma processes is severe loss of high cycle fatigue strength of the parent material due to the process temperature [10,11]. These processes therefore need to be combined with other surface treatments to maintain the fatigue life of the substrate alloy and maximize the benefits of the combined protective surface treatment. Four different coating and lubricant combinations were tested for evaluating the effect of coatings on the fretting fatigue life of Ti-6242 alloy. CuNiIn + $MoS_2$ coating combination is currently being applied to the compressor, hence it was used as the baseline for this investigation.

**Experimental Procedures**

*High-Temperature Pin-On-Disk Wear Test*

A high-temperature tri-pin-on-disk sliding wear tester was used for a preliminary tribological analysis and screening of surface treated titanium alloy substrates. All sliding wear tests were conducted at 454 °C in air. However, because of the wear experienced by the pin and the resulting increase in the contact diameter during the tests, the net force on each pin was incrementally increased after each wear cycle in order to maintain an average contact pressure of 80-90 MPa between the pins and the disks. The tests were run at 120 rpm to maintain a pin velocity of approximately 0.050 m/s throughout the test. In order to ensure that the measured data reflected steady-state, wear behavior, an initial wear-in period was used to remove any large surface asperities on the pin and disk. Once the wear-in was completed and steady-state (linear) behavior observed, the weight loss of each disk was then measured at intervals of approximately 200 revolutions. The resulting wear data was then converted to steady-state data, volumetric wear rate per sliding distance, using linear regression.

Both the pins and the disks received shot peening treatment before testing. The surface treatments were applied only on the disks after shot peening. A comprehensive wear mode analysis was conducted using microscopy. Optical microscopy was performed on both the wear surface and through the base material extending from the wear surface. During this wear mode analysis, the wear disks were sectioned and the wear track and interior examined using microscopy to help determine the underlying wear mechanisms.

*Fretting Fatigue Test*

The fretting fatigue tests were carried out using an MTS servo-hydraulic test system, (Figure 1). The rig is a self-supporting system that uses a four-bolt frame to apply a contact pressure through two elliptical bridges. The bridges were aligned using ball bearings between the plates of the frame, and the average contact pressure was measured using an aluminum cylinder load cell. Strain gauges mounted on the specimen were used to ensure homogenous contact pressure by adjusting the bolt loads to equalize the strain readings. Once the initial load was applied and balanced using the strain gauges, the dog-bone shaped specimens were fretted and the bridge loads were adjusted at regular intervals until equilibrium was reached. Fretting fatigue tests were conducted at elevated temperatures by heating the specimen gauge section using quartz lamps mounted on the MTS frame.

The test conditions were selected to partially simulate the surface interaction of the blade/disk couple. The fatigue specimens were machined from Ti-6242 titanium alloy used in the compressor section of the aircraft engines. Optical microscopy and Scanning Electron Microscopy (SEM) were carried out on the fretted specimens for fretting damage analysis.

Figure 1 - *Fretting Fatigue Test Rig*

**Influence of Surface Treatment on Wear at Elevated Temperature**

Using the experimental procedure described in the previous section, the steady-state wear behavior of surface treated and baseline untreated Ti-6242 titanium alloy were measured. Wear testing in the dry sliding mode was conducted through the wear-in range and continued through the steady-state range. Wear rates were expressed as volume of material removed in mm$^3$ as a function of sliding distance in meters. The testing was carried out in two phases.

In phase-1 only the ion implanted and the platinum IBAD specimens were tested in order to evaluate the effect of ion treatments on the wear of titanium alloy at high temperature. The platinum IBAD treated specimens did not have any lubricant topcoat in phase-1. The wear rate data is shown in tabular form in (Table 2) and graphically in (Figure 2).

Table 2 – *Average Wear Rates for Various Ion Treatments*

| Ion Treatments | Wear Rate (mm$^3$/m) |
|:---:|:---:|
| Ti | 0.0224 |
| PtIBAD | 0.0119 |
| Ni-1 | 0.0148 |
| Ni-2 | 0.0142 |
| Cr-1 | 0.0157 |
| Cr-2 | 0.0129 |

*Note: 6 tests per condition, 9 tests for PtIBAD*

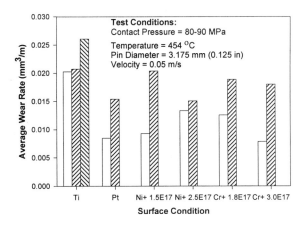

Figure 2 – *Sliding Wear Test Data for Various Ion Treatments*

Typically, steady state wear rate values for the ion treatments were all found to be in the 0.012 – 0.022 mm³/m range. The platinum IBAD treated specimens performed better than the ion-implanted samples. However, there was no significant improvement in wear performance seen for either of the surface treatments relative to the untreated shot peened titanium surfaces. This is most likely a result of the shallow depth of the surface treatments that precluded any significant improvements in wear performance after the initial wear-in period. Without the added wear resistance of the ion treatments, the wear data only reflected the presumably improved performance of work hardening and/or compressive residual stresses induced by the shot peening. The relatively linear wear data without any sudden increase in slope does suggest, however, that the full depth of the shot peen effected zone was not reached during the tests. These results are found to be in consistent with the work carried out earlier on fretting measurements of ion implanted shot peened titanium alloy at room temperature [5]. The results in that study also indicated that the contribution of ion implantation process in extending the fretting life is relatively insignificant when compared to the shot peening process. There does not appear to be much potential for the ion implantation treatment in extending the fretting fatigue life of titanium alloy components with rough surfaces.

In phase-2 wear tests were carried out for five surface treatments, including the platinum IBAD (PtIBAD/OL) treatment with lubricant top coating. Wear rates for each surface conditions are given in (Table 3) and are shown in (Figure 3).

Table 3 – *Average Wear Rates for Various Surface Treatments*

| Surface Treatments | Wear Rate (mm³/m) |
| --- | --- |
| NiFeCr/SFL1 | 0.0048 |
| PtIBAD/OL | 0.0031 |
| OCT/OL | 0.0045 |
| CuNiIn/DAG | Relatively small wear rate |
| NiMo/SFL2 | Wear Rate too high to collect any substantial data |

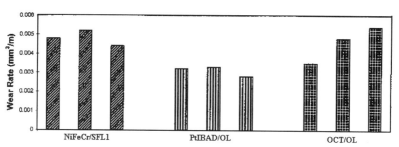

Figure 3 – *Pin-on-Disk Wear Test Data for Surface Treated Ti-6242 Alloy*

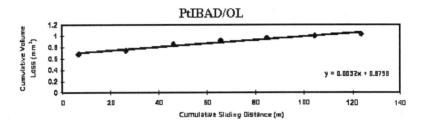

Figure 4 – *Typical Cumulative Volume Loss vs. Cumulative Sliding Distance Plot*

## Influence of Surface Treatment on Fretting Fatigue at Elevated Temperature

Fretting fatigue test results are given in (Figure 5). The CuNiIn/DAG coating

`DISK: Ti-6242, BLADE: Ti-6242, 454 °C, 5 Hz, R=0.25`

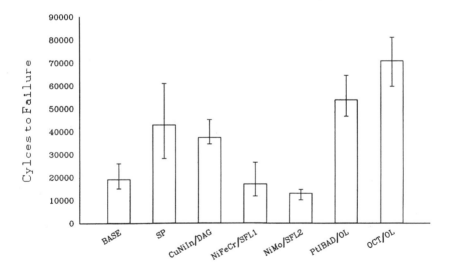

Figure 5 – *Fretting Fatigue Test Data for Surface Treated Ti-6242 Alloy*

configuration was used as the baseline to compare the relative performance of the other
surface treatments evaluated in this study. The improvement factor of all suggested
solutions in the test matrix is given in (Table 4). The PtIBAD/OL showed a marginal
improvement over the OCT/OL treatment and over a 40% improvement over the
CuNiIn/DAG coating. The OCT/OL treatment showed an almost 90% improvement over

the CuNiIn/DAG coating and about 3.7 times improvement over the parent material. Both NiFeCr/SFL1 and NiMo/SFL2 did not show any significant improvement in fretting fatigue life over the bare Ti-6242 and much lower fretting fatigue life than the baseline CuNiIn/DAG coating configuration.

Table 4 – *Relative Performance of Various Surface Treatments to CuNiIn/DAG for Improving Fretting Fatigue Life of Ti-6242 Alloy*

| Surface Treatments | Improvement Factor |
| --- | --- |
| **CuNiIn/DAG** | **1.00** |
| OCT/OL | 1.89 |
| PtIBAD/OL | 1.43 |
| Ti | 1.15 |
| NiFeCr/SFL1 | 0.45 |
| NiMo/SFL2 | 0.34 |

*Fretting Fatigue Damage*

Macro examination of the fracture surfaces showed that failure occurred almost invariably from one of the fretted regions and usually near the outer edge of the fretting faces, (Figure 6). The fracture surface of each specimen revealed extensive crack growth

Figure 6 – *Fretted Surface of Fretting Fatigue Test Specimen for CuNiIn/DAG*

regions, perpendicular to the applied load. Rapid failure occurred in ductile tension followed by quasi-cleavage under a partial bending of the specimen as the failure surface expanded and opened. Surface profiles were taken for all surface treated specimens using a Taylor Hobson Formtalysurf 120L (laser pick-up) profilometer. While the extremely small scale involved and the variability of the results makes specific conclusions problematic, some general observations can be made. The surface treatment with the greatest damage, fretting depth, to both bridges and specimens was the CuNiIn/DAG configuration that is currently used by the OEM. The bare metal alloy, Ti, PtIBAD/OL and OCT/OL specimens all fell within the scatter of each other at the minimum depth of damage range. The remaining coated specimens all had significantly greater depth of damage and ranged from the worst mechanical test results to the middle. The bridge damage all fell within scatter of each other. While this suggests a trend between decreasing fretting depth and increasing fretting cycles survived, no definite conclusion can be made with the data obtained to date.

Optical microscopy was used initially to determine the severity of fretting damage. It was also used to classify and document the types of damage (crack location, failure mechanisms, etc.), (Figure 7). The fracture surface of each of these specimens was

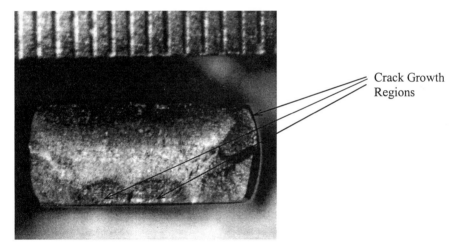

Crack Growth Regions

Figure 7 – *Optical Micrograph Of A Fretted Specimen Showing Typical Failure Modes*

examined to determine the initiation points of the crack growth and resulting failure, using a scanning electron microscope, (Figure 8). By tracing the cracks back to their origin the cause of failure can often be determined. In all the cases under study, the failure predominately originated due to fretting. In general, the fracture surfaces of all the test specimens showed galling, material transfer, smearing, scaling and cracking. In addition, fatigue striations were located on the crack growth surfaces indicating the fact that the failure did occur ultimately due to fatigue crack growth. The failure patterns and the damage on the wear pads indicated that the fretting rig produced representative fretting fatigue damage as seen in service.

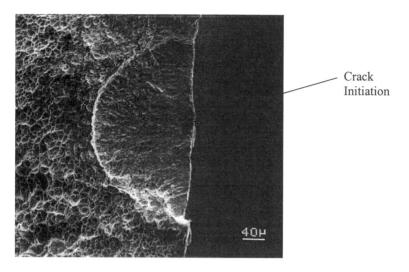

Figure 8 - *Secondary Crack Growth, Fracture Surface*

*Residual Stress Analysis*

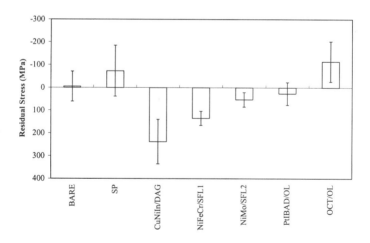

Figure 9 – *Residual Stress Measurements for Surface Treated Ti-6242 Specimens*

Surface residual stress tests were conducted on the failed specimens to try and determine a correlation between residual stress and cycles to failure. A Rigaku D/MAX-

2BX X-ray diffractometer was used for these measurements. The measurements were conducted by continuous scan at 0.3 °/min with a data acquisition interval of 0.02. The results indicate that there is very little relationship between residual stress and cycles to failure, (Figure 9). Also, the measurements themselves vary between measurements on the same specimen. However, there is some indication that certain processes may have detrimental affects on the compressive residual stress caused by the initial peening. Coatings such as CuNiIn, NiFeCr and NiMo tend to cause a tensile surface layer on the substrate [12]. This is usually compensated for by pre-peening the surface. However, in conjunction with the high temperature testing the compressive residual stress may have been completely relieved.

## Comparison of Pin-On-Disk Wear Test Results to Fretting Fatigue Test Results

An interesting difference in test results was observed between the wear test data and the fretting fatigue test data. The ranking of the various surface treatments evaluated in this study during the two tests is as follows:

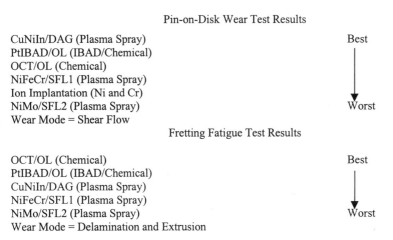

Pin-on-Disk Wear Test Results

CuNiIn/DAG (Plasma Spray)                                    Best
PtIBAD/OL (IBAD/Chemical)
OCT/OL (Chemical)
NiFeCr/SFL1 (Plasma Spray)
Ion Implantation (Ni and Cr)
NiMo/SFL2 (Plasma Spray)                                     Worst
Wear Mode = Shear Flow

Fretting Fatigue Test Results

OCT/OL (Chemical)                                            Best
PtIBAD/OL (IBAD/Chemical)
CuNiIn/DAG (Plasma Spray)
NiFeCr/SFL1 (Plasma Spray)
NiMo/SFL2 (Plasma Spray)                                     Worst
Wear Mode = Delamination and Extrusion

The differences in the ranking in the two tests, especially for the first three surface treatments, can be attributed to different test conditions and correspondingly different wear modes of the surface treated specimens in the two test conditions. In the case of the pin-on-disk sliding wear test, the motion is continuous and oxide particles are swept from the wear track. The low coefficient of friction, hence the low wear rate of CuNiIn/DAG coating configuration can be attributed to its self-lubricating behavior at temperature. As indicated earlier, the fretting fatigue contains a fatigue component as well as a wear component in the failure process. The pin-on-disk wear tests do not consider this fatigue component. Based on this, coatings like CuNiIn tends to perform very well in the wear tests due to their low coefficient of friction at temperature. However, as it can be seen from the fretting fatigue test results, once the fatigue component is added to the test, its performance drastically decreases. CuNiIn does not form adherent protective oxide film under fretting fatigue condition. Abrasion by these oxidized debris and particle generation via delamination of the coating results in adhesion between the contacting surfaces leading to fretting fatigue failure. For the other two surface treatments, OCT/OL and PtIBAD/OL, due to the shallow depth of the surface treated layers, compared to the CuNiIn/DAG coating configuration, no significant improvement in the wear performance was observed after the surface treated layer was removed in the initial wear-in period. It is likely that the surface treated layers in these cases were removed along with any asperities during the initial wear-in period. Whereas in the fretting fatigue tests, there was significant improvement in the fretting fatigue life of the OCT/OL and PtIBAD/OL treated titanium surfaces. This can be attributed to their ability to increase the fatigue life of the parent metal by generating a diffuse metal-oxide interface and a thin tenacious surface oxide, respectively. It is likely that these lubricating thin oxide layers contributed towards reduction in adhesion between the two contacting surfaces and minimized generation of oxide debris under fretting fatigue test conditions, resulting in an enhancement in the fretting fatigue life. In these two surface treatments, a subsurface hardening mechanism via precipitation of fine oxides in the metal matrix might have prevented cracking and subsequent galling of the alloy surfaces.

The conflicting ranking between the first three surface treatments, in both tests, illustrates the potential dangers of using sliding wear tests to rank surface treatments for fretting wear/fatigue amelioration. It is possible that the surface treatments, which performed poorly in the sliding wear tests, may offer greater protection for the small displacement characteristics of fretting wear.

## Conclusions

The OCT/OL provided the best fretting fatigue resistance to the Ti-6242 titanium alloy at elevated temperatures. The resistance was almost twice that of the current conventional anti-fretting coating solution of CuNiIn/DAG and four times that of the base material. The PtIBAD/OL treatment also improved the fretting fatigue life over the shot peened base material by 25%. As expected, a significant contrast was observed in the ranking between the high temperature wear tests results and high temperature fretting fatigue test results for CuNiIn/DAG coating configuration, OCT/OL treatment and PtIBAD/OL treatment. Conflicting ranking in the wear and fretting fatigue test data illustrates the potential dangers of using sliding wear tests to rank coating /surface treatments for fretting fatigue amelioration. Fretting wear tests using realistic fretting conditions such as small displacements characteristic of fretting wear should be used to screen surface treatments for fretting fatigue life improvement as opposed to the conventional sliding wear tests. Finally, the fretting fatigue test is the most realistic test for evaluating the influence of surface treatments and coatings on the fretting fatigue life of materials. For evaluating fretting wear characteristics, wear testing under realistic fretting conditions is recommended.

**Acknowledgements**

The authors would like to acknowledge the financial support provided by the Department of National Defense Canada. Special thanks are extended to C. Adams of Orenda Aerospace Corporation, Institute of Aerospace Research National Research Council of Canada and A.J. Freimanis of Pennsylvania State University. Finally, the authors thank R. Thamburaj of Orenda Aerospace Corporation for his support through the entire course of this work.

**References**

[1] Buckley, D.H., "Effect of Crystal Structure on Fretting," *Paper 13, NATO-AGARD Meeting*, Structures and Materials Panel, Munich, Germany, Oct. 11-12, 1974.

[2] Johnson, R.L., and R.C. Bill, "Fretting in Aircraft Turbine Engines, " *AGARD-CP-161*, 1975.

[3] Bill, R.C., "Materials Evaluation under Fretting Conditions," *ASTM STP 780*, American Society for Testing and Materials, Philadelphia, PA, 1982, pp. 165-182.

[4] Waterhouse, R.B. and Iwabuchi, "The Compositional Properties of Surface Films Formed During the High Temperature Fretting of Titanium Alloys," *Proceedings of the JSLE International Tribology Conference*, Tokyo, Japan, July 8-10, 1985.

[5] Chakravarty S., Andrews R.G., Patnaik P.C. and Koul, A.K., "The Effect of Surface Modification on Fretting Fatigue in Ti Alloy Turbine Components," *Journal of Metals 47, 4*, 1995, pp. 31-35.

[6] Hamdy, M.M. and Waterhouse, R.B., "The Fretting Fatigue Behavior of a Nickel-based Alloy (IN 718) at Elevated Temperatures," *Proceedings of Wear of Materials*, April 16-18, 1979, Dearborn ASME, New York, 351.

[7] Dearnaley, J., *IEEE Transaction of Nuclear Science NS-28 (2)*, 1981, pp. 1808-1810.

[8] Fujishiro, S. and Eylon, D., "Improvement of Ti Alloy Fatigue Properties by Platinum Ion Plating," *Metallurgical Transactions A*, Volume 11, August 1980, pp. 1259-1263.

[9] Fujishiro, S. and Eylon, D., "Improved High Temperature Mechanical Properties of Titanium Alloys by Pt. Ion Plating," *Thin Solid Films 54*, 1978, pp. 309-315.

[10] Levy, M. and Morrosi, J.L., Technical Report 76-4, Army Materials and Mechanics Research Center, 1976.

[11] Groves, M.T., NASA Report CR-134537, 1973.

[12] R.B. Waterhouse, "Fretting Fatigue," Applied Science Publishers Ltd., 1981, pp. 221-240.

# Applications

Taizo Makino,[1] Miyuki Yamamoto,[1] and Kenji Hirakawa[2]

## Fracture Mechanics Approach to the Fretting Fatigue Strength of Axle Assemblies

**REFERENCE:** Makino, T., Yamamoto, M., and Hirakawa, K., "**Fracture Mechanics Approach to the Fretting Fatigue Strength of Axle Assemblies**," *Fretting Fatigue: Current Technology and Practices, ASTM STP 1367*, D. W. Hoeppner, V. Chandrasekaran, and C. B. Elliott, Eds., American Society for Testing and Materials, West Conshohocken, PA, 2000.

**ABSTRACT:** The objective of the present paper is to evaluate the fatigue crack growth behavior in press-fitted axles using a fracture mechanics approach and to predict the fatigue strength regarding crack propagation ($\sigma_{w2}$). The relationship between nominal bending stress ($\sigma_n$) and non-propagating crack length in press-fitted axles is also discussed. Rotating bending fatigue tests were conducted on the induction hardened and quench-tempered axles of 38 and 40 mm in diameter. The equation for $\Delta K$ was formulated from the result of FEM analyses in which the micro-profile at the contact edge was taken into consideration. The threshold stress intensity factor range $\Delta K_{th}$ for small cracks was estimated from the crack size measured after the fatigue tests by using a modified stress ratio effect at fully compressed stress reversals due to high compression residual stress. $\sigma_{w2}$ and the relationship between $\sigma_n$ and non-propagating crack length were predicted by using the above mentioned $\Delta K$ and $\Delta K_{th}$. The predicted $\sigma_{w2}$ and non-propagating crack length were in good agreement with the experimental values.

**KEYWORDS:** fretting fatigue, fracture mechanics, axle assemblies, fatigue strength, small crack growth behavior, non-propagating crack, FEM analysis, stress intensity factor, induction hardening

Press-fitted axle assemblies for railroad vehicles are typical mechanical components that suffer from fretting fatigue [1,2]. Therefore, the fatigue properties of press-fitted axles have been intensively investigated by many researchers [2,3]. As a result, it is known that the fatigue strength regarding crack propagation ($\sigma_{w2}$) can be greatly improved by induction hardening, while the fatigue strength regarding crack initiation ($\sigma_{w1}$) could hardly be improved[3]. However, it was difficult to judge whether initiated cracks could propagate or not, i.e., to analytically predict the fatigue strength $\sigma_{w2}$ of press-fitted axles with many types of residual stress distributions, though some fracture mechanics approaches were adopted for phenomenal comprehension. The difficulty in $\sigma_{w2}$ prediction probably exists in the complex stress distribution at crack initiation sites

[1]Research Engineer and Senior Research Engineer, respectively, Sumitomo Metal Industries, Ltd., 1-8 Fuso-cho, Amagasaki city, Hyogo, JAPAN.

[2]Professor, Kyushu University, 6-1-1 Hakozaki, Higashi-ku, Fukuoka city, Fukuoka, JAPAN.

and small crack growth behavior at fully compression stress reversals due to high compression residual stress.

The fracture mechanics approach is applied to evaluate the fatigue strength of this press-fitted axles. As a result of a previous report, it was clarified that the crack initiation sites corresponded to the maximum stress points obtained by FEM analyses in which contact edge profile was considered[4]. Therefore the generated stresses at the crack initiation sites were accurately described.

The objective of the present paper is to precisely evaluate the fatigue crack growth behavior in press-fitted axles using a fracture mechanics approach and to enable the prediction of the fatigue strength regarding crack propagation ($\sigma_{w2}$). The relationship between the nominal bending stress ($\sigma_n$) and non-propagating crack length in press-fitted axles is also discussed.

The rotating bending fatigue tests were conducted on induction hardened and quench-tempered axles of 38 and 40 mm diameter. Induction hardened axles have two kinds of residual stress distributions. It was experimentally found that $\sigma_{w2}$ was 110MPa for the quench-tempered axle, 270MPa for a kind of induction hardened axle and more than 320MPa for another kind of induction hardened axle.

The stress states of press-fitted axles during fatigue tests were evaluated by FEM analyses in which the microprofile at the contact edge was taken into consideration. The relationships between crack depth and the stress intensity factor range ($\Delta K$) were calculated by using an equation for semi-elliptical cracks under stress states obtained from the FEM analyses.

The threshold stress intensity factor range ($\Delta K_{th}$) for small cracks was estimated from the above mentioned experimental and analytical results, therein, especially for the induction hardened axle, the stress ratio effect was modified by taking into consideration the fully compressed stress reversals due to high compression residual stress. $\sigma_{w2}$ and non-propagating crack length were predicted using the comparison between $\Delta K$ and $\Delta K_{th}$. The predicted $\sigma_{w2}$ and non-propagating crack length were in good agreement with the experimental value.

**Material and Specimen**

The material used in the present paper is 0.38% carbon steel (JIS S38C) used for Shinkansen electric vehicle axles. The chemical composition is given in Table 1. The steel was oil-quenched at 860 °C for 30 minutes and tempered at 590 °C for one hour. The mechanical properties after this heat treatment are given in Table 2.

The configuration of the test specimen is shown in Figure 1. The axle is press-fitted to the boss as in Figure 1. Classification of the test specimen is shown in Table 3. One quench-tempered (QT) axle and two kinds of induction hardened (IH04, IH20) axles were utilized in the present study. The QT axle was machined to the final shape after the above mentioned heat treatment. Thus, the QT axle has a hardness of approximately Hv200. Whereas IH04 and IH20 axles were roughly machined and induction hardened, and finally machined for finishing to be press-fitted. The hardness distribution in the radial direction of the IH04 and IH20 axles is shown in Figure 2. The axial residual stress distribution in the radial direction is shown in Figure 3. The IH04 axle comprises a thinner hardened layer, defined as over 400Hv, of 0.4mm in depth than the IH20 axle in which the hardened layer is 2.0mm in depth. A High compressive residual stress is generated in both the IH04 and IH20 axles, but the peak value of the IH20 axle is higher than that of the IH04 axle.

The boss is made of heat treated high carbon steel (JIS SSW- 1) which is used for the railroad wheel. The chemical composition and mechanical properties of the boss are also given in Table 1 and Table 2. The fitting interference of the specimen was determined in order to maintain the nominal fitting pressure of approximately 70 MPa.

Table 1–*Chemical composition*

|  |  |  |  |  | mass% |
| --- | --- | --- | --- | --- | --- |
|  | C | Si | Mn | P | S |
| Axle(S38C) | 0.38 | 0.28 | 0.76 | 0.009 | 0.008 |
| Boss(SSW-1) | 0.63 | 0.18 | 0.74 | 0.017 | 0.005 |

Table 2–*Mechanical properties after heat treatment*

|  | Y.P., MPa | T.S., MPa | El., % | R.A., % |
| --- | --- | --- | --- | --- |
| Axle(S38C) | 433 | 679 | 25.8 | 70.4 |
| Boss(SSW-1) | 361 | 779 | 19.5 | 34.7 |

Table 3–*Classification of test specimen*

| Class. | Axle diameter, mm | Case depth, mm | Surface hardness, Hv | Surface residual stress, MPa | Peak of compressive residual stress, MPa |
| --- | --- | --- | --- | --- | --- |
| QT | 40 | – | 200 | – | – |
| IH04 | 38 | 0.4 | 520 | –600 | –600 |
| IH20 | 40 | 2.0 | 600 | –600 | -820 |

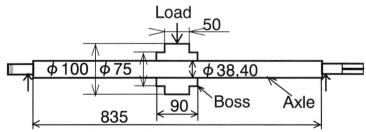

Figure 1 – *The specimen configuration*

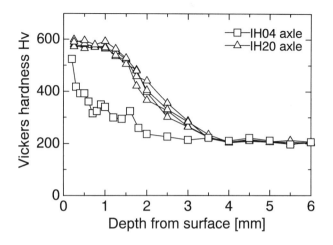

Figure 2 – *Vickers hardness distribution of IH04 and IH20 axles*

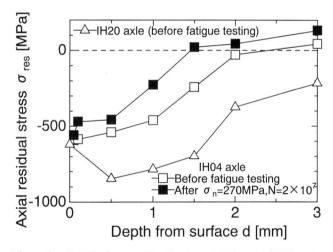

Figure 3 – *Residual stress distribution of IH04 and IH20 axles*

**Fatigue Test**

*Test Method*

The rotational bending fatigue tests were conducted on QT, IH04 and IH20 axles. In these tests, specimens were bent at 3 points by a hydraulic servo control loading system as shown as the arrows in Figure 1 and rotated by an AC motor. After the fatigue tests, the fitted surface of the run-out axles were inspected by the scanning electron and optical microscopes. In this inspection, the size and site of the initiated cracks and contact edge profiles were measured. In addition, the residual stress distribution was measured on some run-out IH04 and IH20 axles in order to investigate the relaxation behavior of the residual stress.

*Test Results*

The S-N diagram is shown in Figure 4. $\sigma_n$ shown in Figure 4 is defined as the bending moment divided by the section modulus of the axle. It was found that $\sigma_{w2}$ was 110MPa for the QT axle, 270MPa for the IH04 axle and more than 320MPa for the IH20 axle. Fretting fatigue cracks were detected on all axles subjected to the stress reversals near below $\sigma_{w2}$ in the fatigue tests. A photograph of the fretting fatigue crack initiated on the IH20 axle is shown in Figure 5. This crack was detected with the fretting corrosion. Crack depths were investigated on some specimens by inspection of the cross section or artificially breaking the crack surface. As a result, the aspect ratio was found to average approximately 0.3.

The relationship between $\sigma_n$ and the surface length of cracks observed in the run-out specimens is shown in Figure 6. The nominal bending stresses of the IH04 and IH20 axles are higher than that of the QT axle, compared at a certain crack length. This means that the IH04 and IH20 axles has a high propagation resistance for small cracks.

The residual stress distribution of the IH04 axle after the fatigue test is also shown in Figure 3. The residual stress of the IH04 axle was slightly relieved after the fatigue tests. This relaxation behavior of residual stress was also indicated in the IH20 axle.

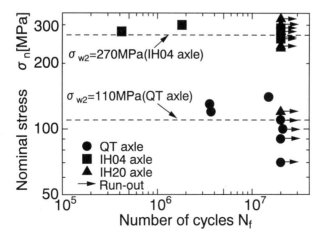

Figure 4 – *S-N diagram*

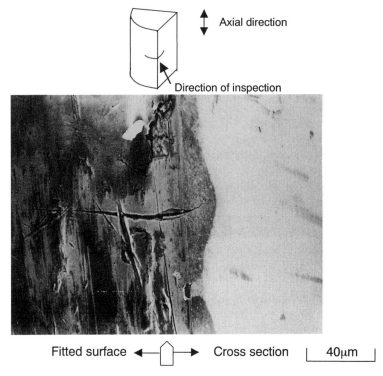

Figure 5 – *Fretting fatigue crack initiated on press-fitted surface of IH20 axle* ($\sigma_n$=300MPa, N=2×10[7])

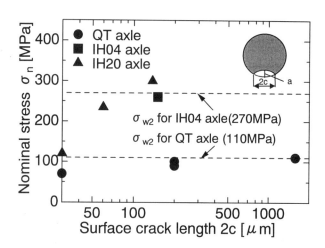

Figure 6 – *Relationship between nominal bending stress and surface length of cracks*

## Stress States of Press-fitted Axles

FEM analyses regarding the fatigue tested press-fitted axles were conducted with ABAQUS. The finite element mesh used in the FEM analyses is shown in Figure 7. The interface elements were arranged on the press-fitted contact surface between boss and axle. The friction coefficient of the interface elements is assumed to be 0.6.

In these analyses, the contact edge microprofile, which is called 'Rounding' in the authors' previous reports[4,5], was taken into consideration by changing the coordinate of mesh nodes. The shape of 'Rounding' and crack initiation site is schematically shown in Figure 8. The details of the analysis results were indicated in the authors previous reports[4,5]. The following results were found from these analyses.

- Maximum axial stress increased and its location shifted from the contact edge to the inside with increasing 'Rounding'.
- Crack initiation sites corresponded well to the maximum stress location calculated by the FEM analysis when considering 'Rounding'.
- The maximum axial stress was estimated as $\sigma_n + 70$[MPa] from the result of the FEM analyses where the deviation caused by the variety of 'Rounding' shapes in each specimen was taken into consideration.

Based on the above-mentioned results, the axial stress distribution in the radial direction is estimated as shown in Figure 9 and the following equations.

When $0 < d \leq D/20$,

$$\sigma_{ax} = \sigma_n + 70 - 20(0.1\sigma_n + 70)d/D$$

When $D/20 \leq d$,

$$\sigma_{ax} = \sigma_n - 2\sigma_n d/D$$

$$\left.\begin{array}{c} \\ \\ \end{array}\right\} (1)$$

Number of elements: 6 964, Number of nodes: 8 065

Figure 7 – *Finite element mesh*

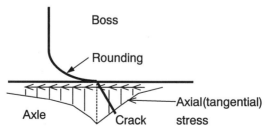

Figure 8 – *Schematic illustration of 'Rounding' and crack initiation site*

where
$d$ = depth from press-fitted surface at tensile side
$D$ = axial diameter (40mm)
$\sigma_{ax}$ = axial stress, MPa
$\sigma_n$ = nominal bending stress, MPa

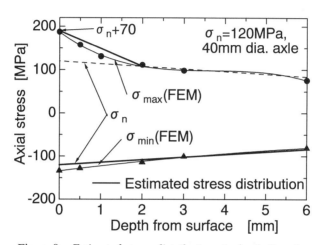

Figure 9 – *Estimated stress distributions in depth direction*

## Evaluation of Crack Propagation Ability using Fracture Mechanics Approach

*Stress Intensity Factor of Cracks Observed in Run-out Specimens*

The relationships between the crack size and the stress intensity factor range ($\Delta K$) were calculated as the following equations using Newman-Raju's equation[6] for a semi-elliptical crack under the stress distributions defined by Eq (1). Cracks are oblique to the axle surface, however the effect of the mixed mode is small according to the comparison between Mode I and Mode II stress intensity factors. Accordingly only Mode I was considered in this evaluation.
When the aspect ratio $a/c$=0.3, $a<D/20$,

$$K_{max}=1.103(\sigma_n+70-1.256\sigma_n a/D-879.2a/D)\sqrt{(\pi a/1.2)} \tag{2}$$

$$K_{min}=-1.103(1-1.256a/D)\,\sigma_n\sqrt{(\pi a/1.2)} \tag{3}$$

$$\Delta K=K_{max}-K_{min} \tag{4}$$

where

*a* = crack depth,
*K* = stress intensity factor
Suffix:   *max* = maximum value,   *min* = minimum value

By using Eq (2)-(4), the stress intensity factor range $\Delta K$ of cracks observed in a run-out specimen was calculated. The relationships between $\Delta K$ and crack depth are shown in Figure 10. In the present paper, it is assumed that cracks observed in run-out specimens are non-propagating when the fatigue test is finished. $\Delta K$ plotted on Figure 10 is then equivalent to the threshold stress intensity factor range, $\Delta K_{th}$, regarding crack growth. $\Delta K_{th}$ of all the axles increased as the crack depth increased, i.e., there is the dependency of $\Delta K_{th}$ on the crack depth like small crack growth behavior in non-fretting fatigue. Furthermore, $\Delta K_{th}$ of the IH04 and IH20 axles are larger than that of the QT axle.

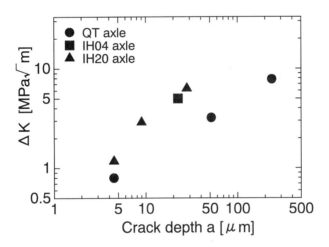

Figure 10 – *Relationship between crack depth and $\Delta K$*

*Threshold Stress Intensity Factor Range of QT Axle*

The relationship between $\Delta K_{th}$ of QT axle and the crack depth seems to be linear on the full-logarithm graph of Figure 10. Thus, this relationship was formulated into the following equation.
When $a < a_{cr}$,

$$\Delta K_{th} = \Delta K_{th\infty} (a/a_{cr})^{\alpha} \tag{5}$$

where
$\Delta K_{th\infty}$=the threshold stress intensity factor range for a large crack
$a_{cr}$ =critical crack length

α=constant

$\Delta K_{th}$ of the QT axle calculated using Eq (5) is shown in Figure 11 as the bold broken line. $a_{cr}$ is defined as the crack length where dependency of $\Delta K_{th}$ on crack depth begins to appear. This was obtained from the intersection point between the relationship $\Delta K$–$a$ of QT axle at $\sigma_{w2}$ and $\Delta K_{th\infty}$. As a result, $a_{cr}$=0.26mm was derived from Figure 11. α was determined to be 0.55 by parameter-fitting of the plotted $\Delta K_{th}$ data in Figure 11.

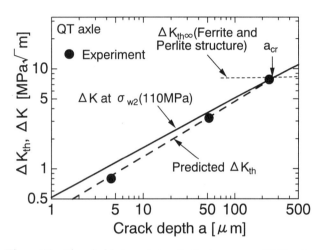

Figure 11– *Threshold stress intensity factor range of QT axle*

*Threshold Stress Intensity Factor Range of IH04 and IH20 axle*

The critical crack length $a_{cr}$ of the IH04 and IH20 axles was acquired by the same procedure as mentioned above. $\Delta K_{th}$ of the IH04 and IH20 axles are shown in Figure 12. In this figure, plotted $\Delta K_{th}$ of the IH04 and IH20 axles have the same tendency. It is caused by the fact that there is no difference of $K_{res}$ between IH04 and IH20 axles at the plotted crack depth. The $\sigma_{w2}$ without residual stress was then assumed to be equal to that of the QT axle because $\sigma_{w2}$ of the fretting fatigue has little dependency on hardness [7]. $\Delta K_{th}$ of the IH04 and IH20 axles without residual stress is shown as the solid line in Figure 12. $a_{cr}$ of the IH04 and IH20 axles was determined as 0.09mm, which was smaller than that of the QT axle, because $\Delta K_{th\infty}$ of the Martensite structure was smaller than $\Delta K_{th\infty}$ of the Ferrite and Perlite structures.

Since compressive residual stress was generated by induction hardening in the IH04 and IH20 axles, Eq (6) should be modified to include the effect of compressive residual stress. The effect of residual stress on fatigue strength is generally equivalent to the effect of mean stress. It is then evaluated by using the stress ratio $R$. However, the evaluation by R has a limit, because the denominator of the $R$, i.e., the addition of residual stress and the maximum stress becomes negative when the compressive residual stress is as large as the IH04 and IH20 axles. Therefore, $\Delta K_{th\infty}$ was defined by the following equation using $R'$ instead of $R$ where the denominator of $R$ was less than 0.

$$\Delta K_{th\infty,R<-1}=\Delta K_{th\infty,R=-1}\{(1-R')/2\}^{\gamma}$$

$$R'=\frac{K_{min}+K_{res}-\beta}{K_{max}+K_{res}+\beta} \qquad \left.\rule{0pt}{40pt}\right\} \quad (6)$$

Here, $K_{res}$ is the stress intensity factor generated as residual stress, which is calculated by Kopsov's equation[8]. Moreover, $\beta$ and $\gamma$ are material constants. These were determined as $\beta$=3.9MPa$\sqrt{m}$ and $\gamma$=0.4, which is obtained by parameter-fitting the plotted $\Delta K_{th}$ data in Figure 12 using Eq (5) and (6).

The two bold dot-dash-lines in Figure 12 indicate the $\Delta K_{th}$ values of the IH04 and IH20 axles calculated by Eq (5) and (6). These estimated $\Delta K_{th}$ respectively correspond to the plotted experimental data of IH04 and IH20 axle. It is interpreted that the approximation by Eq (5) and (6) is appropriate.

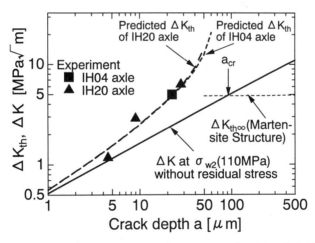

Figure 12– *Threshold stress intensity factor range of IH04 and IH20 axles*

*Prediction of Fatigue Strength and Non-propagating Crack Length, and Comparison Between Predicted and Experimental Values*

In this section, the fatigue strength ($\sigma_{w2}$) and non-propagating crack length of the QT, IH04 and IH20 axles were predicted according to the following procedure. First, the relationships between $\Delta K$ and crack depth are calculated using Eq (2)-(4). Next, the relationships between $\Delta K_{th}$ and crack depth are calculated using Eq (5) and (6). Here, the residual stress distributions measured after the fatigue tests at $\sigma_n$=270MPa for IH04 axle and $\sigma_n$=280MPa for IH20 axle, e.g., shown in Figure 3, are used in calculating $K_{res}$ over 270MPa for the IH04 axle and 280MPa for the IH20 axle in nominal bending stress. Last, the intersection point between the above mentioned two relationships can be the non-propagating crack depth. If this non-propagating crack depth become infinity, i.e., the above mentioned two relationships do not intersect with each other, it is deduced that complete failure can occur. Therefore, the maximum nominal bending stress where the intersection point exists can be $\sigma_{w2}$.

The prediction of $\sigma_{w2}$ and the non-propagating crack length was conducted on the QT, IH04 and IH20 axles. The predicted value is shown in Figure 13 as the relationship between the nominal bending stress and crack length. Here, the crack length 2c in Figure 13 was converted from crack depth as the aspect ratio $a/c$=0.3. It is found from Figure 13 that the predicted non-propagating crack lengths of the QT, IH04 and IH20 axles are in good agreement with the experimental value.

The comparison between the experimental and predicted $\sigma_{w2}$ is shown in Table 4. The predicted $\sigma_{w2}$ of QT and IH04 axles well correspond to experimental value. Whereas, the predicted $\sigma_{w2}$ of the IH20 axle seems too high. It is, however, possible that $\sigma_{w2}$ of the IH20 axle becomes lower than the predicted value indicated in Table 4, because the relaxation behavior of the residual stress in high nominal stress equivalent to $\sigma_{w2}$ may be more significant than the behavior considered in this prediction. Therefore, the predicted $\sigma_{w2}$ of the IH20 axle may closer approach to experimental value. Furthermore, the IH20 axle may fractured from the inside under the predicted $\sigma_{w2}$. In an authors' previous paper[9], a fracture from the inside, i.e., Fish-eye fracture was observed below the hardened layer in 40mm diameter axle with the 2.5mm case depth at 382MPa in nominal

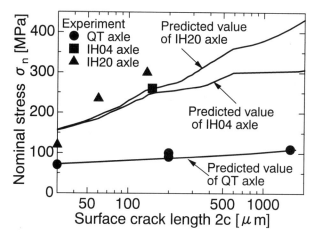

Figure 13 – *Comparison between predicted and experimental non-propagating crack length*

Table 4 – *Comparison between predicted and experimental* $\sigma_{w2}$

| Class. | Experimental $\sigma_{w2}$, MPa | Predicted $\sigma_{w2}$, MPa |
|---|---|---|
| QT | 110 | 110 |
| IH04 | 270 | 310 |
| IH20 | >320 | 540 |

bending stress.

In this paper, $\sigma_{w2}$ was predicted under the assumption that the fretting fatigue crack propagates from the fitted surface under the residual stress distribution measured after the fatigue test. If the relaxation behavior of the residual stress distribution and crack growth behavior from the inside can be predicted, the accuracy of the proposed prediction equation will be improved. Therefore, the prediction of the relaxation behavior of residual stress distribution and the crack growth behavior from the inside should be studied in the future.

**Conclusion**

In order to evaluate the fatigue crack growth behaviors in press-fitted axles using a fracture mechanics approach and to enable the prediction of $\sigma_{w2}$ and the non-propagating crack length in press-fitted axles, rotating bending fatigue tests and FEM analyses were conducted. As a result, the following results were found.

(1) $\sigma_{w2}$ was 110MPa for the QT axle, 270MPa for the IH04 axle and more than 320MPa for the IH20 axle. The IH04 and IH20 axles had a high resistance to small crack propagation.
(2) On the basis of the FEM analysis considered for the contact edge microprofile, the equations for axial stress distribution in the radial direction were estimated.
(3) The $\Delta K_{th}$ values of the QT, IH04 and IH20 axles, which were obtained from the crack size measured after the fatigue tests and the equation for $\Delta K$, increased with increasing crack depth. That is, there was the dependency of $\Delta K_{th}$ on the crack depth like small crack growth behavior during non-fretting fatigue.
(4) Crack size dependency of $\Delta K_{th}$ was formulated by using $a_{cr}$, which was defined as the crack length where the dependency of $\Delta K_{th}$ on crack depth appeared. $a_{cr}$ of the QT axle was found to be 0.26mm. $a_{cr}$ of the IH04 and IH20 axles was found to be 0.09mm.
(5) A new parameter, $R'$, which was a modified stress ratio, was introduced into the equation of $\Delta K_{th}$ mentioned in (4) in order to consider the fully compressed stress reversals due to high compression residual stress.
(6) $\sigma_{w2}$ and the non-propagating crack length were predicted for the QT, IH04 and IH20 axles by comparing $\Delta K$ with $\Delta K_{th}$. The predicted non-propagating crack lengths of all axles and the $\sigma_{w2}$ values of the QT and IH04 axles were in good agreement with the experimental values.
(7) It is possible that $\sigma_{w2}$ of the IH20 axle becomes lower than the predicted, because the relaxation behavior of the residual stress in high nominal stress equivalent to $\sigma_{w2}$ may

be more significant than the behavior considered in this prediction.

## References

[1] Hoger, O.J., "Influence of Fretting Corrosion on the Fatigue Strength of Fitted Members," *ASTM STP144*, ASTM, Philadelphia, (1962), pp.40-51.

[2] Nishioka, K., Nishimura, S., Hirakawa, K., Tokimasa, K. and Suzuki, S. "Fracture Mechanics Approach to the Strength of Wheelsets," 6th International Wheelset Congress, vol.2, U.S.A, (1978), pp.2-4-1-15.

[3] Hirakawa, K. and Toyama, K., Influence of surface residual stresses on fatigue crack initiation of press-fitted axle assemblies," *Fretting Fatigue, ESIS 18*, Mechanical Engineering Publications,   (1994), pp.461-473.

[4] Makino, T., Yamamoto, M. and Hirakawa, K. (in Japanese), "Effect of Contact Edge Profile on Fretting Fatigue Crack Initiation Site on Press-Fitted Axle," *Transactions of the Japan Society of Mechanical Engineers A*, 63-615, (1997), pp.2312-2317.

[5] Makino, T., Yamamoto, M. and Hirakawa, K., "Fracture Mechanics Approach to the Fretting Fatigue Strength of Axle Assemblies," International Conference on Materials and Mechanics '97, (1997), pp.33-38.

[6] Newman, J. C. Jr. and Raju, I. S., "An Empirical Stress-Intensity Factor Equation for the Surface Crack", *Engineering Fracture Mechanics*, 15, pp.185-192.

[7] Nishioka, K. and Hirakawa, K., "Fundamental Investigation of Fretting Fatigue – Part 6. Effects of Contact Pressure and Hardness of Materials," *Transactions of the Japan Society of Mechanical Engineers*, 15-80, (1972), pp.135-144.

[8] Kopsov, I.E., "Stress Intensity Factor Solution for a Semi-elliptical Crack in an Arbitrarily Distributed Stress Field," *International Journal of Fatigue*, 14, No.6, (1992), pp.399-402.

[9] Yamamoto, M., Makino, T. and Hirakawa, K. (in Japanese), "Influence of Residual Stress Distributions on Fretting Fatigue Strength," Proceedings of the 1995 Annual Meeting of JSME/MMD, Vol.B, (1995), pp.455-456.

T. N. Farris,[1] M. P. Szolwinski,[2] and G. Harish[3]

## Fretting in Aerospace Structures and Materials

---

**REFERENCE:** Farris, T. N., Szolwinski, M. P., and Harish, G., **"Fretting in Aerospace Structures and Materials,"** *Fretting Fatigue: Current Technology and Practices, ASTM STP 1367,* D. W. Hoeppner, V. Chandrasekaran, and C. B. Elliott, Eds., American Society for Testing and Materials, West Conshohocken, PA, 2000.

**ABSTRACT:** Fretting, the deleterious and synergistic combination of wear, corrosion, and fatigue phenomena driven by the partial slip of nominally clamped surfaces, has been linked to severe reductions in service lifetimes of a myriad of contacting components, including bearings, turbine blades and mechanically fastened joints—both structural and biological. This paper serves to frame the aggregate of economic, operational, and technical developments responsible for engendering renewed interest in fretting fatigue of critical structural elements of both commercial and military aerospace systems, including riveted primary structure and the blade/disk pair in jet propulsion plants. A collection of both empirical evidence of fretting-induced componential degradation and an overview of results from recent investigations conducted by the authors serves to motivate the need for design-oriented metrics that can be used to ensure the structural integrity and safe operation of both current and future aerospace systems.

**KEYWORDS:** fretting fatigue, aging aircraft, structural integrity, riveted aircraft structure, aircraft engines, aluminum alloys, titanium alloys

### Fatigue and Structural Integrity of Aerospace Systems

Historically, the community of individuals and organizations responsible for the design, inspection, and maintenance of aeronautical and aerospace vehicles have been cognizant of the potential fretting can have in driving system failures. Yet events from the past decade of air and space travel, including the well-publicized catastrophic in-flight disintegration of a passenger plane fuselage, have refocused efforts in understanding the fundamentals of fretting damage mechanisms in aerospace structures and materials.

---

[1]Professor and Head, and [3]Research Assistant, respectively, School of Aeronautics & Astronautics, Purdue University, 1282 Grissom Hall, West Lafayette, IN 47907-1282.
[2]Assistant Professor, Department of Mechanical Engineering, Aeronautical Engineering, and Mechanics, Rensselaer Polytechnic Institute, 110 8th Street, Jonsson Engineering Center, Troy, NY 12180-3590.

To begin forging the link between fretting damage and structural integrity of aerospace systems, it is important to first review the paradigm shifts in the design and operation of both civilian and military aircraft sparked initially by several unexpected crashes of both American and British transports in the late 1940s and early 1950s. Each of these incidents was attributed to fatigue cracks introduced by cyclic loading of key metallic structures, including wing spar caps and components in the pressurized fuselage section. Prior to these incidents aircraft structures were designed and manufactured with primarily static strength and stiffness in mind, a fact echoed by an official from NACA, the National Advisory Committee for Aeronautics (the predecessor of the American space agency, NASA), who stated that interest in fatigue from cyclic loads, "...was low because there had been no service experience to demonstrate that fatigue of the airframe was a serious problem" [1].

Initial attempts to account for fatigue damage nucleated by cyclic loads were steered by an approach dubbed "safe life" design. The safe life of an aircraft was based on laboratory fatigue lives of components subjected to applied load waveforms chosen to be representative of in-flight conditions. A factor of safety of four was then applied to the observed cycles to failure of the test articles to account for variability in both materials and manufacturing quality of in-service airframe components. This safe life approach was the basis for the first Aircraft Structural Integrity Program (ASIP) adopted by the United States Air Force (USAF) in 1958. By providing design and test methods for the prediction and prevention of structural fatigue failures, ASIP was designed to preclude structural failure of in-service and future weapons systems.

A second series of catastrophic losses of F-111, F-5, B-52, and T-30 aircraft in the 1960s focused scrutiny on the inability of the safe life approach to account for the use of relatively brittle material in component subjected to high stresses. The resulting intolerance to defects introduced by either manufacturing processes or in-service damage was amplified by the inability to inspect and detect small cracks in many critical structural components. This renewed attention initiated a second paradigm shift in the approach to dealing with the relationship between material, manufacturing, and service-introduced defects; and the fatigue performance of airframes.

Enter the philosophy of "fail safe" or damage tolerant design. A damage tolerant design acknowledges the presence of variability in initial manufacturing quality or damage accumulated during service. Such designs must be able to maintain their structural integrity or be "tolerant" of these defects or damage between scheduled inspection intervals. The USAF embraced the damage tolerant philosophy in 1975 and continues to rely on its tenets for the current basis of its ASIP program. The methodology was also incorporated into design of civilian aircraft, as evidenced by the fail-safe approach adopted in the design of the primary structure of the DC-10 pressurized fuselage shell [2]. In the two decades following the official adoption of the damage tolerant approach, advances in the understanding of basic fatigue mechanisms, damage tolerant structural design, non-destructive inspection techniques, and a comprehensive tracking of individual aircraft usage have provided the United States Air Force with confidence in this philosophy of ensuring the structural integrity of its fleet.

*Widespread Fatigue Damage and Aging Aircraft*

Successful, robust, and cost-effective implementation of the damage tolerant approach hinges on two elements: the capability to inspect critical structural elements and the ability to blend predictive tools with laboratory and operational service data to schedule appropriate inspection intervals. The aftermath of the Aloha Airlines flight 243 incident, in which a portion of the passenger compartment disintegrated during a short flight between two of the Hawaiian islands, forced the aviation community to refocus the procedure developed to ensure the structural integrity of aircraft—civilian and military alike. The image captured moments after a miraculous landing of injured passengers amid the remnants of a shredded fuselage, which was beamed onto living room television screens and splashed through the popular press of the flying public, represents a watershed event in civilian aviation that symbolizes growing concern regarding the structural integrity of so-called "aging aircraft." The term "aging aircraft" is a moniker used to refer to the burgeoning number of both civilian and military airframes in service that have surpassed their originally intended operational lifetimes. As reflected in the statistics presented in Figure 1, the average age of key defense systems in the USAF fleet will increase steadily over the next few decades [*3*], an increase attributable to post-Cold War cutbacks in defense spending. Similar trends are present in commercial fleets, as the deregulation of the United States domestic passenger air transportation system has lead to stiff, often cutthroat, competition between the larger air carriers and regional carriers.

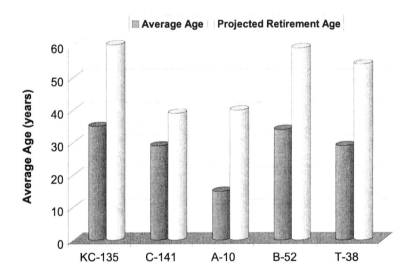

FIG. 1 — *A distribution of average ages for various weapon systems in the United States Air Force fleet, as of 1997. Many of these aircraft have already surpassed their initially intended operational life cycles.*

Expert review of the watershed Aloha Airlines incident attributed the sudden disintegration of the passenger cabin to the sudden linking of multiple undetected cracks at and around rivet holes in the metallic panels comprising the skin of the pressurized fuselage. This in-flight catastrophe unveiled the potential threat to airframe structural integrity proffered by the interaction and uncontrolled linkup of seemingly small and often-undetectable cracks in riveted primary structure. This type of fatigue damage, often referred to as widespread fatigue damage (WFD), is characteristic of the large population of aging aircraft. Widespread fatigue damage poses a current dual-tiered threat to the basis of the damage tolerant design philosophy as (1) it often defies detection by pragmatic non-destructive inspection of riveted structure, and (2) the knowledge base on the nucleation of this type of damage is only beginning to mature. Consequently, a critical need exists currently to understand the detailed relationship between the factors that control the development of WFD, such as aircraft manufacturing processes (*vis-à-vis* rivet installation) and corrosion, and the subsequent fatigue performance of riveted aircraft components.

*Linking Fretting and Widespread Fatigue Damage*

This is a daunting task, though, as the bulk of the knowledge of the mechanisms associated with the nucleation of widespread fatigue damage is primarily anecdotal. Direction in the current research effort into fretting fatigue has drawn on qualitative data from both periodic teardown inspections of aging aircraft structure [4] and laboratory simulations of common riveted aircraft lap splice joints [5, 6]. In each case, the investigators archived evidence of severe fretting damage and multiple fretting fatigue cracks near the rivet/skin interface and at the skin/skin interface or faying surfaces of riveted joints. Of particular interest is a comment summarizing an inspection of a lap joint panel from a service aircraft [6]:

> Visually, without magnification, the panel appeared to be serviceable and corrosion free. There is no reason to believe that this panel would have been inspected further or replaced in a periodic inspection. However, close optical inspection showed the presence of a black substance indicative of fretting at all but two of the rivet holes in the panel...[and upon analysis with microscopy], cracks were found in the fretting debris in each hole.

Similar damage has been observed in riveted lap joints manufactured under controlled conditions in a series of laboratory tests designed to link riveting process parameters and fretting fatigue failures. These observations are reported in greater detail elsewhere in this volume [7].

A review of two early compendia on fatigue in metallic structures [8, 9] reveals that designers were acutely aware of the role fretting can have in driving system failures, since each devoted a full chapter to fretting fatigue. In spite of its seemingly obvious relevance, relatively little work isolating the nucleation of cracks under the influence of fretting contact has been undertaken until recently. Instead, research was focused primarily on either reporting the effect of fretting on the reduction in total fatigue life under both in-service and laboratory conditions or suggesting design palliatives for minimizing the impact of fretting fatigue.

Many early workers [*10-12*] concentrated their efforts toward understanding fretting fatigue in steel with experimental studies designed to correlate applied fretting loads and relative displacements at the contacting surfaces with reductions in the overall fatigue strength of test specimens. The exception to these early efforts are fractographic studies of small cracks that developed in 7075 aluminum alloy subjected to fretting conditions [*13*].

*Fretting in Powerplant Components*

Fretting damage is not confined to only riveted aircraft structure. According to recent estimates [*14*], around one in six of all in-service "mishaps" attributed to high-cycle fatigue (HCF) in USAF engine hardware can be linked to damage induced by both fretting and galling. (Galling generally refers to the wear damage associated with gross relative motion on a much larger scale than the small-amplitude relative motion associated with fretting contacts.) The catastrophic repercussions of the unexpected in-flight failure of critical engine components due to the propagation of fretting fatigue damage nucleated at the mating surfaces of engine components are quite obvious. The threat posed by this near-surface fatigue damage is amplified by the potential for rapid growth of small fatigue cracks by high frequency, low-amplitude loads driven by unsteady aerodynamic forces on the engine blade airfoil. The severe stress concentration at the edge of contact allows the small amplitude loads to drive unrestrained growth of near-surface fatigue damage nucleated by the highly localized cyclic stresses and strains arising from the fretting contact.

Combating this damage through regular maintenance and inspection activities is a costly venture, though, with an estimated \$20 million US [*14*] exhausted annually to remedy fretting-related degradation of engine components. A recent collaborative initiative among the United States Air Force and its domestic aircraft engine vendors has sparked additional interest in fretting fatigue in titanium and other advanced engine alloys. The emerging research program is both a *de facto* effort to understand the role fretting plays in nucleating fatigue damage in current titanium and superalloy engine hardware components and a forward-looking campaign targeted at developing design-based tools to be used in the development of next-generation powerplants. Of primary interest is the intimate contact between the dovetail segment of the blade and notch section of the disk (Figure 6), including the interaction between low-cycle and high-cycle loads on the blade/disk pair. Fretting at this dovetail notch location has attracted the interest of previous researchers, including [*15*].

**Recent Fretting Fatigue Research**

With the clear threat that small and relatively undetectable fretting fatigue cracks pose to damage tolerance and the ensuing structural integrity of airframe components, a strong motivation exists to develop a quantitative, mechanics-based understanding of fretting crack nucleation in aircraft aluminum alloys, providing designers and fleet managers with an important tool for ensuring and maintaining the safe and economically viable operation of both current and future civilian and military airframes. Ideally, the approach developed for airframe aluminum alloys could be applied to engine alloys and the

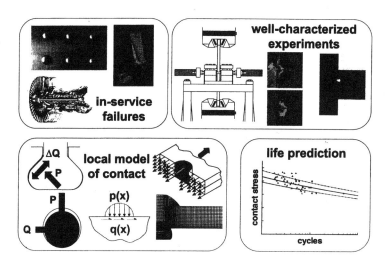

FIG. 2 — *A visual overview of the current approach taken in understanding the role of fretting in the nucleation of cracks in aluminum riveted lap joint structures and engine components.*

associated engine hardware geometries. Figure 2 summarizes the approach adopted by the authors to extend results from basic research studies focused on developing fundamental insight into the mechanics of fretting fatigue to prediction of the fatigue performance of aerospace systems.

This research has used observations from in-service and laboratory failures of riveted lap joints [4-6, 16, 17] to guide a series of experimental and modeling efforts targeted at achieving an accurate understanding of the near-surface conditions driving fretting damage at and around the rivet/hole interface. Realization of this goal required:

- formulating a clear understanding of the state of stress and strain arising from the normal and frictional tractions at and around the rivet/hole interface;
- constructing a well-characterized experimental setup to generate controlled and monitored fretting conditions;
- characterizing the intimately related tribology of partial slip contacts;
- linking the near-surface conditions and the basic fatigue response of the material to predict fretting fatigue crack nucleation; and
- synthesizing this knowledge into a validated design-based criterion for riveted aircraft structure

*Mechanics of Fretting Fatigue*

Pursuit of these objectives began with modeling both the surface tractions and resulting near-surface stresses associated with partial slip of both cylinder-on-flat [18, 19] and spherical-on-flat [20] contacts through the use of both analytical and finite element models. In particular, the finite element model [19] was constructed to address the effects of a cyclic bulk stress in the specimen on both the distribution of stick and slip at the interface and the near-surface stresses. Such a bulk stress is present in many common types of fretting fatigue test fixtures and structural components, and as first illustrated by Nowell and Hills [21], can result in a reversal of slip direction for monotonic loading.

A cursory review of the mechanics of partial slip contacts reveals immediately the intimate dependence of the near-surface conditions on the frictional characteristics at the interface. Several researchers [22-25] have reported linked an increase in tangential force magnitudes recorded during fretting experiments to an evolution of friction coefficient under partial slip conditions. By exercising a well-characterized fretting test setup [26] this evolution has been captured directly in a common aerospace aluminum alloy from both in-situ measurements of near-surface contact temperatures [27] via a non-invasive thermal imaging technique and a series of experiments designed to quantify the rapid increase and eventual stabilization of friction coefficient in the partial slip regime [28]. Additional results from this approach, including experimental validation of the near-surface conditions associated with the partial slip of cylinder-on-flat contacts, can be found elsewhere in this volume [29].

With an accurate characterization of the near-surface conditions during partial slip conditions, attention was focused on forging a link between the cyclic contact stresses and strains and fretting fatigue crack nucleation. A few research efforts have attempted to correlate a mechanics-based understanding of partial slip contacts and fretting crack nucleation [30, 31]. Others have suggested the use of fretting fatigue parameters comprising products of interfacial shear, tangential stress and slip amplitude [15]. Szolwinski and Farris have offered [18] and validated [26] an approach for predicting fretting crack nucleation through application of a multiaxial fatigue life parameter that links the near-surface cyclic stresses and strains to the number of cycles required to nucleate a crack along a critical plane. This approach is particularly attractive as it relies on readily available uniaxial strain-life data to characterize material fatigue response in predicting the location and orientation of fretting fatigue cracks observed at the edge of contact, coincident with a tensile peak in tangential stress associated with the tangential loading.

In continuing the pursuit of design-based criteria for fretting fatigue damage in aerospace systems, additional modeling and experimental efforts by the authors have focused on understanding the contact at and around the rivet/hole interface in lap joint structures. Results from these and other [17, 32] efforts have identified critical areas for current and future research efforts into fretting fatigue, areas which can bridge the gap between the foundational understanding of the mechanics, tribology, and fatigue of fretting contacts in reduced contact configurations and fretting fatigue failures in riveted structural components.

In particular, results from a three-dimensional shell model of a riveted single lap splice structure elucidate the mechanics of load transfer and closely tied potential for fretting fatigue damage in riveted joints [33]. The model resolves the contact interactions both between the rivet and hole and at the faying surface or interface of the joined layers. As shown in Figure 3, transfer of the applied bulk load through the joint is shared between frictional action at the faying surface and contact between the hole and plate. While the model does not include the effects of interference between the rivet and hole, a qualitative understanding of the load transfer mechanism can be gleaned from its results.

Upon initial application of load to the joint with a "slip fit" rivet, the load is transferred solely through the frictional tractions at the interfacial or faying surface. As the load increases, the plates slip with respect to each other, bringing the rivet and hole into contact. (Note that it is this partial slip to small-amplitude sliding behavior at the faying surfaces that leads to the fretting damage observed in riveted lap joints mentioned earlier.) Thus, increases in clamping constraint by the installed rivet result in more frictional load transfer at the faying surfaces, while decreases in clamping effect an increased loading at the interface between the rivet and hole periphery.

A fundamental question to those designing and manufacturing riveted aircraft structure centers on which of these conditions results in longer fatigue lifetimes. Should fastening techniques or installation methods that yield increased clamping constraint be employed, such that a larger percentage of the applied bulk load is transferred at the faying surface? Or will the increased severity of the fretting conditions and contact stresses at the faying surface result in more aggressive nucleation of fatigue damage away from the rivet hole?

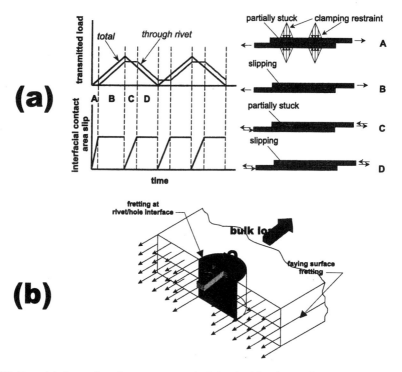

FIG. 3 — (a) A graphical representation of the dual load-transfer mechanism in a riveted lap joint gleaned from a three-dimensional shell model of a riveted lap joint. With these results in mind, (b) highlights the critical regions for the nucleation of widespread fatigue damage by fretting at and around the rivet/hole interface in riveted aircraft structure.

In providing answers to these queries, focus shifted to linking the manufacturing process of rivet installation to the subsequent degradation of fatigue performance of riveted joints by fretting fatigue. One component of this research involved modeling a quasi-static, squeeze-force controlled riveting process with the finite element method to determine the residual stresses induced during rivet installation. Unlike work by several workers examining the effect of residual stresses around rivet holes on fatigue crack propagation [34-36], no assumptions of arbitrary uniform hole expansion or through-thickness clamping restraint were made. Instead, the model included contact both between the rivet and plates and at the faying surface or interface between the two plates during and after rivet installation. Also incorporated was the contact between the riveting tools—a fixed upper curved set and flat bucking surface—and the rivet. Additional details of the model, including characterization of the rivet hardening behavior and experimental validation of the model can be found elsewhere [28].

Figure 4 presents model results for a maximum squeeze force of 19 kN applied to a universal head rivet (MS2047AD6-6) and two bare 2024-T3 aluminum alloy plates (2.3 mm thickness). Included are contours of radial, tangential, and von Mises residual stresses, a profile of expansion of the hole edge through the thickness of the plate, and a deformed mesh after unloading, highlighting the configuration of the driven head. Note the marked variation of hole expansion and residual stress through the thickness of the

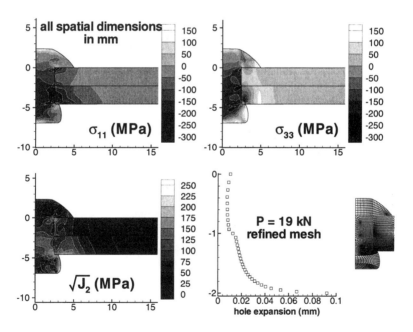

FIG. 4 — *Results from a finite element analysis with refined mesh for a maximum squeeze force of 19.0 kN. Note the through-thickness variation of hole expansion and residual stresses, particularly the zone of tensile hoop stress in the lower sheet, away from the rivet/hole interface.*

plates, further elucidating the shortcomings of the assumption of uniform hole expansion.

While more specific analytical and experimental quantitative correlation between the residual stresses and fatigue performance are presented elsewhere in this volume [7], scrutinizing the distribution of residual hoop stress, $\sigma_{33}$, provides some qualitative insight into this link: the plastic deformation associated with the hole expansion near the lower plate results in a zone of compressive residual hoop stress near the rivet/hole interface, while a compensating zone of tensile residual hoop stress is induced away from the hole edge. Thus, growth of cracks that initiate at the hole edge due to the fretting contact between the rivet and hole would be retarded, while the propagation of cracks formed at the faying surface away from the hole due to the frictional load transfer between the plates would be accelerated.

Figure 5, a plot of fatigue lifetimes of twelve single lap joints manufactured with varied and controlled squeeze forces, reflects a strong correlation between the squeeze force process parameter (and thus residual stresses) and fatigue lifetime. While as expected, fatigue life decreased with an increase in applied load, joint life increased with increases in maximum squeeze force. Indications from post-mortem optical and fractographic inspection of the joints revealed a distinct relationship between the location of fretting damage incurred and the rivet process parameters. Additionally, the authors have been successful at comparing life predictions based on the multiaxial fatigue life parameter mentioned earlier with observed fatigue lifetimes of riveted joints [7].

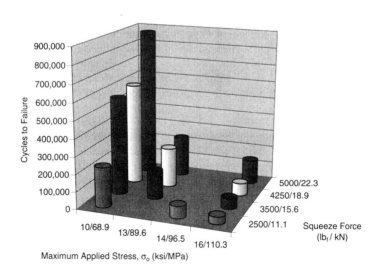

FIG. 5 — *Presentation of results from a series of single-lap joint fatigue tests manufactured with a controlled squeeze force technique. The observed cycles to failure for each test are plotted against the riveting process parameter and applied bulk stress, $R = \sigma_{min}/\sigma_{max} = 0.03$.*

*Blade/Disk Dovetail Connections*

The geometric details at the edges of contact in blade/disk dovetail connections (Figure 6) result in highly localized peaks in the pressure distribution and related sharp gradients in the near-surface stresses at the edges of contact. As such peaks and gradients are difficult to resolve accurately with the finite element method, solution of the singular integral equations governing elastic contact offer an attractive and tenable option for determination of the contact conditions. In support of this approach, Figure 7 presents a plot of non-dimensional pressure for a case of a nominally flat pad loaded against a flat, elastic halfspace with both a normal load and clockwise moment. This distribution was determined via a general series expansion approach presented by Barber [*37*] for solution of singular integral equations and formulated specifically by Szolwinski, *et al.* [*38*] for the case of a nominally flat punch with rounded edges. This approach reproduces efficiently the closed-form results of Ciaverella, *et al.* [*39*]. It is imperative to realize that the sharp gradient in the tangential stress that dominates the nucleation of fretting fatigue damage in cylindrical contacts is even sharper in the dovetail joint. Thus any effort aimed at characterizing fretting fatigue of engine hardware must examine the edge of contact in detail.

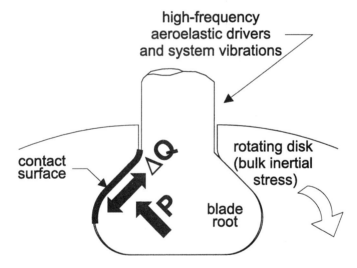

FIG. 6 — *An illustration of the interaction between a blade and disk in rotating jet engine component, highlighting the applied loads and contact interface.*

## Closure

The aggressive damage mechanisms associated with fretting pose a palpable threat to the structural integrity and safe, economical operation of aerospace vehicles. From the nucleation of widespread fatigue damage in riveted lap joint structure to the initiation and rapid propagation of edge-of-contact cracks at the blade/disk pair in jet engines, the sharp

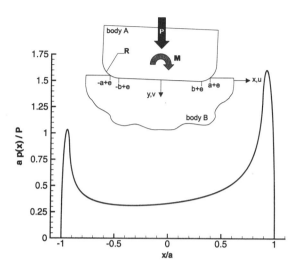

FIG. 7 — *A plot of non-dimensional pressure for a nominally-flat indenter with rounded edges loaded onto an elastic halfspace with a normal load, P, and clockwise moment, M. For the case presented, P = 1.58 kN/mm, M = 10.2 N·m, a = 1.68 mm, E = 110 GPa, ν = 0.3.*

near-surface gradients of stress associated with the partial slip of contacting surfaces can severely degrade the fatigue performance of critical structural elements and mechanical systems.

This document has served to touch on research accomplishments from an intensive effort designed to relate fundamental insight into the mechanics and tribology to the subsequent material fatigue response to develop a predictive methodology for fretting crack nucleation. Additional work served to characterize the local contact conditions at and around the rivet/hole interface in riveted lap joints, including the residual stress field left by the process of rivet installation and the dichotomous nature of the load transfer in the assembled lap joint. The details of the rivet installation process were shown to be critical in the subsequent fatigue performance of joints tested under laboratory conditions representative of those witnessed by fuselage lap joints.

Current and future areas of interest are focused on accurate determination of the mechanics and tribology of fretting contacts of nominally flat geometries representative of the blade/disk pair in rotating jet engine components. Key contributions can be made in linking the extremely sharp near-surface gradients of stress at the edges of contact to fatigue crack nucleation in materials such as titanium and superalloys used commonly in these components.

While fretting and fretting fatigue are not problems that can be eliminated entirely from mechanical systems, it is hoped that this work and future efforts targeted at understanding the aggressive damage mechanisms associated with partial slip contacts will offer insight into managing and minimizing its potential threat to degrading product performance. Emphasis must be placed on transitioning both basic and applied research efforts to cost-effective improvements in design approaches, manufacturing processes and inspection routines for increased performance of mechanical systems.

**Acknowledgments**

This work was supported in part through a multi-disciplinary aging aircraft University Research Initiative F49620-98-1-0293 from the United States Air Force Office of Scientific Research (AFOSR) designed to address basic concerns facing fleets of aging aircraft and a University of Dayton Research Institute (UDRI) subcontract RSC98003 as part of the United States Air Force (USAF) high cycle fatigue (HCF) program.

**References**

[1]    Kuhn, P., "Fatigue Engineering in Aircraft," *Fatigue in Aircraft Structures*, A. M. Freudenthal, Ed., Academic Press, New York, 1956, pp. 295-316.

[2]    Swift, T., "Development of the Fail-Safe Design Features of the DC-10," *Damage Tolerance in Aircraft Structures, ASTM STP 486*, American Society for Testing and Materials, Philadelphia, 1971, pp. 164-214.

[3]    Committee on Aging of U. S. Air Force Aircraft, "Aging of U. S. Air Force Aircraft," National Research Council, Washington, DC, 1997.

[4]    Piascik, R. S. and Willard, S. A., "The Characteristics of Fatigue Damage in the Fuselage Riveted Lap Splice Joint," NASA/TP-97-206257, NASA Langley Research Center, Hampton, VA, 1997.

[5]    Müller, R. P. G., "An Experimental and Analytical Investigation on the Fatigue Behavior of Fuselage Riveted Lap Joints," Ph.D., Delft University of Technology, The Netherlands, 1995.

[6]    Hoeppner, D. W., Elliot, C. B., III, and Moesser, M. W., "The Role of Fretting Fatigue on Aircraft Rivet Hole Cracking," DOT/FAA/AR-96/10, Federal Aviation Administration, Salt Lake City, UT, 1996.

[7]    Szolwinski, M. P., Harish, G., McVeigh, P. A., and Farris, T. N., "Experimental Study of Fretting Crack Nucleation in Aerospace Alloys with Emphasis on Life Prediction," *Fretting Fatigue: Current Technology and Practices, ASTM STP 1367*, D. W. Hoeppner, V. Chandrasekaran, and C. B. Elliott, Eds., American Society for Testing and Materials, West Conshohocken, PA, 1999, this volume.

[8]    Harris, W. J., *Metallic Fatigue*, Pergamon Press, New York, NY, 1961.

[9]    Heywood, R. B., *Designing Against the Fatigue of Metals*, Jarrold and Sons, Norwich, Great Britain, 1962.

[10]    Waterhouse, R. B. and Taylor, D. E., "The Initiation of Fatigue Cracks in a 0.7% Carbon Steel by Fretting," *Wear*, Vol. 17, 1971, pp. 139-147.

[11]    Nishioka, K., Nishimura, S., and Hirakawa, K., "Fundamental Investigations of Fretting Fatigue, Part 1," *Bulletin of JSME*, Vol. 11, No. 45, 1968, pp. 437-445.

[12] Endo, K. and Goto, H., "Initiation and Propagation of Fretting Fatigue Cracks," *Wear*, Vol. 38, 1976, pp. 311-324.

[13] Alic, J. A., Hawley, A. L., and Urey, J. M., "Formation of Fretting Fatigue Cracks in 7075-T7351 Aluminum Alloy," *Wear*, Vol. 56, 1979, pp. 351-361.

[14] Thomson, D., "The National High Cycle Fatigue (HCF) Program," in *3rd National Turbine Engine High Cycle Fatigue (HCF) Conference*, San Antonio, TX, Universal Technology Corp., 1998, CD-ROM proceedings.

[15] Ruiz, C., Boddington, P. H. B., and Chen, K. C., "An Investigation of Fatigue and Fretting in a Dovetail Joint," *Experimental Mechanics*, Vol. 24, No. 3, 1984, pp. 208-217.

[16] Hartman, A., "The Influence of Manufacturing Procedures on the Fatigue Life of 2024-T3 Alclad Riveted Single Lap Joints," NLR TR 68072 U, Nationaal Lucht-En Ruimtevaartlaboratorium (NLR), The Netherlands, 1968.

[17] Iyer, K., Hahn, G. T., Bastias, P. C., and Rubin, C. A., "Analysis of Fretting Conditions in Pinned Connections," *Wear*, Vol. 181-183, 1995, pp. 524-530.

[18] Szolwinski, M. P. and Farris, T. N., "Mechanics of Fretting Fatigue Crack Formation," *Wear*, Vol. 198, 1996, pp. 93-107.

[19] McVeigh, P. A. and Farris, T. N., "Finite Element Analysis of Fretting Stresses," *Journal of Tribology*, Vol. 119, No. 4, 1997, pp. 797-801.

[20] Szolwinski, M. P. and Farris, T. N., "Fretting Fatigue Crack Initiation: Aging Aircraft Concerns," in *35th AIAA/ASME/ASCE/AHS/ASC Structures, Structural Dynamics, and Materials Conference-Part 4*, Hilton Head, NC, AIAA, 1994, pp. 2173-2179, AIAA-94-1591-CP.

[21] Nowell, D. and Hills, D. A., "Mechanics of Fretting Fatigue Tests," *International Journal of Mechanical Sciences*, Vol. 29, No. 5, 1987, pp. 355-365.

[22] Endo, K., Goto, H., and Fukunaga, T., "Behaviors of Frictional Force in Fretting Fatigue," *Bulletin of the JSME*, Vol. 17, No. 108, 1974, pp. 647-654.

[23] Rooke, D. P. and Courtney, T. J., "The Effect of Final Friction Coefficient on Fretting Fatigue Waveforms," *Fatigue and Fracture of Engineering Materials and Structures*, Vol. 12, No. 3, 1989, pp. 227-236.

[24] Vincent, L., "Materials and Fretting," *Fretting Fatigue*, R. B. Waterhouse and T. C. Lindley, Eds., Mechanical Engineering Publications, London, 1994, pp. 323-337.

[25] Hills, D. A. and Nowell, D., *Mechanics of Fretting Fatigue*, Kluwer, Dordrecht, Netherlands, 1994.

[26] Szolwinski, M. P. and Farris, T. N., "Observation, Analysis and Prediction of Fretting Fatigue in 2024-T351 Aluminum Alloy," *Wear*, Vol. 221, No. 1, 1998, pp. 24-36.

[27] Szolwinski, M. P., Harish, G., Farris, T. N., and Sakagami, T., "In-Situ Measurement of Near-Surface Fretting Contact Temperatures in an Aluminum Alloy," *Journal of Tribology*, Vol. 121, No. 1/2, 1999, pp. 11-19/340.

[28] Szolwinski, M. P. and Farris, T. N., "Linking Riveting Process Parameters to the Fatigue Performance of Riveted Aircraft Structures, " *Journal of Aircraft*, 1999, accepted for publication.

[29] Harish, G., Szolwinski, M. P., Harish, G., Farris, T. N., and Sakagami, T., "Evaluation of Fretting Stresses Through Full-Field Temperature Measurements," *Fretting Fatigue: Current Technology and Practices, ASTM STP 1367*, D. W. Hoeppner, V. Chandrasekaran, and C. B. Elliott, Eds., American Society for Testing and Materials, West Conshohocken, PA, 1999, this volume.

[30] Petiot, C., *et al.*, "An Analysis of Fretting-Fatigue Failure Combined with Numerical Calculations to Predict Crack Nucleation," *Wear*, Vol. 181-183, 1995, pp. 101-111.

[31] Fellows, L. J., Nowell, D., and Hills, D. A., "On the Initiation of Fretting Fatigue Cracks," *Wear*, Vol. 205, 1997, pp. 120-129.

[32] Fawaz, S. A., "Fatigue Crack Growth in Riveted Joints," Ph.D. Thesis, Delft University, The Netherlands, 1997.

[33] Harish, G. and Farris, T. N., "Shell Modeling of Fretting in Riveted Lapjoints," *AIAA Journal*, Vol. 36, No. 6, 1998, pp. 1087-1093.

[34] Beuth, J. L. and Hutchinson, J. W., "Fracture Analysis of Multi-Site Cracking in Fuselage Lap Joints," *Computational Mechanics*, Vol. 13, 1994, pp. 315-331.

[35] Park, J. H. and Atluri, S. N., "Fatigue Growth of Multiple-Cracks Near a Row of Fastener-Holes in a Fuselage Lap Joint," *Computational Mechanics*, Vol. 13, 1993, pp. 189-203.

[36] Fung, C. P. and Smart, J., "An Experimental and Numerical Analysis of Riveted Single Lap Joints," *Journal of Aerospace Engineering*, Vol. 208, 1994, pp. 79-90.

[37] Barber, J., *Elasticity*, Kluwer, Dordrecht, The Netherlands, 1992.

[38] Szolwinski, M. P., Harish, G., and Farris, T. N., "The Development and Validation of Design-Oriented Metrics for Fretting Fatigue in Titanium Engine Components," in *IMECE Symposium on Mechanical Behavior of Advanced Materials*, Anaheim, CA, American Society of Mechanical Engineers, 1998, MD-Vol. 84, pp. 11-18.

[39] Ciavarella, M., Hills, D. A., and Monno, G., "The Influence of Rounded Edges on Indentation by a Flat Punch," *Proceedings of the Institution of Mechanical Engineers, Part C, Journal of Mechanical Engineering Science*, Vol. 212, No. 4, 1998, pp. 319-328.

Ky Dang Van[1] and M. Habibou Maitournam[2]

## On a New Methodology for Quantitative Modeling of Fretting Fatigue

**REFERENCE:** Dang Van, K., and Maitournam, M. H., **"On a New Methodology for Quantitative Modeling of Fretting Fatigue,"** *Fretting Fatigue: Current Technology and Practices, ASTM STP 1367,* D. W. Hoeppner, V. Chandrasekaran, and C. B. Elliott, Eds., American Society for Testing and Materials, West Conshohocken, PA, 2000.

**ABSTRACT:** A new intrinsic methodology for the prediction of fretting fatigue failure of a structure is presented. It is based, first, on the evaluation of local relevant thermomechanical parameters by new thermoelastoplastic computational methods (direct cyclic method) and, second, on the systematic use of the Dang Van multiaxial high-cycle fatigue criterion. For the validation of this proposal, numerical simulations of fretting fatigue tests on a particular experimental setup considered as a structure are performed. The resulting prediction of the experimental fretting fatigue map in relation to plastic and fatigue material properties is good.

**KEYWORDS:** fretting fatigue, fretting wear, high-cycle fatigue, numerical methods

## Presentation

Fretting is a major problem for industrial components in contact. It is defined as the surface damage induced by small-amplitude oscillatory displacements between metal components in contact. This damage can either be wear or crack nucleation, depending on the prescribed forces or the displacements amplitude. Many experimental results have been obtained recently. Vingsbo and Soderberg [1] and Vincent et al. [2] have established a test methodology based on fretting maps. These maps give the material response fretting map (MRFM) (no damage, crack nucleation or wear) according to the running condition fretting map (RCFM) (partial slip and gross regime). They are very useful for a qualitative understanding of damage phenomena. However, the results obtained cannot be applied for another configuration (with different solid geometries and material properties) and consequently they are

---

[1]Professor, Laboratoire de Mécanique des Solides, Ecole Polytechnique, 91128 Palaiseau, France.

[2]Research Engineer, L.M.S., Ecole Polytechnique, 91128 Palaiseau, France.

not directly applicable to an industrial component.

Thus, the use of an *intrinsic methodology* to predict fretting is essential to be able to transpose laboratory tests results to real applications on mechanical structures. For this purpose, we propose an original approach which can be summarized as shown on the flow chart in Figure 1. Because of problems arising from contacts between solids, many difficulties must be overcome.

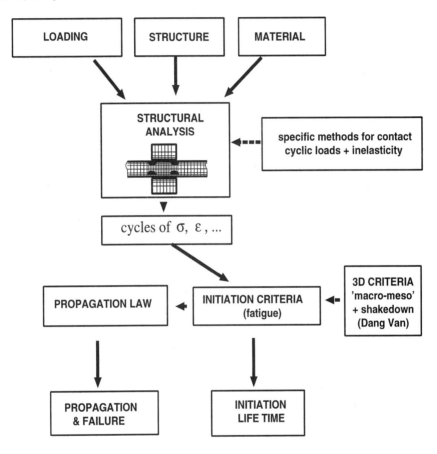

Figure 1 — *General methodology*

- (i) The initial inputs are geometries and nominal loading of the structure. One must compute the local relevant thermomechanical parameters (temperature, stress, total and plastic strain cycles, their evolutions and eventually their stabilized states) in the regions where fretting cracks may occur. The use of classical finite elements methods is inadequate, because of:

    – the type of loading, which is either a moving contact loading or a contact fixed in space but varying in time;

  – plastic flow which generally occurs during the first cycles even in the case of apparent elastic regime (elastic shakedown);
  – their flexibility and accuracy are poor despite of very high computational time;
- (ii) the local stress and strain fields are completely tridimensional with no fixed directions during their evolution. This necessitate the use of fatigue criteria which are able to deal with complex multiaxial loading situations.

To solve the first problem, we have developed a new efficient finite element method. This method, called the *direct cyclic method*, is based on an original scheme of integration. It is used to evaluate the inelastic state of structures subjected to a cyclic loading. The stabilized state (elastic or plastic shakedown) can directly be obtained.

We then use the multiaxial fatigue criterion proposed by Dang Van [3-5]. Its main attribute is that it can be easily identified by classical fatigue uniaxial laboratory tests like repeated tension or torsion. This criterion is essentially based on elastic shakedown hypothesis at all scales of materials description (macroscopic and mesoscopic scales).

To validate this proposal we simulate fretting fatigue tests on a particular experimental set-up considered as a structure. Numerical predictions are compared to experimental fretting map. This work complete the simulations initialized in two previous papers [6-7].

## The Computational Problem and Its Main Difficulties

  Classical approaches to fretting fatigue are based:
- either on the use of nominal parameters characterizing the loading (for example the normal force, the amplitude of oscillatory displacement) reported on the usual fretting-map;
- or, more rarely on the evaluation of the resulting local stress cycles under elastic assumptions.

However, a more careful examination of the problem shows that plastic deformation may occur locally, which changes the local mechanical parameters. Depending on the level of the load, different situations arise:
- purely elastic behaviour for very small loads;
- elastic shakedown: plastic deformation occurs but stabilizes after a certain number of loading cycles; after this period, the response is purely elastic;
- plastic shakedown: the stabilized cycle is a fixed plastic one;
- ratchetting: there is no stabilization of the plastic strain.

These phenomena depend not only on the level of the loading, but also on the geometries of the solids in contact, which influence the local distribution of stress and strain (structural effect). The calculation of this redistribution is generally a very difficult problem due to a great number of nonlinearities related to:
- frictional contact problems with partial or total slip;
- inelastic material behaviour;

• moving nature of obstacles, etc.

Nevertheless, one can use some existing computational codes to estimate the contact characteristics (area, pressure and shear distributions, slip or stick zones, etc). These computations are most of the time difficult to perform. However in the case of elastic shakedown, elastic contact computations are sufficient to estimate the former quantities and give a good idea of the running conditions. We shall use these results as limit loading conditions to evaluate the stress and strain fields in the bulk of the solids in order to check the danger of crack initiation.

In a more general situation, it is only possible to simulate some elastoplastic cycles, because the calculations are very lengthy.

In the following, we are more interested in fretting fatigue. To avoid difficult numerical computations for deriving contact characteristics, a particular contact system corresponding to a cylinder moving on a plane surface is considered. In this case, closed form solutions for elastic behaviour are available. It corresponds for instance to an experimental set-up used by Petiot and al [6], and represented in Figure 2. These experiments will be numerically simulated hereafter.

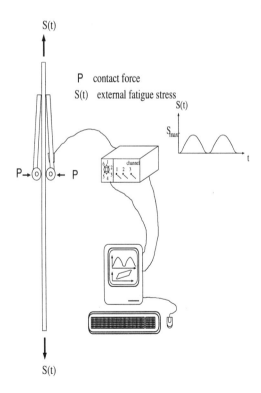

Figure 2 — *Fretting-fatigue setup (Petiot and al [6])*

**Oscillating Cylinder on a Half-space: Contact Characteristics**

The cylinder considered is in contact with a plane under normal constant compressive force $P$ and is subjected to repeated alternating tangential force $T(t)$. In order to calculate the elastoplastic limit response of the material in fretting, we describe first the loading cycle. Although the aim of the numerical method is to evaluate the inelastic state, the normal pressure and the shear tractions are assumed to be respectively given by the elastic theory of Hertz and the theory of Mindlin [8-9]. This approximation is valid when the stabilized state is elastic (elastic shakedown). In the case of plastic shakedown, the contact pressures are in fact modified during the cycles but the hypothesis of elastic contact pressures is maintained.

The contact width is assumed to be sufficiently small in comparison with the dimensions of the two bodies. The constant normal pressure $p(x)$ and the contact area are assumed to be given by Hertz theory:

$$p(x) = p_0 \sqrt{1 - \frac{x^2}{a^2}}$$

where $p_0$ is the maximum normal pressure and $a$ the contact half width. The tangential force $T(t)$ describes an alternate loading cycle of amplitude $2T_{max}$. Two cases are studied:

- partial slip for $T_{max} < \mu P$,
- full slip for $T_{max} = \mu P$.

The description of the loading cycles and the corresponding shear distributions are respectively represented on Figures 3 and 4. For more details, one can refer to Maouche et al. [7], and Hills and Nowell [10]. The cycles of loading just described are prescribed as external forces for the determination of the stabilized state with the new numerical method which is now recalled.

**Direct Cyclic Method**

The direct cyclic method allows the direct determination of the asymptotic response (i.e. stabilized mechanical state) of a structure subjected to a general cyclic loading without an incremental treatment of the whole loading history.

The principle of the finite element method is based on the two following procedures: (i) large time incremental method; (ii) research of the solution in the space of periodic responses, which means that we seek directly for mechanical fields (stresses, strains, plastic strains, etc) which are cyclic, (i.e. which have the same value at the end of the loading cycle as at the beginning of the cycle).

This method can be applied to any kind of elastoplastic constitutive equation; it can predict elastic or plastic shakedown regimes, and even ratchetting.

The details and the applicability of the method were discussed in previous papers [9, 11] on two dimensional examples. In these examples, the surface pressure distributions shown on Figures 3 and 4 are used. The considered loading parameters are $p_0/k = 3.5, \mu = 0.3, T_{max}/P = 0.25$ and $p_0/k = 2., \mu = 0.6, T_{max}/P = 0.6$;

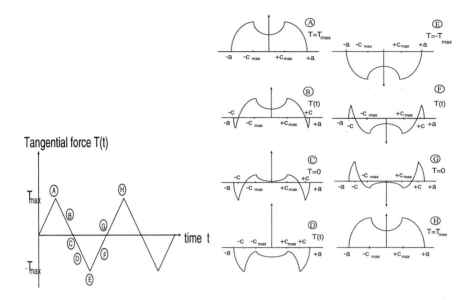

Figure 3 — *Loading cycle with no full sliding and the corresponding shear traction distributions* [7]

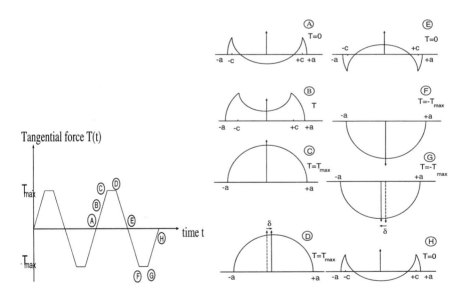

Figure 4 — *Loading cycle with full sliding and the corresponding shear traction distributions* [7]

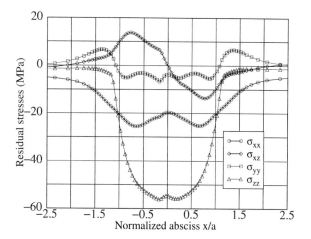

Figure 5 — *Residual stresses at the surface in the case of stick regime*

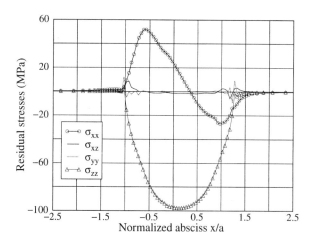

Figure 6 — *Residual stresses at the surface in the case of full sliding*

$k = 159$ MPa is the shear resistance of the considered linear kinematic hardening elastic-plastic von Mises material (Young's modulus: 207GPa, Poisson's coefficient: 0.3, Hardening modulus: 45 GPa). They correspond respectively to stick slip regime with elastic shakedown and gross slip regime with plastic shakedown happens. In Figures 5 and 6 we show the final residual stresses (i.e. stresses remaining after the removal of the loading, once the stabilized state is reached) obtained. In all cases, the CPU time is much lower than by using incremental method.

## The Fatigue Criterion

The used high-cycle fatigue criterion was initially proposed by Dang Van [3-5]. It is based on a multiscale approach in which it is assumed that elastic shakedown happens before crack initiation. In this approach, two scales are considered: (i) a macroscopic scale characterized by an arbitrary elementary volume surrounding the point where fatigue analysis is made and representing for instance an element of finite element mesh; it is the usual scale considered by engineers; (ii) a mesoscopic scale corresponding to subdivision of the previous volume; the stress tensor at this scale results from the macroscopic one and the local residual stresses due to local inelastic deformation.

Thanks to the shakedown assumption at the local scale, it is possible to estimate the local stress cycle from the macroscopic stress cycle. The criteria is then expressed as an inequality related to the mesoscopic stresses at all instants $t$ of the cycle, so that damaging loading can be precisely characterized. The criterion used is expressed as:

$$\max_{t}\{\tau(t) + ap(t)\} \leq b$$

where $\tau(t)$ and $p(t)$ are the instantaneous mesoscopic shear stress and hydrostatic stress, $a$ and $b$ are material constants, which can be determined by two different classical fatigue tests.

Practically, the fatigue resistance of a structure is checked point by point, using two ways.

- The first one is the representation of the loading path $(p(t), \tau(t))$ at each point in the $(p, \tau)$ diagram. In this diagram, two constants $a$ and $b$ define a safety domain (no fatigue cracks) which is the region below the line $(\tau + ap = b)$. If the loading path at each point is entirely in the safety domain, there is no fatigue crack, otherwise fatigue damage occurs.
- The second one is the evaluation at each point of the quantity $\alpha = \max_{t}\{(\tau(t) - ap(t) - b)/b\}$. Positive value of $\alpha$ means occurrence of fatigue crack.

These two representations are used in the section devoted to numerical analysis of fretting to interpret in terms of fatigue, the results of simulations of a particular fretting setup. The experimental study is presented in the next section.

## Experimental Study of Fretting

In this section, we present the experimental results obtained by Petiot and al [6]. The experimental setup used is presented in Figure 2. Two cylindrical fretting pads (diameter 10 mm) are clamped against the two surfaces of a flat uniaxial fatigue specimen tested under constant amplitude loading at a frequency of 20 Hz. The pads are made of 100C6 steel and the fatigue specimen is made of 3Cr-MoV steel. The mechanical properties are given below:

| Material | Yield strength MPa | Tensile strength MPa | Young modulus MPa | Hardness Hv |
|----------|--------------------|----------------------|--------------------|-------------|
| 3Cr-MoV  | 980 | 1140 | 215 | 360 |
| 100C6    | 1700 | 2000 | 210 | 62 |

The prescribed oscillation between the pads are linked to the prescribed oscillatory fatigue stress $S(t)$ in the specimen. For a maximum stress $S_{max}$=500 MPa, the amplitude of displacement, $\delta$, is 0.55 $\mu m$. The flexible beams are equipped with strain gauge in order to measure the clamping force $P$ between pads and specimen and the friction force related to the displacements accommodation. The variations of the tangential force $T(t)$ are recorded for each fatigue cycle and plotted as function of fretting fatigue stress $S(t)$ (fretting fatigue loops). By varying the operating parameters $(P, S_{max})$, three regimes are established:

- Stick regime: Fretting fatigue loops keep a non evolutionary closed shape. Loops are quite linear during the test. The macroscopic displacement between the contacting surfaces is mainly accommodated by elastic deformation in the near surface of the two components. No damage (wear or crack nucleation) appears during the $10^7$ cycles of the test.
- Mixed stick-slip regime: Loops present an elliptical closed shape. There is partial slip and fatigue crack nucleation observed at the edges of the contact.
- Gross slip regime: Loops present a trapezoidal shape. Full slip occurs between the two contacting surfaces. In this regime, particles detachment is observed. The different regimes are obtained for different varying parameters $(P, S_{max})$ summarized in the map shown in Figure 7.

Our aim in the next section is to establish a numerical material response fretting fatigue map and to compare it with the experimental one presented in Figure 7. The principle of this comparison is the following: for each point of Figure 7, we use the corresponding experimental data $(P, S_{max}, T_{max})$ to perform numerical calculation of the stabilized stress (or plastic strain) cycle and then we apply the fatigue criterion to predict crack occurrence.

## Numerical Analysis and Prediction of Wear and Crack Nucleation

This section is devoted to the numerical simulation of the experimental setup presented in the previous section. The prediction of damage mechanisms shown in Figure 7 requires first the calculation of the stress history in the stabilized state. Secondly, Dang Van multiaxial fatigue criterion is applied if the stabilized state is elastic shakedown.

The finite element method (direct cyclic method) described previously is used here to simulate the set-up and to calculate the stresses history in the stabilized state.

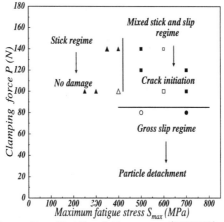

Figure 7 — *Fretting fatigue map (MRFM)*

The specimen is modeled as a half space subjected to a constant normal force $P$ and a varying tangential force $T(t)$ and a fatigue stress S(t) varying linearly with $T(t)$. The material is elastoplastic with a kinematical hardening (hardening modulus C=30 GPa) and with properties given previously.

Four simulations are performed; they correspond to the following four experimental points reported on the Material Response Fretting Map (Figure 7):

- $P$=140 N, $S_{max}$=350 MPa and $T_{max}$=53 N (the biggest filled triangle),
- $P$=100 N and $S_{max}$=400 MPa (the empty triangle),
- $P$=100 N, $S_{max}$=600 MPa and $T_{max}$=80 N (the biggest empty square),
- $P$=80 N, $S_{max}$=500 MPa and $T_{max}$=64 N (the empty circle).

$P$ and $S_{max}$ are the prescribed parameters and $T_{max}$ is measured in the test. The first two points are in the stick regime; the response of material is most of the time purely elastic and no damage is observed. The third point is in the mixed stick-slip regime where crack nucleation is observed. The fourth one is the gross slip regime where wear is observed.

*Numerical Calculation of Stress and Strain cycles*

The loading parameters ($P$, $S_{max}$, $T_{max}$) are used to analytically determine the contact characteristics (length, pressures distributions as shown in Figures 3 and 4). FEM analysis, using the direct cyclic method, is then performed to calculate the stabilized mechanical cycle. The mesh is refined under the contact surface as shown in Figure 8. In this figure, the zone with rectangular elements has a width of $2a$, $a$ varying from 50 $\mu m$ to 80 $\mu m$ in our present applications.

In the case of loading in gross slip regime, plastic shakedown is obtained numerically. This regime is represented by a closed cycle of plastic deformation and leads to

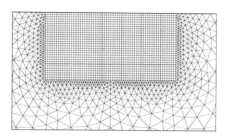

Figure 8 — *Refined mesh under the contact surface (total width of this zone between 100 μm and 200 μm depending on the load)*

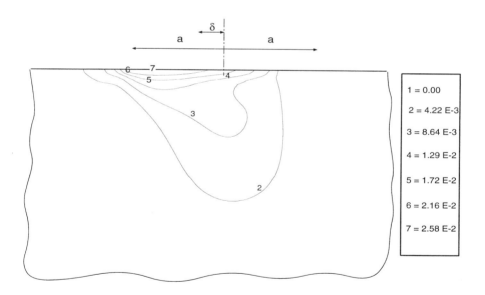

Figure 9 — *Contours of equivalent plastic strain in the case of full sliding regime, at the instant when the tangential force is zero*

low cycle fatigue. Experimentally, wear is observed in this regime. Thus, a connection between wear and the low cycle fatigue properties can be made, confirming the works of Kim and Ludema [12] and Johnson [13]. The numerical method used gives the plastic strain amplitude in the stabilized state. Figure 9 shows the contours of equivalent plastic strain in this case.

*Numerical Prediction of Fatigue*

Since the calculated response of material is respectively purely elastic and elastic shakedown in stick regime and mixed stick-slip regime, high cycle fatigue is concerned. The stress cycle through the contact is multiaxial. Dang Van multiaxial fatigue crack nucleation criterion is used to predict fatigue cracks. For each loading case (a point of Figure 7, corresponding to given values of $P$, $S_{max}$ and $T_{max}$), the stabilized stress cycle calculated as described in the previous section is used to determine at each point of the structure, the mesoscopic loading path consisting of the mesoscopic shear $\tau(t)$ and the hydrostatic pressure $p(t)$. At each point, this loading path is then compared to the fatigue properties of the material (material line) obtained from torsion $t$ and bending $f$ fatigue tests ($t$=380 MPa and $f$=594 MPa for 30NCD16 steel). The most critical point is located at the surface on the edge of the contact. The most critical loading path ($\tau$, $p$) for each case is plotted in Dang Van's fatigue diagram shown in Figure 10. In the stick regime, the two triangles of Figure 7 we simulate ($P$=140 N, $S_{max}$=350 MPa, $T_{max}$=53 N and $P$=100 N, $S_{max}$=400 MPa) give loading path which are beneath the fatigue line material; so, no damage occurs as observed experimentally. In mixed stick-slip regime, the simulated square of Figure 7 ($P$=100 N, $S_{max}$=600 MPa and $T_{max}$=80 N) leads to a loading path which intersects the fatigue line material, meaning crack nucleation. The contours of the Dang Van criterion $\alpha = \max_t \frac{\tau(t) - ap(t) - b}{b}$ are plotted on figure 11. A positive value of $\alpha$ means crack initiation.

All the obtained numerical results are brought together in Figure 10; a result obtained by Petiot and al [6] who used a different simulation for the calculation of the stress cycles is added. The numerical predictions of crack initiation are in total agreement with the experimental observations reported in Figure 7.

**Conclusion**

Quantitative prediction of risk of fretting fatigue on a structure is of prime importance for many structures. However, this problem is so difficult that until now, no predictive method is available. The problem has been studied by engineers empirically; long and expensive experimental tests have to be performed to be sure to avoid fretting fatigue.

We propose a new methodology for studying the fretting phenomena based first on the evaluation of the local mechanical parameters by efficent inelastic numerical methods, and second on the use of multiaxial fatigue criterion which can be easily identified by simple tests, independant of contact phenomena.

To check the validity of our proposal, we simulate the fretting fatigue tests per-

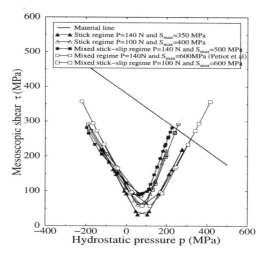

Figure 10 — *Loading paths in Dang Van's diagram in stick regime and mixed stick-slip regime: safe paths are below the material line*

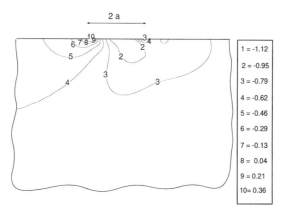

Figure 11 — *Contours of Dang Van criterion* $\alpha = \max_t \frac{\tau(t) - ap(t) - b}{b}$ *for* $P = 100N$ *and* $S_{max} = 600MPa$

formed by Petiot and al. [6], and compare the numerical predictions to the experimental results. By this method, we are able to distinguish the three differents regimes observed experimentally:

- Stick regime: it corresponds to pure elasticity and no damage occurs;

- Mixed stick-slip regime; in this case, the regime is elastic or elastic shakedown; fatigue is observed; the used multiaxial fatigue criterion predicts correctly crack initiation;

- Gross slip regime: wear is observed.

To obtain results, we must evaluate the stabilized state (elastic or plastic shakedown regimes) which is obtained directly by an original scheme of integration, if the contact characteristics (i.e. normal and tangential stress on the surface at any time of the cycle) are known. For real applications with general contact geometry, this last requirement remains a difficult problems, which is not straightforward. However, we have verified that in our example, elastic shakedown hypothesis corresponding to fatigue regime is valid; it is then possible to use some existing classical FEM codes to estimate these contact characteristics. Some applications to real industrial structures are currently being studied by this new methodology.

## References

[1] Vingsbo, O. and Soderberg S., "On Fretting Maps", *Wear*, Vol. 126, 1988, pp. 131-147.

[2] Vincent, L., Berthier, Y., and Godet, M., "Testing Methods in Fretting Fatigue: a Critical Appraisal", *Standardisation of Fretting Fatigue Test Methods and Equipment*, ASTM STP 1159, M. Helmi Attia and R.B. Waterhouse, Eds., American Society for Testing and Materials, Philadelphia, 1992, pp. 33-48.

[3] Dang Van, K., Griveau, B., and Message, O., "On a New Multiaxial Fatigue Limit Criterion: Theory and Application", *biaxial and multiaxial fatigue*, M.W. Brown and K. Miller, Eds., EGF Publication 3, 1982, pp.479-496.

[4] Dang Van, K., "Macro-Micro Approach in High-Cycle Multiaxial Fatigue", *Advances in multiaxial fatigue*, ASTM STP 1991, D.L. McDowell and R. Ellis, Eds., American Society for testing and Materials, Philadelphia, 1993, pp. 120-130.

[5] Dang Van, K., "Introduction to Fatigue Analysis in Mechanical Design by the Multiscale Approach", *High-Cycle Metal Fatigue in the Context of Mechanical Design*, K. Dang Van and I. Papadoupoulos, Eds, CISM Courses and Lectures No. 392, 1999, Springer-Verlag, pp. 57-88.

[6] Petiot, C., Vincent, L., Dang Van, K., Maouche, N., Foulquier, J., and Journet, B., "An Analysis of Fretting-Fatigue Failure Combined with Numerical Calculations to Predict Crack Nucleation", *Wear*, Vol. 181-183, 1995, pp. 101-111.

[7] Maouche, N., Maitournam, H.M., and Dang Van, K., "On a New Method of Evaluation of the Inelastic State due to Moving Contacts", *Wear*, Vol. 203-204, 1997 pp. 139-147.

[8] Mindlin, R.D., Compliance of Elastic Bodies in Contact, *J.Appl.Mech*, Vol.16, 1949, pp. 259-268.

[9] Johnson, K.L., *Contact Mechanics*, Cambridge University Press, 1985.

[10] Hills, D.A., and Nowell, D., *Mechanics of fretting fatigue*, Kluwer Academic Publishers, 1994.

[11] Maitournam, M.H., "Finite Elements Applications Numerical Tools and Specific Fatigue Problems", *High-Cycle Metal Fatigue in the Context of Mechanical Design*, K. Dang Van and I. Papadoupoulos, Eds., CISM Courses and Lectures No. 392, 1999, Springer-Verlag, pp. 169-187.

[12] Kim, K., and Ludema, K.C., "A Correlation Between Low Cycle Fatigue and Scuffing Properties of 4340 Steel", *Wear*, Vol. 117, 1995 pp. 617-621.

[13] Johnson, K.L., "Contact Mechanics and Wear of Metals", *Wear*, Vol. 190, 1995 pp. 162-170.

# Author Index

# Subject Index